Texts and
Monographs
in Physics

W. Beiglböck
J. L. Birman
E. H. Lieb
T. Regge
W. Thirring
Series Editors

J.M. Jauch   F. Rohrlich

The Theory of Photons and Electrons

The Relativistic Quantum Field Theory
of Charged Particles with Spin One-half

Second Expanded Edition

With 64 Figures

Springer-Verlag
Berlin Heidelberg New York 1976

J. M. Jauch †

F. Rohrlich
Department of Physics, Syracuse University, Syracuse, NY 13210, USA

Editors

Wolf Beiglböck
Institut für Angewandte Mathematik
Universität Heidelberg
Im Neuenheimer Feld 294
D-6900 Heidelberg 1.
Fed. Rep. of Germany

Joseph L. Birman
Department of Physics, The City College
of the City University of New York,
New York, NY 10031, USA

Tullio Regge
Istituto di Fisica Teorica
Università di Torino, C. so M. d'Azeglio, 46
I-10125 Torino, Italy

Elliott H. Lieb
Department of Physics
Joseph Henry Laboratories
Princeton University
Princeton, NJ 08540, USA

Walter Thirring
Institut für Theoretische Physik
der Universität Wien, Boltzmanngasse 5
A-1090 Wien, Austria

Second corrected printing 1980

ISBN 3-540-07295-0 Springer-Verlag Berlin Heidelberg New York
ISBN 0-387-07295-0 Springer-Verlag New York Heidelberg Berlin

Library of Congress Cataloging in Publication Data. Jauch, Josef Maria, 1914–1974. The theory of photons and electrons. (Texts and monographs in physics) Includes indexes. I. Quantum electrodynamics. I. Rohrlich, F., joint author. II. Title. QC680.J38   1975   530.1′43   75-8890

This work is subject to copyright. All rights are reserved, whether the whole or part of the material is concerned, specifically those of translation, reprinting, reuse of illustrations, broadcasting, reproduction by photocopying machine or similar means, and storage in data banks. Under § 54 of the German Copyright Law, where copies are made for other than private use, a fee is payable to "Verwertungsgesellschaft Wort", Munich.

© 1955, 1976 J. M. Jauch and F. Rohrlich
Printed in Germany

The use of registered names, trademarks, etc in this publication does not imply, even in the absence of a specific statement, that such names are exempt from the relevant protective laws and regulations and therefore free for general use.

Offset printing: Weihert-Druck GmbH, 6100 Darmstadt
Bookbinding: Konrad Triltsch, Graphischer Betrieb, Würzburg
2153/3130-54321

PREFACE TO THE FIRST EDITION

Since the discovery of the corpuscular nature of radiation by Planck more than fifty years ago the quantum theory of radiation has gone through many stages of development which seemed to alternate between spectacular success and hopeless frustration. The most recent phase started in 1947 with the discovery of the electromagnetic level shifts and the realization that the existing theory, when properly interpreted, was perfectly adequate to explain these effects to an apparently unlimited degree of accuracy. This phase has now reached a certain conclusion: for the first time in the checkered history of this field of research it has become possible to give a unified and consistent presentation of radiation theory in full conformity with the principles of relativity and quantum mechanics. To this task the present book is devoted.

The plan for a book of this type was conceived during the year 1951 while the first-named author (J. M. J.) held a Fulbright research scholarship at Cambridge University. During this year of freedom from teaching and other duties he had the opportunity of conferring with physicists in many different countries on the recent developments in radiation theory. The comments seemed to be almost unanimous that a book on quantum electrodynamics at the present time would be of inestimable value to physicists in many parts of the world.

However, it was not until the spring of 1952 that work on the book began in earnest. It was perhaps fortunate that the first-named author did not foresee all the difficulties, nor estimate correctly the extent of the project. Had he done so, he might have been discouraged from the beginning. These difficulties unfolded gradually as the project grew.

During the fall of 1953 the second author (F. R.) joined the staff of the physics department of the State University of Iowa. He had at that time just completed a set of lecture notes on a course in quantum electrodynamics which he had given during the spring term 1953 at Princeton University. Considerable demand for these notes and a suggestion to publish them brought about the logical development that the two authors should join hands. The resultant cooperation has not only been a rich personal experience and a bond of lasting friendship but, in addition, we think that the product of this cooperation is a book which is much better than either one of us could have written separately.

It was clear from the beginning that the whole approach to quantum electrodynamics had to be re-evaluated in the light of the latest developments. The true significance of this recent progress cannot be properly assessed if it is seen only as a grafting on the old tree. Indeed, one of the most significant aspects of this development is the formal simplification and the resultant clearer physical picture of many old problems in radiation theory.

It would have been very tempting to apply the general theory to problems in meson field theory. We have withstood this temptation. Meson theory exists only in fragments at the present time and even these fragments are subject to considerable doubt when they are studied in the light of experimental evidence.

We have even omitted the study of the electromagnetic interaction of scalar particles, although one of us (F. R.) has been particularly interested in this field. It adds little of *basic* theoretical interest. A complete treatment would involve considerable duplication and the discussion of applications alone would not conform with the purpose of this book.

Thus we have restricted ourselves entirely to the interaction of electrons and photons. But even within this restricted field completeness could not be attempted. For instance, we have not discussed exhaustively the comparison of the experiments with the theory. We have completely ignored the practical problems of energy loss, straggling, and shower production. Many important topics of radiation theory are hardly mentioned. Fortunately, many of the omitted topics are found in other sources, notably in the well-known work by Heitler and in the excellent summary by Bethe and Ashkin (Segrè's *Experimental Nuclear Physics*, Vol. I). On the whole, we have tried to emphasize topics complementary to this standard work. Some duplication could not, of course, be avoided.

We are particularly sorry that we had to omit a full discussion of the relativistic two-body problem. Mathematical difficulties have so far prevented a completely satisfactory treatment of this problem. A thorough discussion would have increased considerably the size of this book and would have further delayed its publication. Consequently, the level shift in positronium and related problems had to be omitted, too.

On the other hand, we believe that we can present a few original contributions within the restricted topic we have elected to discuss. There are several new features both in form and content.

As to form, we have departed from a long-standing tradition of basing the theory on a classical model and of introducing the quantization via the Hamiltonian formalism. The classical analogy and the canonical quantization formed a most fruitful and far-reaching heuristic principle for non-relativistic quantum mechanics. But it was only a crutch. For a quantum theory in full accord with the principle of relativity the canonical quantization procedure is more of a hindrance than a help. Now that we have learned to develop the quantum theory in the four-dimensional continuum we can leave this crutch behind. We have considered the quantum theory as the basic theory. The classical theory is contained in this general theory as a special case, valid within a certain approximation. The fundamental dynamical laws are obtained from the general Action Principle and they are contained entirely in the structure of the Lagrangian. We are fully aware of some of the difficulties connected with this approach. Several of these are so far not completely resolved. They are mentioned in the text. But we believe that in spite of these difficulties this approach has much in its favor.

In content the reader will find some original contributions: the theory of the S-operator, the treatment of time reversal, the substitution law, the positronium selection rules, the ordering theorem, the infrared divergence, the treatment of polarization, the natural line width, and several other minor items.

The organization of the work can be seen at a glance from the accompanying chart, where the numbers indicate the chapters. There is an additional chapter

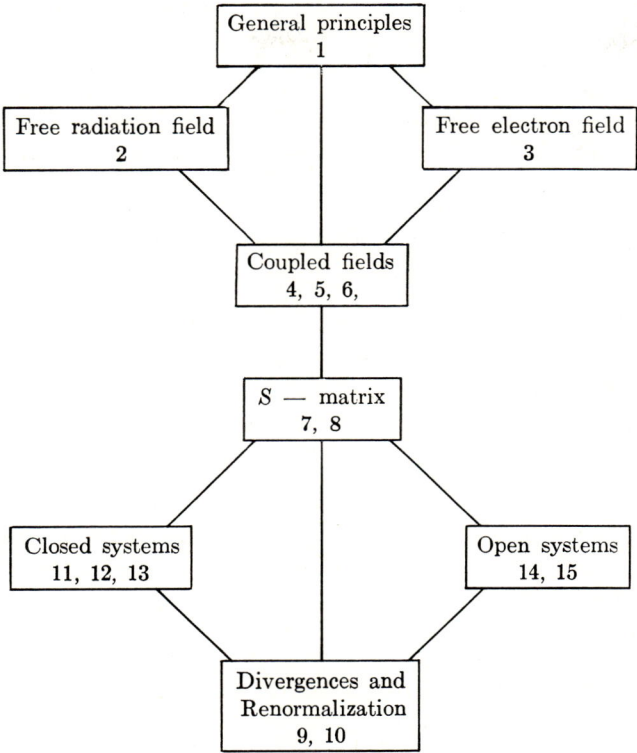

(16) dealing with special problems. We have made a conscious effort to separate those properties of the photon-electron field which have their origin in the invariance under symmetry transformation. We believe these properties to be of fundamental significance not likely to be changed in future developments of the theory.

A major part of the book is concerned with the presentation of the general theory. Applications of the theory are given in four chapters (11, 12, 13, and 15), and are selected mainly for the purpose of illustrating the most important consequences of the theory. The links between theory and applications are the two chapters on the S-matrix.

Many of the applications were treated a long time ago. But the reader will find some new methods of handling the old problems. On the other hand, a few of the examples included have been discussed only very recently and some of them are even incomplete and are left as a challenge to the reader.

In a theory with such an elaborate formal apparatus there is great danger that the physical ideas may be lost in the mathematical manipulations. In order to alleviate this situation, we have relegated many of the more elaborate developments to an appendix. On the whole, we have tried to be as complete as possible in the deductive reasoning and we have always tried to find the most elegant presentation. Elegance, in our opinion, is not just an irrelevant luxury in a scientific theory. We consider it a principle of economy of the greatest pedagogical value and we have found great satisfaction in revealing the intrinsic beauty and perfection of this theory.

Our notation and terminology do not always follow the traditional line. If notation is entirely left to tradition in a rapidly expanding field like this, it grows like weeds in a garden. We have tried hard to bring some consistency into the picture without uprooting the entire tradition. It is, of course, quite impossible to have a system of symbols which is unambiguous; the number of existing alphabets would not be large enough, even if we included the Russian alphabet. Therefore, we have confined ourselves to the Latin and Greek letters of various types. The unavoidable overlap is so arranged that it will not cause confusion.

We also found it desirable to bring some order into the terminology. For instance, we have followed the recommendation of the International Union of Physics adopted at the 1947 conference in Cracow of using the word *electron* as a general term for both the positive and negative variety of this particle. The latter are separately named as the *positon* and the *negaton*.

Another departure from an inconsistent tradition is our distinction between *representation* and *picture*. We have followed Dirac's usage of these words. A *representation* refers to a particular choice of a base-vector system. A *picture* describes the time dependence of the state vectors or the operators. Thus, we speak about the Schroedinger, Heisenberg, and interaction *picture*. It is possible to distinguish these *pictures* without specifying the *representation*.

If this book were without errors it would be a miracle. We are not so much concerned with an occasional error of sign, or of a factor i or 2, although we have made every effort to avoid such mistakes. But we have been seriously concerned with misconceptions or misunderstandings of other research. In several cases we have pointed out errors in the literature. We hope that thereby we are able to save our colleagues unnecessary troubles with a number of papers. We trust that the criticized authors will take our comments in the constructive spirit in which they are offered.

Our selection of references is not entirely systematic and does not claim to be complete, but reflects strongly our personal preferences and judgments. In general, we have tried to give references to papers which contain a new idea or a new result within the restricted topics treated in this book. We have not succeeded in following every author to the most subtle points of his paper and we apologize for any shortcomings on our part in giving credit to authors. The preparation of a complete list of references in this field is a forbidding task. Research which was published after the closing date of the manuscript could not be included. We are confident that this will not detract from the main objective of this book, *viz.*, to stimulate and intensify the search for a deeper understanding of the fundamental physical laws.

We have received valuable comments and suggestions from many colleagues too numerous to mention separately. We are particularly indebted to Professor G. Wentzel, who read and criticized the first six chapters, and to Professor F. J. Dyson, who read critically Chapters 9 and 10.

The burden of writing has been eased considerably by the competent and patient help of Miss Agnes Costello, who typed and retyped the many different versions of the entire manuscript. Finally, we thank Dr. A. Lenard for his part in the proofreading.

J. M. Jauch
F. Rohrlich

Iowa City, Iowa
December, 1954

PREFACE TO THE SECOND EDITION

Last August Josef Jauch and I were making plans to join forces again in order to prepare a second edition of our book. His permanent residence in Geneva, Switzerland, and mine in Syracuse, New York, had permitted few occasions in recent years for us to meet and exchange ideas. Our plan was to meet in Syracuse in mid-September. But this was never to come about. Josef died suddenly at the end of August, to the great shock of his many friends and collaborators.

This second edition is therefore entirely my own doing. However, our points of view have been very similar so that I believe that many of the decisions I had to make would have found his approval.

The first edition of *The Theory of Photons and Electrons* was finished almost exactly twenty years ago. That is a long time by the standards of present day research progress. What was the first presentation in book form of important recent progress twenty years ago has now become standard graduate text material, and some of it is even taught on the advanced undergraduate level. The author of a new edition must take this into account and thus faces a difficult task: he must "bring the subject matter up-to-date" as the phrase goes, but at the same time he cannot change a basic graduate text into the much more advanced text it was twenty years earlier without rewriting the entire book. That would have been unreasonable and uncalled for.

In fact, while a lot of new material can be added, from a fundamental aspect not very much has changed during this time; only our point of view has become much more sophisticated, especially mathematically. We now have a much deeper mathematical understanding of quantum electrodynamics, especially due to the work of axiomatic quantum field theorists; but we have still not solved the basic problem of formulating the theory in a clean mathematical way, not even with all the complicated and highly sophisticated limiting procedures presently used to justify the results of naive renormalization theory in simpler quantum field theories and in lower dimensionality. The hopes and aspirations indicated in the outlook of twenty years ago (Section 16–5) remain valid today.

I have thus decided to leave the level of mathematical rigor basically the same but to provide the necessary stimulation and references for further study for those who want to gain a deeper mathematical insight. In addition, many excellent texts have been published in the intervening years which include more recent material and different points of view. There is clearly nothing gained in repeating what has been done well already.

Thus emerged the plan for the present edition. The previous presentation is left intact apart from minor corrections. Only Sections 11–4 (pair production in the field of an electron) and 15–8 (Delbrück scattering) have been partially rewritten and expanded. (The pagination and cross references thus remain unchanged.) The bulk of the new material is collected in additional chapters in the form of a supplement.

The supplement consists of five chapters. The first summarizes very briefly the various formulations of quantum field theory to the extent that they are relevant to quantum electrodynamics; the second reviews the new developments in renormalization theory. Both parts contain a sufficiently extensive bibliography for further study. The next two chapters deal with coherent states and the infrared problem as it is now understood. Much more detail is given in these two parts because of important recent progress that leads to a much deeper understanding of the physical electron with its complicated "photon cloud." The last part is devoted to a comparison of the theory with precision experiments, supplementing the chapters on the anomalous magnetic moment and on radiative corrections to atomic energy levels.

In this supplement the guiding principles have been to complement rather than duplicate the existing text book literature, to limit the book strictly to photons and electrons, to provide all the necessary reference material, and to bring the reader up to the most recent results wherever possible. Of course, a rather subjective judgment has to be exercised here, first in the selection of subject matter, and then in the choice of references.

Some interesting developments were left out because they must at present still be considered speculative, and at the same time they would require a great deal of background material for their presentation. Into this category belong the attempt to explain the mass of the electron as resulting from symmetry breaking of a massless (and thus probably conform-invariant) quantum electrodynamics, and the attempt to explain the fine structure constant as the solution of an equation that (under certain assumptions only) is necessary for the internal consistency of the theory. Other new developments were omitted because they have relatively little bearing on the already established results in quantum electrodynamics, although their great importance for nonelectromagnetic interactions should not be underestimated. And finally there were considerations of space and time limitations and of costs, *viz.*, to make this book available again soon and at minimum possible price after it has been out of print for several years.

In giving references an effort was made to give credit fairly and otherwise to give those references that I found most instructive. If I have not fully succeeded I want to apologize to those authors who were slighted.

A few modest original contributions can be found in the treatment of coherent states for an infinite number of degrees of freedom (Supplement 3) and in the presentation of the soft photon limit (Supplement 4). The emphasis is always on a better physical description even where only a formal mathematical treatment is available, and to avoid clumsy formalisms which may mask the beauty and basic simplicity of the theory.

Over the years I have received many letters advising me of errors, misprints, and other deficiencies in the previous edition. I am most grateful for all these and I have made corrections wherever possible within the indicated limitations. I also want to thank my colleagues, Aiyalam P. Balachandran and Marcel Wellner for reading parts of the supplement and for their helpful comments. But most importantly I am indebted to Ivan T. Todorov, Daniel Zwanziger, and Thomas Fulton for their criticism of respectively Chapters 1 and 2,

Chapters 3 and 4, and Chapter 5 of the supplement. Their first hand knowledge of recent work in these research fields and their depth of perception contributed greatly. In the last chapter I was also aided considerably by Barry N. Taylor's recent work which he kindly made available to me. Of course, none of these people can be held responsible for the final version of the supplement.

Last but not least I want to thank our physics librarian Irene Karolewski for her continued help in locating various references for me.

F. Rohrlich

Syracuse, New York
August, 1975

CONTENTS

PREFACE TO THE FIRST EDITION v

PREFACE TO THE SECOND EDITION xi

CHAPTER 1 GENERAL PRINCIPLES 1
  1–1 The natural unit system 1
  1–2 Some fundamental notions of the special theory of relativity . 2
  1–3 Some basic notions of quantum mechanics 5
  1–4 Localizability 7
  1–5 Observables of a field 8
  1–6 Canonical transformations 10
  1–7 Lorentz transformations as canonical transformations . . 11
  1–8 The action principle 14
  1–9 The equation of motion 18
  1–10 Momentum operators 19
  1–11 Conservation laws 20
  1–12 Commutation rules 22

CHAPTER 2 THE RADIATION FIELD 25
  2–1 The classical field equations 25
  2–2 The associated boundary value problem 27
  2–3 A Lagrangian for the radiation field 28
  2–4 Quantization of the radiation field 28
  2–5 Momentum operators for the radiation field . . . 31
  2–6 Plane wave decomposition of the radiation field . . . 35
  2–7 Explicit representations of the field operators . . . 37
  2–8 The spin of the photon 40
  2–9 Definition of the vacuum 47

CHAPTER 3 RELATIVISTIC THEORY OF FREE ELECTRONS . . 50
  3–1 The field equations for the one-particle problem . . . 50
  3–2 The associated boundary value problem 51
  3–3 Relativistic invariance of the field equations . . . 52
  3–4 The bilinear covariants 53
  3–5 A Lagrangian for the spinor field 56
  3–6 Quantization 56
  3–7 Momentum operators 59
  3–8 Plane wave decomposition 60
  3–9 Explicit representation of the field operators . . . 65
  3–10 The definition of the vacuum 67

CHAPTER 4 INTERACTION OF RADIATION WITH ELECTRONS . 69
  4–1 The field equations 69

4–2	Commutation rules for the interacting fields	73
4–3	The interaction picture	76
4–4	Measurability of the fields	78

CHAPTER 5 INVARIANCE PROPERTIES OF THE COUPLED FIELDS . . . 82

5–1	Proper Lorentz transformations	82
5–2	Gauge transformations	83
5–3	Space inversion	86
5–4	Time inversion	88
5–5	Charge conjugation	94
5–6	Scale transformations	96

CHAPTER 6 SUBSIDIARY CONDITION AND LONGITUDINAL FIELD . . . 97

6–1	The covariant Coulomb interaction	97
6–2	The subsidiary condition and the construction of the state vector	100
6–3	The Gupta method	103
6–4	Gauge-independent interaction	110
6–5	Radiation fields with finite mass	113

CHAPTER 7 THE S-MATRIX . . . 117

7–1	Preliminary definition of the S-matrix	117
7–2	The wave matrix	120
7–3	The wave operator	126
7–4	Integral representation of the wave operator	129
7–5	Definition of the S-matrix	131
7–6	Invariance properties of the S-matrix	136

CHAPTER 8 EVALUATION OF THE S-MATRIX . . . 144

8–1	The iteration solution	144
8–2	The Feynman-Dyson diagrams	146
8–3	Diagrams in momentum space	151
8–4	Closed loops	159
8–5	The substitution law	161
8–6	Lifetimes and cross sections	163
8–7	Evaluation of the S-matrix in the Heisenberg picture	167

CHAPTER 9 THE DIVERGENCES IN THE ITERATION SOLUTION . . . 170

9–1	Historical background	171
9–2	Classification of divergences	173
9–3	The vacuum fluctuations	176
9–4	The self-energy of the electron	178
9–5	The self-energy of the photon	188
9–6	The vertex part	197

CHAPTER 10 RENORMALIZATION . . . 203

10–1	The primitive divergences	203
10–2	Irreducible and proper diagrams	206

10–3	Separation of divergences in irreducible parts	210
10–4	Separation of divergences in reducible parts	211
10–5	Mass renormalization	219
10–6	Charge renormalization	219
10–7	Wave function renormalization	221
10–8	Sufficiency proof	222
10–9	Regulators	223

CHAPTER 11 THE PHOTON-ELECTRON SYSTEM 228

11–1	Compton scattering	229
11–2	Double Compton scattering	235
11–3	Radiative corrections to Compton scattering	241
11–4	Pair production in photon-electron collisions	247

CHAPTER 12 THE ELECTRON-ELECTRON SYSTEM 252

12–1	Møller scattering	252
12–2	Bhabha scattering	257
12–3	Bremsstrahlung in electron-electron collisions	261
12–4	Annihilation of free negaton-positon pairs	263
12–5	Positronium; selection rules	274
12–6	Positronium annihilation	282

CHAPTER 13 THE PHOTON-PHOTON SYSTEM 287

13–1	Photon-photon scattering as part of a diagram	287
13–2	Photon-photon scattering cross sections	292
13–3	Pair production in photon-photon collision	299

CHAPTER 14 THEORY OF THE EXTERNAL FIELD 302

14–1	The external field approximation	303
14–2	The bound interaction picture	306
14–3	Commutation rules	308
14–4	The electron propagation function	312
14–5	The S-matrix in the external field approximation	318
14–6	Renormalization	321
14–7	Cross sections and energy levels	322

CHAPTER 15 EXTERNAL FIELD PROBLEMS 327

15–1	Coulomb scattering	327
15–2	Radiative corrections to Coulomb scattering	332
15–3	The magnetic moment of the election	342
15–4	Energy levels in hydrogen-like atoms	345
15–5	Radiative transitions between bound states	361
15–6	Bremsstrahlung	364
15–7	Pair production and annihilation	373
15–8	Delbrück and Rayleigh scattering	379

CHAPTER 16 SPECIAL PROBLEMS 390

16–1	The infrared divergences	390

16–2	Radiation damping in collision processes	405
16–3	The natural line width of stationary states	408
16–4	The self-stress of the electron	410
16–5	Outlook	415

MATHEMATICAL APPENDIX

APPENDIX A1 THE INVARIANT FUNCTIONS 419

A1–1	The homogeneous delta-functions	420
A1–2	The inhomogeneous delta-functions	420
A1–3	Relations between the Δ-functions	421
A1–4	Integral representations	423
A1–5	Explicit expressions	424
A1–6	The S-functions	424

APPENDIX A2 THE GAMMA-MATRICES 425

A2–1	Various representations	425
A2–2	The matrices A, B, and C	429
A2–3	The amplitudes of the plane wave solutions	431
A2–4	A theorem on the traces of γ-matrices	436
A2–5	Spin sums	438
A2–6	Polarization sums	439

APPENDIX A3 A THEOREM ON THE REPRESENTATION OF THE EXTENDED LORENTZ GROUP BY IRREDUCIBLE TENSORS 441

APPENDIX A4 THE ORDERING THEOREM 445

A4–1	The ordering theorem for commuting fields	445
A4–2	The ordering theorem for anticommuting fields	449
A4–3	A generalization of the ordering theorem	451
A4–4	The ordering of chronological products	451

APPENDIX A5 ON THE EVALUATION OF CERTAIN INTEGRALS . . 454

A5–1	Convergent integrals	455
A5–2	Divergent integrals	457
A5–3	The integral for the photon self-energy part	460
A5–4	The integral for the electron self-energy part	463

APPENDIX A6 A LIMITING RELATION FOR THE δ-FUNCTION . . 464

APPENDIX A7 THE METHOD OF ANALYTIC CONTINUATION . . 465

A7–1	The Bohr-Peierls-Placzek relation	465
A7–2	The principle of limiting distance	468
A7–3	The fundamental theorem on analytic continuation	469
A7–4	Applications	473

APPENDIX A8 NOTATION 476

Supplement for the Second Edition

supplement S1 Formulations of Quantum Electrodynamics . . 479
- S1–1 Lagrangian QFT 480
- S1–2 Axiomatic QFT 481
- S1–3 Locality, covariance, and indefinite metric 483
- S1–4 Lehmann–Symanzik–Zimmermann and related formalisms . . 486
- S1–5 Null plane QED 490
- References 492

supplement S2 Renormalization 496
- S2–1 Dyson–Salam–Ward renormalization 496
- S2–2 Bogoliubov–Parasiuk–Hepp–Zimmermann renormalization . . 499
- S2–3 Analytic renormalization 501
- References 502

supplement S3 Coherent States 503
- S3–1 A finite number of degrees of freedom 503
- S3–2 Coherent states of the radiation field 506
- S3–3 Application to ordering theorems 509
- References 511

supplement S4 Infrared Divergences 513
- S4–1 Dollard's discovery 513
- S4–2 A new picture 514
- S4–3 The asymptotically modified fields 519
- S4–4 The new S-matrix 525
- References 529

supplement S5 Predictions and Precision Experiments . . . 530
- S5–1 The anomalous magnetic moment 530
- S5–2 The hyperfine structure of the hydrogen ground state . . 531
- S5–3 The Lamb–Retherford shift in hydrogen 533
- S5–4 Energy levels in positronium 535
- S5–5 Muonium hyperfine structure 536
- References 537

Author Index 541

Subject Index 547

CHAPTER 1

GENERAL PRINCIPLES

This chapter begins with a short discussion of the system of units used throughout this book (Section 1–1). After a brief review of the theory of special relativity (Section 1–2) and quantum mechanics (Section 1–3), the concept of the localizable field is introduced (Sections 1–4 and 1–5). We next proceed to define canonical transformations for a field (Section 1–6) and apply the concept to the transformations induced by the inhomogeneous Lorentz group (Section 1–7). This leads us to the ten integrals of momentum and angular momentum.

The specific dynamic properties of a field are then derived from the Action Principle (Section 1–8) and it is shown how one obtains from this principle the field equations, explicit expressions for the ten integrals of momentum and angular momentum, the conservation laws, and some information regarding the commutation rules for the field variables (Sections 1–9, 1–10, 1–11, and 1–12). These results are sufficiently general to comprise any of the known linear localizable field theories. The specific dynamic characteristics are contained in the specific properties of the Lagrangian which is used for the Action Principle, and the type of quantization.

1–1 The natural unit system. The purpose of any physical theory, such as the one under discussion in this book, is to set up a mathematical system of physical laws which enables us to correlate given empirical phenomena. A physical theory is considered satisfactory if we can make with it quantitative predictions of physical data. Such data usually involve the measurements of certain quantities which are expressed in a system of fundamental units. The choice of these units is to some extent arbitrary with respect to magnitude as well as with respect to kind. Such a choice of units will therefore be guided primarily by considerations of convenience.

A great simplification is introduced in quantum electrodynamics if we use as the fundamental units certain physical quantities which are constants of nature. Two constants of this sort which occur in the present theory are c, the velocity of light, and $\hbar$, Planck's constant divided by 2π. We choose these constants as two of the fundamental units of our system. As the third unit we use the centimeter as the conventional and arbitrary unit of length. The advantages for this particular choice will become more evident later on. Here we shall just mention that this choice is guided by the fact that the theory under discussion arises from an intimate fusion of the special theory of relativity characterized by the constant c and the quantum theory characterized by $\hbar$. At the present time there exists no theory which involves in its fundamental laws a universal length, which would make a natural choice of the unit of length. The need for such a theory involving a fundamental length has been the subject of much speculation in the past but it seems safe to say that we are far from understanding the role of such a unit in existing theories.

The natural system of units here adopted is easily converted into the conventional cgs system if we express the units in the latter system in terms of the former. Thus, if we denote by $(f(c,\hbar))$ the value of the function $f(c,\hbar)$ in the cgs system [for instance, $(c) \simeq 3 \times 10^{10}$, $(\hbar) \simeq 10^{-27}$, etc.], the conversion formulas have the form

$$1 \text{ cm} = 1 \text{ cm},$$

$$1 \text{ gm} = \left(\frac{c}{\hbar}\right) \text{cm}^{-1} c^{-1} \hbar, \tag{1-1}$$

$$1 \text{ sec} = (c) \text{ cm } c^{-1}.$$

For instance, the unit of mass in the fundamental system is $\text{cm}^{-1} c^{-1} \hbar$ and the value of mass in these units is converted into its value in grams by multiplication with $(\hbar/c)$. The electric charge has the dimension $c^{1/2} \hbar^{1/2}$ in the fundamental system and in this system the value of a charge is converted into its value in cgs electrostatic units by multiplication with $(c\hbar)^{1/2}$.

It is perhaps worth pointing out that the unit of charge in the cgs system is usually defined with the help of Coulomb's law. The proportionality factor in this law is, of course, arbitrary and can be chosen for convenience. We shall adopt here the so-called rationalized systems of units of charge in which the charge of the electron has the approximate value

$$|e| = \sqrt{\frac{4\pi}{137}} c^{1/2} \hbar^{1/2}. \tag{1-2}$$

The advantage of this choice is that all factors π are eliminated from the fundamental laws which we need to discuss.

1-2 Some fundamental notions of the special theory of relativity. An important feature of the theory to be developed in this book is its complete accord with the principles of the special theory of relativity. The full significance of this formal requirement has only recently been brought to the fore with the realization that the clear-cut separation of the physical effects from the meaningless divergencies can only be carried out in the covariant form of the theory.

The central notion in the mathematical formulation of the relativity theory is the four-vector of space and time which we denote by x. The *contravariant* components [defined by their transformation properties, see below, Eq. (1–6)] of this vector we denote by x^μ ($\mu = 0, 1, 2, 3$). Here x^0 stands for the time-component and the x^i ($i = 1, 2, 3$) represent the three space-components.

The transition to the *covariant* components x_μ is effected with the help of the fundamental metric tensor $g_{\mu\nu}$, defined by

$$g_{11} = g_{22} = g_{33} = -g_{00} = 1. \tag{1-3}$$

All other components vanish.

$$x_\mu = g_{\mu\nu} x^\nu. \tag{1-4}$$

In (1–4) and always in the following we use the "summation convention," that is, repeated indices in the lower and upper position are summed.

The equation for the light cone is conveniently expressed with the invariant scalar product x^2:

$$x^2 \equiv x^\mu x_\mu = 0. \tag{1-5}$$

The kinematics of the special theory of relativity is contained in the general transformation law for four-vectors.* The components of a four-vector referred to two different inertial systems with the same origin are related by a homogeneous proper Lorentz transformation, which is defined as the real linear transformation which leaves the left side of (1–5) invariant,

$$x'^\mu = a^\mu{}_\nu x^\nu; \quad a^\mu{}_\nu a^{\lambda\nu} = g^{\mu\lambda}, \tag{1-6}$$

and which, in addition, satisfies

$$a^\mu{}_\nu \text{ real,}$$

$$\det(a^\mu{}_\nu) > 0, \tag{1-7}$$

$$a^0{}_0 > 0.$$

The "proper" Lorentz transformations are to be distinguished from the "special" transformations which are obtained if the last two conditions (1–7) are omitted. We single out two transformations of this kind, the space inversion σ:

$$\sigma: \quad x'^i = -x^i, \quad x'^0 = x^0, \tag{1-8}$$

---

* See, for instance, A. Einstein, *The Meaning of Relativity*, Princeton University Press, 1945, especially p. 30 ff. For a complete discussion of the theory of relativity, see any of the standard works, e.g., W. Pauli, Relativitätstheorie, in *Encyklopädie der Math. Wiss.*, vol. 5, part 2, p. 543 (1920).

and the time inversion τ:

$$\tau: \quad x'^i = x^i, \quad x'^0 = -x^0. \tag{1-9}$$

For these special transformations,

$$\det(a^\mu{}_\nu) = -1. \tag{1-10}$$

The "extended" Lorentz group is obtained by adjoining these special transformations to the proper Lorentz transformations. If we consider coordinate transformations which also involve displacements, we arrive at the inhomogeneous Lorentz transformations L:

$$L: \quad x'^\mu = a^\mu{}_\nu x^\nu + a^\mu, \tag{1-11}$$

where a^μ is a four-vector independent of x. We shall often write relation (1–11) or (1–6) in the abbreviated form

$$x' = Lx. \tag{1-12}$$

In the relativistic field theories, one studies the properties of a set of functions $\phi_r(x)$ of the position vector x which transform according to a linear transformation law under proper homogeneous Lorentz transformations:

$$\phi_r'(Lx) = S_r{}^s \phi_s(x), \tag{1-13}$$

where the subscripts r and s label the components. This transformation is completely general, valid for spinors, tensors, etc.

The transformation coefficients depend, of course, on the transformation L and they give rise to a representation of the group of homogeneous Lorentz transformations.

The simplest example of a field is the scalar field $\varphi(x)$. (Since there is only one component, we can omit the index r.) It transforms according to the scalar representation

$$\varphi'(Lx) = \varphi(x). \tag{1-14}$$

A special type of vector field is obtained from it by the process of differentiation:

$$\partial_\mu \varphi(x) \equiv \frac{\partial \varphi(x)}{\partial x^\mu}. \tag{1-15}$$

From a general vector field $a_\mu(x)$ one can similarly construct a skew-symmetrical tensor field of second rank:

$$f_{\mu\nu}(x) = \partial_\mu a_\nu(x) - \partial_\nu a_\mu(x), \tag{1-16}$$

$$f_{\mu\nu}(x) = -f_{\nu\mu}(x). \tag{1-17}$$

The principle of the special theory of relativity requires that all fundamental physical laws be covariant under Lorentz transformations. A correct physical law must therefore be expressed by a covariant equation. We call an equation covariant if both sides of the equation have the same transformation properties.

1-3 Some basic notions of quantum mechanics. The essential difference between quantum mechanics and classical mechanics is the entirely different concept of what constitutes the "state" of a physical system* in the two theories. In quantum mechanics the state of a system is described by a vector ω in a finite or infinite dimensional linear vector space. Different state vectors ω_a, ω_b can be multiplied in a scalar product (ω_a,ω_b) with the following properties:

$$(\omega_a,\lambda\omega_b) = \lambda(\omega_a,\omega_b), \quad (\lambda = \text{complex number}),$$

$$(\omega_a,\omega_b)^* = (\omega_b,\omega_a),$$

$$(\omega_a,\omega_b + \omega_c) = (\omega_a,\omega_b) + (\omega_a,\omega_c),$$

$$(\omega,\omega) > 0 \quad \text{for } \omega \neq \text{null vector}.$$

(1-18)

Two vectors ω_1, ω_2 are said to be orthogonal if $(\omega_1,\omega_2) = 0$.

Linear transformations in the vector space are described by linear operators A which have the following properties:

$$\omega \rightarrow \omega' = A\omega,$$

$$A(\omega_a + \omega_b) = A\omega_a + A\omega_b,$$

$$A\lambda\omega = \lambda A\omega.$$

(1-19)

Linear operators give rise to the important concept of the expectation value $\langle A \rangle$ with respect to ω.

$$\langle A \rangle = (\omega,A\omega). \tag{1-20}$$

If two operators A,B stand with respect to each other in the relation

$$AB\omega = BA\omega = \omega \quad \text{for all } \omega, \tag{1-21}$$

we call them inverse to each other and write

$$B = A^{-1}. \tag{1-22}$$

In quantum mechanics, there are two classes of operators which play special roles. They are the Hermitian and the unitary operators. They are defined most easily in terms of the Hermitian conjugate A^* of an

---

* For greater detail, see P. A. M. Dirac, *The Principles of Quantum Mechanics*, 3rd ed., Oxford University Press, 1947.

operator.† A^* is said to be the Hermitian conjugate of A if for any ω

$$(A^*\omega,\omega) = (\omega,A\omega). \qquad (1\text{--}23)$$

A Hermitian operator then is one which is equal to its own Hermitian conjugate:

$$A^* = A. \qquad (1\text{--}24)$$

An operator is unitary if its Hermitian conjugate is its inverse:

$$A^* = A^{-1}. \qquad (1\text{--}25)$$

The importance of the Hermitian operators in quantum mechanics stems from the fact that observable quantities of a physical system are represented by such operators.

A vector ω is said to be an eigenvector of A if

$$A\omega = a\omega, \qquad (1\text{--}26)$$

where a is a number. The following theorems are mentioned without proof.

The eigenvalues and expectation values of Hermitian operators are all real.

The eigenvalues of unitary operators are of magnitude one.

The eigenvectors ω_1, ω_2 associated with two different eigenvalues a_1, a_2, $a_1 \neq a_2$, are orthogonal: $(\omega_1,\omega_2) = 0$ for Hermitian or unitary operators.

For operators in finite dimensional vector spaces one can show further that it is possible to choose a system of mutually orthogonal sets of eigenvectors of a Hermitian or unitary operator which are a complete set. *Complete* means that any other vector of this space is a linear combination of this set.

The extension of this last theorem to infinite dimensional spaces is a mathematical problem of considerable complexity.‡ For the physical interpretation of quantum mechanics it is essential to assume that observables are always represented by Hermitian operators which have this property. This is sometimes called the expansion postulate.

---

† The star operation (*) when applied to an operator in Hilbert space always means Hermitian conjugation. When it is applied to a complex number it means the complex conjugate. For this and similar questions on notation, the reader is referred to Appendix A8.

‡ See J. v. Neumann, *Mathematische Grundlagen der Quantenmechanik*, Springer, Berlin, 1932.

In the following we shall often need the commutator and anticommutator brackets of two operators A, B defined by

$$[A,B] = AB - BA \qquad (1\text{-}27)$$

and

$$\{A,B\} = AB + BA. \qquad (1\text{-}28)$$

In the systematic development of quantum mechanics of atomic systems the commutator brackets are of great importance because they have formal properties which stand in close analogy to the classical Poisson brackets.* They allow a formulation of the dynamical properties of a system which corresponds directly to the classical Hamiltonian form of mechanics.

In the quantum theory of fields this analogy has been of doubtful value. One reason for this is that the Hamiltonian approach is basically nonrelativistic and is therefore not suited for a relativistic field theory. Also, the quantization of spinor fields, such as the matter field to be discussed in Chapter 3, must be carried through with the anticommutator brackets (1-28) and the analogy to classical Poisson brackets breaks down. It is for these reasons that we shall abandon entirely the Hamiltonian formulation of the theory. The above-mentioned difficulties do not occur in the Lagrangian form, and we shall see below that the Lagrangian plays a central role in the generalized Action Principle.

1-4 Localizability. Classically, a field is conceived as a mechanical system which describes physical conditions at any point in the space-time continuum. A typical example is the electromagnetic field. In this case, the six components of the field tensor are needed for a complete description of the physical conditions at any point. Since the field variables are labeled by a continuous variable (*viz.*, the coordinate vector x), the number of degrees of freedom has the power of the continuum.

The quantum-mechanical counterpart of such a classical field is the concept of the localizable field.†

A localizable dynamical variable is a quantum-mechanical operator which describes the physical conditions at one particular point x in space and time. In the quantum theory of radiation, the field variables at one particular point are such operators. Other examples are the energy density and the current density for the matter field. On the other hand, neither the total momentum nor the total charge is localizable, since they describe conditions in an infinite region in space.

---

* P. A. M. Dirac, *loc. cit.*, esp. Chapter IV.

† For the discussion of localizable systems, see P. A. M. Dirac, *Phys. Rev.* **73**, 1092 (1948).

A localizable dynamical system is one for which a complete set of localizable dynamical variables exists. Thus, for instance, a system of point particles is a localizable system, the localizable variables in this case being the position of the particles and their spins, if any. These variables are indeed complete in the sense that a measurement of them is sufficient to determine the state of the system.

A localizable field is such a system. The variables needed for the description of the system are the field variables, which may be denoted by $\phi_r(x)$. In the past, the property of localizability has been a basic assumption in all field theories. The important question arises whether the fields actually occurring in nature are, in fact, localizable. We do not know the answer to this question. Recent attempts to develop theories of nonlocalizable fields are still in a preliminary stage* and have not yet led to any tangible results.

1–5 Observables of a field. In all field theories the connection with empirical facts is established with the help of certain operators called observables. The observables are operators with real eigenvalues and are therefore Hermitian. It is important to note that the field variables do not need to be observables; in fact, they do not even need to be Hermitian. However, in all cases there exist certain functions or functionals of the field variables which represent observables.

In quantum mechanics the concept of a complete set of commuting observables is of fundamental importance.† By measuring the values of such a set of observables in a system the state vector is uniquely determined. It is the common eigenvector of the set associated with the eigenvalues obtained as a result of a measurement of the observable.‡ Although the field variables are not necessarily observables, the description of the field is complete only if it is possible to set up a complete set of commuting observables which are functions (or functionals) of the field variables.§

---

* H. Yukawa, *Phys. Rev.* **80**, 1047 (1950). M. Fierz, *Helv. Phys. Acta* **23**, 412 (1950). D. R. Yennie, *Phys. Rev.* **80**, 1053 (1950); **85**, 877 (1952). C. Bloch, *Progr. Theor. Phys.* **5**, 606 (1950). The difficulty seems to be that noninteracting, nonlocal fields give nothing new because they are equivalent to a superposition of local fields with different spin values, while on the other hand it has not been possible to introduce interactions in a consistent manner.

† P. A. M. Dirac, *Quantum Mechanics*, p. 57.

‡ We assume for simplicity that these are "measurements of the first kind." A repetition of such a measurement will reproduce with certainty the previous result. See W. Pauli, *Handbuch d. Physik*, **24**, 1, p. 152; also L. Landau and R. Peierls, *Z. Physik* **69**, 56 (1931).

§ For the definition of functionals, see V. Volterra, *Theory of Functionals*, Blackie and Son, Ltd., London, 1930.

There may be several independent observables associated with each point x which we do not need to distinguish explicitly. We denote by $\zeta(x)$ the set of observables of the system associated with the point x. Two sets $\zeta(x_1)$, $\zeta(x_2)$ at two different points x_1, x_2 do not necessarily commute with each other, but it follows from general principles of quantum mechanics that two observables located at two points x_1, x_2 with a space-like separation

$$(x_1 - x_2)^2 > 0 \tag{1-29}$$

always commute. The physical content of this statement is the fact that no measurable effect can be propagated faster than the speed of light. The measurement of physical quantities at space-like situated points must therefore be entirely independent of each other. This means in the quantum-mechanical formalism that the observables associated with these quantities must commute.

As a corollary of this conclusion, we note:

If two field quantities do not commute at space-like situated points, at least one of them is not an observable.

A complete set of commuting observables will thus be associated with a three-parameter surface σ which is everywhere space-like; that is, the line joining any two of the points is outside the light cone. Among these surfaces the three-dimensional hyperplanes (hereafter just called planes) are especially easy to deal with. It is quite possible to operate with state vectors arising from the observables on more general space-like surfaces.* Such theories have not led to any physical conclusions which could not also be obtained by the much simpler theories working with planes only. Thus we shall work with planes exclusively.

A general space-like plane is characterized by a unit time-like normal vector n^μ and an "instant" parameter τ related by the equation of the plane

$$\sigma: \quad n \cdot x + \tau = 0, \quad n^2 = -1. \tag{1-30}$$

In (1-30), $n \cdot x = n^\mu x_\mu$ denotes the scalar product of the two four-vectors n, x, a notation which we shall often use in the following.

The dynamical characteristics of the system are contained in the relationship of the state vectors or the observables associated with two dif-

---

* A theory involving more general space-like surfaces was developed for the first time by P. Weiss, *Proc. Roy. Soc.* **A169**, 102 (1938). Later work especially by S. Tomonaga, *Progr. Theor. Phys.* **1**, 27 (1946). J. Schwinger, *Phys. Rev.* **74**, 1439 (1948). K. V. Roberts, *Phys. Rev.* **77**, 146 (1950); *Proc. Roy. Soc.* **A204**, 123 (1950); **A207**, 228 (1951); *Phys. Rev.* **81**, 158 (1951) (L). P. A. M. Dirac, *Can. J. Math.* **2**, 129 (1950); **3**, 1 (1951); *Phys. Rev.* **73**, 1092 (1948). T. S. Chang, *Phys. Rev.* **78**, 592 (1950).

ferent planes σ_1, σ_2. The relationship between the state vectors on two such surfaces is completely arbitrary so long as we do not specify the relation of the observables on different planes. A particularly simple way (which, as we shall see, has several advantages) is to assume that the state vector ω representing the physical state of a system is the same on all planes. This is called the Heisenberg picture. The relationship of the observables on two planes is then fixed by the dynamical properties of the system.

1–6 Canonical transformations. A unitary operator U gives rise to a canonical transformation of a field by the relations

$$\phi_r(x) \rightarrow \phi_r'(x) = U\phi_r(x)U^{-1}, \quad (1\text{--}31)$$

$$UU^* = U^*U = 1. \quad (1\text{--}32)$$

The physical properties of the system described by the transformed field ϕ_r' are exactly the same as the properties of the system described by the original field. Indeed, a canonical transformation (1–31) may always be interpreted as the change induced in the field operators by a coordinate transformation in the space of the state vectors. The physically significant quantities are those which are invariant under these transformations, such as the eigenvalues of operators or the relationship of a state vector to the eigenvectors.

The new operators $\phi_r'(x)$ have the same eigenvalues as the old ones, and if Ω is an eigenvector of $\phi_r(x)$, it follows from (1–31) and (1–26) that

$$\Omega' = U\Omega \quad (1\text{--}33)$$

is an eigenvector of $\phi_r'(x)$.

Because of this property, we may call two fields physically equivalent if they are related by a canonical transformation (1–31).

We obtain infinitesimal canonical transformations if we write U in the form

$$U = 1 + iF, \quad (1\text{--}34)$$

where F is considered an infinitesimal quantity of first order. The unitary condition (1–32) becomes then, if we disregard the higher order contributions,

$$U^*U = (1 - iF^*)(1 + iF) \simeq 1 - i(F^* - F) = 1,$$

or

$$F^* = F. \quad (1\text{--}35)$$

Thus F is Hermitian.

We call F the generator of the infinitesimal canonical transformation U. The iteration of the infinitesimal canonical transformations generates the finite canonical transformations, which may be written as

$$U = 1 + iF + \frac{i^2}{2!}F^2 + \cdots = e^{iF} = \lim_{n \to \infty}\left(1 + \frac{i}{n}F\right)^n,$$

$$U^* = 1 - iF + \frac{(-i)^2}{2!}F^2 + \cdots = e^{-iF} = \lim_{n \to \infty}\left(1 - \frac{i}{n}F\right)^n.$$

(1-36)

In the last two expressions F is considered a finite operator.

The transformed operators ϕ' (omitting for simplicity the index r and the argument x) may be written

$$\phi' = (1 + iF)\phi(1 - iF) \simeq \phi + i[F,\phi]. \tag{1-37}$$

For the increment $\delta\phi = \phi' - \phi$, we thus obtain

$$\delta\phi = i[F,\phi]. \tag{1-38}$$

The finite transformation U generated by F according to (1-36) transforms ϕ into

$$\phi' = e^{iF}\phi e^{-iF} = \phi + i[F,\phi] + \frac{i^2}{2!}[F,[F,\phi]] + \cdots. \tag{1-39}$$

This last result may be written in compact form by introducing the notation

$$[F,\phi]_n \equiv [F,\cdots,[F,[F,\phi]]\cdots], \quad (n \text{ brackets})$$

$$[F,\phi]_0 \equiv \phi,$$

(1-40)

and we obtain for (1-39),

$$\phi' = \sum_{n=0}^{\infty} \frac{i^n}{n!}[F\ \phi]_n. \tag{1-41}$$

1-7 Lorentz transformations as canonical transformations. In Section 1-2 [Eq. (1-13)], we have given the general transformation law for a field $\phi_r(x)$ under Lorentz transformations. We now apply the principle of the special theory of relativity. According to this principle, the original field $\phi_r(x)$ and the transformed field (1-13) describe systems which are physically indistinguishable. The two fields are thus connected by a canonical transformation

$$\phi_r'(x) = U(L)\phi_r(x)U^{-1}(L), \tag{1-42}$$

where

$$\phi_r'(Lx) = S_r{}^s\phi_s(x) \tag{1-43}$$

and L is given by*

$$L: \quad x_\mu \rightarrow x_\mu' = a_\mu{}^\nu x_\nu + a_\mu. \tag{1-11}$$

We observe that the operators $U(L)$ defined by (1–42) form a representation of the Lorentz group:†

$$U(L_2 L_1) = U(L_2) U(L_1). \tag{1-44}$$

Of particular importance are the infinitesimal Lorentz transformations, which may be written

$$\begin{aligned} a_\mu{}^\nu &= g_\mu{}^\nu + \alpha_\mu{}^\nu, \\ a_\mu &= \alpha_\mu, \end{aligned} \tag{1-45}$$

where $\alpha_\mu{}^\nu$ and α_μ are infinitesimals of first order. The relation

$$a^\mu{}_\nu a^{\lambda\nu} = g^{\mu\lambda} \tag{1-46}$$

then leads to

$$\alpha_{\mu\nu} + \alpha_{\nu\mu} = 0. \tag{1-47}$$

The infinitesimal part of the transformation U may be written explicitly

$$U = 1 + iK. \tag{1-48}$$

If we retain the first order terms in K, we may write K as a linear function of the α's:

$$K = \tfrac{1}{2} M^{\mu\nu} \alpha_{\mu\nu} + P^\mu \alpha_\mu. \tag{1-49}$$

Equation (1–49) defines the ten operators $M^{\mu\nu} = -M^{\nu\mu}$ and P^μ which are of fundamental importance in any field theory. They are Hermitian operators and represent observable physical quantities, *viz.*, angular momentum and linear momentum of the system. We shall refer to them for short as the momentum operators, although the P^0 is the operator which in nonrelativistic mechanics is called the energy, and the operators $M^{\mu\nu}$ contain, besides the components usually denoted as angular momenta, the three components $M^{0\mu}$ associated with Lorentz transformations proper (that is, not just rotations in 3-space).

---

* The requirement of a linear transformation law (1–43) for the fields must be relaxed for the time-reversal transformations. For detailed discussion of this point, see Section 5–4.

† Strictly speaking, we obtain from (1–42) only a "ray representation"

$$U(L_2 L_1) = c(L_2, L_1) U(L_2) U(L_1),$$

where $c(L_2, L_1)$ is a numerical factor, since Eq. (1–42) determines $U(L)$ only up to a numerical factor. However, one can show that under fairly general assumptions the "ray representation" can always be changed into a "vector representation" (1–44) by suitable normalization of the $U(L)$. This we shall henceforth do. See E. P. Wigner, *Ann. Math.* **40,** 149 (1939).

1-7] LORENTZ TRANSFORMATIONS AS CANONICAL TRANSFORMATIONS

The momentum operators have some simple properties which we shall now derive. We note first that these operators are not entirely independent of each other. They are restricted by Eq. (1–44), which expresses the fact that the $U(L)$ are a representation of the inhomogeneous Lorentz group. In the theory of Lie groups one shows that this restriction appears in a very simple form* when it is expressed in terms of the operators X_r associated with the infinitesimal transformations of a Lie group (of which the Lorentz group is an example). Then Lie's theorem asserts that the commutators of any two of these operators is a linear combination of these operators:

$$[X_r, X_s] = \sum_t c_{rs}{}^t X_t. \tag{1-50}$$

The coefficients $c_{rs}{}^t$ are called the structure constants of the group because they are independent of the representation of the operators X_r, and may therefore be regarded as a property of the group. They can thus be determined from the composition law of the group elements. For the Lorentz group, the relations (1–50) take the form

$$[P^\mu, P^\nu] = 0,$$

$$-i[M^{\mu\nu}, P^\lambda] = g^{\mu\lambda} P^\nu - g^{\nu\lambda} P^\mu \tag{1-51}$$

$$-i[M^{\mu\nu}, M^{\rho\sigma}] = g^{\mu\rho} M^{\nu\sigma} - g^{\mu\sigma} M^{\nu\rho} + g^{\nu\sigma} M^{\mu\rho} - g^{\nu\rho} M^{\mu\sigma}.$$

We can also obtain a set of commutator relations of the momentum operators with the field operators if we use (1–38) with F replaced by K [(1–49)]:

$$\delta\phi_r = i[K, \phi_r]. \tag{1-52}$$

In this equation we develop the left and right sides in powers of $\alpha_{\mu\nu}$, α_μ and equate the coefficients of the linear terms on both sides. In order to do this for the left side, we need the infinitesimal part of the transformation matrix $S_r{}^s$:

$$S_r{}^s = \delta_r{}^s + \Sigma_r{}^s, \tag{1-53}$$

which we write

$$\Sigma_r{}^s = \tfrac{1}{2} \Sigma_r{}^{s\mu\nu} \alpha_{\mu\nu}, \tag{1-54}$$

where the coefficients $\Sigma_r{}^{s\mu\nu}$ are antisymmetrical in μ and ν. With this notation, the increment of ϕ_r becomes

$$\delta\phi_r(x) = \tfrac{1}{2}[\Sigma_r{}^{s\mu\nu} \phi_s(x) + (x^\mu \partial^\nu - x^\nu \partial^\mu)\phi_r(x)]\alpha_{\mu\nu} - \partial_\mu \phi_r(x) \alpha^\mu. \tag{1-55}$$

---

* See, for instance, L. Pontrjagin, *Topological Groups*, Princeton University Press, 1939, esp. Chapter IX.

If we equate this to the right side of (1–52), we obtain, with (1–49),

$$i[M^{\mu\nu},\phi_r(x)] = \Sigma_r{}^{s\mu\nu}\phi_s(x) + (x^\mu\partial^\nu - x^\nu\partial^\mu)\phi_r(x), \qquad (1\text{–}56)$$

$$i[P^\mu,\phi_r(x)] = -\partial^\mu\phi_r(x). \qquad (1\text{–}57)$$

These last two expressions may also be taken as the defining relations for the momentum operators. They characterize these operators up to an arbitrary additive constant.

Finally we observe that according to the foregoing the existence of these operators is assured provided that our localizable field describes a physical system in accord with the principle of the special theory of relativity. But nothing up to this point gives us any indication as to how these operators can be explicitly constructed in terms of the field variables. Indeed, this construction will depend entirely on the dynamical characteristics of the system, which we have so far not even considered. These dynamical characteristics can be introduced most conveniently by the **generalized Action Principle**.

1–8 The Action Principle. In Section 1–5 we introduced the concept of a complete set of commuting observables on a three-dimensional hyperplane σ. We denote by $\zeta(\sigma)$ such a set of observables. By definition, a measurement of this set will determine the state vector of the system. We may then label the eigenvectors of these observables by the eigenvalues $\zeta'(\sigma)$ of the operator set $\zeta(\sigma)$. These eigenvalues are ordinary real-valued functions of the coordinate vector $x \in \sigma$, and must be clearly distinguished from the operator functions $\zeta(\sigma)$ on the plane σ. The completeness of the set $\zeta(\sigma)$ implies that to every function $\zeta'(\sigma)$ there corresponds exactly one state vector which may therefore be labeled by the function $\zeta'(\sigma)$.

We consider now two different planes σ_0, σ with their associated sets of observables $\zeta(\sigma_0)$, $\zeta(\sigma)$. The dynamical properties of the system are expressed in the law which connects such sets of observables on two different planes. The form in which such a connection will be expressed depends, of course, on the relation of the state vectors on the two different planes σ_0, σ. Thus, in order to express this connection, it is necessary to specify the representation of the state vectors on two different planes. We shall in the following assume the Heisenberg picture. It is characterized by the property that the state vector which represents the physical state of the system is constant. This allows a simplification of the general theory because the dynamical law of the system is expressed entirely by the transformation law which connects the observables on different planes.

For a compact formulation of the general transformation law for operators on two different planes σ_0, σ. we introduce the notation $\zeta_0' = \zeta'(\sigma_0)$

and $\zeta' = \zeta'(\sigma)$. The eigenvectors associated with these functions are then denoted by $\Omega(\zeta_0')$ and $\Omega(\zeta')$, respectively.

The transformation law may be expressed in terms of the transformation matrix $(\zeta_0'|\zeta')$ which transforms the family of eigenvectors $\Omega(\zeta_0')$ into the family $\Omega(\zeta')$:*

$$\Omega(\zeta') = \sum_{(\zeta_0')} (\zeta_0'|\zeta')\Omega(\zeta_0'), \qquad (1\text{--}58)$$

$$\Omega(\zeta_0') = \sum_{(\zeta')} (\zeta_0'|\zeta')^*\Omega(\zeta'), \qquad (1\text{--}58\text{a})$$

$$(\zeta_0'|\zeta') = (\Omega(\zeta_0'),\Omega(\zeta')). \qquad (1\text{--}59)$$

To obtain further information regarding this transformation matrix, we proceed now to introduce the Action Principle. We begin by evaluating the change in the transformation matrix induced by an infinitesimal canonical transformation

$$\delta\phi_r = i[F,\phi_r]. \qquad (1\text{--}60)$$

Such a transformation changes the observables ζ on the plane σ into new observables $\zeta + \delta\zeta$:

$$\delta\zeta = i[F(\sigma),\zeta]. \qquad (1\text{--}61)$$

Such a change is hereafter called a canonical variation, or simply a variation of the field.† Because the variation is canonical, the new operators $\zeta + \delta\zeta$ have the same eigenvalues as the old ones but the state vector corresponding to a given set of eigenvalues ζ', according to (1–33) and (1–34), is changed by an amount

$$\delta\Omega(\zeta') = iF(\sigma)\Omega(\zeta') \qquad (1\text{--}62)$$

It should be remarked that the variation $\delta\zeta$ [(1–61)] does not determine the operator $F(\sigma)$ completely. In fact, we may add to a particular $F(\sigma)$ any operator $G(\sigma)$ which commutes with all the operators $\zeta(\sigma)$. Such operators exist because

---

* The general transformation matrix was introduced by P. A. M. Dirac, *Physik. Z. Sowjetunion* **3**, 64 (1933); later work regarding the transformation matrix, especially by Y. Tanikawa, *Progr. Theor. Phys.* **1**, 12 (1946); *ibid.* **2**, 219 (1947); S. Watanabe, *ibid.* **2**, 71 (1947); R. Utiyama, *ibid.* **2**, 117 (1947).

† It should perhaps be emphasized for clarity that the functional dependence indicated by $F(\sigma)$ is of quite a different nature than, for instance, the previously used $\zeta(\sigma)$. In the latter case, $\zeta(\sigma)$ is an operator function of the variable $x \in \sigma$; in the former case, on the other hand, $F(\sigma)$ merely indicates an operator depending on the four parameters n_μ and τ which characterize the plane σ.

any function of the operators $\zeta(\sigma)$ commutes with all the $\zeta(\sigma)$. The completeness of the operator set $\zeta(\sigma)$ implies that the converse of this is also true, i.e., a $G(\sigma)$ commuting with all $\zeta(\sigma)$ is always a function of the $\zeta(\sigma)$. The transformation (1–62) of the state vectors associated with such a commuting operator merely multiplies each state vector with a purely imaginary number which may be a function of ζ'. Thus (1–62) for this case may be written

$$\delta\Omega(\zeta') = ig(\zeta')\Omega(\zeta'), \tag{1-62a}$$

with $g(\zeta')$ a real-valued function of ζ'. The finite form of this type of transformation is then given by

$$\Omega'(\zeta') = e^{ig(\zeta')}\Omega(\zeta') \tag{1-62b}$$

and it is seen to be simply a transformation of the phases of the state vectors. Such phase transformations in the definition of the eigenvectors have no physical consequences and the ambiguity in the F which is associated with a given $\delta\zeta$ may be ignored.

A similar variation of the operators on σ_0 changes the eigenvectors on σ_0 according to

$$\delta\Omega(\zeta_0') = iF(\sigma_0)\Omega(\zeta_0'). \tag{1-63}$$

The new state vectors thereby obtained give rise to a new transformation matrix which may be written as $(\zeta_0'|\zeta') + \delta(\zeta_0'|\zeta')$. The variation of this matrix is then obtained from (1–59) in the form

$$\delta(\zeta_0'|\zeta') = i(\zeta_0'|\delta W|\zeta'), \tag{1-64}$$

$$\delta W = F(\sigma) - F(\sigma_0). \tag{1-65}$$

We consider now a region in space-time which is bounded by two parallel space-like planes σ_0, σ. A general variation of the field in this region and on its boundary is obtained by giving an arbitrary one-parameter family of operator functions $F(\sigma')$ defined on a one-parameter set of parallel planes σ' between σ_0 and σ. Such a general variation of the field will induce a well-defined variation in all functions or functionals of the field operators.

We are now in a position to formulate the basic *Action Principle:**

There exists an invariant function L (subsequently called the Lagrangian) of the field variables ϕ_r and their first derivatives $\phi_{r\mu} \equiv \partial_\mu \phi_r$ such that the operator δW is obtained by the variation of the action integral

$$W = \int_{\sigma_0}^{\sigma} L\, d^4x. \tag{1-66}$$

---

* J. Schwinger, *Phys. Rev.* **82**, 914 (1951).

Before we discuss the consequences which follow from this principle, we make some preliminary comments.

We have intentionally restricted the function L: it shall not contain derivatives of the field higher than the first. There is no formal difficulty of generalizing the Principle for Lagrangians containing higher order derivatives. We have not done so because the physical interpretation of such theories leads to insurmountable difficulties.*

Next it may be worth pointing out in what respect the principle differs from the usual principle of least action which has always been used in the past as the starting point of the quantum theory of any physical system. It differs in two respects. First, it is an Action Principle for a quantized field. It is thus the quantum-mechanical analogue of the corresponding classical variational principle. Second, it goes beyond the older form of the principle by including variations at the boundary and by postulating a certain connection of the operator δW with these variations at the boundary. By this extension we obtain additional information regarding the dynamical characteristics of the field which in the older theories had to be postulated separately. It was then necessary to show that these additional properties were compatible with the equations of motion. No such problem arises here, since all the basic assumptions are reduced to the one Action Principle. This is a considerable simplification.

The variation at the boundary may be considered as arising from two sources, namely, the variation $\delta_0 \phi_r$ at a fixed point in space-time, and the additional variation $\partial_\mu \phi_r \delta x^\mu$ caused by an infinitesimal displacement of the boundary plane. The total variation at the boundary is then the sum of the two:

$$\delta \phi_r = \delta_0 \phi_r + \partial_\mu \phi_r \delta x^\mu. \tag{1-67}$$

This separation of the variation at the boundary will be convenient later on.

The variations $\delta \phi_r$ do not necessarily commute with the field variables themselves. This fact may give rise to some complication if δW, the variation of W in (1-66), is expressed in terms of $\delta \phi_r$. For instance, if a variable ϕ of a scalar field does not commute with $\delta \phi$ then the variation of ϕ^2 is given by

$$\delta \phi^2 = \delta \phi \phi + \phi \delta \phi. \tag{1-68}$$

In the formula discussed below we shall for convenience ignore this point and pay no attention to the position of the variations $\delta \phi$. The formulas thereby obtained are then merely symbolic expressions of the properly symmetrized quantities such as (1-68).

---

* For complete details on this question the reader is referred to A. Pais and G. Uhlenbeck, *Phys. Rev.* **79**, 145 (1950), where also references to the numerous older publications on this problem can be found.

An explicit expression for the variation of the action integral is obtained if (1–67) is combined with*

$$\delta L = \frac{\partial L}{\partial \phi_r} \delta_0 \phi_r + \frac{\partial L}{\partial \phi_{r\mu}} \delta_0 \phi_{r\mu} \qquad (1\text{–}69)$$

for the variation of L.

After partial integration and substitution of (1–67), we obtain for δW

$$\delta W = \int_{\sigma_0}^{\sigma} \left(\frac{\partial L}{\partial \phi_r} - \partial_\mu \frac{\partial L}{\partial \phi_{r\mu}} \right) \delta_0 \phi_r d^4 x + F(\sigma) - F(\sigma_0), \qquad (1\text{–}70)$$

with

$$F(\sigma) = \int_\sigma \left[\left(\frac{\partial L}{\partial \phi_{r\mu}} \partial^\nu \phi_r - g^{\mu\nu} L \right) \delta x_\nu - \frac{\partial L}{\partial \phi_{r\mu}} \delta \phi_r \right] d\sigma_\mu \qquad (1\text{–}71)$$

and a similar expression for $F(\sigma_0)$.

The expression for $F(\sigma)$ is an integral over the three-dimensional hyperplane. We have used the notation $n_\mu d\sigma = d\sigma_\mu$ for the directed surface element on this plane, where n_μ is defined by (1–30).

By choosing special variations, we can draw various conclusions from the Action Principle. We shall begin with the equation of motion.

1–9 The equation of motion. We obtain the equation which governs the change of the field variables if we choose a variation which vanishes at the boundary planes σ_0, σ and also leaves these planes unchanged. Otherwise the variation is left completely arbitrary. For all such variations the observables ζ_0, ζ at the boundary are unchanged, and consequently the transformation matrix is also unaltered. Thus

$$F(\sigma) = F(\sigma_0) = 0 \qquad (1\text{–}72)$$

and

$$\delta W = \int_{\sigma_0}^{\sigma} \left(\frac{\partial L}{\partial \phi_r} - \partial_\mu \frac{\partial L}{\partial \phi_{r\mu}} \right) \delta_0 \phi_r d^4 x = 0. \qquad (1\text{–}73)$$

This equation is satisfied if the integrand vanishes at every point:

$$\frac{\partial L}{\partial \phi_r} - \partial_\mu \frac{\partial L}{\partial \phi_{r\mu}} = 0. \qquad (1\text{–}74)$$

These are the field equations.

The field equations can be said to follow from the Action Principle only if it is possible to show that (1–74) is a consequence of (1–73). In the calculus of variations for ordinary functions one shows that if an equation

---

* Note that the variation $\delta_0 \phi_{r\mu}$ of $\phi_{r\mu} \equiv \partial_\mu \phi_r$ is defined by $\delta_0 \phi_{r\mu} \equiv \partial_\mu \delta_0 \phi_r$, which may differ from $i[F, \phi_{r\mu}]$ if F varies from one surface to another.

such as (1–73) holds for a sufficiently large class of variations, then the relation (1–74) is a consequence of (1–73). It is very plausible that a similar theorem holds for the corresponding operator equations, but a rigorous proof does not seem to exist. We shall assume the theorem to be true. The other consequences of the Action Principle are independent of this assumption.

We note for future reference that, as a consequence of (1–73), we may write for the general variation (1–70) of W simply

$$\delta W = F(\sigma) - F(\sigma_0). \tag{1-75}$$

1–10 Momentum operators. In Section 1–7 we have shown that in a relativistically invariant field theory there exists a set of ten operators $M^{\mu\nu}$, P^μ which satisfy the relations (1–51), (1–56), and (1–57). We shall now obtain explicit expressions for these operators in terms of the Lagrangian L from the Action Principle. To this end we use a variation in the field induced by an infinitesimal Lorentz transformation. Under such a transformation the plane σ suffers a displacement given by the displacement vector

$$\delta x^\mu = \alpha^\mu{}_\nu x^\nu + \alpha^\mu, \tag{1-76}$$

and according to (1–43), (1–53), and (1–54) the field at the displaced point $x + \delta x$ is $\phi_r + \delta\phi_r$, with

$$\delta\phi_r(x) = \tfrac{1}{2} \Sigma_r{}^{s\mu\nu} \phi_s(x) \alpha_{\mu\nu}. \tag{1-77}$$

With the last two expressions, we obtain for the generating operator $F(\sigma)$ [(1–71)]

$$F(\sigma) = \int_\sigma [T^{\mu\nu}(\alpha_{\nu\rho} x^\rho + \alpha_\nu) - \tfrac{1}{2} \Pi^{r\mu} \Sigma_r{}^{s\nu\rho} \phi_s \alpha_{\nu\rho}] d\sigma_\mu. \tag{1-78}$$

where we have introduced the abbreviations*

$$\Pi^{r\mu} = \frac{\partial L}{\partial \phi_{r\mu}}, \tag{1-79}$$

$$T^{\mu\nu} = \Pi^{r\mu} \partial^\nu \phi_r - g^{\mu\nu} L. \tag{1-80}$$

We write (1–78) in the form

$$F(\sigma) = \tfrac{1}{2} M^{\mu\nu} \alpha_{\mu\nu} + P^\mu \alpha_\mu, \tag{1-81}$$

---

* We shall not discuss here the complications which arise when the operators (1–79) are functionally dependent, leading to subsidiary conditions among the field variables. A problem of this sort occurs in the quantization of the radiation field and will be dealt with in Chapter 2. The general case is discussed by J. Schwinger, *Phys. Rev.* **82** (1951), esp. p. 919. See also P. G. Bergmann and R. Schiller, *Phys. Rev.* **89**, 4 (1953).

with
$$M^{\mu\nu} = \int_\sigma (T^{\rho\mu}x^\nu - T^{\rho\nu}x^\mu - \Pi^{r\rho}\Sigma_r{}^{s\mu\nu}\phi_s)d\sigma_\rho, \tag{1-82}$$

$$P^\mu = \int_\sigma T^{\rho\mu}d\sigma_\rho. \tag{1-83}$$

We now recall that the operator $F(\sigma)$ [(1-81)] is the generating operator for the variation of the field at a point on the boundary σ. According to (1-67), (1-76), and (1-77), this variation is given by

$$\delta_0\phi_r = \tfrac{1}{2}[\Sigma_r{}^{s\mu\nu}\phi_s + (x^\mu\partial^\nu - x^\nu\partial^\mu)\phi_r]\alpha_{\mu\nu} - \partial^\mu\phi_r\alpha_\mu$$

$$= i[F(\sigma),\phi_r]. \tag{1-84}$$

Since this equation must hold for arbitrary values of the ten parameters $\alpha_{\mu\nu}$, α_μ, we obtain the set of equations

$$i[M^{\mu\nu},\phi_r] = \Sigma_r{}^{s\mu\nu}\phi_s + (x^\mu\partial^\nu - x^\nu\partial^\mu)\phi_r, \tag{1-56}$$

$$i[P^\mu,\phi_r] = -\partial^\mu\phi_r, \tag{1-57}$$

which are seen to be identical with (1-56) and (1-57). We have thus obtained with (1-82) and (1-83) explicit expressions for the ten momentum operators in terms of the Lagrangian L and its derivatives.

The tensor $T^{\mu\nu}$ defined by (1-80) is called the *canonical momentum tensor*. We note that $M^{\mu\nu}$ may be split in a natural way into two parts, defined by

$$M^{\mu\nu} = L^{\mu\nu} + N^{\mu\nu}, \tag{1-85}$$

$$L^{\mu\nu} = \int_\sigma (T^{\rho\mu}x^\nu - T^{\rho\nu}x^\mu)d\sigma_\rho, \tag{1-86}$$

$$N^{\mu\nu} = -\int_\sigma \Pi^{r\rho}\Sigma_r{}^{s\mu\nu}\phi_s d\sigma_\rho, \tag{1-87}$$

and we shall show later that these two parts correspond to the separation of the total angular momentum ($M^{\mu\nu}$) of the field into orbital ($L^{\mu\nu}$) and spin ($N^{\mu\nu}$) angular momentum.

We also remark that the equations (1-51) need not be verified separately, since they are a consequence of the representation property (1-44) which is implicit in our definition of the unitary operators $U(L)$.

1-11 Conservation laws. The definitions of the momentum operators (1-82) and (1-83) seem to depend on the surface σ over which the integrals are to be taken. These integrals are, in fact, independent of this surface and this is the content of the conservation laws. We prove these conservation laws by establishing first that the transformation matrix $(\zeta_0'|\zeta')$ is invariant under Lorentz transformations. Indeed, since the transformed eigenvectors Ω' are given by (1–33), we obtain for the new transformation matrix

$$(\zeta_0'|\zeta')' \equiv (U\Omega(\zeta_0'), U\Omega(\zeta'))$$
$$= (\Omega(\zeta_0'), \Omega(\zeta')) = (\zeta_0'|\zeta'), \tag{1-88}$$

where we have made use of the fact that U is unitary. Specializing this result for the infinitesimal Lorentz transformations, we obtain by (1–64)

$$\delta(\zeta_0'|\zeta') \equiv i(\zeta_0'|\delta W|\zeta') = 0, \tag{1-89}$$

or with (1–65)

$$F(\sigma) = F(\sigma_0), \tag{1-90}$$

where $F(\sigma)$ is given by (1–81). Since the parameters $\alpha_{\mu\nu}$, α_μ are arbitrary, this means that each of the ten momentum operators satisfies a separate conservation law:

$$M^{\mu\nu}(\sigma) = M^{\mu\nu}(\sigma_0), \tag{1-91}$$

$$P^\mu(\sigma) = P^\mu(\sigma_0), \tag{1-92}$$

for any two planes σ, σ_0.

The last equations state that the ten momentum operators are integrals of motion.

We can express the conservation laws just obtained in differential form by transforming equations (1–91) and (1–92) with the help of Gauss' theorem applied to a region bounded by the two surfaces σ_0 and σ. This theorem states that for any set of functions $f^\mu(x)$ ($\mu = 0, \cdots, 3$) which vanish sufficiently fast in space-like directions

$$\int_{\sigma_0}^{\sigma} \partial_\mu f^\mu d^4x = -\int_\sigma f^\mu d\sigma_\mu + \int_{\sigma_0} f^\mu d\sigma_\mu. \tag{1-93}$$

If the conservation laws hold, the surface integrals on the right side of (1–93) cancel and since σ_0 and σ are arbitrary, we must have

$$\partial_\mu f^\mu = 0. \tag{1-94}$$

Applying this result to the integrands in (1–82) and (1–83), we obtain the equations

$$T^{\mu\nu} - T^{\nu\mu} + \partial_\rho H^{\rho\mu\nu} = 0, \tag{1-95}$$

$$\partial_\mu T^{\mu\nu} = 0, \tag{1-96}$$

with

$$H^{\rho\mu\nu} = \Pi^{r\rho} \Sigma_r{}^{s\mu\nu} \phi_s = -H^{\rho\nu\mu}. \tag{1-97}$$

Equations (1–95) and (1–96) can also be verified directly. Referring back to the definition (1–80), we see that they are indeed immediate consequences of the field equations (1–74).

Equation (1–95) can be used to define the symmetrical momentum tensor:

$$\theta^{\mu\nu} = T^{\mu\nu} + \partial_\rho G^{\rho\mu\nu}, \tag{1-98}$$

with

$$G_{\rho\mu\nu} = \tfrac{1}{2}(H_{\rho\mu\nu} + H_{\mu\nu\rho} + H_{\nu\mu\rho}). \tag{1-99}$$

The verification of the following properties of this tensor are left to the reader:

$$\theta^{\mu\nu} = \theta^{\nu\mu}, \tag{1-100}$$

$$P^\nu = \int_\sigma \theta^{\mu\nu} d\sigma_\mu, \tag{1-101}$$

$$\partial^\mu \theta_{\mu\nu} = 0. \tag{1-102}$$

1–12 Commutation rules. The Action Principle also contains some relations concerning the commutation rules of the field operators. To extract these relations, we consider an arbitrary variation $\delta\phi_r$ with stationary planes σ_0 and σ. According to (1–71), the generating operator for this case is given by

$$F(\sigma) = -\int_\sigma \Pi^{r\mu} \delta\phi_r d\sigma_\mu, \tag{1-103}$$

and Eq. (1–60) yields

$$\delta\phi_r(x) = i[\phi_r(x), \int_\sigma \Pi^{s\mu}\delta\phi_s d\sigma_\mu], \quad \text{for } x \in \sigma. \tag{1-104}$$

We can draw no further conclusions unless we make some assumptions regarding the commutation properties of $\delta\phi_s$ with ϕ_r. The Action Principle does not contain any information regarding these commutators. We can show, however, that there exist two natural assumptions regarding the commutation rules, which we obtain if we write (1–104) in two different ways. They are based on the following identities for any three operators A, B, and C:

$$[A,BC] = [A,B]C + B[A,C]$$
$$= \{A,B\}C - B\{A,C\}. \tag{1-105}$$

If we substitute for A, B, and C the three operators $\phi_r(x)$, $\Pi^{s\mu}(y)$, $\delta\phi_s(y)$ under the integral in (1–104), we obtain the identities

$$\delta\phi_r(x) = i\int [\phi_r(x), \Pi^{s\mu}(y)]\delta\phi_s(y) d\sigma_\mu + i\int \Pi^{s\mu}(y)[\phi_r(x), \delta\phi_s(y)] d\sigma_\mu$$
$$= i\int \{\phi_r(x), \Pi^{s\mu}(y)\}\delta\phi_s(y) d\sigma_\mu - i\int \Pi^{s\mu}(y)\{\phi_r(x), \delta\phi_s(y)\} d\sigma_\mu. \tag{1-106}$$

From this we see that there exist two simple possibilities of satisfying (1-104):

(a)
$$[\phi_r(x), \delta\phi_s(y)] = 0,$$
$$[\phi_r(x), \Pi^{s\mu}(y)] = -i\delta_r{}^s \delta^\mu(x,y).$$
(1-107a)

(b)
$$\{\phi_r(x), \delta\phi_s(y)\} = 0,$$
$$\{\phi_r(x), \Pi^{s\mu}(y)\} = -i\delta_r{}^s \delta^\mu(x,y).$$
(1-107b)

In the last two expressions we have introduced the δ-function $\delta^\mu(x,y)$ with respect to the plane σ. It has the property

$$\int \delta^\mu(x,y) f(y) d\sigma_\mu = f(x) \qquad (1\text{-}108)$$

for any function $f(x)$.*

The particular choice (1-107a) or (1-107b) for the basic commutation rules implies a restriction for the admissible generating functions F which generate the variations at the boundary. For instance, from (1-107a) follows

$$[\phi_r(x), [F, \phi_s(y)]] = 0. \qquad (1\text{-}109a)$$

Similarly, from (1-107b) follows

$$\{\phi_r(x), [F, \phi_s(y)]\} = 0. \qquad (1\text{-}109b)$$

Thus we are led to the two possibilities of quantization expressed in Eqs. (1-107a) and (1-107b). In order to decide which of the two types of quantizations apply, some further requirement of a physical nature is needed. These two possibilities correspond to the two possibilities for the statistics of identical particles known as Einstein-Bose and Fermi-Dirac statistics. They lead to different physical results which can be tested empirically. It has been known for a long time that systems of identical particles with integral spin (such as photons, for instance) are described by fields quantized according to case (a), while systems of identical particles with half-integer spin (such as electrons) are described by a field quantized in accord with case (b).

The question arises whether this connection between spin and statistics is a general law of nature, and if so, whether it can be derived from other physical principles. This is indeed the case.† Pauli showed that there

---

* It should be clear that the arguments which lead us to the selection of the two types of commutation rules (a) and (b) are merely of *heuristic* value.

† W. Pauli, for spin values $\tfrac{1}{2}$ and 0: *Ann. de l'Inst. H. Poincaré* **6**, 137 (1936); for general spin values: *Phys. Rev.* **58**, 716 (1940); *Progr. Theor. Phys.* **5**, 526 (1950).

are two principles involved:

(1) The total energy of the system must be a positive definite operator such that the vacuum state is the state of lowest energy.
(2) Observables at two points with space-like separation must commute with each other.

It can be shown that the quantization of fields with half-integer spin according to case (a) would violate principle (1), while on the other hand the quantization of fields with integer spin according to case (b) would violate principle (2).*

From these general considerations, it follows, then, that the radiation field (photons) is quantized according to case (a), while the matter field (electrons) satisfies the commutation rules under case (b). This is, of course, in agreement with well-known empirical facts such as the blackbody radiation for photons and the exclusion principle for electrons.

---

* There is also a connection of the quantization with the invariance under time reversal which was noted by Schwinger [*Phys. Rev.* **82,** 914 (1951)]. We do not agree, however, with the conclusion reached in this paper that the commutation rules are unambiguously determined from the Action Principle and the invariance under time reversal. Counter examples were given by J. M. Jauch, *Helv. Phys. Acta* **27,** 89 (1954) and W. K. Burton and B. F. Touschek, *Phil. Mag.* **44,** 161 (1953). We shall discuss the time-reversal transformation in detail in Section 5-4.

CHAPTER 2

THE RADIATION FIELD

The general method of Chapter 1 is now applied to the electromagnetic field. In this chapter we disregard entirely the effect of the sources and study only the quantum-mechanical formalism of the field far removed from the sources, or the radiation field. The field equations are especially simple if they are expressed in terms of the vector potential subject to a subsidiary condition (Sections 2-1, 2-2, and 2-3). This condition introduces a complication in the quantum-mechanical formalism which was not discussed in the general theory of Chapter 1. Apart from this, the general theory can be readily applied to derive the correct expressions for the commutation rules and the momentum operators (Sections 2-4 and 2-5).

We then decompose the fields into plane wave solutions of the field equations (Section 2-6). The amplitudes of these waves have the same commutation rules as the amplitudes of harmonic oscillators and explicit representations are readily constructed (Section 2-7).

We also give an explicit discussion of the photon spin (Section 2-8). Special attention is given to the state vector which represents the photon vacuum (Section 2-9).

2-1 The classical field equations. The electromagnetic field in a space region which is far removed from any sources of the field is called the radiation field. Such a field is classically described by a skew-symmetrical tensor field $f_{\mu\nu}(x)$.

$$f_{\mu\nu}(x) = -f_{\nu\mu}(x), \quad (\mu,\nu = 0, \cdots, 3). \tag{2-1}$$

The field components satisfy the homogeneous Maxwell equations

$$\partial_\lambda f_{\mu\nu} + \partial_\mu f_{\nu\lambda} + \partial_\nu f_{\lambda\mu} = 0, \tag{2-2}$$

$$\partial^\mu f_{\mu\nu} = 0. \tag{2-3}$$

It follows from the last two equations and (2-1) that each component separately satisfies the wave equation

$$\partial^\lambda \partial_\lambda f_{\mu\nu} = 0. \tag{2-4}$$

The last equation is much simpler than the pair (2-2), (2-3) but it cannot be used to replace this pair, since it states less than the equations (2-2) and (2-3). We can, however, define a vector field $a_\mu(x)$ which satisfies the wave equation by writing $f_{\mu\nu}$ as

$$f_{\mu\nu} = \partial_\mu a_\nu - \partial_\nu a_\mu, \tag{2-5}$$

so that (2-2) is identically satisfied. If we write (2-3) in terms of $a_\mu(x)$, we obtain the equation

$$\partial^\mu \partial_\mu a_\nu - \partial_\nu \chi = 0 \tag{2-6}$$

with

$$\chi \equiv \partial^\mu a_\mu. \tag{2-7}$$

We note here that the vector potential, defined by (2-5), is not uniquely determined by (2-5). We may always add to $a_\mu(x)$ the gradient of a scalar $\varphi(x)$ without changing the field variables:

$$a_\mu'(x) = a_\mu(x) + \partial_\mu \varphi(x). \tag{2-8}$$

The field variables $f_{\mu\nu}(x)$ describe directly measurable quantities, namely, the strength of the fields at a point x, while the vector potential must be considered as a mathematical and auxiliary field which can be determined by measurements only to the extent expressed by the *gauge transformations* (2-8). Thus a_μ and a_μ' describe the same physical system.

In spite of this complication, the field $a_\mu(x)$ is easier to work with than the Maxwell field $f_{\mu\nu}(x)$. The freedom in the definition of $a_\mu(x)$ can be used to simplify Eq. (2-6). Indeed, if we choose φ as the solution of the equation

$$\partial^\mu \partial_\mu \varphi + \chi = 0, \tag{2-9}$$

then the field $a_\mu'(x)$ satisfies

$$\chi' \equiv \partial^\mu a_\mu' = 0. \tag{2-10}$$

For this field, Eq. (2-6) reduces to

$$\partial^\mu \partial_\mu a_\nu' = 0. \tag{2-11}$$

In the following we shall always choose φ as explained above, and omit the primes. Our basic equations of the theory are then

$$f_{\mu\nu} = \partial_\mu a_\nu - \partial_\nu a_\mu, \tag{2-12}$$

$$\partial^\lambda \partial_\lambda a_\mu = 0, \tag{2-13}$$

$$\chi \equiv \partial^\mu a_\mu = 0. \tag{2-14}$$

These last three equations are equivalent to the original Maxwell equations (2-2) and (2-3). Equation (2-14) is treated as a subsidiary condition and is referred to as the Lorentz condition in the following. Even with the restriction (2-14) the potentials are not uniquely defined by (2-12), since the *restricted gauge transformations* are still allowed:

for φ satisfying
$$a_\mu \to a_\mu + \partial_\mu \varphi,$$
$$\partial^\mu \partial_\mu \varphi = 0. \tag{2-15}$$

2-2 The associated boundary value problem.

With a partial differential equation such as (2-13) there is associated a boundary value problem which may be formulated as follows. Given $a_\mu(x)$ and $n^\nu \partial_\nu a_\mu(x) \equiv \partial a_\mu(x)$ on a space-like plane σ with normal vector n, find the value of $a_\mu(x)$ at any other point in space. The solution of this problem can be written down immediately, with the help of Green's theorem,* in the following form:

$$a_\mu(x) = \int_\sigma d\sigma_\nu'(\partial'^\nu a_\mu(x')D(x-x') - \partial'^\nu D(x-x')a_\mu(x')), \quad (2\text{-}16)$$

where D is the D-function defined in the mathematical appendix A1. In our present context, the following properties of the D-function are relevant:

$$\partial^\lambda \partial_\lambda D(x) = 0,$$

$$D(x) = 0, \quad (\text{for } x^2 > 0), \quad (2\text{-}17)$$

and

$$\partial_\mu D(x) = \delta_\mu(x), \quad (\text{for } x^2 > 0), \quad (2\text{-}18)$$

where $\delta_\mu(x)$ is the δ-function associated with the surface σ which has already been used in Chapter 1:

$$\int \delta_\mu(x)f(x)d\sigma^\mu = f(0). \quad (2\text{-}19)$$

The proof that (2-16) is the solution of the boundary value problem is now obtained as follows. The integral (2-16) is of the form $\int d\sigma_\nu' f^\nu(x')$ with $f^\nu(x)$ satisfying $\partial_\nu f^\nu = 0$. Thus its value is independent of the plane σ. By choosing a plane σ which passes through x, we find that (2-16) becomes identical with (2-19).

We note that from (2-16) it follows that the Lorentz condition (2-14) is satisfied at any point x if it is satisfied at a point on the plane together with its normal derivative. In other words, the equations

$$\chi(x) = 0, \quad \partial\chi(x) = 0, \quad (\text{for } x \in \sigma) \quad (2\text{-}20)$$

imply

$$\chi(x) = 0, \quad (\text{for all } x). \quad (2\text{-}21)$$

---

* See, for instance, H. Jeffreys and B. S. Jeffreys, *Methods of Mathematical Physics*, 2nd ed., Cambridge University Press, 1950, p. 195 ff. The theorem used in the present case is only a slight generalization of the usual form of this theorem to the case of four dimensions with the metric of the Lorentz group.

2-3 A Lagrangian for the radiation field. According to the general theory of Chapter 1, the Lagrangian of a field theory is to be constructed in such a way that the Euler-Lagrange equations (1-74) of the associated variational problem are identical with the field equations. This condition is, of course, not quite sufficient to determine the Lagrangian. The remaining ambiguities are of no physical consequence. The actual choice of the Lagrangian which we adopt here is made primarily on the basis of simplicity.*

If we assume for L the expression

$$L = -\tfrac{1}{2} \partial_\mu a_\nu \partial^\mu a^\nu, \tag{2-22}$$

we obtain

$$\frac{\partial L}{\partial a_\mu} = 0, \tag{2-23}$$

and the conjugate variables are

$$\pi^{\lambda\mu} \equiv \frac{\partial L}{\partial(\partial_\mu a_\lambda)} = -\partial^\mu a^\lambda. \tag{2-24}$$

The field equations (1-74) are then

$$-\partial_\mu \pi^{\lambda\mu} = \partial_\mu \partial^\mu a^\lambda = 0, \tag{2-25}$$

and are seen to be identical with (2-13). The subsidiary condition (2-14) cannot be obtained as a field equation and must be imposed as a separate condition on the field variables. This circumstance requires some special considerations in the quantum theory, which were not included in the general theory of Chapter 1.

2-4 Quantization of the radiation field.† The quantum theory of the radiation field is obtained by interpreting the field variables $f_{\mu\nu}(x)$ as

---

* The Lagrangian adopted here corresponds to the method of quantization developed by E. Fermi, *Rendiconti d. R. Acc. dei Lincei* (6) **9**, 881 (1929). For a discussion of alternative Lagrangians, see, for instance, W. Heisenberg and W. Pauli, *Z. Physik* **56**, 1 (1929) and **59**, 169 (1930); P. A. M. Dirac, V. A. Fock, B. Podolsky, *Physik. Z. Sowjetunion* **2**, 468 (1932).

† The quantum theory of the radiation field was developed by P. A. M. Dirac, *Proc. Roy. Soc.* **112**, 661 (1926); **114**, 243, 710 (1927). P. Jordan and W. Pauli, *Z. Physik* **47**, 151 (1928). W. Heisenberg and W. Pauli, *Z. Physik* **56**, 1 (1929); **59**, 169 (1930).

Hermitian operators which represent the localized observables of the system. It follows that the field variables must commute at two different points which are separated by a space-like interval. Consequently, the radiation field must be quantized according to Eq. (1–107a). We write for the commutators of the potentials

$$[a_\lambda(x), a_\mu(x')] = C_{\lambda\mu}(x,x'). \tag{2–26}$$

Since

$$\pi^{\nu\mu} = -\partial^\mu a^\nu, \tag{2–24}$$

it follows from (1–107a) and (2–18) that

$$[a_\lambda(x), \pi^{\nu\mu}(x')] = -[a_\lambda(x), \partial'^\mu a^\nu(x')] = -ig_\lambda{}^\nu \partial^\mu D(x - x').$$

From (2–26), by differentiation with respect to x', we obtain

$$[a_\lambda(x), \partial'^\mu a^\nu(x')] = \partial'^\mu C_\lambda{}^\nu(x,x').$$

Comparing the two expressions, we see that

$$C_{\lambda\nu}(x,x') = -ig_{\lambda\nu} D(x - x') + \text{const.} \tag{2–27}$$

The possible constant of integration must vanish because the $a_\lambda(x)$ commute for points x, x' with a finite space-like interval. Thus we obtain for the potentials the commutation rules

$$[a_\lambda(x), a_\mu(x')] = -ig_{\lambda\mu} D(x - x'). \tag{2–28}$$

The field equations (2–13) together with the quantum conditions (2–28) form the basic equations of the theory.

It must be noted, however, that the quantized field thus obtained is not yet equivalent to the Maxwell field. The reason is that the field equations are equivalent to Maxwell's equations only if the field variables are further restricted by the subsidiary condition (2–14). In the classical theory it was sufficient to impose the subsidiary condition (2–14) as an additional condition for the field variables, by assuming (2–20) on a particular space-like surface. This procedure can no longer be used in the quantum theory. The operator χ cannot be equal to the zero operator, since it has nonvanishing commutators with other field operators, for instance with $a_\lambda(x)$:

$$[a_\lambda(x), \chi(x')] = i\partial_\lambda D(x - x'). \tag{2–29}$$

It follows that the field equations (2–3) cannot be satisfied as operator equations.

This difficulty can be avoided by imposing the subsidiary condition not as a condition on the operators, but as a condition on the state vectors.

That is, instead of the operator relation (2–14), we require only the weaker relation

$$\chi(x)\omega = 0 \tag{2-30}$$

for all state vectors ω which describe a Maxwell field.

The consequences of assumption (2–30) are that the Maxwell equations (2–3) are satisfied for the expectation values. Thus, instead of (2–3), we find the weaker relation

$$\partial^\mu \langle f_{\mu\nu} \rangle \equiv -\partial_\nu \langle \chi \rangle = 0. \tag{2-31}$$

This relation is sufficient to guarantee a classical limit of the theory which is in full agreement with the Maxwell theory for the radiation field.

It is necessary to make sure that state vectors with the property (2–30) actually do exist. We shall see in Section 6–2 how such state vectors may be constructed explicitly. Here we shall merely point out that the relation (2–30) implies that the operators $\chi(x)$ have eigenvalues zero for all x and that ω is a common eigenvector for all these operators. In general, a set of operators has common eigenvectors only if the operators commute:

$$[\chi(x),\chi(x')] = 0, \quad \text{for all } x, x'. \tag{2-32}$$

Equation (2–32) is indeed satisfied because

$$[\chi(x),\chi(x')] = -i\partial_\mu \partial'^\mu D(x - x')$$
$$= i\partial_\mu \partial^\mu D(x - x') = 0. \tag{2-33}$$

This expression would not vanish if the field had a finite rest mass. We see from this that state vectors with the property (2–30) can be assumed to exist only for a field with rest mass zero.

The relation (2–32) is, of course, not sufficient to assure the existence of a state vector with eigenvalue zero. We can, however, refer to a general theorem which states that if two Hermitian operators A, B have a commutator which is a c-number different from zero

$$[A,B] = ic1, \quad (c \text{ real}), \tag{2-34}$$

then the two operators have all the eigenvalues from $-\infty$ to $+\infty$.

For the proof, let ω_a be an eigenfunction of A corresponding to the eigenvalue a:

$$A\omega_a = a\omega_a.$$

The state vector

$$\omega' = e^{-i\lambda B}\omega_a, \quad (\lambda \text{ real}), \tag{2-35}$$

is then again an eigenvector of A associated with the eigenvalue $a + \lambda c$, for we have*

$$A\omega' = Ae^{-i\lambda B}\omega_a = [A, e^{-i\lambda B}]\omega_a + e^{-i\lambda B}A\omega_a$$
$$= (\lambda c + a)\omega'. \tag{2-36}$$

Thus, by choosing $\lambda = -(a/c)$ we obtain an eigenvector $\omega' = \omega_0$ for which

$$A\omega_0 = 0.$$

For the two operators, we may choose $a_\lambda(x)$ and $\chi(x')$ which, according to (2–29), have a c-number for their commutator. Thus the theorem applies and we can for each x construct an eigenvector with the property (2–30).

We should mention, however, that, although the state vectors ω with the property (2–30) are seen to exist, they are in general not normalizable, as they belong to a continuum of eigenvalues.

This situation is exactly analogous to the corresponding situation in ordinary wave mechanics where the eigenfunctions of the canonically conjugate variables q and p are also not normalizable. Just as in wave mechanics, it is possible to approximate such eigenfunctions by sequences of normalizable wave function (wave packets) so that it will be possible to construct the state vector which satisfies (2–30) as a sequence of normalizable vectors (see Section 6–2).

2–5 Momentum operators for the radiation field. The ten operators P^μ, $M^{\mu\nu}$ may be readily constructed for the radiation field by the method of Chapter 1. The formulas needed are (1–80), (1–82), and (1–83). From (1–80), we obtain

$$T^{\mu\nu} = T^{\nu\mu} = -\partial^\mu a^\lambda \partial^\nu a_\lambda + \tfrac{1}{2}g^{\mu\nu}\partial_\lambda a_\rho \partial^\lambda a^\rho. \tag{2-37}$$

It follows from (1–83) that

$$P^\mu = \int_\sigma (-\partial a^\lambda \partial^\mu a_\lambda + \tfrac{1}{2}n^\mu \partial_\lambda a_\nu \partial^\lambda a^\nu)d\sigma. \tag{2-38}$$

---

* The simple formal algebraic procedure used for this "proof" ignores all the complications arising from the singular character of unbounded Hermitian operators in Hilbert space. For a discussion of the mathematical problems involved, the reader is referred to J. v. Neumann, *Math. Annalen* **102**, 49 (1930). The difficulty in our heuristic proof arises from the implied assumption that the state vector $\omega' = e^{-i\lambda B}\omega_a$ belongs to the domain of definition of the operator A. This need not be the case. Simple counter examples occur frequently in ordinary wave mechanics. For instance, let $A = \varphi$ be the azimuth angle and $-i(\partial/\partial\varphi) = B = L_3$ the angular momentum in the 3-direction. Then $[A,B] = i$, but the eigenvalues of L_3 are known to be discrete. A similar situation prevails for the operators $A = r > 0$ and $B = -i\partial/\partial r$. The operators which occur in the field theory under discussion actually do have the desired regularity property and thus the "proof" applies, but it is well to keep in mind its limitations.

To construct the operators $M^{\mu\nu}$, we need the transformation coefficients $\Sigma_r{}^{s\mu\nu}$ defined in (1–54). Since r and s are now vector indices, we have

$$\Sigma^{\rho\sigma\mu\nu} = g^{\rho\mu}g^{\sigma\nu} - g^{\sigma\mu}g^{\rho\nu}. \tag{2-39}$$

The tensor corresponding to (1–97) then becomes

$$h^{\kappa\mu\nu} \equiv \pi^{\rho\kappa}\Sigma_\rho{}^{\sigma\mu\nu}a_\sigma, \tag{2-40}$$

$$h^{\kappa\mu\nu} = -\partial^\kappa a^\mu a^\nu + \partial^\kappa a^\nu a^\mu. \tag{2-41}$$

This leads to

$$M^{\mu\nu} = L^{\mu\nu} + N^{\mu\nu}, \tag{2-42}$$

$$L^{\mu\nu} = \int_\sigma (T^{\rho\mu}x^\nu - T^{\rho\nu}x^\mu)d\sigma_\rho, \tag{2-43}$$

$$N^{\mu\nu} = \int_\sigma (\partial^\kappa a^\mu a^\nu - \partial^\kappa a^\nu a^\mu)d\sigma_\kappa \tag{2-44}$$

for the total (M), the spin (N), and the orbital (L) angular momentum of the radiation field. It follows from the general theory of Chapter 1 that the operators (2–38) and (2–42) satisfy the characteristic equations (1–51), (1–56), and (1–57) of the momentum operators. It is instructive to verify that these equations are a mathematical consequence of the definitions (2–38), (2–42), (2–43), (2–44), and the commutation rules (2–28). This is left to the reader.

At this point, it is necessary to discuss the problem which arises from the fact that the vector potential is not determined by the field tensors but allows the gauge transformations (2–8). The expressions which we have obtained for the momentum operators contain the vector potential explicitly and are not invariant under such transformations. How then is it possible to ascertain that these expressions represent physically observable quantities?

The answer to this question is connected with the observation that in quantum mechanics the prediction of physical measurements is obtained from expectation values of operators. Thus it is not necessary that the operators themselves be invariant under gauge transformations, but only that the expectation values be invariant, and this only for state vectors which satisfy the subsidiary condition (2–30).

The invariance of the expectation values of the momentum operators can be proved most easily if we make use of the fact that gauge transformations are canonical transformations. It is sufficient to show this for the infinitesimal gauge transformations.

We consider the transformation of adding a C– number gradient

$$a_\lambda(x) \rightarrow a_\lambda(x) + \partial_\lambda\varphi(x), \tag{2-45}$$

where $\varphi(x)$ is considered an infinitesimal of first order. The increment of the $a_\lambda(x)$ may thus be written as

$$\delta_\varphi a_\lambda(x) = \partial_\lambda \varphi(x). \tag{2-46}$$

The generating operator F for this transformation is given by*

$$F = \int_\sigma (\partial^\mu \varphi(x)\chi(x) - \varphi(x)\partial^\mu \chi(x))d\sigma_\mu. \tag{2-47}$$

The integral is taken over an arbitrary plane σ. Since the divergence of the integrand vanishes, its value is independent of the plane.

For the commutator of F with $a_\lambda(x)$, we obtain

$$i[F, a_\lambda(x)] = \partial_\lambda \int_\sigma (\partial'^\mu \varphi(x')D(x - x') - \varphi(x')\partial'^\mu D(x - x'))d\sigma_\mu'. \tag{2-48}$$

This is the same integral which was evaluated in (2–16) and we have thus

$$i[F, a_\lambda(x)] = \partial_\lambda \varphi(x) = \delta_\varphi a_\lambda(x). \tag{2-49}$$

This result shows that the gauge transformations are indeed canonical transformations.

The variation of the momentum operator under a gauge transformation is now obtained as

$$\delta_\varphi P^\nu = i[F, P^\nu]. \tag{2-50}$$

This commutator can be evaluated most easily if we recall that P^ν is the generating operator of displacement [see (1–57)]. Thus

$$\delta_\varphi P^\nu = \int_\sigma (\partial^\mu \varphi(x)\partial^\nu \chi(x) - \varphi(x)\partial^\mu \partial^\nu \chi(x))d\sigma_\mu. \tag{2-51}$$

For state vectors which satisfy the subsidiary condition (2–30) the expectation value of this expression vanishes. Thus we have shown that the expectation value $\langle P^\nu \rangle$ is gauge invariant.

It is possible to express this expectation value in terms of manifestly gauge-invariant quantities by defining the "classical" energy-momentum tensor:

$$\theta_{\mu\nu} = -f_{\mu\rho}f_\nu{}^\rho + \tfrac{1}{4}g_{\mu\nu}f_{\rho\sigma}f^{\rho\sigma}. \tag{2-52}$$

The divergence of this tensor is not zero. Instead, with the help of the field equations, one obtains for it the value†

$$\partial^\mu \theta_{\mu\nu} = f_\nu{}^\rho \partial_\rho \chi. \tag{2-53}$$

---

* The expression given here is the relativistic generalization of the corresponding operator given by W. Heisenberg and W. Pauli, *Z. Physik* **59**, 169 (1930) especially Eq. (20) on p. 173.

† In the classical theory the right side of Eq. (2–53) vanishes, but this is not the case in the quantized theory.

However, the expectation value of this expression vanishes. One can thus evaluate $\langle \int \theta^{\mu\nu} d\sigma_\mu \rangle$ in any special coordinate system, for instance in the one for which $n^\mu = (1, 0, 0, 0)$. Disregarding terms which are space divergences and dropping terms which involve χ, we obtain

$$\left\langle \int \theta^{\mu\nu} d\sigma_\mu \right\rangle = \langle P^\nu \rangle. \tag{2-54}$$

By similar methods, we prove the further result

$$\left\langle \int (\theta^{\rho\mu} x^\nu - \theta^{\rho\nu} x^\mu) d\sigma_\rho \right\rangle = \langle M^{\mu\nu} \rangle. \tag{2-55}$$

In the last expression we find that the total angular momentum no longer appears separated into an orbital and a spin part. This separation is, in fact, impossible in a gauge-invariant manner. This raises the question of how much physical reality one may associate with the notion of the photon spin. This question will be discussed in Section 2–8, where we shall see that the photon spin still has, in a certain restricted sense, a gauge-invariant and therefore physical meaning.*

---

* The transformation which leads to the expressions (2–54) and (2–55) involves the dropping of surface terms at infinity, as we have indicated in the text. This procedure is entirely correct only if the state vector which represents the state of the system satisfies the condition that the expectation values of the fields A_μ vanish sufficiently fast at infinity. It is plausible that any real physical state can always be approximated with any desirable accuracy by states which satisfy this condition.

However, it is well to remember that the state of the system which corresponds to a plane monochromatic wave (that is, an eigenstate of the momentum operator P_μ) is not of this sort. This gives rise to a peculiar paradox which has been the subject of many papers. This can be seen, for instance, by calculating the angular momentum density in the direction of propagation for a circularly polarized wave, with the expression (2–55). It turns out to be 0, contrary to the value obtained from (2–42). It can be shown that the difference is properly accounted for by the omitted surface terms. For more detail the reader is referred to the following: W. Heitler, *Quantum Theory of Radiation*, 3rd edition, Oxford University Press, 1954. A. W. Conway, *Proc. Roy. Irish Acad.* **50**, 115 (1936). W. Heitler, *Proc. Cambridge Phil. Soc.* **32**, 112 (1936). The problem occurs in classical electrodynamics: E. Henriot, *Mémorial des Sciences Physique* **30**, Paris, Gauthier-Villars; M. J. Humblet, *Physica* **10**, 585 (1943); as well as in meson theory: J. Serpe, *Physica* **8**, 748 (1941); L. Rosenfeld, *Bull. de l'Acad. Roy. de Belgique* **28**, 562 (1942). Some of the earlier discussions are given in A. Sommerfeld, *Atombau und Spektrallinien*, 5th ed., p. 684; M. Abraham, *Physik. Z.* **15**, 914 (1914); P. S. Epstein, *Ann. Physik* **44**, 593 (1914); H. Busch, *Physik. Z.* **15**, 455 (1914); J. M. Poynting, *Proc. Roy. Soc.* **82**, 565 (1909). A detailed discussion of the problem is also given by L. de Broglie, *Mécanique Ondulatoire du Photon et Théorie Quantique des Champs*, Paris, Gauthier-Villars, 1949, Chapter VI, p. 65.

2-6 Plane wave decomposition of the radiation field.

The potentials of the radiation field can be developed in a four-dimensional Fourier integral:

$$a_\mu(x) = \frac{1}{(2\pi)^2} \int b_\mu(k) e^{ik \cdot x} d^4k. \tag{2-56}$$

We recall the significance of the invariant scalar product $k \cdot x = k^\mu x_\mu$ for the phase factor in the exponential. The transformation (2-56) can be inverted to give

$$b_\mu(k) = \frac{1}{(2\pi)^2} \int a_\mu(x) e^{-ik \cdot x} d^4x. \tag{2-57}$$

The fact that $a_\mu(x)$ is Hermitian, $a_\mu{}^*(x) = a_\mu(x)$, implies that

$$b_\mu{}^*(k) = b_\mu(-k). \tag{2-58}$$

The field equation (2-13) becomes an algebraic equation in the operators b_μ:

$$k^2 b_\mu(k) = 0. \tag{2-59}$$

Thus the operator $b_\mu(k)$ can be different from zero only if $k^2 = 0$. This implies that it is of the form

$$b_\mu(k) = \delta(k^2) c_\mu(k). \tag{2-60}$$

It follows from (2-60) and the well-known property of the δ-function,

$$\delta(k^2) = \frac{1}{2\omega}(\delta(k^0 - \omega) + \delta(k^0 + \omega)),$$

$$\omega = +\sqrt{k_1{}^2 + k_2{}^2 + k_3{}^2} \equiv |\mathbf{k}|, \tag{2-61}$$

that the integration over k^0 can be carried out. The result is conveniently expressed in terms of the operators

$$a_\mu(\mathbf{k}) = \frac{1}{\sqrt{4\pi\omega}} c_\mu(\mathbf{k}, \omega), \qquad a_\mu{}^*(\mathbf{k}) = \frac{1}{\sqrt{4\pi\omega}} c_\mu(-\mathbf{k}, -\omega)$$

in the form

$$a_\mu(x) = \frac{1}{(2\pi)^{3/2}} \int \frac{d^3k}{\sqrt{2\omega}} (a_\mu(\mathbf{k}) e^{ik \cdot x} + a_\mu{}^*(\mathbf{k}) e^{-ik \cdot x}), \tag{2-62}$$

which shows that an invariant decomposition of the field operators exists:

$$a_\mu(x) = a_\mu{}^{(-)}(x) + a_\mu{}^{(+)}(x),$$

$$a_\mu{}^{(-)}(x) = \frac{1}{(2\pi)^{3/2}} \int \frac{d^3k}{\sqrt{2\omega}} a_\mu(\mathbf{k}) e^{ik \cdot x}, \tag{2-63}$$

$$a_\mu{}^{(+)}(x) = \frac{1}{(2\pi)^{3/2}} \int \frac{d^3k}{\sqrt{2\omega}} a_\mu{}^*(\mathbf{k}) e^{-ik \cdot x}. \tag{2-64}$$

The invariance of this decomposition under Lorentz transformations is not obvious from the form of (2–62), but it follows from the manner in which this decomposition was derived.

This decomposition can be carried out with Schwinger's method of complex integration, and without the use of the Fourier transformation, as follows.* We start from the integral formula

$$\frac{1}{2\pi i} \int_P \frac{e^{i\alpha\tau}}{\tau} d\tau = \begin{cases} 1, & \text{for } \alpha > 0. \\ 0, & \text{for } \alpha < 0. \end{cases}$$

The path P is a contour along the real axis in the complex τ-plane avoiding the pole at the origin, $\tau = 0$, by a detour into the negative imaginary half-plane. If we apply this integral formula to the expressions (2–63) and (2–64), we obtain

$$\frac{1}{2\pi i} \int_P \frac{d\tau}{\tau} a_\mu(x + \tau n) = a_\mu^{(+)}(x)$$

$$\frac{1}{2\pi i} \int_P \frac{d\tau}{\tau} a_\mu(x - \tau n) = a_\mu^{(-)}(x),$$

where n is a time-like constant four vector which points into the future ($n^0 > 0$). The form of the last two expressions shows that under all proper Lorentz transformations the operators $a_\mu^{(+)}$ and $a_\mu^{(-)}$ transform separately among themselves.

We mention that the decomposition (2–62) can be obtained for any function and we shall also be using it for the invariant D-function to obtain the D_+ and D_- functions:†

$$D_\pm = \frac{1}{2\pi i} \int_P \frac{d\tau}{\tau} D(x \mp \tau n).$$

It is not difficult to find the commutation relations for the operators $a_\mu^{(+)}(x)$ and $a_\mu^{(-)}(x)$. From (2–28) follows

$$[a_\mu^{(+)}(x), a_\nu^{(-)}(y)] = \left(\frac{1}{2\pi i}\right)^2 \int_{P'} \frac{d\tau'}{\tau'} \int_{P''} \frac{d\tau''}{\tau''} [a_\mu(x + \tau' n), a_\nu(y - \tau'' n)]$$

$$= -ig_{\mu\nu} \left(\frac{1}{2\pi i}\right)^2 \int_{P'} \frac{d\tau'}{\tau'} \int_{P''} \frac{d\tau''}{\tau''} D(x - y + (\tau' + \tau'')n).$$

---

* J. Schwinger, *Phys. Rev.* **75**, 651 (1949). Note that our definition of the superscripts (+) and (−) differs from that of Schwinger.

† An alternative definition of the function $D_\pm$ is given in the mathematical appendix, Section A1–1.

If we choose the paths P' and P'' such that

$$\text{Im}\,\tau' < \text{Im}\,\tau'' < 0,$$

the new paths P and P_σ obtained from the transformation

$$\tau = \tau' + \tau'', \quad \sigma = \tfrac{1}{2}(\tau' - \tau'')$$

will be determined. The integration over σ can be performed and yields

$$\frac{1}{2\pi i}\int_{P_\sigma}\frac{\tau\,d\sigma}{\sigma^2 - \tfrac{1}{4}\tau^2} = \frac{1}{2\pi i}\int_{P_\sigma}\left(\frac{d\sigma}{\sigma - \tfrac{1}{2}\tau} - \frac{d\sigma}{\sigma + \tfrac{1}{2}\tau}\right) = 1.$$

Therefore,

$$[a_\mu^{(+)}(x), a_\nu^{(-)}(y)] = -ig_{\mu\nu}D_-(x - y). \tag{2-65}$$

In a similar way, we obtain

$$[a_\mu^{(+)}(x), a_\nu^{(+)}(y)] = [a_\mu^{(-)}(x), a_\nu^{(-)}(y)] = 0. \tag{2-66}$$

With the Fourier transform of (2–65), substituting (2–63), (2–64), and (A1–22), we find the commutation rules for the Fourier amplitudes in the form

$$[a_\mu(\mathbf{k}), a_\nu(\mathbf{k}')] = [a_\mu^*(\mathbf{k}), a_\nu^*(\mathbf{k}')] = 0, \tag{2-67}$$

$$[a_\mu(\mathbf{k}), a_\nu^*(\mathbf{k}')] = g_{\mu\nu}\delta(\mathbf{k} - \mathbf{k}'). \tag{2-68}$$

This result shows that the operators $a_\mu(\mathbf{k})$ have the familiar properties of the creation and destruction operators of the harmonic oscillator problem. There is a peculiar difference, however, for the components with $\mu = \nu = 0$. Since $g_{00} = -1$, formula (2–68) gives

$$[a_0(\mathbf{k}), a_0^*(\mathbf{k}')] = -\delta(\mathbf{k} - \mathbf{k}'), \tag{2-69}$$

which shows that in this particular case the roles of the two operators are interchanged; a_0 is the creation operator and a_0^* the destruction operator.

2–7 Explicit representations of the field operators. The commutation rules (2–67) and (2–68) can serve as a convenient starting point for constructing an explicit representation of the field operators. To simplify the notation it is useful to introduce the device of enclosing the radiation field in a finite volume V in the form of a cube with edge length L. If we impose periodic boundary conditions, the wave number vector $\mathbf{k}$ is discrete and is given by

$$k_r = \kappa n_r, \quad (r = 1, 2, 3),$$

$$n_r = 0, \pm 1, \pm 2, \cdots, \tag{2-70}$$

$$\kappa = \frac{2\pi}{L}.$$

Every formula of the preceding section remains correct provided we carry through the substitutions

$$\frac{1}{(2\pi)^{3/2}} \int d^3k \cdots \to \frac{1}{\sqrt{V}} \sum_{\mathbf{k}} \cdots,$$

$$\delta(\mathbf{k} - \mathbf{k}') \to \delta_{\mathbf{k}\mathbf{k}'} \equiv \delta_{n_1 n_1'} \delta_{n_2 n_2'} \delta_{n_3 n_3'},$$

$$\frac{1}{(2\pi)^{3/2}} \int d^3x \cdots \to \frac{1}{\sqrt{V}} \int_V d^3x \cdots$$

We now regard the indices $\mu, \mathbf{k}$ as one single index. Since operators with different indices commute, we can discuss the operator referring to one particular index and then construct the whole family of operators with the method of the Kronecker product.† We may therefore omit this index in the following discussion. The problem is thus simplified to that of constructing the irreducible representation of an operator a which satisfies the operator relation

$$[a, a^*] = 1. \tag{2-71}$$

The solution of this problem is well known from the quantum theory of the harmonic oscillator problem.‡ It may be summarized in the following two statements:

(1) A representation of a is given by the matrix

$$a = \begin{pmatrix} 0 & 1 & 0 & 0 & \cdot \\ 0 & 0 & \sqrt{2} & 0 & \cdot \\ 0 & 0 & 0 & \sqrt{3} & \cdot \\ \cdot & \cdot & \cdot & \cdot & \cdot \end{pmatrix}. \tag{2-72}$$

(2) Every other irreducible representation is related to (2–72) by an S-transformation: SaS^{-1}.

It is not difficult to verify that (2–72) satisfies (2–71). More complex is the proof that it is the only representation in the precise sense stated under (2). We shall forego this proof here, and merely state that the

---

† For the definition of the Kronecker product see, for instance, F. D. Murnaghan, *The Theory of Group Representations*, Baltimore, Johns Hopkins Press, 1938, p. 68.

‡ For details the reader is referred to the book by P. A. M. Dirac, *Principles of Quantum Mechanics, loc. cit.*, p. 136 ff., where the operator corresponding to a is denoted by η.

theorem involved is a special case of a theorem by v. Neumann† on the representation of canonical operators, to which the reader is referred in this connection.

We may write (2–72) in matrix notation by labeling the rows and columns of a with an integer N ($= 0, 1, 2, \cdots$):

$$(N|a|N') = \sqrt{N+1}\, \delta_{N+1,N'}. \tag{2–73}$$

We also note the matrix representations of the operators which follow from (2–73):

$$\begin{aligned}(N|aa^*|N') &= (N+1)\delta_{NN'}, \\ (N|a^*a|N') &= N\delta_{NN'}.\end{aligned} \tag{2–74}$$

An alternative way to express these results, which is often used, is to introduce the eigenfunctions $\omega(N)$ of the operator a^*a associated with the eigenvalue N,

$$a^*a\omega(N) = N\omega(N). \tag{2–75}$$

From (2–73) we then obtain the relations

$$\begin{aligned}a\omega(N) &= \sqrt{N}\, \omega(N-1), \\ a^*\omega(N) &= \sqrt{N+1}\, \omega(N+1).\end{aligned} \tag{2–76}$$

The number N, which mathematically is the eigenvalue of a^*a, has a simple physical interpretation in the quantum theory of radiation which is brought out more clearly if we express the momentum operators P^μ in terms of the operators a, a^*. This can be done by substituting (2–62) for $a_\mu(x)$ in (2–38). The result of the straightforward calculation is

$$P^\mu = \int d^3k\, \tfrac{1}{2} k^\mu [a^\lambda(\mathbf{k})a_\lambda^*(\mathbf{k}) + a_\lambda^*(\mathbf{k})a^\lambda(\mathbf{k})]. \tag{2–77}$$

With the notation

$$n(\mathbf{k}) = \tfrac{1}{2}[a^\lambda(\mathbf{k})a_\lambda^*(\mathbf{k}) + a_\lambda^*(\mathbf{k})a^\lambda(\mathbf{k})], \tag{2–78}$$

this last expression may be written

$$P^\mu = \int d^3k\, k^\mu n(\mathbf{k}). \tag{2–79}$$

---

† J. v. Neumann, *Math. Ann.* **104**, 570 (1931). The extension of the uniqueness theorem to an infinite set of operators a_i ($i = 1, 2, \cdots$) is only possible if the additional assumption is made that there exists a state vector ω_0 with the property $\sum_i a_i^* a_i \omega_0 = 0$. Without this assumption, there exist nondenumerably many inequivalent representations. This was recently proved by L. Gårding and A. S. Wightman, *Proc. Nat. Acad.* **40**, 617 and 622 (1954). See also A. S. Wightman and S. S. Schweber, *Phys. Rev.* **98**, 812 (1955).

The representation (2–77) for the total energy and momentum of the radiation shows that the quantized radiation field is mathematically equivalent to a system of independent harmonic oscillators each of which is quantized according to the usual procedure of quantum mechanics. In the harmonic oscillator problem, the operator

$$n = \tfrac{1}{2}(aa^* + a^*a) \tag{2-80}$$

represents the total energy of the system which has, according to (2–74), the eigenvalues $N + \tfrac{1}{2}$ ($N = 0, 1, \cdots$). In addition to the above interpretation, we may also consider the operator $n - \tfrac{1}{2}$ as the photon number associated with a definite degree of freedom labeled by indices $\mu,\mathbf{k}$. Its eigenvalues are called *occupation numbers*.

The operator $L^{\mu\nu}$ (2–43) does not have such a simple interpretation because it cannot be written as a sum of individual contributions for each index $\mu,\mathbf{k}$. This is still possible, however, for the spin operator $N^{\mu\nu}$ which, according to (2–44) and (2–62), may be written as

$$N_{\mu\nu} = i\int [a_\mu{}^*(\mathbf{k})a_\nu(\mathbf{k}) - a_\nu{}^*(\mathbf{k})a_\mu(\mathbf{k})]d^3k. \tag{2-81}$$

2–8 The spin of the photon. The spin of the photon is usually assumed to be 1. This statement is in need of elaboration in view of the fact, pointed out in connection with Eq. (2–55), that the separation of the total angular-momentum operator into spin and orbital parts is not gauge invariant. Since only gauge-invariant quantities are observable, the concept of the photon spin, as a physical quantity, is obscure.

The simplest way to examine this point is to investigate the operator for the spin angular momentum in the plane wave representation (2–81). The operator has the six components of an antisymmetrical tensor of second rank. The angular momenta proper are the three space components of this tensor:

$$N_{ij} = i(a_i{}^*a_j - a_j{}^*a_i). \tag{2-82}$$

This expression is not gauge invariant. However, its projection in the direction of $\mathbf{k}$ is invariant and therefore has a physical meaning. In order to write this operator most conveniently, we introduce a special coordinate system in which the transverse components are denoted by a_1, a_2 and the longitudinal component by a_3. Gauge transformations affect only the latter and leave the former invariant. Consequently, the expression

$$N_{12} = i(a_1{}^*a_2 - a_2{}^*a_1) \tag{2-83}$$

is gauge invariant and is therefore an observable physical quantity. According to our definition, it is the spin of the photon in the direction of propagation.† We may describe this situation by the statement that the spin of the photon is always oriented in the direction of propagation of the photons. The spin orthogonal to the direction of propagation does not exist. The photons have this property in common with particles of arbitrary spin and zero rest mass.‡

In order to discuss the properties of the operator (2–83), it is convenient to introduce the following set of operators Σ_μ on an equal footing:

$$\begin{aligned}\Sigma_1 &= a_1{}^*a_2 + a_2{}^*a_1, \\ \Sigma_2 &= i(-a_1{}^*a_2 + a_2{}^*a_1), \\ \Sigma_3 &= a_1{}^*a_1 - a_2{}^*a_2, \\ \Sigma_0 &= a_1{}^*a_1 + a_2{}^*a_2.\end{aligned} \qquad (2\text{–}84)$$

These four operators may be more conveniently written by introducing the two-component "spinor" operators a:

$$a = \begin{pmatrix} a_1 \\ a_2 \end{pmatrix}, \qquad (2\text{–}85)$$

and the corresponding "spin" matrices σ_μ, with $\sigma_0 = 1$. The operators (2–84) are then given by§

$$\Sigma_\mu = (a^*, \sigma_\mu a), \quad (\mu = 0, \cdots, 3). \qquad (2\text{–}86)$$

We note here that all four of these quantities are gauge invariant but only two of them have so far a simple physical interpretation:

$$\Sigma_2 = -N_{12}, \quad \Sigma_0 = n - 1.$$

We shall show now that these four operators are the quantum-mechanical analogue of the four Stokes parameters in the classical description of the transverse vector wave. These parameters were introduced by Stokes in 1852. They are the most convenient mathematical characterization for the state of polarization of a plane wave but they have received little

---

† That this quantity is observable was demonstrated by R. A. Beth, *Phys. Rev.* **50**, 115 (1936).

‡ This was shown in general by M. Fierz, *Helv. Phys. Acta* **13**, 95 (1940).

§ The close relation of the mathematical description of the polarization states with the spin theory of Pauli was previously noted by P. Jordan, *Z. Physik* **44**, 292 (1927).

attention until very recent times.† Since none of the standard texts on classical electromagnetic theory makes any mention of these parameters, we shall here briefly summarize some of their most important properties.

A transverse vector wave in a definite state of polarization is given by

$$E_1 = \mathcal{E}_1 \cos \omega t,$$
$$E_2 = \mathcal{E}_2 \cos (\omega t + \alpha). \tag{2-87}$$

The three real parameters $\mathcal{E}_1$, $\mathcal{E}_2$, α characterize the state of polarization. An equivalent description is obtained by introducing the notation

$$E = c + c^*, \tag{2-88}$$

where

$$E = \begin{pmatrix} E_1 \\ E_2 \end{pmatrix}, \quad c = \begin{pmatrix} c_1 \\ c_2 \end{pmatrix}, \tag{2-89}$$

$$c_1 = \tfrac{1}{2}\mathcal{E}_1 e^{i\omega t}, \quad c_2 = \tfrac{1}{2}\mathcal{E}_2 e^{i(\omega t + \alpha)}. \tag{2-90}$$

Unitary transformations in the two-dimensional space of the "spinors" c induce real orthogonal transformations in the three variables

$$s_\mu = (c^*, \sigma_\mu c), \quad (\mu = 1, 2, 3). \tag{2-91}$$

The three quantities s_μ are the Stokes parameters of the wave. In terms of the original parameters, they are given as

$$s_1 = \tfrac{1}{2}\mathcal{E}_1\mathcal{E}_2 \cos \alpha,$$
$$s_2 = \tfrac{1}{2}\mathcal{E}_1\mathcal{E}_2 \sin \alpha, \tag{2-92}$$
$$s_3 = \tfrac{1}{4}(\mathcal{E}_1{}^2 - \mathcal{E}_2{}^2).$$

A fourth dependent parameter $s_0 = -s^0$ may be introduced by

$$s_0 = (c^*, c) \equiv (c^*, \sigma_0 c), \quad (\sigma_0 = 1). \tag{2-93}$$

The four parameters together satisfy the relation

$$s_\mu s^\mu = 0 \quad \text{or} \quad s_0 = \sqrt{s_1{}^2 + s_2{}^2 + s_3{}^2}. \tag{2-94}$$

---

† The original paper by Stokes is G. G. Stokes, *Trans. Camb. Phil. Soc.* **9**, 399 (1852). It is reprinted in *Mathem. and Physical Papers*, Cambridge University Press, London, 1901, Vol. 3, p. 233. One of the first to rediscover the elegance of this treatment was P. Soleillet, *Ann. phys.* **12**, 23 (1929). More recent papers in which use is made of the Stokes parameters are F. Perrin, *J. Chem. Phys.* **10**, 415 (1942); H. Mueller, Report No. 2 of OSRD Project OEMsr-576, Nov. 15, 1943; R. Clark Jones, *J. Opt. Soc. Am.* **37**, 107, 110 (1947); U. Fano, *ibid.* **39**, 859 (1949); D. L. Falkoff and J. E. Macdonald, *J. Opt. Soc. Am.* **41**, 862 (1951). The formalism was most effectively used by S. Chandrasekhar in his elegant solution of the problem of radiative transfer in atmospheres: *Radiative Transfer*, Oxford, Clarendon Press, 1950.

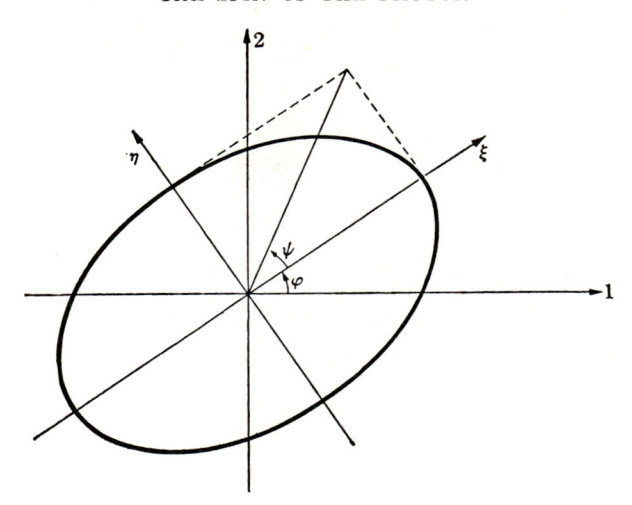

FIG. 2-1. The polarization ellipse for a plane monochromatic wave.

The quantity s_0 represents half the average intensity of radiation in the wave:

$$I = \tfrac{1}{2}(\mathcal{E}_1^2 + \mathcal{E}_2^2) = 2s_0. \tag{2-95}$$

If ξ, η are the semi-axes of the polarization ellipse and if φ is the angle of inclination of the major axis with the 1-direction, then the following relations hold:

$$\tan 2\varphi = \frac{s_1}{s_3},$$

$$\tan 2\psi = \frac{s_2}{\sqrt{s_1^2 + s_3^2}}, \tag{2-96}$$

where

$$\tan \psi = \frac{\eta}{\xi}.$$

These relations show that the parameters characterize the state of polarization.

The Stokes parameters are closely related to the parameters which were introduced by Poincaré for the description of the state of polarization only.* Poincaré introduces a complex number $z = u + iv$ (u,v real) which is related to the Stokes vector by

$$u = \frac{s_1}{s_0 - s_3}, \quad v = \frac{s_2}{s_0 - s_3}. \tag{2-97}$$

---

* H. Poincaré, *Théorie Mathématique de la Lumière*, Vol. 2, p. 275 (1892).

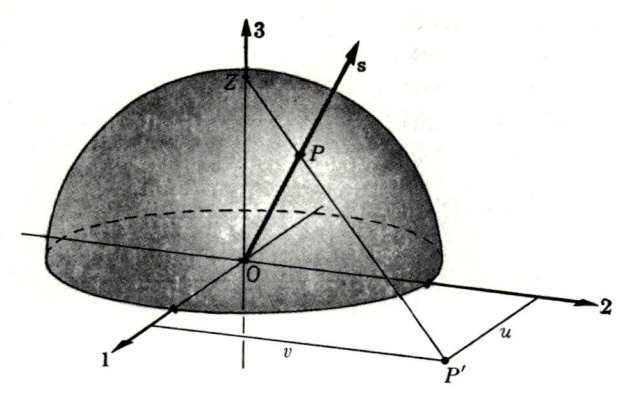

FIG. 2-2. The relationship between the Stokes vector **s** and the Poincaré parameters u and v.

This number is obtained by stereographic projection of the intersection of the Stokes vector with the unit sphere into the complex z-plane passing through the origin of the sphere.

In order to describe a partially polarized wave, it is convenient to follow Stokes, and introduce the concept of *opposite polarization*. Two waves are said to be opposite in polarization if their respective Stokes vectors **s**, **s**′ point in opposite directions:

$$\mathbf{s}' = \lambda \mathbf{s} \text{ with } \lambda < 0. \tag{2-98}$$

Their respective Poincaré parameters are then in the relation

$$z' = -\frac{1}{z^*}. \tag{2-99}$$

According to the theorem of Fresnel and Arago, two waves in opposite states of polarization do not interfere, and the Stokes vector of their incoherent superposition is given by

$$S_\mu = s_\mu + s_\mu'. \tag{2-100}$$

For the resultant Stokes vector we have, from the triangle inequality,

$$S_\mu S^\mu \leq 0. \tag{2-101}$$

The quantity

$$\kappa = \frac{S}{S_0} \quad (S = \sqrt{S_1^2 + S_2^2 + S_3^2}) \tag{2-102}$$

is called the degree of polarization ($0 \leq \kappa \leq 1$).

The particular case of unpolarized light is described by $\kappa = 0$. It is obtained as the incoherent superposition of two oppositely polarized waves with the same intensity.

After this digression into the realm of classical physics, let us return to the quantum-mechanical description of the polarization. The quantum-mechanical operators which correspond to the amplitudes c are the absorption operators a, and consequently the Stokes parameters (2–91) and (2–93) become the "Stokes operators" Σ_μ defined by (2–86).

They satisfy the following relations as a consequence of the commutation rules (2–67) and (2–68):

$$[\Sigma_1, \Sigma_2] = 2i\Sigma_3,$$

$$[\Sigma_2, \Sigma_3] = 2i\Sigma_1, \qquad (2\text{–}103)$$

$$[\Sigma_3, \Sigma_1] = 2i\Sigma_2.$$

$$[\Sigma_i, \Sigma_0] = 0, \quad (i = 1, 2, 3). \qquad (2\text{–}104)$$

$$\Sigma_1^2 + \Sigma_2^2 + \Sigma_3^2 = \Sigma_0(\Sigma_0 + 2). \qquad (2\text{–}105)$$

The last equation replaces the classical relation (2–94). The commutation rules (2–103) are, apart from a factor two, those of angular momentum operators. As Hermitian operators, the Σ_μ may be considered as the generators of a group of symmetry transformations which is locally isomorphic to the three-dimensional rotation group and which leaves the operator Σ_0 invariant [cf. (2–104)]. The eigenvectors of Σ_0 belong thus to linear vector spaces of degenerate eigenvalues which under the transformation of the symmetry group transform according to some irreducible representation of this group.

Let $N = 2j$ be one of the eigenvalues of Σ_0, and let ω_N be an eigenvector:

$$\Sigma_0 \omega_N = N\omega_N, \quad (N = 0, 1, \cdots). \qquad (2\text{–}106)$$

The operators Σ_i, when considered in the subspace of the ω_N (N fixed), satisfy, according to (2–105) and (2–106), the operator relation

$$\tfrac{1}{4}(\Sigma_1^2 + \Sigma_2^2 + \Sigma_3^2) = j(j+1). \qquad (2\text{–}107)$$

The three operators Σ_i belong thus to the representation D_j of the rotation group.* The eigenvalues of the operators $\tfrac{1}{2}\Sigma_i$ in these subspaces are j, $j-1$, $\cdots$, $-j$, just as in the case of ordinary angular momenta. In particular, for $N = 1$ the eigenvalues of $-\Sigma_2$ (spin in the propagation direction) are equal to ± 1. This is the reason for the assertion that the spin of the photon is 1.

---

* See Van der Waerden, *Die Gruppentheoretische Methode in der Quantenmechanik*, Springer, 1932, especially p. 61. The existence of this symmetry group for the two-dimensional oscillator problem was previously noted by J. M. Jauch and E. L. Hill, *Phys. Rev.* **57**, 641 (1940), especially section C.

It should be emphasized, however, that it would be entirely wrong to associate these transformations which are generated by Σ_i ($i = 1, 2, 3$) with any rotations in ordinary space. The transformations originate in the "spinor" space associated with the manifold of polarization states of the photons.

The noncommutability of the quantum-mechanical analogue of the Stokes parameters precludes the simultaneous measurement of the physical quantities represented by these operators.

The concept of partial polarization can also be formulated in the quantum-mechanical description. A partially polarized system of photons is described by a statistical mixture of states described by a density matrix ρ.† The density matrix is positive definite, Hermitian, and of trace 1:

$$\rho^+ = \rho, \quad \text{Tr}\, \rho = 1. \tag{2-108}$$

The operator ρ^+ denotes the Hermitian conjugate in the "spinor" space. (See also Appendix A8.) The relationship to the classical description is established by equating the expectation values of the Stokes operators to the classical Stokes parameters. Thus we have

$$\text{Tr}\,(\rho \Sigma_\mu) = S_\mu. \tag{2-109}$$

Just as in the classical case, we have for a general density matrix

$$S \leq S_0, \tag{2-110}$$

but this inequality is much harder to prove than the corresponding inequality in the classical case. We can also show that the equality sign holds only for pure states. Thus we may define the degree of polarization for a statistical assembly of photons by

$$\kappa = \frac{S}{S_0} \quad (0 \leq \kappa \leq 1). \tag{2-111}$$

The concept of the state of opposite polarization can also be defined in quantum mechanics. With any given state ω one can associate a state ω' of opposite polarization such that the expectation value of the Stokes operators have opposite signs for the two states:

$$\langle \Sigma_j \rangle = -\langle \Sigma_j \rangle', \quad (j = 1, 2, 3). \tag{2-112}$$

The transformation $\omega \to \omega'$ is uniquely defined by the condition (2-112) and is given by

$$\omega \to \omega' = U\omega^*, \tag{2-113}$$

---

† For the definition of the density matrix see, for instance, R. C. Tolman, *The Principles of Statistical Mechanics*, Oxford University Press, 1938, especially p. 327 ff.

where the operator U has the property*

$$U^+U = 1, \qquad (2\text{-}114)$$

$$U^+\Sigma_j U = -\tilde{\Sigma}_j. \qquad (2\text{-}115)$$

The sign $\sim$ denotes the transposed operator. Just as in the classical case, we can also define an arbitrary state of partial polarization by setting up special density matrices which have only two diagonal matrix elements with respect to the two states ω, ω' of opposite polarization.

2–9 Definition of the vacuum. The vacuum state of a radiation field is most conveniently defined as the state associated with the lowest value of the energy P^0, which is the state corresponding to the values $n_1 = 0$, $n_2 = 0$ of the occupation numbers. The expression (2–78) gives for each Fourier component a zero-point value of the vacuum energy of the amount k^0. When integrated over all the values of $\mathbf{k}$, this would lead to an infinite energy of the vacuum, a result which cannot have any physical significance. The operators P^μ must therefore be redefined in such a way that their vacuum expectation values vanish. This can be done most conveniently by defining what we shall call in the following the *ordered product* of operators.

The ordered product of a sum of emission and absorption operators is one in which all the absorption operators are placed to the right and the emission operators to the left in each term. The vacuum expectation value of an ordered product is always zero.

In the following it will be useful to have a definition of the ordered product directly in x-space without going to the momentum space. This can be done with the decomposition of the operators given in (2–62), (2–63), and (2–64). If we have two operators $a(x)$, $b(x)$ with their decomposition

$$\begin{aligned} a(x) &= a^{(+)}(x) + a^{(-)}(x), \\ b(x) &= b^{(+)}(x) + b^{(-)}(x), \end{aligned} \qquad (2\text{-}116)$$

then the ordered product, in the following denoted by $(a(x)b(y))_\text{ord}$, is

$$\begin{aligned}(a(x)b(y))_\text{ord} = a^{(+)}(x)b^{(+)}(y) &+ b^{(+)}(y)a^{(-)}(x) \\ &+ a^{(+)}(x)b^{(-)}(y) + a^{(-)}(x)b^{(-)}(y),\end{aligned} \qquad (2\text{-}117)$$

and it is related to the ordinary product $a(x)b(y)$ by

$$(a(x)b(y))_\text{ord} = a(x)b(y) + [b^{(+)}(y), a^{(-)}(x)]. \qquad (2\text{-}118)$$

---

* This transformation is simply the quantum-mechanical time-reversal transformation. This corresponds exactly to the classical interpretation of opposite polarization.

The new definition of energy and momentum can now be given by writing the tensor $T^{\mu\nu}(x)$ as

$$T^{\mu\nu}(x) = -(\partial^\mu a^\lambda(x)\partial^\nu a_\lambda(x))_{\text{ord}} + \tfrac{1}{2}g^{\mu\nu}(\partial_\lambda a_\rho(x)\partial^\lambda a^\rho(x))_{\text{ord}}. \quad (2\text{--}119)$$

In a similar manner, we define all the quadratic expressions occurring in P^μ and $M^{\mu\nu}$. One verifies easily that the commutation rules (1–51) are thereby not affected.

With this new definition, the vacuum has the property that all expectation values of the ten quantities P^μ, $M^{\mu\nu}$ vanish:

$$\langle P^\mu \rangle_0 = 0, \quad \langle M^{\mu\nu} \rangle_0 = 0. \quad (2\text{--}120)$$

The vacuum state characterized by (2–120) can also be described as the state with no transverse photons present.

In the special coordinate system for each Fourier component used in connection with (2–83) this could be expressed by

$$a_1(\mathbf{k})\omega_0 = 0, \quad a_2(\mathbf{k})\omega_0 = 0, \quad \text{(for all } \mathbf{k}\text{)}. \quad (2\text{--}121)$$

It is easily seen that the two characterizations (2–120) and (2–121) of the vacuum are equivalent. Indeed, (2–120) is a consequence of (2–121) by the definition of the ordered product. If, on the other hand, (2–121) were not a consequence of (2–120), there would exist a component $\mathbf{k}$ such that

$$\omega_0' \equiv a_1(\mathbf{k})\omega_0 \neq 0. \quad (2\text{--}122)$$

Because of the commutation properties

$$[P^\mu, a_1(\mathbf{k})] = -k^\mu a_1(\mathbf{k}) \quad (2\text{--}123)$$

and (2–120), we then find that

$$\begin{aligned}
\langle P^\mu \rangle_{0'} &\equiv (a_1(\mathbf{k})\omega_0, P^\mu a_1(\mathbf{k})\omega_0) \\
&= \langle a_1^*(\mathbf{k}) P^\mu a_1(\mathbf{k}) \rangle_0 \\
&= -k^\mu \langle a_1^* a_1 \rangle_0 = -k^\mu (\omega_0', \omega_0').
\end{aligned} \quad (2\text{--}124)$$

This means the state ω_0' is one for which the energy would have a negative expectation value, which is impossible. Consequently, $\omega_0' = 0$, in agreement with (2–121).

We can also formulate the vacuum condition in a general manner without reference to a special coordinate system. To this end, we introduce the *transverse field* operators b_μ defined by

$$b_\mu(\mathbf{k}) = a_\mu(\mathbf{k}) - n_\mu \kappa^\lambda a_\lambda(\mathbf{k}) - \kappa_\mu n^\lambda a_\lambda(\mathbf{k}) - \kappa_\mu \kappa^\lambda a_\lambda(\mathbf{k}) \quad (2\text{--}125)$$

with

$$\kappa^\lambda = \frac{k^\lambda}{k \cdot n}, \quad (2\text{--}126)$$

such that

$$\kappa^2 = 0, \quad \kappa \cdot n = 1; \quad (2\text{--}127)$$

the $b_\mu^*(k)$ are then defined in a similar manner. With these definitions, we verify that
$$\kappa \cdot b = n \cdot b = 0, \qquad (2\text{--}128)$$
which justifies the name *transverse field*.

The vacuum state is now defined by the relations
$$b_\mu(\mathbf{k})\omega_0 = 0, \quad \text{(for all } \mathbf{k}) \qquad (2\text{--}129)$$
which are equivalent to (2–121).

We can transcribe this definition entirely into x-space if we define the field $b_\mu(x)$ corresponding to the Fourier components $b_\mu(\mathbf{k})$. We define for this purpose the differential operator $\partial^{-1}\partial^\lambda$, which is defined by its equivalent expression κ^λ (2–126) in momentum space ($\partial \equiv \partial^\lambda n_\lambda$) and obtain for the *transverse field* in x-space
$$b_\mu(x) = a_\mu(x) - n_\mu \partial^{-1}\partial^\lambda a_\lambda(x) - \partial^{-1}\partial_\mu n^\lambda a_\lambda(x) - \partial^{-2}\partial_\mu \partial^\lambda a_\lambda(x). \qquad (2\text{--}130)$$
This field satisfies identically
$$\partial^\mu b_\mu(x) = n^\mu b_\mu(x) = 0. \qquad (2\text{--}131)$$
The vacuum state is now defined by
$$b_\mu^{(-)}(x)\omega_0 = 0, \quad \text{(for all } x) \qquad (2\text{--}132)$$
together with (2–30). Because of (2–30) the last equation can also be written as
$$a_\mu^{(-)}(x)\omega_0 = \partial^{-1}\partial_\mu n^\lambda a_\lambda^{(-)}(x)\omega_0. \qquad (2\text{--}133)$$

The definition (2–133) of the vacuum state ω_0 seems to imply that ω_0 depends explicitly on n. This, however, is not the case. The following theorem can be shown to hold. If ω satisfies (2–133) and (2–30), then it satisfies an equation of the form (2–133) for arbitrary unit vectors n^λ.[†] This shows that the vacuum state is actually independent of the orientation of the plane.

[†] F. J. Belinfante, *Phys. Rev.* **76**, 228 (1949). Also F. Coester and J. M. Jauch, *Phys. Rev.* **78**, 149 (1950).

CHAPTER 3

RELATIVISTIC THEORY OF FREE ELECTRONS

In this chapter, the general principles which were developed in the first chapter are now applied to the spinor field which describes the system of electrons. The field equation is the equation of Dirac (Section 3–1). As a linear partial differential equation of first order the values of the field variables are determined by the values of these variables on a space-like surface (Section 3–2). The field equation is relativistically invariant (Section 3–3). The field variables themselves do not represent observables directly, but one can construct bilinear expressions of the field variables which have the transformation properties of tensors (Section 3–4). Some of these have a simple physical interpretation.

The quantization of the spinor field follows the pattern of the general theory (Sections 3–5 and 3–6) and the momentum operators can readily be constructed (Section 3–7). We then proceed to the plane-wave representation of the spinor field and derive the commutation rules for the plane-wave amplitudes (Section 3–8). Explicit representations of these operators can easily be constructed and the uniqueness of these representations can be proved (Section 3–9). Finally, we define the state vector for the vacuum (Section 3–10).

3–1 The field equations for the one-particle problem. According to Dirac,[*] the wave function which describes an electron is a four-component spinor $\psi_\rho(x)$ which transforms under proper Lorentz transformations according to the irreducible spinor representation of the Lorentz group.[†] It satisfies the homogeneous Dirac equation

$$(\partial + m)\psi = 0, \quad (\partial \equiv \gamma^\mu \partial_\mu). \tag{3-1}$$

Here the γ^μ are a set of 4×4 matrices operating on the spinor indices of ψ. Matrix product notation is used in (3–1) and ψ represents a column spinor:

$$\psi = \begin{pmatrix} \psi_1 \\ \psi_2 \\ \psi_3 \\ \psi_4 \end{pmatrix}. \tag{3-2}$$

---

[*] P. A. M. Dirac, *Proc. Roy. Soc.* **117**, 610; **118**, 341 (1928).

[†] For the discussion of this representation see, for instance, B. L. van der Waerden, *Die gruppentheoretische Methode in der Quantenmechanik*, Berlin, 1932, p. 78 ff.

The four matrices γ^μ satisfy the anticommutation rules

$$\{\gamma^\mu,\gamma^\nu\} = 2g^{\mu\nu}. \tag{3-3}$$

It follows from (3-3) and (3-1) that the spinors ψ always satisfy the second order wave equation

$$(\partial^\mu\partial_\mu - m^2)\psi = 0. \tag{3-4}$$

This equation is obtained by operating with $(\partial - m)$ on the left and using (3-3).

The constant m is the reciprocal Compton wavelength of the electron ($mc/\hbar$ in cgs units). The fact that Planck's constant enters into the definition of this m indicates that the spinor wave function ψ does not really represent a classical field. It is for this reason that the quantization procedure to be discussed below is usually referred to as "second quantization."

The relations (3-3) characterize the γ_μ sufficiently (cf. Section A2-1). The physical content of the theory can be extracted from these relations alone. We shall therefore never introduce any specific representation of the γ_μ but shall formulate the theory in such a way that all the results involving the γ_μ are derived directly from (3-3).

3-2 The associated boundary value problem. The field equations lead naturally to a boundary value problem which may be stated as follows:

Given $\psi(x)$ on a space-like plane with normal vector n^ν. Find the value of $\psi(x)$ at any other point in space.

The solution of this problem can be given in the form of a definite integral by using the function $S(x)$ defined in the mathematical appendix (Section A1-6):

$$\psi(x) = \int_\sigma d\sigma_\mu' S(x - x')\gamma^\mu\psi(x'). \tag{3-5}$$

To prove (3-5), we notice that the integrand is of the form

$$\psi(x) = \int_\sigma d\sigma_\mu' f^\mu(x') \tag{3-6}$$

with

$$f^\mu(x') = S(x - x')\gamma^\mu\psi(x'), \tag{3-7}$$

and $f^\mu(x')$ satisfies

$$\partial_\mu' f^\mu(x') = 0 \tag{3-8}$$

because of (3-1) and

$$\partial^\mu S\gamma_\mu = \gamma_\mu\partial^\mu S \equiv \partial S = -mS, \tag{3-9}$$

which follows from Eq. (A1-30). The integral (3-5) is, by Gauss' theorem, independent of the surface and may therefore be evaluated on a plane σ

passing through the point x. The surface may in particular be chosen such that $n^\mu = (1, 0, 0, 0)$. The integral (3–5) now reduces to

$$\int d\sigma' S(x - x')\gamma_0 \psi(x') = \int d\sigma' \partial_0 \Delta(x - x')\psi(x') = \psi(x).$$

We see that the function $\psi(x)$ is determined everywhere if it is given on a space-like surface σ.

3–3 Relativistic invariance of the field equations. The relativistic invariance of the field equations amounts to the statement that under a Lorentz transformation L the spinor function $\psi(x)$ transforms according to

$$x' = Lx, \quad \psi'(x') = S\psi(x), \tag{3-10}$$

and in such a manner that the new spinor $\psi'(x')$ considered as a function of x' satisfies Eq. (3–1) with the same γ_μ:

$$(\gamma^\mu \partial_\mu' + m)\psi'(x') = 0. \tag{3-11}$$

This statement is an immediate consequence of the theorem regarding the uniqueness of the representation of the γ's (*cf.* Appendix, Section A2–1). To show this, we write for the transformed coordinates

$$L: \quad x_\mu' = a_\mu{}^\nu x_\nu, \quad x_\nu = a^\mu{}_\nu x_\mu'. \tag{3-12}$$

and obtain for (3–1) in the new variables

$$(\gamma^\mu a_{\nu\mu}\partial'^\nu + m)\psi = 0. \tag{3-13}$$

Since

$$\partial_\mu x_\nu' = a_{\nu\mu}, \tag{3-14}$$

we may write this

$$(\gamma'^\mu \partial_\mu' + m)\psi = 0, \tag{3-15}$$

where

$$\gamma'^\mu = a^\mu{}_\nu \gamma^\nu, \tag{3-16}$$

and ψ is considered as a function of x' through the intermediary relation (3–12). We observe now that the set of matrices γ'^μ satisfies again the relation (3–3),

$$\{\gamma'^\mu, \gamma'^\nu\} = 2g^{\mu\nu}, \tag{3-17}$$

and is therefore related to the original set γ^μ by a similarity transformation:

$$\gamma'^\mu = S^{-1}\gamma^\mu S. \tag{3-18}$$

The relation (3–18) defines $S(L)$ only up to an arbitrary factor. This factor is further restricted to a $\pm$ sign if we require that the $S(L)$ form a representation of the Lorentz group. We obtain thus the two-valued spinor representation, in agreement with our previous assumption as to the transformation properties of ψ. Equation (3–15) now becomes, after multiplying with S from the left,

$$(\gamma^\mu \partial_\mu' + m)\psi'(x') = 0, \tag{3-19}$$

with

$$\psi'(x') = S\psi(x). \tag{3-20}$$

We have thus established that (3–11) is a consequence of (3–1).

An explicit expression for the matrix S can be obtained with the method of the infinitesimal Lorentz transformations. Let

$$a_\nu{}^\mu = g_\nu{}^\mu + \alpha_\nu{}^\mu, \tag{3-21}$$

where the $\alpha_\nu{}^\mu$ are infinitesimals of first order. The matrix S will then appear in the form

$$S = 1 + \Sigma, \tag{3-22}$$

with

$$\Sigma = \tfrac{1}{2}\Sigma^{\mu\nu}\alpha_{\mu\nu}. \tag{3-23}$$

The defining equation (3–18) leads then to the relation

$$[\gamma^\rho, \Sigma^{\mu\nu}] = g^{\rho\mu}\gamma^\nu - g^{\rho\nu}\gamma^\mu, \tag{3-24}$$

with the obvious solution

$$\Sigma^{\mu\nu} = \tfrac{1}{2}\gamma^\mu\gamma^\nu. \tag{3-25}$$

The finite transformation S generated by the infinitesimal one (3–22) is then given by

$$S = e^\Sigma = \lim_{n \to \infty} \left(1 + \frac{1}{n}\Sigma\right)^n. \tag{3-26}$$

3–4 The bilinear covariants. The physical observables associated with electrons are obtained from certain bilinear expressions of ψ which transform according to some irreducible tensor representation of the Lorentz group. It is easiest to start with the construction of the invariant. According to Eq. (A2–33) in the mathematical appendix, the transformation matrix S satisfies

$$S^+ A = A S^{-1}, \tag{3-27}$$

where A is defined by

$$-\gamma_\mu{}^+ = A\gamma_\mu A^{-1}, \quad A^+ = A. \tag{3-28}$$

It follows from these expressions that the adjoint spinor defined by

$$\bar{\psi} = \psi^+ A \tag{3-29}$$

satisfies

$$\partial_\mu \bar{\psi} \gamma^\mu - m\bar{\psi} = 0. \tag{3-30}$$

In the following, this and similar equations involving the row-spinor $\bar{\psi}$ will be written by introducing inverted differential operators $\overleftarrow{\partial}_\mu$ operating on the function *before* it. The contracted operator $\rlap{/}{e} = \gamma^\mu e_\mu$ is also convenient. Equation (3–30) may then be written

$$\bar{\psi}(\rlap{/}{e} - m) = 0. \tag{3-31}$$

In (3–30) the adjoint spinor $\bar{\psi}$ stands on the left of the matrix γ^μ because the operation of Hermitian conjugation has changed the column spinor into a row spinor. From Eq. (A2–33), it follows that

$$K_1 = \bar{\psi}\psi \tag{3-32}$$

is an invariant. From the defining equations (3–16) and (3–18), we obtain immediately the general result that

$$K_t = \bar{\psi}\Gamma_t\psi, \quad (t = 1, \cdots, 5), \tag{3-33}$$

is a set of quantities which transform as a scalar, vector, tensor of second rank, axial vector, or pseudoscalar, according to whether $t = 1, 2, 3, 4,$ or 5. The Γ_t introduced here stand for the set of matrices defined in (A2–5) of the mathematical appendix.

We note here that (3–29) and (3–33) are not the only possibilities for forming bilinear covariants with the spinor ψ. Indeed, a glance at Eq. (A2–35) of the mathematical appendix shows that we could have defined the row spinor

$$\hat{\psi} = \psi^\sim B\gamma_5 \tag{3-34}$$

with B defined by

$$\gamma_\mu{}^\sim = B\gamma_\mu B^{-1},$$

and we would have found that

$$\hat{\psi}' = \hat{\psi}S^{-1} \tag{3-35}$$

since by (A2–35),

$$S^\sim B\gamma_5 = B\gamma_5 S^{-1}. \tag{3-36}$$

The set of quantities

$$\hat{K}_t = \hat{\psi}\Gamma_t\psi \tag{3-37}$$

has then the same transformation properties as the corresponding set K_t. This possibility is intimately connected with the existence of the charge conjugate transformation

$$\psi \to \psi^c = C^*\psi^*, \tag{3-38}$$

where C is the matrix defined by

$$\gamma_\mu{}^* = C\gamma_\mu C^{-1},$$
$$C^*C = 1. \tag{3-39}$$

The charge conjugate spinor ψ^c satisfies the same equation as the original spinor ψ:

$$(\partial + m)\psi^c = 0, \tag{3-40}$$

and because of [*cf.* (A2–34)]

$$S^*C = CS \tag{3-41}$$

it has the same transformation property as ψ. With the help of the relation derived in the mathematical appendix (A2–32):

$$A = C^+B\gamma_5, \tag{3-42}$$

we find immediately

$$\bar{\psi} = \hat{\psi}^c$$

and

$$\hat{\psi} = \bar{\psi}^c. \tag{3-43}$$

Thus the covariants (3–37) may also be written as

$$\hat{K}_t = \bar{\psi}^c \Gamma_t \psi. \tag{3-44}$$

These alternate forms of covariant expressions do not seem to have any physical significance in radiation theory. They may be of importance, however, in theories of elementary particle interactions, in particular for the correct description of a universal Fermi interaction for particles of spin $\frac{1}{2}$.†

The expressions K_t may be generalized for two independent spinor functions ψ, φ to

$$K_t = \bar{\psi} \Gamma_t \varphi. \tag{3-45}$$

In many calculations, the following relations are useful:

$$(\bar{\psi}\Gamma_t \varphi)^* = -\bar{\psi}^c \Gamma_t \varphi^c = \zeta_t \bar{\varphi} \Gamma_t \psi, \tag{3-46}$$

where ζ_t is the sign function defined by

$$\Gamma_t^+ = \zeta_t A \Gamma_t A^{-1} \tag{3-47}$$

and tabulated in Table A2–1 of the mathematical appendix. We note that (3–46) is correct only in the c-number theory so far discussed. In the quantized theory, attention must be paid to the position of noncommuting factors.

In radiation theory, the current density is of special importance:

$$j_\mu = i\bar{\psi}\gamma_\mu \psi. \tag{3-48}$$

Because of (3–1) and (3–30), it satisfies the continuity equation

$$\partial^\mu j_\mu = 0, \tag{3-49}$$

and because of (3–46) it is real,

$$j_\mu^* = j_\mu. \tag{3-50}$$

---

† Cf. E. R. Caianiello, *Nuovo Cimento* **8**, 534 (1951); *ibid.* **8**, 749 (1951). C. N. Yang, J. Tiomno, *Phys. Rev.* **79**, 495 (1950). The use of the $\hat{\psi}$-spinor facilitates especially the construction of interactions with symmetry properties, such as the interaction of Wigner and Critchfield. C. L. Critchfield and E. P. Wigner, *Phys. Rev.* **60**, 412 (1941). C. L. Critchfield, *Phys. Rev.* **63**, 417 (1943).

3-5 A Lagrangian for the spinor field.

The field equations (3-1) and (3-30) can be derived from a Lagrangian:

$$L = -\bar{\psi}(\partial + m)\psi, \quad (\partial \equiv \gamma_\mu \partial^\mu). \tag{3-51}$$

Inserting this expression into Eq. (1-74), we obtain the equation for the spinor $\bar{\psi}$:

$$\bar{\psi}(\overleftarrow{\partial} - m) = 0, \quad (\bar{\psi}\overleftarrow{\partial} \equiv \partial_\mu \bar{\psi}\gamma^\mu). \tag{3-30}$$

By conjugation then follows

$$(\partial + m)\psi = 0. \tag{3-1}$$

The canonically conjugate spinor to ψ, according to (1-79), is defined by

$$\pi_\mu = -\bar{\psi}\gamma_\mu. \tag{3-52}$$

We note here that the Lagrangian (3-51) has, in addition to its relativistic invariance, a further invariance property—the phase transformation of the field operators:

$$\psi(x) \rightarrow \psi'(x) = e^{i\varphi}\psi(x). \tag{3-53}$$

Here φ denotes an arbitrary constant c-number. Under this transformation,

$$\bar{\psi}(x) \rightarrow \bar{\psi}'(x) = e^{-i\varphi}\bar{\psi}(x), \tag{3-54}$$

and consequently L (3-51) remains unaffected.

3-6 Quantization.

One of the fundamental properties of electrons is the fact that they obey the exclusion principle. According to the discussion in Section 1-12, the quantization of the electron field will be accomplished in accord with case (b) of that section.*

According to (1-107b) and (3-52), we have for the anticommutators at two *space-like* situated points x,x':

$$\{\psi(x),\bar{\psi}(x')\gamma_\mu\} = i\delta_\mu(x,x'). \tag{3-55}$$

The Hermitian conjugate of this equation is

$$\gamma_\mu\{\psi(x),\bar{\psi}(x')\} = i\delta_\mu(x,x'). \tag{3-56}$$

---

* The quantization of the electron field was given by P. Jordan and E. Wigner, *Z. Physik* **47**, 631 (1928), and V. Fock, *Z. Physik* **75**, 622 (1932). The interaction of electrons was discussed by P. Jordan and O. Klein, *Z. Physik* **45**, 751 (1927). Of the more recent work, we mention especially J. G. Valatin, *J. de Phys. et le Radium* **12**, 131, 542, 607 (1951).

We can obtain the anticommutators for arbitrary points x, x' by using Eq. (3–5). Thus, for instance,

$$\{\psi(x), \bar{\psi}(x')\} = \int d\sigma_\mu'' S(x - x'') \gamma^\mu \{\psi(x''), \bar{\psi}(x')\}. \tag{3–57}$$

If we choose the plane of integration in the last equation to pass through x', we can substitute (3–56) and obtain

$$\{\psi(x), \bar{\psi}(x')\} = \int S(x - x'') d\sigma_\mu'' i\delta^\mu(x'', x'). \tag{3–58}$$

With the characteristic property (1–108) of the δ-function, this simplifies to*

$$\{\psi(x), \bar{\psi}(x')\} = iS(x - x'). \tag{3–59}$$

This anticommutation relation is, of course, not sufficient for the quantization of the ψ-field. We must also know† the anticommutator $\{\psi(x), \psi(x')\}$. The anticommutator $\{\bar{\psi}(x), \bar{\psi}(x')\}$ will then follow from it.

The simplest way to obtain a value for such an anticommutator is to postulate that it be invariant under phase transformations, as is Eq. (3–59). This requirement is necessary in order to obtain a satisfactory description of the interaction of electrons with photons (see Chapter 4). It then follows that

$$\{\psi(x), \psi(x')\} = 0, \tag{3–60}$$

$$\{\bar{\psi}(x), \bar{\psi}(x')\} = 0. \tag{3–61}$$

It is interesting to see how much is implied for the anticommutation rules of the ψ at different space-time points without the condition of invariance under phase transformations. One can show that the anticommutation rules (3–59), together with the condition of relativistic invariance, determine to a large extent the anticommutators between $\psi(x)$ and $\psi(x')$,‡ and that instead of (3–60) and (3–61), one has always an equation of the form

$$\{\psi(x), \hat{\psi}(x')\} = i\rho S(x - x').$$

---

* We remind the reader that (3–59) and similar expressions are matrix equations with respect to the spinor indices. Thus (3–59), for instance, reads explicitly

$$\{\psi_\rho(x), \bar{\psi}_\sigma(x')\} = iS_{\rho\sigma}(x - x').$$

† In the paper by J. M. Jauch, quoted below, it is shown that Eq. (3–59) implies that $\{\psi(x), \psi(x')\}$ is a c-number. It then follows that $[\psi(x), \psi(x')]$ cannot be a c-number, so that there exists no commutator of this form compatible with (3–59), but only an anticommutator.

‡ J. M. Jauch, *Helv. Phys. Acta* **27**, 89 (1954). Y. Takahashi, *Nuovo Cimento* **1**, 414 (1955).

Here

$$\dot{\psi} = \bar{\psi}^c = \psi^{\sim} B \gamma_5.$$

The phase of ψ can always be so chosen that ρ is real. In the latter case, we have

$$0 \leq \rho \leq 1.$$

The case $\rho = 1$ is especially interesting because it follows then that $\psi^c = \psi$. Such a field, which is equal to its own charge conjugate, is called a Majorana field.† We merely mention these fields here for the sake of completeness, but we shall work exclusively with Eqs. (3–60) and (3–61).

We note that according to the remark made in Section 1–6 the field operators ψ and $\bar{\psi}$ are not observables because they do not commute on a space-like surface. However, the bilinear expressions discussed in Section 3–4 do commute at two different points on such a surface. Thus we find, for instance, for the current operator (3–48), by using (3–59), the expression

$$[j_\mu(x), j_\nu(x')] = -i\bar{\psi}(x)\gamma_\mu S(x - x')\gamma_\nu \psi(x') + i\bar{\psi}(x')\gamma_\nu S(x' - x)\gamma_\mu \psi(x),$$

which clearly vanishes if the points x, x' are separated by a space-like interval and $x \neq x'$.

The commutation rules (3–59) are invariant under the operation of charge conjugation (3–38). Indeed, we obtain‡

$$\{\psi^c(x), \bar{\psi}^c(x')\} = \{C^*\psi^*(x), (C^*\psi^*(x'))^+ A\}$$

$$= C^*\{\psi^*(x), \psi^{\sim}(x')\} C^{\sim} A.$$

The expression in curly brackets on the right side of the above equation is

$$\{\psi^*(x), \psi^{\sim}(x')\} = -i(\partial^* - m)A^{*-1}\Delta(x - x').$$

Combining this with

$$A^{*-1}C^{\sim}A = -C, \tag{3–62}$$

which follows from relations (A2–32) in the mathematical appendix, we obtain, with (3–39),

$$\{\psi^c(x), \bar{\psi}^c(x')\} = iS(x - x'). \tag{3–63}$$

---

† E. Majorana, *Nuovo Cimento* **14**, 171 (1937); G. Racah, *ibid.* **14**, 322 (1937). W. H. Furry, *Phys. Rev.* **54**, 56 (1938); also discussed in the review article by W. Pauli, *Rev. Mod. Phys.* **13**, 203 (1941), Part II, Section 3c.

‡ The notation used here needs some comment. As long as we discussed the c-number theory the ψ^* meant simply the complex conjugate of the numbers ψ. In the quantized theory ψ^* always means the Hermitian conjugate operator. The operation ψ^+ is then used for the Hermitian conjugate and transposed spinor. The transposition operation for the spinor index is thus explicitly separated from the transposition operation inherent in the Hermitian conjugate operation.

3-7 Momentum operators.

The general method of Chapter 1 allows us to construct the operators P^μ and $M^{\mu\nu}$ which represent the mechanical properties of the field. According to (1-80), we obtain for the tensor $T^{\mu\nu}$:

$$T^{\mu\nu} = -\bar\psi\gamma^\mu\partial^\nu\psi + g^{\mu\nu}\bar\psi(\partial + m)\psi. \tag{3-64}$$

For a ψ which satisfies the field equation (3-1), the last term can be omitted and we obtain simply

$$T^{\mu\nu} = -\bar\psi\gamma^\mu\partial^\nu\psi. \tag{3-65}$$

To obtain the symmetrical tensor, we need the quantity $H^{\rho\mu\nu}$ (1-97) which, because of (3-25), is given by

$$H^{\rho\mu\nu} = -\tfrac{1}{2}\bar\psi\gamma^\rho\gamma^\mu\gamma^\nu\psi. \tag{3-66}$$

Finally, with the definition (1-99), we obtain

$$\partial_\rho G^{\rho\mu\nu} = -\tfrac{1}{2}\bar\psi\gamma^\nu\partial^\mu\psi + \tfrac{1}{2}\bar\psi\gamma^\mu\partial^\nu\psi. \tag{3-67}$$

The symmetrical tensor $\theta^{\mu\nu}$ is thus given by

$$\theta^{\mu\nu} = -\tfrac{1}{2}(\bar\psi\gamma^\mu\partial^\nu\psi + \bar\psi\gamma^\nu\partial^\mu\psi). \tag{3-68}$$

For the total energy and momentum, we obtain

$$P^\mu = -\int \bar\psi\gamma^\rho\partial^\mu\psi \, d\sigma_\rho, \tag{3-69}$$

and for the total angular-momentum operator,

$$M^{\mu\nu} = \int (-\bar\psi\gamma^\rho\partial^\mu\psi x^\nu + \bar\psi\gamma^\rho\partial^\nu\psi x^\mu + \tfrac{1}{2}\bar\psi\gamma^\rho\gamma^\mu\gamma^\nu\psi)d\sigma_\rho. \tag{3-70}$$

We can again decompose this expression into orbital and spin angular momentum [cf. Eq. (1-85)]:

$$M^{\mu\nu} = L^{\mu\nu} + N^{\mu\nu}, \tag{3-71}$$

with

$$L^{\mu\nu} = \int (-\bar\psi\gamma^\rho\partial^\mu\psi x^\nu + \bar\psi\gamma^\rho\partial^\nu\psi x^\mu)d\sigma_\rho, \tag{3-72}$$

$$N^{\mu\nu} = \tfrac{1}{2}\int (\bar\psi\gamma^\rho\gamma^\mu\gamma^\nu\psi)d\sigma_\rho. \tag{3-73}$$

We see from (3-73) that the axial vector K_4 (3-33) represents the spin angular momentum density of the spinor field. The general theory of Chapter 1 assures us that these quantities have the right commutation properties (1-56) and (1-57). It is instructive to verify these commutation rules directly by utilizing (3-59).

3-8 Plane wave decomposition. The spinor field ψ can be developed in a Fourier series:

$$\psi(x) = \frac{1}{(2\pi)^2} \int \varphi(p) e^{ip \cdot x} d^4p, \qquad (3\text{-}74)$$

with the inversion

$$\varphi(p) = \frac{1}{(2\pi)^2} \int \psi(x) e^{-ip \cdot x} d^4x. \qquad (3\text{-}75)$$

The field equation (3–1) leads to*

$$(i\boldsymbol{p} + m)\varphi(p) = 0. \qquad (3\text{-}76)$$

It is useful to introduce the operators

$$\Lambda_{\pm}(p) = \frac{1}{2m}(m \pm i\boldsymbol{p}) \qquad (3\text{-}77)$$

with the properties

$$\Lambda_{\pm}(p)\Lambda_{\mp}(p) = \frac{1}{4m^2}(p^2 + m^2) = 0 \qquad (3\text{-}78)$$

$$\Lambda_{+}(p) + \Lambda_{-}(p) = 1, \qquad (3\text{-}79)$$

$$\Lambda_{\pm}^2 = \Lambda_{\pm}. \qquad (3\text{-}80)$$

Equation (3–76) shows that the spinor $\varphi(p)$ has the form

$$\varphi(p) = \delta(p^2 + m^2) \Lambda_{-}(p) \chi(p) \qquad (3\text{-}81)$$

with arbitrary $\chi(p)$. Substituting this into (3–74) and integrating over p^0, we obtain

$$\psi(x) = \frac{1}{(2\pi)^2} \int \frac{d^3p}{2\epsilon} [\Lambda_{-}(p)\chi(p)e^{ip \cdot x} + \Lambda_{+}(p)\chi(-p)e^{-ip \cdot x}]. \qquad (3\text{-}82)$$

Here ϵ is the positive square root,

$$p^0 \doteq \epsilon \equiv +\sqrt{\mathbf{p}^2 + m^2}. \qquad (3\text{-}83)$$

The operators $\Lambda_{\pm}$ have the characteristic properties of projection operators of rank two. Thus, they project the four-dimensional spinor space of the χ into two-dimensional subspaces. We can express this explicitly if we introduce two basis vectors u_r, v_r ($r = \pm$) in each of these subspaces defined, for instance, as certain linearly independent solutions of the equations

$$\Lambda_{+} u_r = 0, \quad \Lambda_{-} v_r = 0, \quad (r = \pm), \qquad (3\text{-}84)$$

---

* The use of bold-faced italic letters for the product of a γ-matrix with a four-vector and other notations are summarized in Appendix A8.

with p^0 given by Eq. (3-83). If we then substitute in the expressions in (3-82) the linear combinations

$$\Lambda_-(p)\chi(p) = \sqrt{8\pi m\epsilon} \sum_r a_r(\mathbf{p})u_r(\mathbf{p}),$$

$$\Lambda_+(p)\chi(-p) = \sqrt{8\pi m\epsilon} \sum_r b_r^*(\mathbf{p})v_r(\mathbf{p}),$$

(3-85)

we obtain the development

$$\psi(x) = \frac{1}{(2\pi)^{3/2}} \sum_r \int d^3p \sqrt{\frac{m}{\epsilon}} \{a_r(\mathbf{p})u_r(\mathbf{p})e^{ip\cdot x} + b_r^*(\mathbf{p})v_r(\mathbf{p})e^{-ip\cdot x}\}. \quad (3\text{-}86)$$

Since this is equivalent to the obviously invariant expression (3-74), we have obtained with (3-86) an invariant decomposition into two parts,

$$\psi^{(-)}(x) = \frac{1}{(2\pi)^{3/2}} \sum_r \int d^3p \sqrt{\frac{m}{\epsilon}} a_r(\mathbf{p})u_r(\mathbf{p})e^{ip\cdot x},$$

$$\psi^{(+)}(x) = \frac{1}{(2\pi)^{3/2}} \sum_r \int d^3p \sqrt{\frac{m}{\epsilon}} b_r^*(\mathbf{p})v_r(\mathbf{p})e^{-ip\cdot x},$$

(3-87)

$$\psi(x) = \psi^{(-)}(x) + \psi^{(+)}(x), \quad (3\text{-}88)$$

and a similar expression for

$$\bar{\psi}(x) = \bar{\psi}^{(+)}(x) + \bar{\psi}^{(-)}(x). \quad (3\text{-}88')$$

With the method of complex integration used in the analogous problem of the radiation field, we may represent this decomposition in a manner independent of the Fourier integral representation by the formula

$$\psi^{(\pm)}(x) = \frac{1}{2\pi i} \int_P \psi(x \pm \tau n) \frac{d\tau}{\tau}, \quad (3\text{-}89)$$

where the path P avoids the pole $\tau = 0$ by a detour into the negative imaginary half-plane.

The preceding expression (3-89) is useful for establishing the following relations:

$$\bar{\psi}^{(+)} = \overline{\psi^{(-)}}, \quad \bar{\psi}^{(-)} = \overline{\psi^{(+)}}. \quad (3\text{-}90)$$

To determine the operators a_r, b_r completely, it is necessary to choose a suitable normalization for the spinors u_r, v_r. We shall adopt the definition and normalization of the u_r, v_r given in the mathematical appendix, Eq. (A2-71):

$$\bar{u}_r u_s = \delta_{rs}, \quad \bar{v}_r v_s = -\delta_{rs}, \quad (r,s = \pm), \quad (3\text{-}91)$$

and (A2–74):
$$\sum_r u_r \bar{u}_r = \Lambda_-, \quad \sum_r v_r \bar{v}_r = -\Lambda_+. \tag{3-92}$$

These equations, together with Eq. (3–79), yield
$$\sum_r (u_r \bar{u}_r - v_r \bar{v}_r) = 1. \tag{3-93}$$

The commutation rules for the operators a_r, b_r are most easily obtained if we establish first those for the parts $\psi^{(\pm)}$ of the spinor ψ. Since from (3–89) and (3–59) follows
$$\{\psi^{(+)}(x), \bar{\psi}^{(+)}(x')\} = \{\psi^{(-)}(x), \bar{\psi}^{(-)}(x')\} = 0, \tag{3-94}$$
we obtain
$$\{a_r{}^*(\mathbf{p}), b_s{}^*(\mathbf{p}')\} = 0, \tag{3-95}$$
and
$$\{a_r(\mathbf{p}), b_s(\mathbf{p}')\} = 0. \tag{3-96}$$

On the other hand, we have
$$\{\psi^{(-)}(x), \bar{\psi}^{(+)}(x')\} = iS_+(x - x').$$

If we substitute the Fourier integrals (3–87) and (3–88) in this equation and use Eq. (A2–74), we obtain
$$\frac{1}{(2\pi)^3} \sum_{r,r'} \iint d^3p\, d^3p' \frac{m}{\sqrt{\epsilon \epsilon'}} \{a_r(\mathbf{p}), a_{r'}{}^*(\mathbf{p}')\} u_r(\mathbf{p}) \bar{u}_{r'}(\mathbf{p}') e^{ip \cdot x - ip' \cdot x'}$$
$$= \frac{1}{(2\pi)^3} \int \frac{m}{\epsilon} \Lambda_-(p) e^{ip \cdot (x - x')} d^3p.$$

On identification of the integrands, we find
$$\{a_r(\mathbf{p}), a_{r'}{}^*(\mathbf{p}')\} = \delta_{rr'} \delta(\mathbf{p} - \mathbf{p}'). \tag{3-97}$$

In the same way, we establish
$$\{b_r(\mathbf{p}), b_{r'}{}^*(\mathbf{p}')\} = \delta_{rr'} \delta(\mathbf{p} - \mathbf{p}'). \tag{3-98}$$

We note from the general principles of Chapter 1 that we obtain no further information regarding the commutation rules. In particular, brackets of the form $\{a, b^*\}$ do not occur in the set so far obtained, (3–95) to (3–98). Such brackets arise from commutation rules involving two ψ's at different space-time points instead of a ψ and $\bar{\psi}$ as hitherto considered. In fact, Eqs. (3–60) and (3–61) lead to the anticommutators
$$\{a_r(\mathbf{p}), b_{r'}{}^*(\mathbf{p}')\} = 0, \quad \{a_r(\mathbf{p}), a_{r'}(\mathbf{p}')\} = 0, \quad \{b_r(\mathbf{p}), b_{r'}(\mathbf{p}')\} = 0, \tag{3-99}$$
and their Hermitian conjugates.

The physical significance of the operators a and b can be brought to light more explicitly if we calculate the total energy, momentum, and charge of the field in terms of the operators a and b. If we use expression (3–69) just as it stands, we obtain

$$P^\mu = \sum_r \int d^3 p\, p^\mu [a_r{}^*(\mathbf{p}) a_r(\mathbf{p}) - b_r(\mathbf{p}) b_r{}^*(\mathbf{p})]. \tag{3–100}$$

In the derivation of this last expression, use is made of formulas such as

$$i\bar{u}_r(\mathbf{p})\gamma^0 u_{r'}(\mathbf{p}) = i\bar{v}_r(\mathbf{p})\gamma^0 v_{r'}(\mathbf{p}) = \delta_{rr'} \frac{\epsilon}{m} \tag{3–101}$$

and

$$\bar{u}_r(\mathbf{p})\, \gamma^0 v_{r'}(-\mathbf{p}) = \bar{v}_r(\mathbf{p})\, \gamma^0 u_{r'}(-\mathbf{p}) = 0, \tag{3–102}$$

which follow from the defining equations (3–84) and the normalization (3–91). The expression for the total charge operator, in a similar manner, becomes

$$Q = i \int d\sigma_\mu \bar{\psi}\gamma^\mu \psi$$

$$= -\sum_r \int d^3 p \{ a_r{}^*(\mathbf{p})a_r(\mathbf{p}) + b_r(\mathbf{p})b_r{}^*(\mathbf{p}) \}. \tag{3–103}$$

The expressions (3–100) and (3–103) show a dissymmetry with respect to the operators a and b. This has as a consequence that the vacuum expectation values of these expressions are infinite. The situation is analogous to the situation encountered in Chapter 2 in the discussion of the radiation field, and it can be remedied by using ordered products. We define the ordered product of two field operators as

$$(\psi(x)\bar{\psi}(y))_{\text{ord}} \equiv \psi(x)\bar{\psi}(y) - \{\psi^{(-)}(x), \bar{\psi}^{(+)}(y)\}. \tag{3–104}$$

As a result of this definition, all creation operators appear on the left and all destruction operators on the right. The sign differs, however, from the corresponding expression (2–118). The definition (3–104) provides a sign change for every interchange of an emission-absorption operator, a provision which is necessary if the ordered product of two field operators is to differ from the ordinary product only by a c-number. We postulate now that all physically meaningful integrals such as the energy, momentum, and total charge shall be defined as ordered products. In this way, we obtain for the expressions corresponding to (3–100) and (3–103),

$$P^\mu = \sum_r \int d^3 p\, p^\mu \{ a_r{}^*(\mathbf{p})a_r(\mathbf{p}) + b_r{}^*(\mathbf{p})b_r(\mathbf{p}) \}, \tag{3–105}$$

$$Q = \sum_r \int d^3p \{ -a_r^*(\mathbf{p})a_r(\mathbf{p}) + b_r^*(\mathbf{p})b_r(\mathbf{p}) \}. \qquad (3\text{--}106)$$

The expressions thereby obtained exhibit a symmetry with respect to charge conjugation. In fact, since charge conjugation is equivalent to the interchange of the operators a and b [*cf.* Eq. (A2–76)], we see that

$$P^{c\mu} = P^\mu \qquad (3\text{--}107)$$

and

$$Q^c = -Q. \qquad (3\text{--}108)$$

This last relation justifies the use of the term "charge conjugation."

The expression for the charge could also have been obtained from the current expression which is antisymmetrized with respect to the operation of charge conjugation. We have thus an equation of the form

$$i(\bar\psi \gamma^\mu \psi)_{\text{ord}} = \frac{i}{2}(\bar\psi \gamma^\mu \psi - \bar\psi^c \gamma^\mu \psi^c). \qquad (3\text{--}109)$$

It is not possible to write (3–105) in a similar manner as a symmetrical expression in x-space. The use of the ordered product is thus a more general method of eliminating the undesirable vacuum expectation values. The definition (3–109) of the current density assures that its vacuum expectation value is zero,

$$\langle j_\mu \rangle_0 = 0. \qquad (3\text{--}110)$$

This equation is not satisfied by j_μ as defined in Eq. (3–48) unless the operator product is ordered. In the following, we shall understand all expressions for physical quantities which involve products of ψ-functions as ordered products, even if they are not explicitly indicated as such with the notation used in (3–104). The form of Eqs. (3–105) and (3–106) makes the following interpretation possible. We denote by

$$n^-(\mathbf{p}) = \sum_r a_r^*(\mathbf{p}) a_r(\mathbf{p}) \qquad (3\text{--}111)_-$$

the operator for the particle number of negative charge, and by

$$n^+(\mathbf{p}) = \sum_r b_r^*(\mathbf{p}) b_r(\mathbf{p}) \qquad (3\text{--}111)_+$$

the operator for the particle number of positive charge and of momentum p^μ. The choice of the sign of the charge here adopted is, of course, conventional and arbitrary. In order to interpret the dependence on the index r, we calculate the spin operator in the direction of propagation $\mathbf{p}$, which is defined as the direction of the momentum of the particles. We obtain this operator by first evaluating the Fourier transform of the spin operator (3–73):

$$N^{\mu\nu} = \int N^{\mu\nu}(\mathbf{p}) d^3p. \qquad (3\text{–}112)$$

We consider now just the space components $\mu = i$, $\nu = k$, and define the spin vector $\mathbf{N}(\mathbf{p})$ with the components

$$N^l(\mathbf{p}) = N^{ik}(\mathbf{p}); \qquad (3\text{–}113)$$

i, k, l are a cyclic permutation of 123. The operator in question is then defined as

$$N(\mathbf{p}) = \frac{1}{|\mathbf{p}|}(\mathbf{N}(\mathbf{p}) \cdot \mathbf{p}) \qquad (3\text{–}114)$$

and it corresponds to the operator (2–83) defined for photons. The operator $\mathbf{N}(\mathbf{p})$ is complicated because it contains terms in a^*a, b^*b as well as a^*b^* and ab. But the operator (3–114) which represents the spin in the direction of propagation is simple and it can easily be evaluated by using the formulas (A2–58) and (A2–71) of the appendix. The result is

$$N(\mathbf{p}) = \tfrac{1}{2}(a_+^* a_+ - a_-^* a_-) + \tfrac{1}{2}(b_+^* b_+ - b_-^* b_-). \qquad (3\text{–}115)$$

From this expression, we see that the quantities

$$n_r^{(-)} = a_r^* a_r, \qquad (3\text{–}116)_-$$
$$(r = \pm)$$
$$n_r^{(+)} = b_r^* b_r, \qquad (3\text{–}116)_+$$

are the operators of negative (positive) particles with spins in the direction of propagation $\tfrac{1}{2}r$. We note especially that in order to obtain this interpretation it was necessary to define the u_r, v_r in accord with Eq. (A2–56). In the following we shall use the words *negatons* and *positons* for designating the two kinds of particles.

3–9 Explicit representation of the field operators. We have obtained in (3–97), (3–98), and (3–99) a set of commutation rules which must be satisfied for the amplitudes of the plane wave solutions. But we do not yet know whether operators with these properties do in fact exist. The problem is more complicated than the corresponding problem encountered in the theory of the radiation field. In that case we could refer to the quantum mechanics of the harmonic oscillator, which gives explicit matrix representations of the operators in question. Since there is no such correspondence in the present case, the construction of such operators must be carried out with purely algebraic methods.

To simplify the notation, we make the indices which label the plane waves a countable set by enclosing the physical system in a finite region in space [cf. Eqs. (2–70)]. We then label each component by a single index r which may also be used for labeling both the amplitudes a and b.

The mathematical problem is then to construct a countable set of matrices a_r which satisfy the relations

$$\{a_r, a_s{}^*\} = \delta_{rs}, \quad \{a_r, a_s\} = 0. \tag{3-117}$$

The solution of this problem was given by Jordan and Wigner.† Following a mathematical construction by Clifford, we construct first the 2×2 matrices:

$$\alpha = \begin{pmatrix} 0 & 1 \\ 0 & 0 \end{pmatrix}, \quad \alpha^* = \begin{pmatrix} 0 & 0 \\ 1 & 0 \end{pmatrix}, \quad \beta = \begin{pmatrix} 1 & 0 \\ 0 & -1 \end{pmatrix}, \tag{3-118}$$

which satisfy the relations

$$\{\alpha, \alpha^*\} = 1, \quad \{\alpha, \alpha\} = 0$$
$$\beta = 1 - 2\alpha^*\alpha. \tag{3-119}$$

We now order the indices r in a definite but otherwise arbitrary manner and then form by the method of the direct (or Kronecker) product the following matrices:

$$\alpha_r = 1 \times 1 \times \cdots \times \alpha \times 1 \times \cdots,$$
$$\beta_r = 1 \times 1 \times \cdots \times \beta \times 1 \times \cdots. \tag{3-120}$$

In (3–120) the factors α, β on the right occur at the rth place. One then easily verifies that the matrices

$$a_r = \alpha_r \prod_{q=1}^{r-1} \beta_q,$$

$$a_r{}^* = \alpha_r{}^* \prod_{q=1}^{r-1} \beta_q \tag{3-121}$$

satisfy the required commutation rules.

Much more subtle is the proof that this construction is essentially unique. More precisely: if a_r' and a_s' belong to any other irreducible set of matrices which satisfy the relations (3–117), then this set is related to the original set by an S-transformation:

$$\{a_r', a_s'{}^*\} = \delta_{rs}, \tag{3-117}'$$

$$a_r' = S^{-1} a_r S. \tag{3-122}$$

---

† *Loc. cit.* on page 56.

For finite sets of operators this theorem is a special case of a well-known theorem on representations of Clifford algebras and the reader is referred to the literature.†

For the physical application the uniqueness property of these matrices is of great importance because it implies that for irreducible representations every transformation which leaves the commutation rules invariant is a canonical transformation.

The eigenvalues of the operators $a_r{}^*a_r$ are seen to be 0 and 1. This result is in accord with the exclusion principle of Pauli, according to which individual states cannot be occupied more than once.

3-10 The definition of the vacuum. Just as in the case of the radiation field, we may define the vacuum as the state associated with the lowest value of the energy P^0. According to (3-105) the energy is a positive definite Hermitian matrix which assumes its lowest value if the individual occupation numbers a^*a and b^*b are equal to zero. This state may also be characterized by

$$\psi^{(-)}(x)\omega_0 = 0,$$

$$\text{(for all } x\text{)}, \tag{3-123}$$

$$\psi^{c(-)}(x)\omega_0 = 0,$$

which is equivalent to

$$a_r(\mathbf{p})\omega_0 = 0,$$

$$\text{(for all } r,\mathbf{p}\text{)}. \tag{3-124}$$

$$b_r(\mathbf{p})\omega_0 = 0,$$

---

† The uniqueness was proved by Jordan and Wigner with a group theoretical method, *loc. cit.* on page 56. The most concise proof known to the authors was given by A. A. Albert, *Mathematical Reviews* **10**, 180 (1949). For more details see H. C. Lee, *Ann. of Math.* (2) **49**, 760 (1948), especially Section II, case A. For infinite sets of operators the representation which we have constructed is not the only possible representation. In order to assure the uniqueness of the representation of infinite sets of operators a_i it is necessary to make the additional assumption that there exists a state vector with the property

$$\sum_i a_i{}^*a_i\omega_0 = 0.$$

This is very similar to the corresponding case for the commutation rules of Bose statistics. It can be shown that without this assumption there exists an infinite set of inequivalent irreducible representations. The above condition has a simple physical interpretation. We shall see below that the state vector ω_0 represents the vacuum state of the system. The condition under discussion means therefore, in physical terms, that the physical system should contain among its possible states the vacuum state. We are indebted to Professor Wightman for informing us of this theorem. See footnote reference on page 39.

The expectation value of the energy for this state is then

$$\langle P^0 \rangle_0 = (\omega_0, P^0 \omega_0) = 0, \tag{3-125}$$

since

$$P^0 \omega_0 = 0. \tag{3-126}$$

If the system is irreducible, the conditions (3–123) or (3–124) determine the state vector uniquely, provided it is normalized. Since the operators representing physical quantities are all ordered operators, we note that for all such operators F, we have

$$\langle F \rangle_0 = 0; \tag{3-127}$$

in particular,

$$\langle P^\mu \rangle_0 = \langle M^{\mu\nu} \rangle_0 = \langle Q \rangle_0 = 0. \tag{3-128}$$

Another example of an equation of type (3–127) was encountered previously in Eq. (3–110).

We emphasize that in the quantized theory for electrons there is no difficulty arising from negative energy states. This difficulty exists only in a one-particle theory (c-number theory). In the quantized theory the negative energy solutions do not occur because we have interpreted the charge conjugate solutions $v = u^c$ as the states of the field with positive energy and positive electric charge.

CHAPTER 4

INTERACTION OF RADIATION WITH ELECTRONS

The two fields discussed in the previous two chapters are now allowed to interact. The interaction term appears in the field equations as a coupling term which involves both fields. The possibilities of this term are severely restricted by the requirement that the coupled fields must, in the classical limit, satisfy Maxwell's equations and the postulate of relativistic invariance (Section 4–1). The commutation rules for the Heisenberg fields are then discussed (Section 4–2). This is done in a formally covariant manner by introducing a family of uncoupled fields which satisfy the free-field commutation rules. We then introduce the interaction picture in which the field operators satisfy the free-field equations. The coupling of the two fields is then entirely contained in the time variation of the state vector (Section 4–3). This chapter is concluded with a brief discussion of the measurability of the field variables (Section 4–4).

4–1 The field equations. In the last two chapters we have developed the quantum theory of two separate systems, the radiation field and the electron field. These systems were considered free and independent of each other. The radiative processes arise from the interaction of the two systems. An interaction implies that observable quantities, such as momentum and energy, for instance, can be transferred from one system to the other, thereby giving rise to all kinds of radiative processes. The coupling of the systems can be achieved by postulating the existence of a coupling term in the field equations which depends on the field variables of both fields.

If no further restrictions were assumed, we would have a great deal of freedom to construct such coupling terms. An obvious restriction which must be satisfied by any field theory stems from the requirement of the relativistic invariance of the field equations. But even with this restriction there exist in general a large number of possible interactions.* Further conditions are obtained only from requirements which are to some extent arbitrary. One such requirement which immediately reduces the number of possible interaction terms to one is the following: the field equations shall be modified only by terms which are of lowest order in the field vari-

---

* The possible physical implications of such more general interaction terms were studied for the one-particle theory by L. L. Foldy, *Phys. Rev.* **87**, 688 (1952). A special case was also discussed by F. J. Belinfante, *Phys. Rev.* **73**, 641 (1948).

ables and contain no derivatives. This means that the interaction term will be linear in the potentials A_μ and bilinear (for reasons of relativistic invariance) in the operators Ψ.* The final test for the correctness of the interaction term, however, will not be found in such requirements of simplicity but in its agreement with observation. It is a fact (and the major portion of the remainder of this book will be devoted to elaboration of this point) that the interaction term thereby obtained describes correctly the interaction of the radiation field with the electron field.

One of the first of these consequences of the theory of interacting fields is the fact that the theory has a classical limit which is in complete agreement with the classical theory of the radiation field interacting with charges and currents.

The classical electromagnetic field is described by the skew-symmetrical tensor field $F_{\mu\nu}(x)$:

$$F_{\mu\nu} = -F_{\nu\mu}, \tag{4-1}$$

which satisfies the system of field equations

$$\partial_\lambda F_{\mu\nu} + \partial_\mu F_{\nu\lambda} + \partial_\nu F_{\lambda\mu} = 0, \tag{4-2}$$

$$\partial^\mu F_{\lambda\mu} = J_\lambda. \tag{4-3}$$

The coupling is indicated by the term J_λ in (4-3) which describes the classical current distribution. The equations (4-2) are identically satisfied if we express $F_{\mu\nu}$ in terms of the vector potential A_μ,

$$F_{\mu\nu} = \partial_\mu A_\nu - \partial_\nu A_\mu. \tag{4-4}$$

We can always make use of the freedom in the definition of A_μ by choosing A_μ in such a way that†

$$\chi \equiv \partial^\mu A_\mu = 0. \tag{4-5}$$

For this choice of A_μ the equations (4-3) are equivalent to

$$\partial^\mu \partial_\mu A_\lambda = -J_\lambda. \tag{4-6}$$

The last three equations are the fundamental equations of the classical radiation field interacting with the current density J_λ.

Of course, the classical equations are complete only if they are coupled with an equation of motion for the current density. Such an equation of

---

* In the following we use capital letters for the field variables which satisfy the inhomogeneous field equations, while the small letters are used, as before, for the free fields.

† For more details compare the corresponding discussion in Section 6-2.

motion would depend on the mechanical properties of the carriers of the electric charge and, therefore, involves further assumptions regarding these properties. Classical theories of this sort have only a limited validity and look rather artificial because of the considerable arbitrariness regarding the fundamental assumptions involved.

The most interesting of these classical models is the classical theory of a point charge which was developed by H. A. Lorentz* and has been subsequently investigated by many authors.† The guiding idea in much of this work is the hope that a satisfactory solution of the problems arising from the quantum mechanical treatment of the interaction of charges with radiation might be obtained by first finding a solution of the corresponding classical problem for a point electron. Up to the present time this hope has not been fulfilled. Whereas it is possible to construct a theory of classical point particles which is free from the more obvious defects of the original Lorentz theory, this has not been of any help in solving the more profound problems which remain in the quantum theory of radiation. Moreover, the classical theory has problems of its own, such as the self-acceleration,‡ which have not yet found a satisfactory solution.

Therefore we shall not pursue these classical models further. We shall instead proceed directly to the quantum mechanical formalism of the interacting fields.

In the quantum theory we retain the equations (4–6) as the field equations for the electromagnetic field, together with the definition (4–4) of the field strengths. But Eq. (4–5) cannot be taken over into quantum theory as an equation between operators, because the operator χ does not commute with the other operators of the fields. We have already encountered this situation in Chapter 2 for the free-field case. We replace Eq. (4–5) by a subsidiary condition on the state vectors

$$\chi\Omega = 0. \qquad (4\text{–}7)$$

This is sufficient to ensure the validity of Maxwell's equations for the expectation values. In order to be sure that condition (4–7) can be satis-

---

* H. A. Lorentz, *The Theory of Electrons*, Leipzig, 1909.

† Some of the more recent works are G. Wentzel, *Z. Physik* **86**, 479 (1933); **87**, 726 (1934). P. A. M. Dirac, *Proc. Roy. Soc.* **167**, 148 (1938); **209**, 291 (1951). M. H. L. Pryce, *ibid.* **168**, 389 (1938). F. Bopp, *Ann. Physik* **38**, 345 (1940); **43**, 565 (1943). *Zeits. f. Naturf.* **1**, 53, 196 (1946); **3**, 564 (1948). W. Wessel, *Z. Physik* **92**, 407 (1934); **110**, 625 (1938). H. Hönl and A. Papapetrou, *Z. Physik* **112**, 65; **114**, 478 (1939); **116**, 154 (1940). E. Kanei, S. Takagi, *Progr. Theor. Phys.* **1**, 43 (1946). R. Utiyama, *ibid.* **3**, 114 (1948). M. Schönberg, *Phys. Rev.* **74**, 738 (1948). A survey of these developments was given by A. Pais, *Developments in the Theory of the Electron*, Institute for Advanced Study, Princeton, 1948.

‡ For a discussion of this and other problems see C. J. Eliezer, *Rev. Mod. Phys.* **11**, 147 (1947).

fied, it is necessary to verify again that

$$[\chi(x),\chi(x')] = 0 \tag{4-8}$$

for all values of x,x'. We shall see that this relation is also true in the case of coupled fields (Section 4-2 below).

The current density J_λ in (4-6) is taken from the current operator in the free-field problem discussed in Chapter 3, and it is given by

$$J_\lambda = -i\frac{e}{2}[\overline{\Psi}\gamma_\lambda\Psi - \overline{\Psi}^c\gamma_\lambda\Psi^c], \tag{4-9}$$

[*cf.* Eq. (3-109)]. The dimensionless numerical factor e (>0) is the charge of the particles described by the spinor field. Its sign is so chosen that the charge of a negaton is $(-e)$. In the system of units used here it has the value $e \simeq \sqrt{4\pi/137}$. We have used the charge symmetrical form of the current density in order to obtain for the vacuum the total charge zero (*cf.* Section 3-10). In the future we shall write more simply

$$J_\lambda = -ie\overline{\Psi}\gamma_\lambda\Psi \tag{4-9a}$$

with the understanding that (4-9a) is to be replaced by the correct expression (4-9) for actual calculations of expectation values. This convention is adopted for all bilinear expressions in Ψ.

We can obtain the additional term J_λ on the right side of (4-6) by an additional term L_1 in the Lagrangian for the coupled fields. The total Lagrangian then appears in the form

$$L = L_0 + L_1, \tag{4-10}$$

$$L_0 = L_{\text{rad}} + L_{\text{el}}, \tag{4-11}$$

$$L_{\text{rad}} = -\tfrac{1}{2}\partial_\mu A_\nu \partial^\mu A^\nu, \tag{4-12}$$

$$L_{\text{el}} = -\overline{\Psi}(\partial + m)\Psi, \tag{4-13}$$

$$L_1 = J_\lambda A^\lambda = -ie\overline{\Psi}A\Psi, \quad (A \equiv \gamma_\mu A^\mu). \tag{4-14}$$

Equation (1-74) of Chapter 1 is then seen to be identical with the field equation (4-6).

The derivation of (4-6) from a Lagrangian has the advantage that from it we obtain also the equations of motion for the spinor field.

By taking the variation of L with respect to $\overline{\Psi}$, we obtain the system

$$(\partial + m)\Psi = -ieA\Psi. \tag{4-15}$$

In a similar manner, varying Ψ, we obtain

$$\overline{\Psi}(\partial - m) = ie\overline{\Psi}A \tag{4-16}$$

which could also have been obtained directly from (4-15) by the transition from Ψ to $\overline{\Psi}$ (Eq. 3–29). We can also derive an equation of motion for the charge conjugate spinor by applying the process of charge conjugation, (3–38), to (4–15) or (4–16). We then obtain

$$(\partial + m)\Psi^c = ieA\Psi^c, \tag{4–17}$$

$$\overline{\Psi}^c(\overleftarrow{\partial} - m) = -ie\overline{\Psi}^c A. \tag{4–18}$$

The last two equations show that the charge conjugate field Ψ^c interacts with the radiation field with the opposite sign of the charge. This justifies the term "charge conjugation."

As a simple consequence of the field equations, we verify the law of the conservation of charge. By multiplying (4–15) from the left with $\overline{\Psi}$ and (4–16) from the right with Ψ and adding the two equations, we obtain the result

$$\partial_\mu(\overline{\Psi}\gamma^\mu\Psi) = 0$$

or

$$\partial_\mu J^\mu(x) = 0. \tag{4–19}$$

As a consequence of (4–19) and (4–6), we also note that the operator χ of (4–5) satisfies the free field equation

$$\partial^\lambda \partial_\lambda \chi = 0. \tag{4–20}$$

4–2 Commutation rules for the interacting fields. The commutation rules for the coupled fields are exactly the same as those for the free fields provided we consider only bracket expressions involving two field variables at two space-like situated points. This follows immediately from the general theory of Chapter 1. There we have shown that the appropriate bracket expressions are given either by (1–107a) or by (1–107b) according to the type of quantization. The canonically conjugate operators are not changed by the presence of the coupling term L_1, since this term does not contain any derivatives of the field operators. Hence the commutation rules for the fields and their conjugate operators remain unchanged on space-like planes.

In the preceding two chapters we have extended these commutation rules for field variables at general points in the four-dimensional continuum. This is no longer possible in the present case. In order to extend these rules for coupled fields it would be necessary to know the solutions of the coupled field equations (4–6) and (4–15). Unfortunately, it has so far been impossible to construct solutions for these equations in closed form.

It is, however, possible to construct an equivalent set of commutator relations at arbitrary points by defining new field operators, which will also be useful in a different context later on.

We define a family of free fields $a_\lambda(x,\sigma)$, $\psi(x,\sigma)$ which depend, aside from the coordinate x, on an arbitrary space-like plane σ. Such a plane is characterized by four real parameters which are contained in the normal vector n of magnitude 1,

$$n^2 + 1 = 0,$$

and the "instant parameter" τ which is related to the equation of the plane by

$$n \cdot x + \tau = 0.$$

A point x which satisfies this equation is said to be coincident with the plane σ and we write for it $x \in \sigma$.

The fields $a_\lambda(x,\sigma)$ and $\psi(x,\sigma)$ are defined by the following properties:*

$$a_\lambda(x,\sigma) = A_\lambda(x), \quad \partial a_\lambda(x,\sigma) = \partial A_\lambda(x), \quad \psi(x,\sigma) = \Psi(x), \quad (4\text{--}21)$$

$$\text{for } x \in \sigma, \quad \partial \equiv \partial^\mu n_\mu.$$

$$\partial^\mu \partial_\mu a_\lambda(x,\sigma) = 0,$$
$$(\partial + m)\psi(x,\sigma) = 0. \tag{4--22}$$

$$[a_\lambda(x,\sigma), a_\mu(x',\sigma)] = -ig_{\lambda\mu}D(x - x'),$$
$$\{\psi(x,\sigma), \bar\psi(x',\sigma)\} = iS(x - x'). \tag{4--23}$$

It is clear that the family of fields is completely determined by these conditions. Equations (4–22) and (4–23) state that they are free fields and (4–21) fixes their boundary values at σ. It is equally clear that a knowledge of the family is equivalent to a knowledge of the solution of the inhomogeneous system (4–6) and (4–15). We merely need to let the point coincide with the plane σ to obtain the values of the Heisenberg fields A_μ, Ψ at that point (4–21).

The commutation rules (4–23) are now the desired four-dimensional generalization of the commutation rules on space-like surfaces only.

They can be used to prove that the operators χ commute at any two points:

$$[\chi(x), \chi(x')] = 0, \quad x, x' \text{ arbitrary.} \tag{4--8}$$

Since $\chi(x)$ satisfies the free-field equation (4–20), it is sufficient to prove the two relations

$$[\chi(x), \chi(x')] = 0, \quad [\partial\chi(x), \chi(x')] = 0, \quad \text{for } x, x' \in \sigma. \tag{4--24}$$

---

* The fields $a_\lambda(x,\sigma)$, $\psi(x,\sigma)$ are the same as the fields introduced by C. N. Yang and D. Feldman, *Phys. Rev.* **79**, 975 (1950), especially Eq. (20). They may be looked upon as a generalization of the Wentzel potentials, G. Wentzel, *Z. Physik* **86**, 479 (1933); also *Quantum Theory of Fields*, Interscience Publishers (1949), especially §18. Note that the $\psi(x,\sigma)$ anticommute on σ.

These equations (4–24) are equivalent to (4–8) because the values of $\chi(x)$, $\partial\chi(x)$ on σ determine the operators $\chi(x)$ at any point x in space [*cf.* the corresponding problem discussed in connection with (2–20) and (2–21)].

For the proof, let us denote by $\phi(x)$ any of the field components and by $\varphi(x,\tau)$ the corresponding free field depending on a one-parameter family of parallel planes. Then, since $\partial_\mu \tau = -n_\mu$,

$$\partial_\mu \phi \equiv \partial_\mu \varphi + \frac{d\varphi}{d\tau}\partial_\mu \tau = \partial_\mu \varphi - n_\mu \dot\varphi \quad \left(\dot\varphi \equiv \frac{d\varphi}{d\tau}\right), \tag{4-25}$$

$$\partial_\lambda \partial_\mu \phi = \partial_\lambda \partial_\mu \varphi - \partial_\lambda n_\mu \dot\varphi - \partial_\mu n_\lambda \dot\varphi + n_\lambda n_\mu \ddot\varphi. \tag{4-26}$$

For the radiation field, we have*

$$\dot a_\mu = 0, \tag{4-27}$$

and from (4–26), with the help of the last relation and the field equation (4–6),

$$\partial_\lambda \chi \equiv \partial_\lambda \partial^\mu A_\mu = \partial_\lambda \partial^\mu a_\mu + n_\lambda n^\mu J_\mu. \tag{4-28}$$

From the last equation, one deduces

$$[\partial_\lambda \chi(x), \chi(x')] = 0, \quad x,x' \in \sigma, \tag{4-29}$$

and from (4–25), one obtains

$$[\chi(x), \chi(x')] = 0, \quad x,x' \in \sigma. \tag{4-30}$$

Thus (4–24) holds and therefore (4–8).

For future use, we also note the relation deduced from (4–28) with the help of the commutation rules for the Ψ:

$$\int_\sigma [\partial_\lambda \chi(x), \Psi(x')] d\sigma^\lambda = e\Psi(x'), \quad x' \in \sigma. \tag{4-31}$$

It is important to note that commutators are needed only between field variables referring to the same plane σ and that the right side of (4–23) is independent of σ.

The last remark leads to a consequence which we shall use later on. It is based on the observation made in the two preceding chapters that the commutation rules of the free fields determine the representation of the field operators up to a similarity transformation. It follows that the various fields $\varphi(x,\sigma)$ (φ standing for either a_λ or ψ) are connected by a unitary transformation. Thus, if σ,σ_0 are two arbitrary planes, then there exists a unitary operator $U(\sigma,\sigma_0)$ such that

$$\varphi(x,\sigma) = U(\sigma,\sigma_0)\varphi(x,\sigma_0)U^{-1}(\sigma,\sigma_0). \tag{4-32}$$

---

* This can be seen most easily in the interaction picture, see Eqs. (4–38) and (4–44) below.

The operator family has the following properties:

$$U(\sigma_0,\sigma_0) = 1, \tag{4-33}$$

$$U^{-1}(\sigma,\sigma_0) = U(\sigma_0,\sigma), \tag{4-34}$$

$$U(\sigma,\sigma_1)U(\sigma_1,\sigma_0) = U(\sigma,\sigma_0), \tag{4-35}$$

which follow directly from the definition (4-32) with a suitable choice of phase factor.

4–3 The interaction picture. The Heisenberg picture, which has been the basis of the discussion up to the present point, has a definite advantage insofar as it is possible to restrict the discussion of the general theory to the field operators. The state vectors do not enter into the dynamical laws of the theory. They are merely used for the calculation of expectation values and matrix elements. However, in the actual calculations of approximate solutions of the field equations it is often more convenient to introduce different descriptions for which the state vectors are no longer constant. Of particular importance in the more recent development of the theory has been the interaction picture, which we shall discuss now.

We begin by defining a one-parameter family of parallel planes $\sigma(\tau)$, $-\infty < \tau < +\infty$, characterized as usual by a unit normal vector n and the equation

$$\sigma(\tau): \quad n \cdot x + \tau = 0. \tag{4-36}$$

A τ-dependent canonical transformation defines a new field $\varphi(x)$ by the relation

$$\varphi(x) = V(\tau)\phi(x)V^{-1}(\tau). \tag{4-37}$$

Since $\tau = \tau(x) = -n \cdot x$ is, by (4-36), a function of x, this new field $\varphi(x)$, considered as a function of x, satisfies a new field equation. From (4-37) and (4-36), we obtain

$$\partial_\mu \varphi(x) = V(\tau)\partial_\mu \phi(x)V^{-1}(\tau) + n_\mu i[H(\tau),\varphi(x)], \tag{4-38}$$

with

$$H(\tau) = i\dot{V}(\tau)V^{-1}(\tau), \tag{4-39}$$

$$\dot{V}(\tau) \equiv \frac{dV(\tau)}{d\tau}. \tag{4-40}$$

The transformation (4-37) of the operators induces a transformation of the state vectors,

$$\omega(\tau) = V(\tau)\Omega. \tag{4-41}$$

The interaction picture is obtained by choosing the operator $V(\tau)$ in such a manner that the field equations for the transformed field are the free-field equations. We shall now determine this operator for the interacting radiation and matter field. Because of Eq. (4-38), the $\psi(x)$ satisfy the free-field equations, provided we choose $H(\tau)$ such that

$$[H(\tau),\psi(x)] = -e n\!\!\!/(x)\psi(x). \tag{4-42}$$

One verifies easily that

$$H(\tau) = ie\int_{\sigma(\tau)} \bar\psi(x) \mathbf{a}(x)\psi(x)d\sigma \tag{4-43}$$

is a solution of (4–42). With this solution $H(\tau)$, the transformed radiation field $a_\lambda(x)$ also satisfies the free-field equation, as one may verify, for instance, as follows. We observe first that

$$[H(\tau),a_\lambda(x)] = 0, \quad x \in \sigma(\tau), \tag{4-44}$$

since $H(\tau)$ contains no derivatives of the field. Consequently,

$$\partial^\mu\partial_\mu a_\lambda(x) = V(\tau)\partial^\mu\partial_\mu A_\lambda(x)V^{-1}(\tau) + i[H(\tau),\partial a_\lambda(x)]. \tag{4-45}$$

The first term on the right side is, by (4–6) and (4–9),

$$V(\tau)\partial^\mu\partial_\mu A_\lambda(x)V^{-1}(\tau) = ie\bar\psi(x)\gamma_\lambda\psi(x); \tag{4-46}$$

the second term on the right side of (4–45) is, by (4–43) and the commutation rules of the free fields,

$$i[H(\tau),\partial a_\lambda(x)] = -e\int \bar\psi(y)\gamma^\mu\psi(y)[a_\mu(y),\partial a_\lambda(x)]d\sigma_y$$

$$= -ie\bar\psi(x)\gamma_\lambda\psi(x). \tag{4-47}$$

Thus, the two terms cancel and we find

$$\partial^\mu\partial_\mu a_\lambda(x) = 0. \tag{4-48}$$

The fact that both Eq. (4–48) and the free-field equation for ψ are satisfied assures that $H(\tau)$ in Eq. (4–43) is unique.

The equation

$$H(\tau) = i\dot V(\tau)V^{-1}(\tau) \tag{4-49}$$

which connects the operator $H(\tau)$ with $V(\tau)$ is called the Schroedinger equation in the interaction picture. It is usually written as an equation for the state vectors $\omega(\tau)$,

$$i\dot\omega(\tau) = H(\tau)\omega(\tau), \tag{4-50}$$

which is, of course, equivalent to (4–49) because of (4–41).

The subsidiary condition (4–7), when transformed into the interaction picture, becomes

$$g(x)\omega(\tau) = 0, \quad x \in \sigma(\tau), \tag{4-51}$$

with

$$g(x) = V(\tau)\chi(x)V^{-1}(\tau) = \partial^\mu a_\mu(x). \tag{4-52}$$

It is important to realize that in (4–51) x must be situated on $\sigma(\tau)$. If x is not on $\sigma(\tau)$, $g(x)\omega(\tau)$ can also be calculated, since $g(x)$ satisfies the free-field equation and therefore, by (2–16), for any x,

$$g(x)\omega(\tau) = \int_{\sigma(\tau)} d\sigma_\nu'(\partial^{\prime\prime\nu}g(x')D(x-x') - \partial^{\prime\prime\nu}D(x-x')g(x'))\omega(\tau). \quad (4\text{–}53)$$

Now, using (4–38) and (4–51), this becomes

$$g(x)\omega(\tau) = -i\int_{\sigma(\tau)} d\sigma'[H(\tau),g(x')]D(x-x')\omega(\tau)$$

$$= \int j^\lambda(x')D(x-x')d\sigma_\lambda'\omega(\tau). \quad (4\text{–}54)$$

Thus we obtain an equivalent form of (4–51),

$$\gamma(x)\omega(\tau) = 0, \quad \text{for all } x,\tau,$$
$$\gamma(x) = \partial^\lambda a_\lambda(x) - \int_\sigma j^\lambda(x')D(x-x')d\sigma_\lambda'. \quad (4\text{–}55)$$

The subsidiary condition (4–51) or (4–55) may finally be replaced by an initial condition which the state vector $\omega(\tau_0)$ must satisfy for one particular value of $\tau = \tau_0$. We obtain it by taking the derivative of $\gamma(x)$ in the direction of the normal for $x \in \sigma_0 \equiv \sigma(\tau_0)$. Since

$$\int_{\sigma_0} j^\lambda(x')\partial D(x-x')d\sigma_\lambda' = j^\lambda(x)n_\lambda,$$

we find the two initial conditions

$$\partial^\lambda a_\lambda(x)\omega(\tau_0) = 0,$$
$$(\partial\partial^\lambda a_\lambda(x) - n_\lambda j^\lambda(x))\omega(\tau_0) = 0, \quad \text{for } x \in \sigma_0. \quad (4\text{–}56)$$

We have in (4–51), (4–55), and (4–56) three equivalent forms of the subsidiary condition in the interaction picture. The last one, (4–56), is especially useful because it shows that the subsidiary condition (4–51) can be satisfied by subjecting the state vector $\omega(\tau_0)$ on a particular surface, $\sigma_0 = \sigma(\tau_0)$, to an initial condition.

4–4 Measurability of the fields. One of the characteristic features of quantum mechanical systems is the restriction which is imposed by the formalism on the measurements of noncommuting observables. If two such observables are represented by the Hermitian operators A, B, and their commutator $C \equiv [A,B]$ is not zero, then a simultaneous measurement of the two physical quantities is restricted in accuracy. Indeed, a measurement of B which follows one of A partly invalidates the measurement of A. A quantitative expression of this physical law is contained in the uncertainty relation, which can be generally expressed in the form

$$\Delta a \Delta b \geq \tfrac{1}{2}\langle C \rangle, \tag{4-57}$$

where $(\Delta a)^2 = \langle (A - \langle A \rangle)^2 \rangle$ is the mean square deviation of the quantity A for the state in question.

The quantum theory of the radiation field leads us to assume certain commutation rules between the potentials and consequently also between the field strength operators $f_{\mu\nu} = \partial_\mu a_\nu - \partial_\nu a_\mu$. Thus, contrary to the classical field concept, the measurability of the fields will be restricted by the commutation rules.

For the quantitative discussion of this restriction one would therefore have to start from the commutation rules for the field strengths which, according to (2–28), may be written in the form

$$[f_{\mu\nu}(x), f_{\kappa\lambda}(x')] = +i(\partial_\mu \partial_\kappa g_{\nu\lambda} + \partial_\nu \partial_\lambda g_{\mu\kappa} - \partial_\mu \partial_\lambda g_{\nu\kappa} - \partial_\nu \partial_\kappa g_{\mu\lambda}) D(x - x')$$

$$\equiv i d_{\mu\nu, \kappa\lambda} D(x - x'). \tag{4-58}$$

Because of the occurrence of the singular functions on the right side of this equation, this form is not very suitable for a discussion of the limitation on actual physical measurements of the field strengths.

In actual measurements one never measures the values of the fields at a given point x in space-time. The measurement of the field, in fact, always consists of the measurement of the momentum transfer on a suitable test body. Such a test body must necessarily have a finite extension V in space and the measurement must consume a certain time T. The field measurements therefore will only yield the value of the fields averaged over a finite region $G = VT$ in space-time. The quantities which are measurable are thus appropriately represented by the operators

$$\bar{f}_{\mu\nu}(G) = \frac{1}{G} \int_G d^4x f_{\mu\nu}(x). \tag{4-59}$$

For these quantities the commutation rules are finite. For instance,

$$[\bar{f}_{\mu\nu}(G), \bar{f}_{\kappa\lambda}(G')] = \frac{i}{GG'} \iint_{GG'} d_{\mu\nu, \kappa\lambda} D(x - x') d^4x d^4x'. \tag{4-60}$$

As a simple consequence of the last equation and the property of the function

$$D_{\mu\nu, \kappa\lambda}(x - x') \equiv d_{\mu\nu, \kappa\lambda} D(x - x')$$

of being antisymmetrical,

$$D_{\mu\nu, \kappa\lambda}(x - x') = -D_{\mu\nu, \kappa\lambda}(x' - x), \tag{4-61}$$

[we use the symmetry properties of the D-function given under (A1–12)], it follows that for two regions G,G' which coincide ($G = G'$), all the field components commute:

$$[\bar{f}_{\mu\nu}(G), \bar{f}_{\kappa\lambda}(G)] = 0. \tag{4-62}$$

Consequently, there is no restriction imposed on the joint measurement of the various field components averaged over the same space-time region.

Similarly, if the regions G,G' are entirely outside each other's light cones, the commutators vanish because of the vanishing of the D-function for space-like distances $x - x'$. This is in conformity with the notion of the limiting property of the velocity of light for the transmission of physical effects from one point to another.

For other relative positions the right side of (4–60) has, in general, a finite value, and in these cases the quantum formalism predicts a limitation of the measurability which finds its quantitative expression in the uncertainty relation (4–57).

In order to make sure that the physical content of the uncertainty relation, when applied to the field strengths, is not empty, it is necessary to establish that there exists, at least in principle, a physical experiment which allows a determination of the field strengths up to the limits imposed by the uncertainty relation. For this purpose it is not necessary to have an experiment which actually can be carried out with existing experimental equipment. It is sufficient to construct a thought experiment which is limited only by the physical laws involved.

It was shown by Bohr and Rosenfeld* in a fundamental paper that such a thought experiment can indeed be constructed. The problem is considerably more complicated than the corresponding discussion in quantum mechanics of the joint measurement of noncommuting observables. Here we encounter, in addition to the expected characteristic quantum mechanical effects which limit the measurements of noncommuting observables, difficulties which have their source entirely in the nature of the measuring device and which seem to limit even the measurements of individual field components irrespective of the value of the other components.

The point of the Bohr-Rosenfeld investigation is to show that these apparent limitations can be overcome entirely if test bodies are used with sufficiently heavy mass and a sufficiently large charge and current density which is uniformly spread over the finite measuring region and if, in addition, suitable compensating devices are introduced for the uncontrollable reactions which seem to limit the accuracy of the field measurements.

In a subsequent paper† Bohr and Rosenfeld extend these considerations

---

* N. Bohr and L. Rosenfeld, *Det Kgl. Danske Videnskab. Selskab.* **12,** No. 8 (1933).
† N. Bohr and L. Rosenfeld, *Phys. Rev.* **78,** 794 (1950).

to the measurement of the charge-current density $j_\mu(x)$. Just as for the field measurement, it is necessary to consider only the measurability of average values of $j_\mu(x)$,

$$\bar{J}_\mu(G) = \frac{1}{G}\int_G j_\mu(x)d^4x, \qquad (4\text{-}63)$$

which then satisfy commutation rules of the form

$$[\bar{J}_\mu(G),\bar{J}_\nu(G')] = \frac{i}{GG'}\iint(-\bar{\psi}(x)\gamma_\mu S(x-x')\gamma_\nu\psi(x')$$
$$+ \bar{\psi}(x')\gamma_\nu S(x-x')\gamma_\mu\psi(x))d^4x'd^4x. \qquad (4\text{-}64)$$

Because of the field equations for the expectation values,

$$\partial^\mu\langle f_{\lambda\mu}\rangle = \langle j_\lambda\rangle,$$

the measurement of the average charge-current over the four-dimensional region G can be replaced by the measurement of the flux of the electromagnetic field through the boundary S of this region G:

$$\langle \bar{J}_\mu(G)\rangle = \frac{1}{G}\int_S \langle f_{\lambda\mu}\rangle d\sigma^\lambda.$$

This determination of the flux can be reduced to the measurement of the electromagnetic field in a thin four-dimensional shell surrounding the region G. It is now possible to construct additional suitable compensating devices for the effect of actual and virtual pair creation on the accuracy of the charge-current measurement which allow in principle a determination of the value of these operators limited only by the uncertainty relation derived from the commutation rules.*

It must be emphasized that the possibility of the measurements of fields and currents by idealized thought experiments, outlined here so briefly, does not give us any clue as to the actual limitations of these measurements, because of the atomic structure of all test bodies. It is not inconceivable that a more profound analysis of the measuring process in which due account is taken of this limitation might well be an important step in the direction of a new and more embracing field theory.

---

* The charge and current fluctuations were discussed by E. Corinaldesi, *Nuovo Cimento* **8**, 494 (1951) and **9**, 194 (1952); also Suppl. **10**, 83 (1953).

CHAPTER 5

INVARIANCE PROPERTIES OF THE COUPLED FIELDS

In this chapter we shall study all the known invariance properties of the field which represents the interaction of radiation with electrons. Let us briefly recall the basic concepts underlying these considerations. The field variables on a particular space-like plane, σ, may be transformed by a unitary operator, U, into a new set of variables on that plane σ by the correspondence

$$\phi(x) \rightarrow \phi'(x) = U\phi(x)U^{-1}. \tag{5-1}$$

If the transformation operator U is independent of σ, then the original and the transformed field describe two systems with identical physical properties. In particular, the commutation rules and the field equations for the two fields are the same. Conversely, any transformation $\phi \rightarrow \phi'$ which leaves the field equations and the commutation rules invariant gives rise to a unitary operator U with the property (5-1). (This follows from the uniqueness of the representations discussed in Chapters 2 and 3.)

Transformations of this sort are first of all furnished by the Lorentz transformations of the fields, and the invariance of the field equations under these transformations is implicit in the principle of relativity. For the case of free fields we have previously discussed the operators U (or rather their infinitesimal parts) associated with the Lorentz transformations and have shown how they can be constructed for the free fields with the help of the Action Principle. There is no difficulty in extending these considerations to the case of the interacting fields (Section 5–1).

In addition to these invariance properties which follow from the restricted principle of relativity (which is implied in the Action Principle) the system exhibits further symmetries which we shall now study also. There are, first of all, the gauge transformations, which can be shown to be canonical transformations (Section 5–2). The symmetry associated with the spatial reflection leads us to the construction of the parity operator (Section 5–3). There is also a symmetry with respect to the reversal of time (Section 5–4), and finally the symmetry which is expressed by the transformation of charge conjugation (Section 5–5). The chapter ends with a few remarks on scale transformations (Section 5–6).

5–1 Proper Lorentz transformations. According to the preceding remarks, the invariance of the field equations and commutation rules under proper Lorentz transformations needs no further proof. In fact formulas (1–82) and (1–83) furnish directly the infinitesimal generators of the unitary operators U. We obtain for the canonical tensor (1-80) the expression

$$T^{\mu\nu} = -\partial^\mu A^\lambda \partial^\nu A_\lambda + \tfrac{1}{2}g^{\mu\nu}\partial_\lambda A_\rho \partial^\lambda A^\rho - \overline{\Psi}\gamma^\mu \partial^\nu \Psi. \quad (5\text{-}2)$$

With it we can construct the momentum operators by using formulas (1-82) and (1-83). The result is given by

$$P^\mu = \int (-\partial^\rho A^\lambda \partial^\mu A_\lambda + \tfrac{1}{2}g^{\rho\mu}\partial_\lambda A_\kappa \partial^\lambda A^\kappa)d\sigma_\rho - \int \overline{\Psi}\gamma^\rho \partial^\mu \Psi d\sigma_\rho. \quad (5\text{-}3)$$

$$M^{\mu\nu} = \int_\sigma (T^{\rho\mu}x^\nu - T^{\rho\nu}x^\mu)d\sigma_\rho + \int_\sigma (\partial^\rho A^\mu A^\nu - \partial^\rho A^\nu A^\mu)d\sigma_\rho$$
$$+ \tfrac{1}{2}\int_\sigma \overline{\Psi}\gamma^\rho \gamma^\mu \gamma^\nu \Psi d\sigma_\rho. \quad (5\text{-}4)$$

We note here that the expressions (5-2), (5-3), and (5-4) do not contain the charge explicitly, although it is, of course, implicitly contained in these expressions through the field equations. It is perhaps of interest to see how the coupling term disappeared from the canonical tensor. In the expression (1-80) for this tensor the coupling term would be present in the part which contains $g_{\mu\nu}L$. L is given by (4-10) and contains three terms, including the coupling term L_1. The term L_{el} (4-13) can be transformed with the help of the field equations (4-15) and it then just cancels the coupling term L_1.

The expressions for the momentum operator for the coupled fields thus have exactly the same form as for the free fields. Since we have previously seen that the commutation rules on space-like planes are also unchanged, it follows immediately that the characteristic commutation properties (1-56) and (1-57) of the momentum operators hold just as in the case of the free fields.

There is one important difference in the two cases. For uncoupled fields the momentum operators referring to each of the fields separately satisfy a conservation law. For the coupled fields this is no longer the case; only the sum of the two contributions together satisfies such a conservation equation. This means, of course, that energy and momentum can be transferred from one system to the other, thus giving rise to the well-known observable radiative processes such as the Compton effect.

5-2 Gauge transformations. The transformation

$$A_\lambda(x) \to A_\lambda'(x) = A_\lambda(x) + \partial_\lambda \Lambda(x),$$
$$\Psi(x) \to \Psi'(x) = \Psi(x)e^{-ie\Lambda(x)}, \quad (5\text{-}5)$$

with $\Lambda(x)$ an arbitrary c-number satisfying the equation

$$\partial^\mu \partial_\mu \Lambda(x) = 0, \quad (5\text{-}6)$$

is called a *gauge transformation*.* It has the property that the new fields satisfy the same field equations and commutation rules as the old ones. Any quantity which is left invariant under this transformation is called *gauge invariant*. The current operator, for instance, is such a quantity.

According to the introductory remarks to this section, gauge transformations are canonical transformations. We shall show this explicitly by constructing the infinitesimal generating operators of these canonical transformations.

For infinitesimal gauge transformations, we may consider $\Lambda(x)$ as an infinitesimal quantity of first order and write (5-5) as

$$\delta A_\lambda(x) = \partial_\lambda \Lambda(x), \quad \delta \Psi(x) = -ie\Psi(x)\Lambda(x). \tag{5-7}$$

We thus need an operator F with the property

$$i[F, A_\lambda(x)] = \partial_\lambda \Lambda(x), \quad i[F, \Psi(x)] = -ie\Psi(x)\Lambda(x). \tag{5-8}$$

The operator F which has this property is given by†

$$F = \int_\sigma (\partial_\mu \Lambda \chi - \Lambda \partial_\mu \chi) d\sigma^\mu. \tag{5-9}$$

We observe first that the operator F is independent of the plane σ, since

$$\partial^\mu(\partial_\mu \Lambda \chi - \Lambda \partial_\mu \chi) = 0$$

because of (5-6) and (4-20). Thus F is an integral. We shall refer to it in the following as the *gauge integral*. It is actually not just one integral but a continuous set, since it depends on the arbitrary functions $\Lambda(x)$ [restricted only by (5-6)].

Next we observe that for state vectors Ω which satisfy the subsidiary condition (4-7)

$$F\Omega = 0. \tag{5-10}$$

In order to verify (5-8), we place the surface through the point x and then make use of relations (4-28) and (4-31) and the commutation rules (4-23).

---

* For reasons of simplicity we have restricted the function $\Lambda(x)$ to be a *c*-number, as is customarily done. It is possible to define canonical gauge transformations with operator functions $\Lambda(x)$. These operators are then restricted only by the condition that the new field variables A_λ', Ψ' have the same commutation rules as the original variables. Compare also Section 6–3.

† This is essentially the same as the operator C in the paper by W. Heisenberg and W. Pauli, *Z. Physik* **59**, 168 (1930), written in covariant notation. See also Z. Koba, T. Tati, S. Tomonaga, *Progr. Theor. Phys.* **2**, 101 (1947), especially p. 110 ff., and Z. Koba, Y. Oisi, M. Sasaki, *ibid.* **3**, 141, 229 (1948).

Thus, since F is shown to be the generating operator of the gauge transformations, we see that the gauge invariance of a quantity is equivalent to the property of commuting with the operator F.

Examples of gauge-invariant operators are the field strengths $F_{\mu\nu}$ and the current density operators J_λ.

We shall define an observable in quantum electrodynamics as an operator which, in a representation in which F is diagonal, has no matrix elements connecting the eigenvalue 0 of F with any other eigenvalue. The physical meaning of this requirement for observables is that the measurement of the physical quantity represented by an observable shall not result in a state vector which violates the subsidiary condition. A consequence of this condition is that the expectation value of an observable for state vectors which satisfy the subsidiary condition is gauge invariant.

The gauge-invariant operators such as $F_{\mu\nu}$ and J_λ are thus obviously observables. But there are operators possible which by themselves are not gauge invariant but which are nevertheless observables in the sense described above. We may therefore distinguish between observables of the first kind which are represented by gauge-invariant operators, and observables of the second kind with gauge-invariant expectation values only. The operators $F_{\mu\nu}$ and J_λ are then observables of the first kind.

We shall now show that the operators P^μ and $M^{\mu\nu}$, (5–3) and (5–4), are observables of the second kind. This statement is mathematically expressed by the conditions

$$[P_\mu, F]\Omega = 0, \tag{5-11}$$

$$[M_{\mu\nu}, F]\Omega = 0, \tag{5-12}$$

where the state vector Ω satisfies (4–7). The proof of these relations is made very easy by the observation that the commutator $[P_\mu, F]$ may be interpreted in two different ways. If F is considered the generating operator, then this commutator represents essentially (that is apart from a numerical factor) the increment of P_μ due to gauge transformations. But if P_μ is considered the generating operator, this same commutator represents essentially the increment induced in F due to translations. Thus we have immediately

$$[P_\mu, F] = i \int_\sigma (\partial_\lambda \Lambda \partial_\mu \chi - \Lambda \partial_\lambda \partial_\mu \chi) d\sigma^\lambda \tag{5-13}$$

and therefore

$$[P_\mu, F]\Omega = 0$$

for state vectors which satisfy (4–7).

In a similar manner, one proves (5–12). In that case the operator $M_{\mu\nu}$ generates the infinitesimal rotations.

We mention also, without elaboration, that it is possible to construct a manifestly gauge-invariant "classical" tensor

$$\theta^{\mu\nu} = -F^{\mu\rho}F^{\nu}{}_{\rho} + \tfrac{1}{4}g^{\mu\nu}F_{\rho\sigma}F^{\rho\sigma} - \overline{\Psi}\gamma^{\mu}(\partial^{\nu} + ieA^{\nu})\Psi \qquad (5\text{--}14)$$

such that

$$\left\langle \int \theta^{\rho\mu} d\sigma_{\rho} \right\rangle = \langle P^{\mu} \rangle, \qquad (5\text{--}15)$$

$$\left\langle \int (\theta^{\rho\mu}x^{\nu} - \theta^{\rho\nu}x^{\mu})d\sigma_{\rho} \right\rangle = \langle M^{\mu\nu} \rangle. \qquad (5\text{--}16)$$

The last equations correspond exactly to equations (2–54) and (2–55). The proof proceeds along lines similar to those outlined for (2–54) and, just as in that case, so here, too, it is necessary to omit terms arising from surface integrals at infinity.*

5–3 Space inversion. The invariance of the field equations under the proper Lorentz transformations has a much more direct intuitive appeal than the invariances to be discussed now under the special transformations σ, τ. [See Eqs. (1–8), (1–9).] The reason for this is, perhaps, the fact that the proper transformations can be connected by a continuous set of such transformations with the identity for which the invariance is obvious. This is not so for the special transformations σ, τ, and $\rho = \sigma\tau$, which give rise to three additional disconnected sets of transformations. It is, however, true that the field equations of quantum electrodynamics do have these additional invariance properties, and it is an interesting question whether the invariance under the special transformations should be postulated as a general principle underlying all physical laws.

We shall first study the effect of the space inversion σ on the fields A_μ, Ψ. The transformed fields $\phi^\sigma(x')$ at the transformed point x' are defined as a certain linear transformation of the original fields $\phi(x)$ at the original point. Here x' is the coordinate

$$x_i' = -x_i, \quad x_0' = x_0, \quad (i = 1, 2, 3)$$

of the transformed point. We shall now show that the linear transformation in question is almost completely determined by the invariance requirement. We first observe that from (4–15) it follows that A_μ must transform in the same way as ∂_μ if (4–15) is to remain invariant. Thus we have immediately

$$A_\mu^\sigma(x') = A_\mu'(x), \qquad (5\text{--}17)$$

where

$$A_i' = -A_i, \quad A_0' = A_0, \quad (i = 1, 2, 3). \qquad (5\text{--}18)$$

---

* Regarding the surface terms at infinity, *cf.* the footnote in Chapter 2, page 34.

For the transform of Ψ, we may write

$$\Psi^\sigma(x') = S\Psi(x), \tag{5-19}$$

where the 4×4 matrix S is determined by the condition that

$$S\gamma_\mu' = \gamma_\mu S, \tag{5-20}$$

where

$$\gamma_i' = -\gamma_i, \quad \gamma_0' = \gamma_0, \quad (i = 1, 2, 3). \tag{5-21}$$

Thus

$$S = \lambda\gamma_0 \tag{5-22}$$

with

$$\lambda^*\lambda = 1. \tag{5-23}$$

The last equation follows from the invariance of Eq. (4–6). To see this, we need the transformation property of the current operator. According to (A2–27),

$$J_\mu^\sigma(x') = -ie\,\lambda^*\lambda\,\overline{\gamma_0\Psi(x)}\gamma_\mu\gamma_0\Psi(x)$$

$$= J_\mu'(x)\lambda^*\lambda. \tag{5-24}$$

Since the transformation property of A_μ is given by (5–17), the invariance of (4–6) requires that $\lambda^*\lambda = 1$ (5–23). The as yet undetermined phase of λ can be further restricted if we fix the arbitrary constant phase factor in the definition of the operators Ψ such that the Ψ transform according to the two-valued representation of the proper Lorentz group. We must then have

$$S^2 = \pm 1$$

or

$$\lambda^2 = \mp 1. \tag{5-25}$$

λ is thus restricted to one of the four values $\lambda = \pm 1, \pm i$. There is no reason to choose any particular one of these four values, but we shall, for simplicity, put $\lambda = i$ in the following, so that $S^2 = +1$.†

It remains to be shown that the commutation rules are also invariant. Since the field equations are invariant, it is sufficient to prove the invariance for commutators at space-like situated points, and this means, accord-

---

† C. N. Yang and T. Tiomno, *Phys. Rev.* **79**, 495 (1950), discuss the possible physical significance of the assumption that different spinor fields transform under space inversion according to the two different possibilities $S^2 = \pm 1$. So long as we consider only one spinor field, however, it is irrelevant which of the two possibilities we adopt. In fact, there is no compelling physical reason to exclude a transformation law under space inversion of a more general type such as $\lambda^2 = e^{i\alpha}$ (α real and arbitrary). The reason for this ambiguity in the definition of the phase is the impossibility of measuring the phase of a spinor field, a point which was especially brought out by G. C. Wick, A. S. Wightman, and E. P. Wigner, *Phys. Rev.* **88**, 101 (1952).

ing to (4–23) that we need to show the invariance for free-field commutation rules only. For the A_μ, this invariance is trivial; for Ψ we need (5–20) and the property $\Delta(x') = \Delta(x)$ of the Δ-function entering in the bracket expression. Thus, if we write $\psi(x)$ for a free field

$$\{\psi^\sigma(x), \bar{\psi}^\sigma(y)\} = -i\gamma_0 S(x' - y')\gamma_0 = iS(x - y). \tag{5–26}$$

Now that the complete invariance is established, we know that there must exist a unitary operator Π with the property

$$A_\mu^\sigma(x) = \Pi A_\mu(x)\Pi^{-1},$$
$$\Psi^\sigma(x) = \Pi\Psi(x)\Pi^{-1}. \tag{5–27}$$

This operator Π is called the parity operator. It follows from the definition of this operator that Π^2 commutes with all the field operators and is therefore a multiple of the unit operator

$$\Pi^2 = c1. \tag{5–28}$$

Since Π is unitary, $|c| = 1$, and we can always normalize the operator Π in such a way that $c = 1$. The eigenvalues of this normalized operator are then ± 1. Since Π leaves the field equations invariant, the operator Π represents an integral of the system and the conservation of the parity leads to observable selection rules in radiative processes.

5–4 Time inversion. The same argument which led us to the construction of the space-inversion transformation can be used to show that there exist no linear time-inversion transformations of the field variables which leave the equations of motion invariant.

We observe first that if Eq. (4–15) is to remain invariant then A_μ must transform in the same way as ∂_μ. In the classification of irreducible tensors given in the appendix (Section A3), this means that A_μ would have to be a vector of kind 0 or a "regular" vector:

$$A_\mu^\tau(x') = A_\mu'(x). \tag{5–29}$$

For the transform of Ψ, we may write

$$\Psi^\tau(x') = R\Psi(x), \tag{5–30}$$

where the 4×4 matrix R is determined by the condition that

$$R\gamma_\mu' = \gamma_\mu R, \tag{5–31}$$

$$\gamma_0' = -\gamma_0, \quad \gamma_i' = \gamma_i, \quad (i = 1, 2, 3). \tag{5–32}$$

Thus

$$R = \mu\gamma_1\gamma_2\gamma_3, \tag{5–33}$$

where μ is a numerical factor of magnitude one, $|\mu| = 1$. The transformation law so far derived is entirely determined by the assumption that it is a linear transformation of the field variables and that it leaves Eq. (4–15) invariant.

We shall now show that this transformation is incompatible with the invariance of Eq. (4–6). Since the differential operator $\partial_\mu \partial^\mu$ is invariant under time inversion, and since we have already shown that A_μ is a regular vector, this assertion amounts to saying that the current operator J_μ is not a regular vector. In fact, we shall now prove that J_μ is a vector of kind 1 under the transformation (5–30).

Indeed, we find

$$J_\mu^\tau(x) = \overline{\Psi^\tau}(x)\gamma_\mu \Psi^\tau(x) = \Psi^+(x')R^+A\gamma_\mu R\Psi(x'). \tag{5-34}$$

By direct inspection, using (5–33) and (A2–27), we obtain

$$R^+A = -AR^{-1}. \tag{5-35}$$

Substituting this into (5–34) and using (5–31), we obtain the result

$$J_\mu^\tau(x) = -J_\mu'(x'), \tag{5-36}$$

which shows that the current vector behaves under this transformation like a pseudovector of kind 1. The equation (4–6) is therefore not invariant under this transformation. With similar methods, one can also prove that the commutation rules for both fields are not invariant. Thus the transformation (5–29) and (5–30) has no invariance properties and is not even canonical.†

Although a linear time-reversal transformation is thus seen not to exist, it is easy to define a nonlinear transformation which does have the desired properties. In order to give this transformation explicitly it is necessary to introduce a new notation for the operation of complex conjugation as contrasted with the Hermitian conjugation which so far has been used exclusively. If Φ is any field quantity, we denote by

Φ^* its Hermitian conjugate,

$\Phi^\sim$ its transpose,

$\Phi^\times$ its complex conjugate.

We have thus the relation

$$(\Phi^\times)^\sim = (\Phi^\sim)^\times = \Phi^*. \tag{5-37}$$

If any of the three operations is carried out on a spinor operator, we consider that the spinor index, as before, is not affected (see note on page 58).

---

† The operator $R = \gamma_1\gamma_2\gamma_3$ was introduced by G. Racah, *Nuovo Cimento* **14**, 322 (1937), in connection with the special Lorentz transformation τ. See also W. Pauli, *Rev. Mod. Phys.* **13**, 221 (1941). It is often referred to in the literature as the Racah time-inversion operator, and contrasted with the Wigner operator discussed below.

If we take the complex conjugate of a product, we obtain

$$(\Phi_1\Phi_2)^\times = \Phi_1{}^\times \Phi_2{}^\times,$$

which is contrasted with the relation

$$(\Phi_1\Phi_2)^* = \Phi_2{}^*\Phi_1{}^*. \tag{5-38}$$

We take now the ×-operation on the system (4–6) and (4–15) and obtain

$$\partial_\mu\partial^\mu A_\lambda{}^\times = -ie\overline{\Psi}^\times \gamma_\lambda{}^*\Psi^\times, \tag{4-6}^\times$$

$$(\partial_\mu\gamma^{\mu*} + m)\Psi^\times = ie\gamma^{\mu*}\Psi^\times A_\mu{}^\times. \tag{4-15}^\times$$

The field variables may be considered as functions of the time reversed position coordinate x_μ':

$$x_0' = -x_0, \quad x_i' = x_i, \quad (i = 1, 2, 3). \tag{5-39}$$

Let us define a 4×4 matrix D by the conditions†

$$\gamma'^{\mu*} = D^{-1}\gamma^\mu D,$$
$$D^*D = -1, \tag{5-40}$$

where

$$\gamma'^0 = -\gamma^0, \quad \gamma'^i = \gamma^i, \quad (i = 1, 2, 3). \tag{5-41}$$

The new field variables Ψ^τ, $A_\mu{}^\tau$ defined by

$$\Psi^\tau(x') = D\Psi^\times(x), \quad A_\mu{}^\tau(x') = A_\mu'{}^\times(x), \tag{5-42}$$

$$A_0' = A_0, \quad A_i' = -A_i, \quad (i = 1, 2, 3), \tag{5-43}$$

then have the property to satisfy again the same field equations as the original ones. For Eq. (4–15) this follows immediately from (4–15)$^\times$ upon substitution of (5–39), (5–40), (5–41), (5–42), and (5–43). To show the invariance of (4–6) also, we need the relation that the matrix AD is antisymmetrical:‡

$$(AD)^\sim = -AD. \tag{5-44}$$

---

† The fact that $D^*D = DD^* = d\cdot 1$ is a multiple of the unit matrix follows from the first of equations (5–40). The proof that actually $d < 0$ is more involved. It is established by a judicious use of the relations (A2–27), (A2–28), and (A2–32) (with $c = 1$). An alternative way is to prove first that d is independent of the representation of the γ-matrices and then to show that it is negative in a particular representation, for instance the standard representation (A2–14). We shall be content with this brief sketch; the details of the proof are left to the reader. The value $d = -1$ in Eq. (5–40) is then obtained by suitable normalization of D.

‡ This follows from the defining equations (5–40) and the relation (A2–32) of the Appendix. In fact, one finds

$$AD = -(B\gamma_0)^*,$$

from which (5–44) follows because of (A2–29).

We obtain for the right side of $(4\text{--}6)^\times$, using (5–44) and $A^+ = A$,

$$\overline{\Psi}^\times(x)\gamma_\lambda{}^*\Psi^\times(x) = \Psi^{\tau+}(x')D^{+-1}A^*\gamma_\lambda{}^*D^{-1}\Psi^\tau(x')$$
$$= \Psi^{\tau+}(x')AD\gamma_\lambda{}^*D^{-1}\Psi^\tau(x') = \overline{\Psi^\tau}(x')\gamma_\lambda{}'\Psi^\tau(x').$$

In view of (5–43) Eq. $(4\text{--}6)^\times$ appears now in the form

$$\partial_\mu\partial^\mu A_\lambda{}^\tau(x) = ie\overline{\Psi^\tau}(x)\gamma_\lambda\Psi^\tau(x) = -J_\lambda{}^\tau(x). \tag{5–45}$$

Thus it is seen to be invariant also. The invariance of the commutation rules may then be proved in a similar manner if use is made of the following properties of the Δ-functions [*cf.* (A1–11) and (1–12) in the Appendix]:

$$\Delta^*(x) = \Delta(x), \quad \Delta(x') = -\Delta(x).$$

The transformation $A_\mu \to A_\mu{}^\tau$, $\Psi \to \Psi^\tau$ is called the time-reversal transformation, and the fields $A_\mu{}^\tau$, Ψ^τ are referred to as the time-reversed fields.† The invariance property, which we have proved, entails the existence of the unitary time-reversal operator T with the property

$$A_\mu{}^\tau(x) = TA_\mu(x)T^{-1},$$
$$\Psi^\tau(x) = T\Psi(x)T^{-1}. \tag{5–46}$$

Instead of transforming the operators, as we have done up to now, one may transform the state vectors. To each state vector Ω can be associated a time-reversed state vector Ω^τ which is so defined that the expectation values of all observables with respect to Ω^τ assume the time-reversed values

$$\langle \zeta(x) \rangle_\tau \equiv (\Omega^\tau, \zeta(x)\Omega^\tau) = (\Omega, \zeta'(x')\Omega) = \langle \zeta'(x') \rangle. \tag{5–47}$$

This condition determines the time-reversed state vector Ω^τ associated with any vector Ω as

$$\Omega^\tau = T^{-1}\Omega^\times. \tag{5–48}$$

In the derivation of this result the relation $\zeta'(x') = \zeta^{\tau\times}(x)$, which holds for all observables, was used. With this definition, the expectation values of observables transform according to a linear transformation law under time-reversal and, insofar as they are irreducible tensors, fall into one of

---

† The time-reversal transformation in quantum mechanics was discussed by E. Wigner, *Göttinger Nachrichten* **31**, 546 (1932) for the nonrelativistic case. The case of relativistic quantum electrodynamics was treated by S. Watanabe, *Phys. Rev.* **84**, 1008 (1951). Our treatment differs from the latter insofar as we discuss here the time-reversal transformation in the Heisenberg picture. We have adopted what Watanabe calls in his paper "standpoint II," which seems to us the proper formulation of this transformation. The "standpoint I" corresponds to a combination of the time-reversal and charge conjugation to be discussed in the following section. It can be shown that the relativistic time-reversal operation here adopted reduces to the Wigner case in the nonrelativistic limit.

the four classes of pseudotensors discussed in the mathematical appendix (*cf.* Section A3).

We note in particular that the current operator is a tensor of class 1. This means that the space part of the current is reversed for both the space and time inversion, while the 0-component (the charge density) is invariant under both these transformations. It follows in particular that the total charge is an absolute invariant under all Lorentz transformations, including the time-reversal transformation. The same transformation properties as for the current hold also for the vector potential. It is also a pseudovector of class 1. As a particular consequence, we note, for instance, that the electric field (the components F_{0k}) reverses itself under space inversion but remains invariant under time inversion. The opposite is true for the magnetic field.

While the transformation $\phi \rightarrow \phi^\tau$ has thus all the properties of the classical time-reversal transformation, it is important to emphasize also the properties which single it out from all the other unitary transformations of the field variables. This singular character is due to the fact that it is, as already mentioned, a nonlinear transformation of the field operators. A consequence is the somewhat unfamiliar transformation law of the operator T if the field operators Φ are subjected to a unitary transformation. To exhibit this transformation law, let $T = T_0$ be the operator obtained in one particular representation and let us transform the field variables with a unitary transformation U. We have

$$(U\phi(x)U^{-1})^\tau = U^\times \phi'^\times(x')(U^\times)^{-1} = U^\times \phi^\tau(x)(U^\times)^{-1}$$
$$= U^\times T_0 \phi(x) T_0^{-1}(U^\times)^{-1} = TU\phi(x)U^{-1}T^{-1}. \quad (5\text{–}49)$$

Thus
$$T = U^\times T_0 U^{-1}. \quad (5\text{–}50)$$

The unitary character of T remains invariant, since

$$TT^* = U^\times T_0 U^{-1} U T_0^* U^\sim = U^\times T_0 T_0^* U^\sim$$
$$= U^\times U^\sim = 1, \quad (5\text{–}51)$$

and similarly for T^*T.

An interesting relation is obtained if we take for U the transformation T_0 itself. T in (5–50) is then the transformation which transforms back from the time reversed to the original field, and according to (5–50) it is given by

$$T = T_0^\times. \quad (5\text{–}52)$$

Now according to (5–40), the resultant Ψ field is restored with the sign reversed,

$$\Psi^{\tau\tau} = -\Psi, \quad (5\text{–}53)$$

while according to (5–43) the resultant A_μ is the same as the original field,

$$A_\mu^{\tau\tau} = A_\mu. \tag{5-54}$$

Thus we find that the operator $T^\times T$ has the property that it commutes with A_μ while it anticommutes with Ψ:

$$[T^\times T, A_\mu] = 0, \tag{5-55}$$

$$\{T^\times T, \Psi\} = 0. \tag{5-56}$$

Consequently, $(T^\times T)^2$ commutes with all the field variables and is therefore a multiple of the unit matrix. Since it is also unitary, it is the unit matrix. Thus $T^\times T$ has eigenvalues ± 1. If Ω is a state vector with eigenvalue $+1$, then it follows that $\Psi\Omega$ belongs to the eigenvalue -1.

The relations (5-55) and (5-56) give rise to a "superselection rule."† If the Hilbert space of a physical system can be decomposed into two subspaces with the properties:

(a) a selection rule holds between state vectors of the two subspaces,
(b) observables have vanishing matrix elements connecting the two subspaces,

then we say that a superselection rule is operative with respect to the two subspaces.

Such a decomposition of the Hilbert space is given by the operator $T^\times T$. We denote by Ω_+ the state vectors which belong to the eigenvalue $+1$, and by Ω_- those vectors which belong to the eigenvalue -1 of this operator.

Because the interaction operator (4-14) is quadratic in the operators Ψ, the state of the system when once in one of these substates will remain in it for all time. Thus we have a selection rule. But we have even a superselection rule, for it follows from (5-47) that for any observable $\zeta(x)$,

$$(\Omega_2^\tau, \zeta(x)\Omega_1^\tau) = (\Omega_2, \zeta'(x')\Omega_1).$$

Thus if we carry out the time-reversal transformation twice,

$$(\Omega_1^{\tau\tau}, \zeta(x)\Omega_2^{\tau\tau}) = (\Omega_1, \zeta(x)\Omega_2).$$

If we choose now for $\Omega_1 = \Omega_+$ and $\Omega_2 = \Omega_-$ and take account of the fact that

$$\Omega_+^{\tau\tau} = \Omega_+, \quad \Omega_-^{\tau\tau} = -\Omega_-,$$

then it follows that

$$(\Omega_+, \zeta(x)\Omega_-) = 0. \tag{5-57}$$

A matrix element for any observable between the two subspaces must vanish.

---

† This concept was introduced by G. C. Wick, A. S. Wightman, and E. P. Wigner, *Phys. Rev.* **88**, 101 (1952).

5-5 Charge conjugation. The charge-conjugate transformation is defined by

$$\Psi^c(x) = C^*\Psi^*(x),$$
$$A_\mu{}^c(x) = -A_\mu(x).$$
(5-58)

Under this transformation, the current operator transforms according to the equation

$$J_\mu{}^c(x) = -J_\mu(x). \quad (5\text{-}59)$$

This equation is correct only if we use the charge-symmetrical expression (4-9) for the current operator. The invariance of the field equation (4-6) is then obvious. In order to show the invariance of (4-15) also, we take the Hermitian conjugate† of (4-15), giving

$$(\gamma_\mu{}^*\partial^\mu + m)\Psi^* = ie\gamma_\lambda{}^*\Psi^*A^\lambda, \quad (4\text{-}15)^*$$

since $A_\lambda{}^* = A_\lambda$. We now substitute the defining equation (A2-27) for the operator C, use $C^*C = 1$, and obtain

$$(\partial + m)\Psi^c = -ieA^c\Psi^c \quad (4\text{-}15)^C$$

This shows that the transformation (5-58) leaves the system (4-6) and (4-15) invariant.

This transformation is also a canonical transformation. To see this, we must show that the commutation rules are invariant. Since the field equations are invariant, it is sufficient to establish the invariance of the commutation rules on a space-like plane, and this means we can discuss the invariance for the free fields (4-23). For the field $a_\mu(x)$ this is again obvious. For $\psi(x)$, we find

$$\{\psi^c(x), \overline{\psi^c}(x')\} = \{C^*\psi^*(x), \psi^\sim(x')C^\sim A\}. \quad (5\text{-}60)$$

From (A2-32) and (A2-29) follows

$$C^\sim A = -A^\sim C = -A^*C.$$

Substituting this into (5-60), we find with the use of the defining equation (A2-27),

$$\{\psi^c(x), \overline{\psi^c}(x')\} = iS(x - x'). \quad (5\text{-}61)$$

We have thus shown that charge conjugation is a canonical transformation. There exists therefore a unitary operator Γ such that

$$A_\mu{}^c = \Gamma A_\mu \Gamma^{-1} = -A_\mu,$$
$$\Psi^c = \Gamma \Psi \Gamma^{-1} = C^*\Psi^*.$$
(5-62)

---

† We remind the reader that the star operation [defined in Section 1-3, Eq. (1-23)] which is used here means complex conjugation for the c-number matrices in spinor space.

It should perhaps be noted that for the invariance of the commutation rule (5–61) under charge conjugation it is essential that the ψ field is quantized according to the case b. Indeed, in the transition from (5–60) to (5–61) we had to make use of the symmetry of the commutator bracket, because the operation of Hermitian conjugation interchanges the order of the noncommuting operators. The principle of invariance under charge conjugation thus furnishes a restriction for the quantization of spinor fields.†

By combining charge conjugation with the time-reversal operation, we obtain a new transformation. The two transformations can be carried out in any order, since they commute. We have for any field $\Phi(A_\mu \text{ or } \Psi)$,

$$\Phi^{c\tau} = \Phi^{\tau c}. \tag{5-63}$$

For the field A_μ this is trivial. For Ψ it follows from the relation

$$DC = C^*D^* = R. \tag{5-64}$$

The operator R has the property

$$\gamma_k = R^{-1}\gamma_k R, \quad -\gamma_0 = R^{-1}\gamma_0 R, \tag{5-65}$$

and is given by $R = \gamma_1\gamma_2\gamma_3$. It is thus identical with the Racah operator (5–33) previously discussed. We have anticipated this result with the notation.

The new canonical transformation thus defined leaves the field equations also invariant and is given explicitly by

$$\Psi^R(x) = R\tilde{\Psi}(x'), \quad A_\mu^R(x) = A_\mu'^{\sim}(x'), \tag{5-66}$$

where

$$A_0' = -A_0, \quad A_k' = A_k, \quad (k = 1, 2, 3),$$

$$x_0' = -x_0, \quad x_k' = x_k, \quad (k = 1, 2, 3).$$

There exists therefore a unitary operator P such that

$$\Phi^R(x) = P\Phi(x)P^{-1}. \tag{5-67}$$

For any Hermitian operator ζ we then have a relation

$$\langle \zeta(x) \rangle_R = \langle \zeta'(x') \rangle \tag{5-68}$$

if we define the state vector Ω^R by

$$\Omega^R = P^{-1}\Omega. \tag{5-69}$$

---

† See S. Watanabe, *Phys. Rev.* **84**, 1008 (1951), esp. Section VIII.

The transformation (5–69) could be used as an alternative description of the time-reversed state vector.† In this description the current and the vector potential transform like regular vectors (vectors of kind 0, see Appendix A3). This corresponds to the fact that classically the time-reversed solution may also be obtained by reversing the electric field, the velocities, the sign of the charge, and the current density but retaining the magnetic field. It is not a natural interpretation, since it contradicts the usual classical notion of the total charge as an absolute invariant under time reversal. We can exclude it by postulating that the total charge be an invariant under all Lorentz transformations.

5–6 Scale transformations. One can verify easily that the system of our fundamental field equations (4–6) and (4–15) is invariant under the "scale transformations" defined as follows:

$$A_\mu^s(sx) = \frac{1}{s} A_\mu(x),$$

$$\Psi^s(sx) = \frac{1}{s^{3/2}} \Psi(x), \quad (5\text{–}70)$$

$$m_s = \frac{1}{s} m,$$

where s is an arbitrary dimensionless real number. The fact that the mass (or rather the reciprocal Compton wavelength in our system of units) is not invariant shows that the transformation $\Phi \to \Phi^s$ is not a canonical transformation. The commutation rules for the fields are indeed not invariant. This is to be expected, since this theory contains the length $1/m$ as an arbitrary constant.

On the other hand, we can see that a theory for which $m = 0$ does have this additional invariance property.

---

† This corresponds to "standpoint I" in the paper of S. Watanabe, *Phys. Rev.* **84**, 1008 (1951).

CHAPTER 6

SUBSIDIARY CONDITION AND LONGITUDINAL FIELD

In this chapter we shall discuss some of the peculiarities of the radiation field which have their origin in the vanishing of the photon mass. We begin with a covariant treatment of the Coulomb interaction and the elimination of the longitudinal and zero components of the field (Section 6–1). The construction of the state vector which satisfies the subsidiary condition requires special precautions (Section 6–2). Alternative solutions of this problem are the method of Gupta (Section 6–3), Valatin's gauge-invariant interaction (Section 6–4), and the radiation field with small finite mass (Section 6–5).

6–1 The covariant Coulomb interaction. In the original formulation of the quantum theory of radiation by Dirac[*] the electromagnetic field was separated into a radiation field and a static Coulomb interaction. The radiation field was then subjected to the usual quantum procedure, while the Coulomb interaction was treated as an unquantized classical interaction potential. In this form the relativistic invariance is, of course, completely destroyed and as a result it is very difficult to prove that the whole scheme is in accord with the principle of relativity. For this reason it is of importance that the Coulomb interaction can be obtained in a relativistic form from the completely covariant theory so far developed.[†]

We prepare the covariant derivation of the Coulomb interaction by studying the effect of a particular τ-dependent canonical transformation on the Schroedinger equation in the interaction picture. Let us define a new state vector $\eta(\tau)$ related to $\omega(\tau)$ by a unitary transformation:

$$\omega(\tau) = e^{i\Sigma(\tau)}\eta(\tau), \qquad (6\text{–}1)$$

where $\Sigma(\tau)$ is a τ-dependent Hermitian operator. Substituting (6–1) into (4–50), we obtain the new Schroedinger equation in the form

$$i\dot{\eta}(\tau) = G(\tau)\eta(\tau), \qquad (6\text{–}2)$$

---

[*] P. A. M. Dirac, *Proc. Roy. Soc.* **112**, 661 (1926); **114**, 243, 710 (1927).

[†] Covariant treatments of this problem are found in J. Schwinger, *Phys. Rev.* **74**, 1439 (1948), esp. Sec. 3. N. Hu, *Phys. Rev.* **76**, 391 (1949); **77**, 150 (1950). F. Coester and J. M. Jauch, *Phys. Rev.* **78**, 149, 827 (1950). Z. Koba, T. Tati. and S. Tomonaga, *Progr. Theor. Phys.* **2**, 198 (1947). S. Hayakawa, Y. Miyamoto, S. Tomonaga, *J. Phys. Soc. Japan* **2**, 172 (1947). Z. Koba, Y. Oisi, M. Sasaki. *Progr. Theor. Phys.* **3**, 141, 229 (1948). J. G. Valatin, *Det. Kong. Danske Vid. Selsk.* **26**, No. 13 (1951).

with
$$G(\tau) = e^{-i\Sigma(\tau)}H(\tau)e^{i\Sigma(\tau)} - ie^{-i\Sigma(\tau)}\frac{d}{d\tau}e^{i\Sigma(\tau)}. \qquad (6\text{-}3)$$

The last expression can be developed in powers of Σ:

$$G = H + \left(\frac{i}{1!}[H,\Sigma] + \dot{\Sigma}\right) + \left(\frac{i^2}{2!}[[H,\Sigma],\Sigma] + \frac{i}{2!}[\dot{\Sigma},\Sigma]\right) + \cdots. \qquad (6\text{-}4)$$

Here we have bracketed together the terms which are of the same order in Σ. We obtain the Coulomb interaction term if we choose for $\Sigma(\tau)$ the operator

$$\Sigma(\tau) = \int d\sigma_\mu j^\mu(x) c(x) \qquad (6\text{-}5)$$

with
$$c(x) = -\partial^{-1} n^\nu a_\nu(x). \qquad (6\text{-}6)$$

The operator ∂^{-1} stands for the inverse of the differential operator $\partial \equiv n^\nu \partial_\nu$ in the direction of the normal. It may be most conveniently defined by developing the function on which it operates into a Fourier integral. The operator ∂^{-1} multiplies each term in this integral by a factor $(ik^\nu n_\nu)^{-1}$. With this expression for $\Sigma(\tau)$, we obtain

$$[H,\Sigma] = -\int_\sigma d\sigma \int_\sigma d\sigma'[j^\nu(x)a_\nu(x), j^\mu(x')n_\mu c(x')]. \qquad (6\text{-}7)$$

Gauss' theorem, together with the continuity equation $\partial^\nu j_\nu(x) = 0$, gives

$$\dot{\Sigma} = -\int d\sigma j^\mu(x) \partial_\mu c(x), \qquad (6\text{-}8)$$

from which we obtain

$$[\dot{\Sigma},\Sigma] = -\int d\sigma \int d\sigma'[j^\mu(x)\partial_\mu c(x), j^\nu(x')n_\nu c(x')]. \qquad (6\text{-}9)$$

The commutators which appear on the right sides of (6-7) and (6-9) can be evaluated, with the result

$$[H,\Sigma] = i\int d\sigma \int d\sigma' j^\nu(x) j^\mu(x') n_\nu n_\mu \partial^{-1} D(x-x'), \qquad (6\text{-}10)$$

$$[\dot{\Sigma},\Sigma] = i\int d\sigma \int d\sigma' j^\nu(x) j^\mu(x') n_\mu \partial_\nu \partial^{-2} D(x-x'). \qquad (6\text{-}11)$$

All the other commutators which appear in the expression (6-4) for G

vanish. Consequently, we obtain

$$G(\tau) = -\int d\sigma j^\mu(x)(a_\mu(x) + \partial_\mu c(x))$$

$$- \int d\sigma \int d\sigma' j^\nu(x) j^\mu(x')(n_\nu n_\mu \partial^{-1} + \tfrac{1}{2} n_\mu \partial_\nu \partial^{-2}) D(x - x'). \quad (6\text{–}12)$$

The second expression on the right may be further simplified if we make use of the relation

$$(\partial_\mu + n_\mu \partial)\partial^{-2} D(x - x') = 0, \quad (6\text{–}13)$$

which holds if x and x' are on the same plane $\sigma(\tau)$. Thus we finally obtain for $G(\tau)$,

$$G(\tau) = -\int d\sigma j^\mu(x)(a_\mu(x) + \partial_\mu c(x))$$

$$- \tfrac{1}{2} \int d\sigma_\nu \int d\sigma_\mu' j^\nu(x) j^\mu(x') \partial^{-1} D(x - x'). \quad (6\text{–}14)$$

The second term on the right is the covariant expression for the Coulomb interaction. This can be seen by evaluating it in the special coordinate system for which $n^\mu = (1, 0, 0, 0)$. Indeed, in this special system $\partial^{-1} D(y)$, with $y = x - x'$, becomes simply [cf. Eq. (A1–23)]

$$\partial^{-1} D(y) = -\frac{1}{(2\pi)^3} \int e^{i\mathbf{k}\cdot\mathbf{y}} \frac{1}{\omega^2} d^3k = -\frac{1}{4\pi} \frac{1}{|\mathbf{y}|}. \quad (6\text{–}15)$$

If we denote the charge density operator by $j^0(\mathbf{x}) = \rho(\mathbf{x})$, we obtain

$$-\tfrac{1}{2} \int d\sigma_\nu \int d\sigma_\mu' j^\nu(x) j^\mu(x') \partial^{-1} D(x - x') = \frac{1}{2} \frac{1}{4\pi} \int \frac{\rho(x)\rho(x')}{|\mathbf{x} - \mathbf{x}'|} d^3x d^3x', \quad (6\text{–}16)$$

which is seen to be the Coulomb energy of the charge distribution $\rho(x)$.

There remains the discussion of the first term on the right side of (6–14). The field $a_\mu + \partial_\mu c$ which interacts with the current is not the transverse field b_μ introduced in Chapter 2, Eq. (2–130). But a comparison with this equation shows that it differs from this field only by a term which involves $g(x) = \partial^\mu a_\mu(x)$. Indeed, because of (6–16), we have simply

$$b_\mu(x) = a_\mu(x) + \partial_\mu c(x) - \partial^{-1}(n_\mu + \partial^{-1}\partial_\mu)g(x). \quad (6\text{–}17)$$

Up to this point we have not yet used the subsidiary condition. We shall now show that the first term of G in (6–14) can be simplified further if we use the fact that $G(\tau)$ operates only on state vectors $\eta(\tau)$ which are restricted by the subsidiary condition. In Section 4–3 we gave this restriction for the state vectors $\omega(\tau)$; we must now express it for the transformed state vectors $\eta(\tau)$. We take, for instance, the form (4–55) which, because of (6–1), may be written as

$$e^{-i\Sigma}\left(\partial^\lambda a_\lambda(x) - \int_\sigma j^\lambda(x')D(x-x')d\sigma_\lambda'\right)e^{i\Sigma}\eta = 0. \qquad (6\text{–}18)$$

Now

$$e^{-i\Sigma}a_\lambda(x)e^{i\Sigma} = a_\lambda(x) + i[a_\lambda(x),\Sigma]$$

$$= a_\lambda(x) + n_\lambda \partial^{-1}\int d\sigma_\nu' j^\nu(x')D(x-x'). \qquad (6\text{–}19)$$

Consequently,

$$e^{-i\Sigma}\partial^\lambda a_\lambda(x)e^{i\Sigma} = \partial^\lambda a_\lambda(x) + \int d\sigma_\nu' j^\nu(x')D(x-x'). \qquad (6\text{–}20)$$

Since the second term in (6–18) commutes with Σ, the transformation $e^{i\Sigma}$ results in a cancellation between this term and the second term in (6–20). It follows that $\eta(\tau)$ satisfies for all τ and x the condition

$$\partial^\lambda a_\lambda(x)\eta(\tau) = 0. \qquad (6\text{–}21)$$

It follows from this result and (6–17) that

$$(a_\mu(x) + \partial_\mu c(x))\eta(\tau) = b_\mu(x)\eta(\tau), \quad \text{for all } x,\tau. \qquad (6\text{–}22)$$

Thus the Schroedinger equation is not affected if we replace in (6–14) the field $a_\mu + \partial_\mu c$ by the transverse field b_μ. Consequently the interaction operator which governs the time development of the state vectors $\eta(\tau)$ is given by

$$G(\tau) = -\int j^\mu(x)b_\mu(x)d\sigma - \tfrac{1}{2}\int d\sigma_\mu \int d\sigma_\nu' j^\mu(x)j^\nu(x')\partial^{-1}D(x-x'). \qquad (6\text{–}23)$$

The Schroedinger equation is

$$i\dot\eta(\tau) = G(\tau)\eta(\tau) \qquad (6\text{–}24)$$

and the subsidiary condition is

$$\partial^\lambda a_\lambda(x)\eta(\tau) = 0, \quad \text{for all } x,\tau. \qquad (6\text{–}25)$$

6–2 The subsidiary condition and the construction of the state vector. If the subsidiary condition (2–30) is written in terms of the momentum variables (2–62), it appears in the form

$$\chi(k)\omega \equiv k^\mu a_\mu \omega = 0, \qquad (6\text{–}26)$$

$$\chi^*(k)\omega \equiv k^\mu a_\mu^* \omega = 0. \qquad (6\text{–}27)$$

These two relations must hold for all k. They are compatible because $\chi(k)$ commutes with $\chi^*(k')$ for all values of k,k'. For with (2–67) and

(2–68), we obtain

$$[\chi(k),\chi^*(k')] = k^\mu k_\mu' \delta(\mathbf{k} - \mathbf{k}') = 0. \tag{6-28}$$

The right side vanishes because $k^\mu k_\mu = 0$.

In order to construct explicitly a state vector which satisfies conditions (6–26) and (6–27), we consider a component for which $k_1 = k_2 = 0$, $k_3 = k^0$. Any other component can be brought into this form by a suitable rotation of the coordinate system. Thus there is no loss of generality if we treat this case only. The part of the state vector which is associated with this component will again be denoted by ω, omitting for simplicity of notation the argument k on which it depends. The complete state vector for all components is then the direct product of all the state vectors for each individual k.

For the longitudinal and 0-component of the field operators, we write

$$a_3 = a, \quad a_0 = b^* \tag{6-29}$$

so that according to (2–68),

$$[a,a^*] = [b,b^*] = 1 \tag{6-30}$$

and

$$[a,b] = [a,b^*] = 0. \tag{6-31}$$

The most general state vector ω referring to two degrees of freedom can be constructed as a linear superposition in the form

$$\omega = \sum_{n,m=0}^{\infty} \frac{1}{\sqrt{n!m!}} c_{nm}(a^*)^n (b^*)^m \omega_0, \tag{6-32}$$

where ω_0 is the simultaneous eigenvector of a^*a and b^*b with eigenvalue 0.

$$a^*a\omega_0 = b^*b\omega_0 = 0. \tag{6-33}$$

The conditions (6–26) and (6–27), in this special coordinate system, become

$$(a + b^*)\omega = 0, \tag{6-26}'$$

$$(a^* + b)\omega = 0. \tag{6-27}'$$

Making use of the commutation rules and the relations (6–33), we may write the last two conditions:

$$\sqrt{n}\, c_{n,m} + \sqrt{m}\, c_{n-1,m-1} = 0, \tag{6-34}$$

$$\sqrt{m}\, c_{n,m} + \sqrt{n}\, c_{n-1,m-1} = 0. \tag{6-35}$$

Nontrivial solutions of this homogeneous linear system exist only if the determinant vanishes, that is, if

$$n = m.$$

Thus
$$c_{nm} = c_n \delta_{nm}, \qquad (6\text{-}36)$$
$$c_n = (-1)^n.$$

The state vector (6-32) appears now in the form
$$\omega = \sum_{n=0}^{\infty} \frac{1}{n!} (-a^*b^*)^n \omega_0 = e^{-a^*b^*} \omega_0. \qquad (6\text{-}37)$$

This state vector is not normalizable, since its norm is given by
$$(\omega,\omega) = \sum_{n=0}^{\infty} \frac{1}{(n!)^2} \langle a^n b^n a^{*n} b^{*n} \rangle_0, \qquad (6\text{-}38)$$
where
$$\langle a^n b^n a^{*n} b^{*n} \rangle_0 \equiv (\omega_0, a^n b^n a^{*n} b^{*n} \omega_0) = (n!)^2. \qquad (6\text{-}39)$$
Thus
$$(\omega,\omega) = \sum_{n=0}^{\infty} 1 = \infty. \qquad (6\text{-}40)$$

This result was expected in view of the remarks made in connection with the subsidiary condition (2-30).

The proper procedure in this case is to define a family of normalizable state vectors depending on a parameter λ, say, which is allowed to approach a limit in such a manner that the subsidiary condition is satisfied in the limit only.† Such a family of state vectors can be defined, for instance, by setting
$$\omega^\lambda = \sqrt{1-\lambda^2} \sum_{n=0}^{\infty} \frac{1}{n!} (-\lambda a^* b^*)^n \omega_0 = \sqrt{1-\lambda^2}\, e^{-\lambda a^* b^*} \omega_0. \qquad (6\text{-}41)$$

The norm of these vectors is given by
$$(\omega^\lambda,\omega^\lambda) = (1-\lambda^2) \sum_{n=0}^{\infty} \lambda^{2n} = 1, \quad \text{for } |\lambda| < 1. \qquad (6\text{-}42)$$

For the left side of the subsidiary condition, we obtain
$$(a + b^*)\omega^\lambda \equiv \alpha = (1-\lambda) b^* \omega^\lambda, \qquad (6\text{-}26)_\lambda$$
$$(a^* + b)\omega^\lambda \equiv \beta = (1-\lambda) a^* \omega^\lambda. \qquad (6\text{-}27)_\lambda$$

The norm of the vectors α,β can be easily calculated. We obtain
$$(\alpha,\alpha) = (\beta,\beta) = (1-\lambda)^2 (\omega^\lambda, bb^* \omega^\lambda) = \frac{1-\lambda}{1+\lambda}. \qquad (6\text{-}43)$$

---

† This procedure was proposed by R. Utiyama, T. Imamura, S. Sunakawa, T. Dodo, *Progr. of Theor. Phys.* **6**, 587 (1951).

It is seen from the last expression that this norm approaches zero with $\lambda \to 1$. Thus the subsidiary condition is satisfied for the limit $\lambda = 1$. With the state vectors ω^λ we can now calculate the expectation value of any operator without any convergence problems. For instance, for the various quadratic expressions of the longitudinal and transverse components of the field operators, we obtain the following expectation values with the state vector ω^λ:

$$\langle a_0^* a_0 \rangle = \langle a_3 a_3^* \rangle = \frac{1}{1 - \lambda^2}, \tag{6-44}$$

$$\langle a_0 a_0^* \rangle = \langle a_3^* a_3 \rangle = \frac{\lambda^2}{1 - \lambda^2}, \tag{6-45}$$

which are seen to be consistent with the commutation rules. The part of the operator n (2–78) which refers to the a_3, a_0 components has thus the expectation value

$$\tfrac{1}{2} \langle a_3 a_3^* + a_3^* a_3 - a_0 a_0^* - a_0^* a_0 \rangle = 0. \tag{6-46}$$

This value is zero and independent of λ and its limit is therefore also 0. Thus we may write

$$\langle n \rangle = \langle \tfrac{1}{2}(a_1^* a_1 + a_1 a_1^*) + \tfrac{1}{2}(a_2^* a_2 + a_2 a_2^*) \rangle. \tag{6-47}$$

We see then that for state vectors which represent a Maxwell field only the transverse components contribute to the expectation value of the total momentum operator. Since gauge transformations do not affect these transverse components, we have obtained a second proof that the expectation values of the momentum operators are gauge invariant [*cf.* Section 5–2, especially Eqs. (5–15) and (5–16)].

This last result also shows that, in spite of the indefinite appearance of Eq. (2–78), the expectation value of n is actually positive definite for state vectors satisfying the subsidiary condition.

6–3 The Gupta method. In the preceding section we have shown that it is possible to construct a state vector which satisfies the subsidiary condition by a limiting process. Various alternative methods have been proposed to circumvent this somewhat awkward situation. Gupta[†] has proposed such a method which makes use of the quantization with an indefinite metric in Hilbert space introduced by Dirac.[‡] The method was further developed by Bleuler and Heitler.[§] We introduce the essential

---

[†] S. N. Gupta, *Proc. Phys. Soc.* **63**, 681 (1950); **64**, 850 (1951).
[‡] P. A. M. Dirac, *Proc. Roy. Soc.* **A180**, 1 (1942); *Comm. Dublin Inst. Advanced Studies*, A., No. 1, 1943. W. Pauli, *Rev. Mod. Phys.* **15**, 175 (1943).
[§] K. Bleuler, *Helv. Phys. Acta* **23**, 567 (1950). K. Bleuler and W. Heitler, *Progr. Theor. Phys.* **5**, 600 (1950).

idea of this procedure by discussing a very simple example first. Let a be an operator and a^* its Hermitian conjugate which satisfies

$$[a,a^*] = -1. \tag{6-48}$$

This relation is usually interpreted by saying that a^* is an absorption operator, while a is an emission operator. As a consequence of (6–48), there exists a state vector ω_0 which belongs to the lowest eigenvalue 0 of aa^* such that

$$\omega' = a^*\omega_0 = 0. \tag{6-49}$$

The reason is that the state vector ω' has the norm zero:

$$(\omega',\omega') = (a^*\omega_0, a^*\omega_0) = (\omega_0, aa^*\omega_0) = 0,$$

and consequently must be the zero vector, because of the positive definite character of the scalar product. In a similar manner, we show that the state vector

$$\omega_1 = a\omega_0 \tag{6-50}$$

belongs to the eigenvalue 1 of aa^*:

$$aa^*\omega_1 = \omega_1. \tag{6-51}$$

The relation (6–48) holds for the 0-component of the operators $a_\mu(k)$ and it is precisely this mixing of emission and absorption operators, which one encounters in the subsidiary condition (6–26)′ and (6–27)′, which causes the state vector to become not normalizable.

The new quantization procedure can now be introduced by a redefinition of the adjoint operation. To a linear operator A is associated an adjoint operator $A\dagger$ which is required to satisfy only

$$(\lambda A)\dagger = \lambda^* A\dagger,$$
$$(A + B)\dagger = A\dagger + B\dagger, \tag{6-52}$$
$$(AB)\dagger = B\dagger A\dagger,$$
$$A\dagger\dagger = A$$

where λ is an arbitrary number. Corresponding to this definition, we define also the adjoint state vector $\omega\dagger$ with similar properties. The scalar product (ω_1,ω_2) of two state vectors is now interpreted to mean‡

$$(\omega_1,\omega_2) = \omega_1\dagger\omega_2 = (\omega_2,\omega_1)^*. \tag{6-53}$$

---

‡ With regard to the notation, we introduce the following convention. A scalar product (ω_1,ω_2) is, as before, written with a comma, the expression $\omega_1\dagger\omega_2$, on the other hand, is conceived as a matrix product. Thus if ω_2 is considered a column vector (that is a matrix with only one column) then $\omega_1\dagger$ is a row vector (a matrix with only one row) since the matrix product of the two is by definition a number.

The greater generality of this scheme is introduced by not requiring the positive definite character of the scalar product. That is, (ω,ω) may be smaller than zero. It follows in particular that we may have state vectors $\omega \neq 0$ for which nevertheless $(\omega,\omega) = 0$.

From the definitions (6–52) and (6–53) follows immediately that

$$(\omega_1, A\omega_2) = (A^\dagger \omega_1, \omega_2) = (\omega_2, A^\dagger \omega_1)^*. \tag{6-54}$$

In particular, if $A^\dagger = A$ then $(\omega, A\omega)$ is real. We call such an operator self-adjoint.

The adjoint operation is determined if we assume the existence of a state vector ω_0 with the property

$$a\omega_0 = 0; \quad (\omega_0, \omega_0) = 1. \tag{6-55}$$

It follows from $[a, a^\dagger] = -1$ that $\omega_1 = a^\dagger \omega_0$ is the eigenvector of $a^\dagger a$ with eigenvalue -1. Indeed, we have

$$a^\dagger a a^\dagger \omega_0 = a^\dagger [a, a^\dagger] \omega_0 = -a^\dagger \omega_0.$$

The norm of this vector is

$$(\omega_1, \omega_1) = (a^\dagger \omega_0, a^\dagger \omega_0) = \omega_0^\dagger a a^\dagger \omega_0$$
$$= \omega_0^\dagger [a a^\dagger] \omega_0 = -(\omega_0, \omega_0) = -1.$$

More generally, we can construct the state vectors

$$\omega_n = \frac{1}{\sqrt{n!}} (a^\dagger)^n \omega_0 \tag{6-56}$$

with the properties

$$(\omega_n, \omega_{n'}) = (-1)^n \delta_{nn'}, \tag{6-57}$$

$$a^\dagger a \omega_n = -n \omega_n. \tag{6-58}$$

The state vectors ω_n with $n = 0, 1, \cdots$, form a complete system of eigenvectors of the operator $a^\dagger a$. The most general state vector ω is a linear superposition of the ω_n:

$$\omega = \sum_n c_n \omega_n.$$

Canonical transformations are obtained in the form

$$a' = U a U^{-1}, \tag{6-59}$$

where the operator U is now no longer unitary but instead satisfies

$$U^\dagger U = 1 = U U^\dagger. \tag{6-60}$$

We call an operator which satisfies this relation a canonical operator. Canonical operators leave the self-adjoint property of operators invariant.

This follows immediately from (6–52). Thus, if

then
$$A' = UAU^{-1} \quad \text{and} \quad U{\dagger}U = 1 = UU{\dagger}.$$
$$A'{\dagger} = UA{\dagger}U^{-1},$$

and the assertion is obviously true. It follows in particular that the scalar product is an invariant under canonical transformations of the state vectors. That is, if $\omega' = U\omega$, then

$$(\omega_1', \omega_2') = (U\omega_1){\dagger}U\omega_2 = \omega_1{\dagger}U{\dagger}U\omega_2 = (\omega_1, \omega_2).$$

It is easy to construct explicit matrix representations of the operators a, $a{\dagger}$.

If we define the matrix $(n|a|n')$ of the operator a in the coordinate system ω_n ($n = 0, 1, 2 \cdots$) by the relation

$$a\omega_n = \sum_{n'} (n'|a|n)\omega_{n'}, \tag{6-61}$$

we see from (6–57) that we can express the matrix elements in terms of the scalar product by

$$(n|a|n') = (-1)^n (\omega_n, a\omega_{n'}). \tag{6-62}$$

If we employ this formula, we find the explicit matrix representation

$$a = \begin{pmatrix} 0 & -1 & 0 & 0 & \cdot \\ 0 & 0 & -\sqrt{2} & 0 & \cdot \\ 0 & 0 & 0 & -\sqrt{3} & \cdot \\ \cdot & \cdot & \cdot & \cdot & \cdot \end{pmatrix}, \tag{6-63}$$

$$a{\dagger} = \begin{pmatrix} 0 & 0 & 0 & 0 & \cdot \\ 1 & 0 & 0 & 0 & \cdot \\ 0 & \sqrt{2} & 0 & 0 & \cdot \\ 0 & 0 & \sqrt{3} & 0 & \cdot \\ \cdot & \cdot & \cdot & \cdot & \cdot \end{pmatrix}. \tag{6-63}{\dagger}$$

It is seen that this representation differs from the usual one by the relative sign of the matrix elements and the fact that a is an emission operator.

The entire formalism of the quantization of this one-dimensional problem can now be developed in complete analogy with the corresponding one-dimensional oscillator problem. The only difference is in the indefinite character of the scalar product. This causes serious difficulty in the interpretation. For suppose we have an operator, representing an observable, with only positive eigenvalues. There exist then, in general, state vectors for which the expectation values of this operator are <0. In the usual interpretation of quantum mechanics this would mean that some of the eigenvalues of A can occur with negative probability. This makes no sense.

In spite of this difficulty, it is possible to carry through this quantization procedure for the case of the electromagnetic field. The reason is that we have to satisfy also the subsidiary condition, and this condition restricts the admissible state vectors in just such a manner that the norm of all the admitted state vectors becomes positive definite.

We shall first discuss this for the free radiation field. The field operators $a_\mu(x)$ are assumed to be self-adjoint, so that their expectation values are always real. We then define the Fourier amplitudes in complete analogy to (2–62):

$$a_\mu(x) = \frac{1}{(2\pi)^{3/2}} \int \frac{d^3k}{\sqrt{2\omega}} \, (a_\mu(\mathbf{k})e^{ikx} + a_\mu\dagger(\mathbf{k})e^{-ikx}). \tag{6-64}$$

The only difference is that now $a_\mu\dagger$ is the adjoint of a_μ and not the Hermitian conjugate. The commutation rules are exactly the same as before. In particular,

$$[a_\mu(\mathbf{k}), a_\nu\dagger(\mathbf{k}')] = g_{\mu\nu}\delta(\mathbf{k} - \mathbf{k}'). \tag{6-65}$$

In order to discuss now the effect of the subsidiary condition, we make the propagation vector $\mathbf{k}$ discrete, consider one particular $\mathbf{k}$, and suppress the argument. We also introduce the coordinate system in which a_1, a_2 are the transverse components, while a_3 is the longitudinal component.

The state vector ω_0 is defined by

$$a_\mu \omega_0 = 0, \quad (\mu = 0, 1, 2, 3). \tag{6-66}$$

The subsidiary condition (2–30) cannot be satisfied in this form. Instead, we can require a weaker condition, namely, using (2–7)

$$\chi^{(-)}(x)\omega = 0, \tag{6-67}$$

which can be satisfied. Although (6–67) is weaker than (2–30) it is still sufficient to ensure the classical Maxwell equations for the expectation values. For this we need only

$$\langle \chi(x) \rangle \equiv (\omega, \chi(x)\omega) = 0 \tag{6-68}$$

for ω which satisfy (6–67). This is indeed the case, since

$$\chi^{(-)}(x)\dagger = \chi^{(+)}(x)$$

and, therefore,

$$(\omega, \chi^{(+)}\omega) = (\chi^{(-)}\omega, \omega) = 0.$$

Condition (6–67) can now be satisfied. We shall write down the explicit solutions of (6–67) for one particular propagation vector $\mathbf{k}$.

Let us introduce $a = a_3$ and $b = a_0$ for the longitudinal and 0-component of the field operators and put $q = a + b$. We ignore the dependence of ω on the transverse variables. For the remaining part of the state vector the subsidiary condition is simply

$$(a + b)\omega \equiv q\omega = 0. \tag{6-69}$$

Since $[q, q\dagger] = 0$ the state vectors which satisfy (6–69) can be generated from ω_0 by defining

$$\omega_N = (q\dagger)^N \omega_0. \qquad (6\text{--}70)$$

We find, then, for all N,

$$q\omega_N = q(q\dagger)^N \omega_0 = [q, (q\dagger)^N]\omega_0 = 0. \qquad (6\text{--}71)$$

One verifies easily that with this method we obtain the complete system of vectors which satisfy condition (6–69). Thus the most general vector of this kind is a linear combination

$$\omega = \sum_N c_N \omega_N \qquad (6\text{--}72)$$

with arbitrary coefficients c_N.

One can verify that all state vectors of this form have a nonnegative norm because

$$(\omega_N, \omega_M) = 0, \quad \text{unless } N = M = 0, \qquad (6\text{--}73)$$

and therefore

$$(\omega, \omega) = |c_0|^2 \geq 0. \qquad (6\text{--}74)$$

In contrast to the theory discussed in Section 6–2, we now obtain a whole family of state vectors which satisfy the subsidiary condition (6–69). The physical meaning of this freedom in the definition of ω is brought to light if we study the canonical transformations which transform from one of these state vectors to another. These transformations are closely associated with the gauge transformations.* In the Fourier amplitudes a, b a gauge transformation is given by

$$\begin{aligned} a \to a' &= a + \lambda, \\ b \to b' &= b - \lambda. \end{aligned} \qquad (6\text{--}75)$$

The transverse components are not affected.

λ is usually assumed to be a c-number, but it may be an operator which is restricted only by the requirement that the transformation is canonical:

$$\begin{aligned} a' &= UaU^{-1}, \\ b' &= UbU^{-1}, \\ U\dagger U &= 1 = UU\dagger \end{aligned} \qquad (6\text{--}76)$$

It follows from conditions (6–75) and (6–76) that

$$[U, q] = [U, q\dagger] = 0. \qquad (6\text{--}77)$$

---

* K. Bleuler, *loc. cit.*, p. 575. Bleuler studies only those gauge transformations arising from $\omega_0 \to \omega_0 + c_1 \omega_1$. In order to obtain all the transformations of the state vectors ω, it is necessary to consider more general gauge transformations.

With the operator U is associated a canonical transformation of the state vector

$$\omega' = U\omega. \qquad (6\text{--}78)$$

It follows from (6–77) that ω' also satisfies the subsidiary condition

$$q\omega' = 0. \qquad (6\text{--}79)$$

Thus it must be of the form

$$\omega' = \sum_N c_N' \omega_N. \qquad (6\text{--}72')$$

In particular, for $\omega = \omega_0$,

$$U\omega_0 = \sum_N d_N \omega_N, \quad \text{with } d_0 = 1, \qquad (6\text{--}80)$$

and for $\omega = \omega_M$,

$$U\omega_M = U(q\dagger)^M \omega_0 = \sum_{N=0}^{\infty} d_N \omega_{N+M}.$$

Consequently, with (6–72),

$$\omega' = U\omega = \sum_{N,M=0}^{\infty} c_M d_N \omega_{N+M} = \sum_{N=0}^{\infty} \left(\sum_{K=0}^{N} c_K d_{N-K} \right) \omega_N. \qquad (6\text{--}81)$$

Comparison of this expression with (6–72') gives, because of the linear independence of the ω_N,

$$c_N' = \sum_{K=0}^{N} c_K d_{N-K}. \qquad (6\text{--}82)$$

If we now give the two state vectors ω, ω', then we can always find a set of coefficients d_N by solving the system of linear equations (6–82). The solution is unique even if $c_0 = 0$. On the other hand, to every set of such coefficients one can associate a canonical transformation which furnishes the gauge transformation (6–75).

The particular set of transformations for which $d_0 = 1$, $d_1 \neq 0$, $d_N = 0$ for $N \neq 0,1$ furnishes the special kind of transformations for which λ is a c-number. These are the transformations which were considered by Bleuler.

The definition of the vacuum is especially simple in this theory. It is described by the uniquely determined state vector for which

$$a_\mu^{(-)}(x)\omega_0 = 0, \quad \text{for all } x. \qquad (6\text{--}83)$$

It is not necessary to introduce the transverse field.

There is no difficulty in extending this theory to the case of the interacting fields. The only difference is that instead of the subsidiary condition (4–55), we now have to require

with
$$\begin{rcases} \gamma^{(-)}(x)\omega(\tau) = 0, \quad \text{for all } x, \tau, \\ \gamma(x) = \partial^\lambda a_\lambda(x) - \int_{\sigma(\tau')} j^\lambda(x')D(x-x')d\sigma_\lambda'. \end{rcases} \quad (4\text{--}55)\dagger$$

The longitudinal field can be eliminated by the same method used in Section 6–1, and the result is identically the same as in that section. We shall not carry out the details of the calculation again for this case.*

This method has many satisfactory features, notably the fact that the subsidiary condition can be satisfied without any limiting process and the vacuum can be defined without reference to the transverse field. These advantages are somewhat weakened by the occurrence of the singular state vectors with vanishing or negative norm for which expectation values become meaningless.

6–4 Gauge-independent interaction. A third method of avoiding the complication of the subsidiary condition is based on an observation of Novobátsky† according to which the subsidiary condition can be avoided if the interaction of the two fields involves only gauge-independent quantities.‡

We start with the Maxwell tensor $f_{\mu\nu}$ which satisfies (in the interaction picture) the free field equations

$$\partial^\mu f_{\mu\nu} = 0, \quad \partial_\lambda f_{\mu\nu} + \partial_\mu f_{\nu\lambda} + \partial_\nu f_{\lambda\mu} = 0. \quad (6\text{--}84)$$

With this field is associated a potential defined by the relation

$$b_\nu = \partial^{-1} n^\mu f_{\mu\nu}. \quad (6\text{--}85)$$

Here the operator $\partial \equiv n^\nu \partial_\nu$ and ∂^{-1} is defined in terms of the Fourier components of the field as the multiplication with $(ik^\nu n_\nu)^{-1}$. The field b_ν then satisfies the following relations:

$$n^\nu b_\nu = 0, \quad \partial^\nu b_\nu = 0, \quad (6\text{--}86)$$

as a consequence of (6–85) and (6–84). It is thus a transverse vector

---

* For details, see K. Bleuler, *loc. cit.*
† K. F. Novobátsky, *Z. Physik* **111**, 292 (1938).
‡ A similar treatment is found in F. J. Belinfante and J. S. Lomont, *Phys. Rev.* **84**, 541 (1951). A covariant formulation was given by G. Valatin, *Kgl. Danske Videnskab. Selskab* **26**, No. 13 (1951), which we follow.

field. It is quantized in accordance with the commutation rules

$[f_{\mu\nu}(x), f_{\lambda\kappa}(x')]$
$$= -i\{g_{\nu\lambda}\partial_\mu\partial_\kappa + g_{\mu\kappa}\partial_\nu\partial_\lambda - g_{\nu\kappa}\partial_\mu\partial_\lambda - g_{\mu\lambda}\partial_\nu\partial_\kappa\}D(x - x'), \quad (6\text{-}87)$$

$$[b_\mu(x), b_\nu(x')] = -id_{\mu\nu}'D(x - x'), \quad (6\text{-}88)$$

$$d_{\mu\nu}' = g_{\mu\nu} - \partial_\mu\partial_\nu\partial^{-2} - n_\mu\partial_\nu\partial^{-1} - n_\nu\partial_\mu\partial^{-1}. \quad (6\text{-}89)$$

We notice that although the field b_μ is not the same as the transverse field b_μ introduced in Chapter 2 [Eq. (2-130)], the commutation rules of the two fields are exactly the same. It follows that the equation of motion for the state vector $\omega(\tau)$,

$$i\dot\omega = H\omega,$$

leads to exactly the same results as the conventional theory provided we define H as the sum of the two terms

$$H = H_1 + H_C, \quad (6\text{-}90)$$

$$H_1 = -\int b_\mu(x) j^\mu(x) d\sigma, \quad (6\text{-}91)$$

and H_C is the Coulomb interaction term given in Eq. (6-16).

Instead of this term H_C, one can now introduce an auxiliary scalar field $q(x)$ which satisfies (in the interaction picture) the field equations

$$\partial^\lambda \partial_\lambda q(x) = 0 \quad (6\text{-}92)$$

and the commutation rules

$$[q(x), q(x')] = iD(x - x'). \quad (6\text{-}93)$$

The unconventional sign in (6-93) makes $q^{(-)}(x)$ an emission operator. The field $q(x)$ is assumed to be dynamically independent of $f_{\mu\nu}$ such that

$$[f_{\mu\nu}(x), q(x')] = 0, \quad \text{for all } x, x'. \quad (6\text{-}94)$$

We can construct with $q(x)$ a vector field $c_\mu(x)$:

$$c_\mu(x) = -\partial_\mu \partial^{-1} q(x) \quad (6\text{-}95)$$

which satisfies the relations

$$[c_\mu(x), c_\nu(x')] = -id_{\mu\nu}''D(x - x'), \quad (6\text{-}96)$$

$$d_{\mu\nu}'' = -\partial_\mu\partial_\nu\partial^{-2}. \quad (6\text{-}97)$$

The curl of this vector field vanishes:

$$\partial_\mu c_\nu - \partial_\nu c_\mu = 0 \quad (6\text{-}98)$$

because of (6–95). From the definition (6–95) follow also the relations

$$n^\nu c_\nu = -q, \tag{6-99}$$

$$\partial^\nu c_\nu = 0, \tag{6-100}$$

and

$$\partial^\lambda \partial_\lambda c_\mu = 0. \tag{6-101}$$

Introducing now the vector potential

$$a_\mu = b_\mu + c_\mu, \tag{6-102}$$

we obtain

$$\partial^\lambda \partial_\lambda a_\mu = 0, \tag{6-103}$$

$$\partial^\mu a_\mu = 0, \tag{6-104}$$

and

$$[a_\mu(x), a_\nu(x')] = -i d_{\mu\nu} D(x - x'), \tag{6-105}$$

with

$$d_{\mu\nu} = d_{\mu\nu}' + d_{\mu\nu}''.$$

The interaction operator is now given by

$$H(\tau) = -\int_{\sigma(\tau)} a_\mu(x) j^\mu(x) d\sigma \tag{6-106}$$

and the Schroedinger equation in the interaction picture is

$$i\dot{\omega} = H\omega.$$

The vacuum state ω_0 is defined as the state with no transverse and scalar photons, i.e., ω_0 satisfies

$$f_{\mu\nu}{}^{(-)}\omega_0 = 0, \quad q^{(+)}\omega_0 = 0. \tag{6-107}$$

It is now possible to show that the effect of the term

$$H_2 = -\int c_\mu(x) j^\mu(x) d\sigma, \tag{6-108}$$

which contains the scalar photons, is exactly the same as that of the Coulomb term. More precisely, if we define the transformed state vector

$$\eta(\tau) = e^{-i\Sigma(\tau)} \omega(\tau) \tag{6-109}$$

with

$$\Sigma(\tau) = \int d\sigma n_\mu j^\mu(x) \partial^{-1} q(x), \tag{6-110}$$

then the new state vector $\eta(\tau)$ satisfies the Schroedinger equation:*

$$i\dot{\eta} = (H_1 + H_C)\eta \tag{6-111}$$

---

* The details of the calculation are similar to the ones given in Section 6–1.

with H_C given by (6–16). The dynamical properties of the system are thus the same as those of conventional quantum electrodynamics, with the state vector restricted by the subsidiary condition. However, no such condition is needed in the present formalism.

6–5 Radiation fields with finite mass. The complication in connection with the subsidiary condition does not occur in a vector field theory which has a finite rest mass (neutral vector meson theory). The reason is that in such a theory the subsidiary condition can be postulated as a field equation. This means that it is possible to choose the Lagrangian in such a way that the field equations for the vector field $\varphi_\mu(x)$ are*

$$\partial^\lambda \partial_\lambda \varphi_\mu - \kappa^2 \varphi_\mu = 0, \quad \partial^\mu \varphi_\mu = 0. \tag{6-112}$$

Since in this theory there is no difficulty with the subsidiary condition, one may ask whether it might not be possible to consider the radiation field as a limiting case of a neutral vector meson field with vanishing rest mass. It is, of course, true that there is no experimental evidence of a finite value of the rest mass of the photon, but it must be remembered that these experiments allow the conclusion of vanishing rest mass only within a certain latitude.

The most important experimental results which have a bearing on the value of the rest mass are the constancy of the propagation velocity of light waves, the Coulomb law for the interaction of two charges, and the blackbody radiation law. De Broglie† has analyzed the significance of the experimental evidence in these cases and finds that a photon mass of 10^{-65} gram or less could certainly not be distinguished by any known experiment from the mass value zero.

The transition to the limiting case of vanishing rest mass is, however, mathematically not quite trivial, as one may see if one writes down the commutation rules for the field operators. For a vector meson field with finite mass κ, they are

$$[\varphi_\mu(x), \varphi_\nu(x')] = -i \left(g_{\mu\nu} - \frac{1}{\kappa^2} \partial_\mu \partial_\nu \right) \Delta(x - x'). \tag{6-113}$$

The second term in parentheses on the right side arises from the fact that $\partial^\mu \varphi_\mu$ must commute with all the field variables, since it is the zero operator.

---

* For details on the vector field with finite rest mass the reader is referred to G. Wentzel, *Quantum Theory of Fields*, Interscience Publishers, New York, 1949, especially Chapter III.

† See, for instance, L. de Broglie, *Mécanique Ondulatoire du Photon et Théorie Quantique des Champs*, Paris, Gauthier-Villars, 1949, especially Chapter V.

These commutation rules become singular in the limit of vanishing rest mass κ.

Another difficulty arises when the interaction with the negaton-positon current is introduced. It can be shown that a consistent theory leads to the occurrence of certain contact terms in the interaction operator $H(\tau)$ which themselves become singular if $\kappa \to 0$.* However, all these difficulties can be overcome, since it can be shown† that the terms which would diverge in the limit $\kappa \to 0$ all cancel and do not contribute to the observable effects, so that this limit exists and is even uniform.

The neutral vector theory can be discussed most conveniently when it is cast into a form due to Stueckelberg‡ and Coester§. In Coester's method one starts with the free fields $a_\mu(x)$ which satisfy the following field equations and commutation rules:

$$\partial^\lambda \partial_\lambda a_\mu(x) - \kappa^2 a_\mu(x) = 0, \tag{6-114}$$

$$[a_\mu(x), a_\nu(x')] = -ig_{\mu\nu}\Delta(x - x'), \tag{6-115}$$

where the Δ-function corresponds to the finite mass value κ and therefore satisfies

$$(\partial^\lambda \partial_\lambda - \kappa^2)\Delta = 0. \tag{6-116}$$

To ensure the validity of Maxwell's equations in the limit of $\kappa = 0$, we impose on the state vector $\omega(\tau)$ the condition

$$\langle \gamma(x) \rangle = 0$$

with

$$\gamma(x) = \partial^\lambda a_\lambda(x) - \int_{\sigma(\tau)} j^\lambda(x')\Delta(x - x')d\sigma_\lambda'. \tag{6-117}$$

Because of the finite value of κ, the condition (6-117) is weaker than the corresponding condition (4-55). One can show this by carrying out first a τ-dependent canonical transformation which transforms the state vectors according to

$$\eta(\tau) = e^{i\Sigma(\tau)}\omega(\tau). \tag{6-118}$$

---

* See F. J. Belinfante, *Phys. Rev.* **76**, 66 (1949); *Progr. Theor. Phys.* **4**, 2 (1949) (in Esperanto). A simple procedure to exhibit these contact terms will be shown below.
† P. T. Matthews, *Phys. Rev.* **76**, 1657 (1949).
‡ E. C. G. Stueckelberg, *Helv. Phys. Acta* **11**, 225 (1938).
§ F. Coester, *Phys. Rev.* **83**, 798 (1951).

The operator $\gamma(x)$ then transforms into

$$\gamma'(x) = e^{i\Sigma(\tau)}\gamma(x)e^{-i\Sigma(\tau)}$$
$$= \gamma(x) + i[\Sigma(\tau),\gamma(x)] + \cdots. \quad (6\text{-}119)$$

If we choose

$$\Sigma(\tau) = \frac{1}{\kappa^2}\int_{\sigma(\tau)}\partial'^\lambda a_\lambda(x')j^\mu(x')d\sigma_\mu', \quad (6\text{-}120)$$

then the remaining commutators in (6–119) all vanish and we obtain

$$i[\Sigma(\tau),\gamma(x)] = \partial^\lambda a_\lambda(x) \equiv g(x). \quad (6\text{-}121)$$

The transformed subsidiary condition is then

$$\langle g(x)\rangle = 0. \quad (6\text{-}122)$$

Because of the commutation rules (6–115), the $g(x)$ satisfy

$$[g(x),g(x')] = i\kappa^2\Delta(x - x'). \quad (6\text{-}123)$$

The condition (6–122) can now be satisfied by requiring

$$g^{(+)}(x)\eta(\tau) = 0. \quad (6\text{-}124)$$

In the case of vanishing mass, the condition (6–124) implies $g^{(-)}(x)\eta(\tau) = 0$, and therefore $g(x)\eta(\tau) = 0$. In the present case ($\kappa \neq 0$), this is no longer true, and relation (6–124) simply means the absence of the field quanta associated with the field $g(x)$ (scalar photons). For the validity of Maxwell's equations for vanishing mass it is sufficient to satisfy the weaker condition (6–124), which does not cause any difficulty with the non-normalizable state vector.

The new state vectors $\eta(\tau)$ satisfy a modified Schroedinger equation

$$i\dot\eta(\tau) = G(\tau)\eta(\tau), \quad (6\text{-}125)$$

with

$$G(\tau) = e^{i\Sigma(\tau)}H(\tau)e^{-i\Sigma(\tau)} - \dot\Sigma - \frac{i}{2}[\Sigma,\dot\Sigma] \quad (6\text{-}126)$$

and

$$H(\tau) = -\int j^\mu(x)a_\mu(x)d\sigma. \quad (6\text{-}127)$$

The evaluation of the commutators in (6–126) is straightforward and gives

$$G(\tau) = -\int j^\mu(x)a_\mu(x)d\sigma - \frac{1}{2\kappa^2}\int n^\mu j_\mu(x)n^\nu j_\nu(x)d\sigma. \quad (6\text{-}128)$$

Consider now a new field

$$\varphi_\mu(x) = a_\mu(x) - \frac{1}{\kappa^2}\partial_\mu g(x). \quad (6\text{-}129)$$

The field $\varphi_\mu(x)$ satisfies the equations (6–112) and (6–113) and represents therefore a neutral vector meson field in the conventional quantization. The second term in (6–128) is the contact term which was previously mentioned.

The theory is thus equivalent to the customary theory of a neutral vector meson. The forms (6–114) and (6–115) have the advantage that the limiting process $\kappa \to 0$ is easily carried out. Coester shows that all observable quantities (namely, expectation values and transition probabilities) go uniformly into the values for $\kappa = 0$.*

A disadvantage of the method is the fact that the theory has no physical meaning until the limit $\kappa \to 0$ is taken. Also, there exist no gauge transformations so long as $\kappa \neq 0$.

In using this method, great care must be taken to include in the calculations the effect of the longitudinal neutral vector mesons, since otherwise one would arrive at inconsistencies. For this purpose it is important to note that the emission probability of a longitudinal meson of momentum **k** is

$$\frac{\kappa^2}{\mathbf{k}^2 + \kappa^2} \qquad (6\text{–}130)$$

times the emission probability of a transverse meson (F. Coester, *loc. cit.*). Otherwise there is no difficulty in the use of this procedure. Its formal convenience far outweighs the above objections, and its usefulness will be amply demonstrated, especially in Sections 15–2 and 15–4.

---

* See F. Coester, *loc. cit.*, esp. Sections III and IV.

CHAPTER 7

THE S-MATRIX

The study of the S-matrix forms the central part of this book. It is the major goal to which the previously discussed theoretical developments lead, and it is the starting point for most applications.

In this chapter we give the general formulation of S-matrix theory and introduce the basic notions and definitions. The following chapters are dedicated to a more detailed study of the difficulties which arise from one particular explicit solution of the S-matrix problem which recently proved most successful, i.e., the *iteration solution*. We want to emphasize the very special conditions under which this solution is justifiable. Although no explicit general solution is known, it is the task of the first few chapters to clearly expose the difficulties so that the iteration solution which we are forced to use in the later chapters can be put in its proper place.

The following five sections are concerned with the definition and properties of the S-matrix. In Section 7–1 a preliminary definition is given in terms of the transformation matrix which, in the interaction picture, transforms the remote past into the remote future. The introduction of Møller's *wave matrix* in Section 7–2 leads to a formulation in terms of the *wave operator* which is independent of the pictures (Section 7–3). The wave operator is then represented explicitly by an integral in terms of the operators H_0 and H (Section 7–4). In Section 7–5 the S-operator is defined rigorously in terms of the wave operators. The resulting S-matrix is shown to be equivalent to other definitions of this quantity. In particular, the equivalence of the time-dependent and stationary definitions is discussed and the range of validity of the preliminary definition is exhibited. The general invariance properties of the S-matrix are dealt with in Section 7–6.

7–1 Preliminary definition of the S-matrix. In Section 4–3 we introduced the unitary operator $V(\tau)$ which establishes the connection between the Heisenberg fields $\Phi(x)$ and the free fields $\varphi(x)$:

$$\varphi(x) = V(\tau)\Phi(x)V^{-1}(\tau). \tag{7-1}$$

The operator $V(\tau)$ was then shown to satisfy the Schroedinger equation

$$i\dot{V}(\tau) = H(\tau)V(\tau), \tag{7-2}$$

where the interaction operator is given in terms of the free fields $\varphi(x)$:

$$H(\tau) = ie\int_{\sigma(\tau)} d\sigma \bar{\psi}(x)\gamma^\lambda \psi(x)a_\lambda(x). \tag{7-3}$$

Actually, the solution of (7–2) is determined only if we specify the initial conditions for the operator $V(\tau)$. We can define the solution $V(\tau,\tau_0)$ with the condition that

$$V(\tau_0,\tau_0) = 1 \tag{7-4}$$

is the unit operator. $V(\tau,\tau_0)$ is then the operator which transforms the state vectors $\omega_0 = \omega(\tau_0)$ defined on $\sigma_0 = \sigma(\tau_0)$ into the state vectors $\omega(\tau)$. The free fields $\varphi(x,\tau_0)$ associated with this operator are then those fields which coincide with the Heisenberg fields for $x \in \sigma_0$:

$$\varphi(x,\tau_0) = V(\tau,\tau_0)\Phi(x)V^{-1}(\tau,\tau_0), \tag{7-5}$$

that is, they are the same fields introduced in (4–21) in connection with the commutation rules of the Heisenberg fields.

The interaction operator $H(\tau,\tau_0)$ is then the operator (7–3) expressed in terms of the fields $\varphi(x,\tau_0)$:

$$H(\tau,\tau_0) = ie\int_{\sigma(\tau)} d\sigma \bar{\psi}(x,\tau_0)\gamma^\lambda \psi(x,\tau_0)a_\lambda(x,\tau_0), \tag{7-6}$$

and the operator $V(\tau,\tau_0)$ satisfies

$$i\frac{dV(\tau,\tau_0)}{d\tau} = H(\tau,\tau_0)V(\tau,\tau_0). \tag{7-7}$$

The solution of Eq. (7–7) which satisfies the initial condition (7–4) can also be expressed as the solution of an integral equation. Indeed, one verifies easily that an operator $V(\tau,\tau_0)$ with the property

$$V(\tau,\tau_0) = 1 - i\int_{\tau_0}^{\tau} d\tau' H(\tau',\tau_0)V(\tau'.\tau_0) \tag{7-8}$$

also satisfies Eqs. (7–7) and (7–4).

There is a simple connection between the operators $U(\sigma,\sigma_0) \equiv U(\tau,\tau_0)$ defined by (4–32) and $V(\tau,\tau_0)$ defined by (7–5). From the definition of U,

$$\varphi(x,\tau) = U(\tau,\tau_0)\varphi(x,\tau_0)U^{-1}(\tau,\tau_0), \tag{7-9}$$

it follows by comparison with (7–5) that, independently of τ_0,

$$U(\tau,\tau_0)V(\tau_1,\tau_0) = V(\tau_1,\tau). \tag{7-10}$$

By specializing in this equation $\tau = \tau_1$ we obtain from it

$$U(\tau_1,\tau_0)V(\tau_1,\tau_0) = 1,$$

or with (4–34)

$$V(\tau_1,\tau_0) = U(\tau_0,\tau_1). \tag{7-11}$$

Since therefore

$$V(\tau,\tau_0) = V^{-1}(\tau_0,\tau),$$

and consequently

$$\frac{dV(\tau,\tau_0)}{d\tau_0} = -V^{-1}(\tau_0,\tau)\frac{dV(\tau_0,\tau)}{d\tau_0}V^{-1}(\tau_0,\tau),$$

we obtain for $V(\tau,\tau_0)$ a Schroedinger equation with respect to the initial instant τ_0:

$$i\frac{dV(\tau,\tau_0)}{d\tau_0} = -V(\tau,\tau_0)H(\tau_0,\tau). \quad (7\text{--}12)$$

We mention also that the operator family $U(\tau,\tau_0)$ or $V(\tau_0,\tau)$ is intimately connected with the general transformation matrix introduced in Section 1–8. The matrix element $(\zeta_0'|\zeta')$ of the transformation matrix is, according to (1–59), given by

$$(\zeta_0'|\zeta') = (\omega_0(\zeta_0'),\omega(\zeta')) = (\omega_0(\zeta_0'),V(\tau,\tau_0)\omega_0(\zeta'))$$

or

$$(\zeta_0'|\zeta') = (\omega_0(\zeta_0'),U(\tau_0,\tau)\omega_0(\zeta')). \quad (7\text{--}13)$$

A knowledge of the operator family $U(\tau,\tau_0)$ thus implies a complete solution of all the problems of the interacting radiation and matter field. In most physical applications of the theory such a complete knowledge of the development in time is not necessary. Indeed, the majority of experiments involve the measurements of cross sections for radiative processes or the transition probabilities in spontaneous transitions. Such information can be obtained by the observations of the state of a system at two different instants τ_0,τ separated by a finite interval of time which for all practical purposes can be taken as infinitely large. The theory contains this information in the transformation matrix which corresponds to the values $\tau_0 = -\infty$, $\tau = +\infty$ of the instant parameter.

In this manner, we are led to consider the operator*

$$S = \lim_{\substack{\tau \to +\infty \\ \tau_0 \to -\infty}} V(\tau,\tau_0) = \lim_{\substack{\tau \to +\infty \\ \tau_0 \to -\infty}} U(\tau_0,\tau). \quad (7\text{--}14)$$

This operator S furnishes the probability $P_{\zeta_0' \to \zeta'}$ for the transition $\zeta_0' \to \zeta'$, according to the formula

$$P_{\zeta_0' \to \zeta'} = |(\zeta'|S|\zeta_0')|^2. \quad (7\text{--}15)$$

In connection with this last expression we have used the language corresponding to discrete states, which must be appropriately modified if the variables ζ' label a continuum of states. The matrix $(\zeta'|S|\zeta_0')$ is called the S-matrix or scattering matrix. As a matrix it always refers to a special coordinate system in the Hilbert space. It is often more convenient to refer to the general S-operator (7–14).

---

* E. C. G. Stueckelberg, *Helv. Phys. Acta* **17**, 3 (1943).

The definition (7–14) of the operator S has been the basis of nearly all calculations in field theory. This definition is only suitable, however, if the implied double limit exists. If this limit does not exist, then this definition of the S-operator is inadequate.

It is possible to give a more rigorous definition of the S-operator which does not depend on the existence of the limit implied in Eq. (7–14). If the limit exists, this definition is actually equivalent to Eq. (7–14). But it also holds in more general cases where Eq. (7–14) fails to give sensible results.

We note that, according to Eq. (7–14), the operator S is unitary. This is clear, provided the limit exists, since from

$$U^*(\tau,\tau_0)U(\tau,\tau_0) = U(\tau,\tau_0)U^*(\tau,\tau_0) = 1$$

follows

$$S^*S = SS^* = 1 \tag{7-16}$$

in the limit $\tau \to +\infty$, $\tau_0 \to -\infty$. On the other hand, if this limit does not exist, the proof of this property of S must be based on the more general definition.

7–2 The wave matrix. The S-matrix was originally introduced by Wheeler[†] in connection with the problem of nuclear reactions. Subsequently, Heisenberg[‡] in a series of papers emphasized the significance of the S-matrix in a possible future theory in which the transformation matrix $U(\tau,\tau_0)$ for finite values of τ,τ_0 is not assumed to exist. In these papers the S-matrix is defined in terms of the stationary states solution of the Schroedinger equation. We shall proceed in the same manner.

We assume that there exists an operator in the form $H_0 + H$ which represents the total energy of the system and which is independent of the instant parameter τ. The operator H_0 represents the energy of the constituent parts, while H stands for the interaction energy. This assumption is, of course, quite far-reaching, but it is consistent with the form of quantum electrodynamics so far developed. The separation of the total energy $H_0 + H$ into two parts is to some extent arbitrary, depending on what we wish to define as the "free" particles of the system.

The particular τ-independent operator introduced here is the energy operator in the *Schroedinger picture*. This picture is commonly used in ordinary wave mechanics for the discusson of stationary states. In a relativistic field theory it is not so important, partly because it is not

---

[†] J. A. Wheeler, *Phys. Rev.* **52**, 1107 (1937).

[‡] W. Heisenberg, *Z. Physik* **120**, 513, 673 (1943); *Z. Naturf.* **1**, 608 (1946). Also C. Møller, *Kgl. Danske Videnskab. Selskab* **23**, No. 1 (1945) (part I); *ibid.* **22**, No. 19 (1946) (part II).

adapted to an invariant formalism, and partly because a systematic approximation technique for the S-operator is simpler in the interaction or Heisenberg picture.

We consider the state vectors Ω which are solutions of the eigenvalue problem

$$(H_0 + H)\Omega = E\Omega. \tag{7-17}$$

Here E is the total energy of the system. The set of eigenvalues E of the operator $H_0 + H$ are restricted by the condition of normalizability of the eigenvectors Ω (suitably generalized in case E belongs to the continuum) and we must assume that the totality of such eigenvectors form a complete system.*

In the following discussion it is important to know something about the eigenvalue spectrum of the two operators H_0 and $H_0 + H$. We shall make certain assumptions regarding this spectrum, which are satisfied for an important class of scattering problems. We shall assume that

(1) every eigenvalue of H_0 belongs to the continuum,
(2) the continuum part of the spectrum $H_0 + H$ coincides with the spectrum of H_0,
(3) the discrete eigenvalues E_α of $H_0 + H$ are all smaller than the continuum values.

These three conditions are actually satisfied for many scattering problems in wave mechanics. According to condition (2) there exists an eigenfunction ω of H_0 for every continuum eigenvalue E of $H_0 + H$:

$$H_0 \omega = E\omega$$

and vice versa. Condition (3) avoids the complication which results from the mixture of discrete and continuous eigenvalues. It states that if E is any eigenvalue of $H_0 + H$ which belongs to the continuum and E_α is one of the discrete eigenvalues, then

$$E_\alpha < E. \tag{7-18}$$

The properties of the operators H_0 and $H_0 + H$ depend to some extent on the way in which the splitting of $H_0 + H$ into the two operators H_0 and H is carried out. If the free fields are defined as in Chapters 2 and 3, then condition (1) is valid in quantum electrodynamics.

---

* The property here formulated is the physicist's manner of stating that the operator $H_0 + H$ is hypermaximal. The precise mathematical formulation of this property is found in v. Neumann, *Mathematische Grundlagen der Quantenmechanik*, Dover Publications, 1943, esp. Chapter II. We shall continue to use the heuristic language of the physicist. No difficulties arise if its limitations are clearly kept in mind. Whether the operator $H_0 + H$ in quantum electrodynamics has actually the required property is an open question.

Condition (2), on the other hand, is not satisfied. The interaction produces a shift of the continuum levels. It is then necessary to modify the definition of the S-operator.* We shall not discuss this modification at this time, since it introduces additional complications which we believe are peripheral to the problem under discussion.†

The third condition is also not satisfied. Bound states can exist, for instance, in an external field, or if we have two kinds of particles with opposite charge. The energy levels of these bound states may very well fall into the continuum region. This is the case, for instance, for the energy levels of the positronium atom (cf. Section 12–6) although these are not stable states. They always decay into two or more photons by annihilation. However, the assumption of condition (3) is not really a serious limitation. We have adopted it here merely for the sake of having a definite and sufficiently simple situation. The theory of the S-operator can be formulated more generally. The necessary modifications in the presence of external fields will be further discussed in Chapter 14. Our discussion of the general properties of the S-operator is thus necessarily incomplete. Nevertheless, certain properties can be obtained under these conditions which are of more general validity, and which give some insight into the general situation.‡

The energy values E of H_0 are, in general, highly degenerate, and in order to specify the state vectors completely it is necessary to introduce additional variables. In general reaction processes these variables will be the various momentum and spin (or polarization) values of the participating particles. We shall introduce a single symbol q for all of these variables. They can be visualized as the eigenvalues of a complete set of commuting observables Q, all of which commute with H_0. In general, the q are a point in a multidimensional continuum together with a set of discrete numbers for the spin and polarization states. We write therefore

$$H_0 \omega_q = E(q) \omega_q. \tag{7-19}$$

We shall always work with state vectors ω_q which are normalized according to

$$(\omega_q, \omega_{q'}) = \delta(qq'). \tag{7-20}$$

---

* This point is discussed by J. Pirenne, *Phys. Rev.* **86**, 395 (1952).

† This problem is discussed in full detail in Chapters 9 and 10.

‡ The assumptions which we have explicitly made here are essentially implicit in nearly all discussions on the S-matrix (see, for instance, C. Møller, *loc. cit.*). A rigorous theory of the S-matrix which is applicable without modifications to the quantum theory of fields has never been given. See also H. E. Moses, *Nuovo Cimento* **1**, 103 (1955).

The symbol $\delta(qq')$ is a generalized δ-function with the property

$$f(q) = \int \delta(qq')f(q')dq'. \tag{7-21}$$

The integration in Eq. (7-21) must be interpreted as a generalized summation process involving integration over the continuous range of values and summation over the discrete indices which make up the symbol q.

We need a similar characterization of the solutions Ω of Eq. (7-17). The operators Q in general do not commute with the total energy operator $H_0 + H$. We can nevertheless label the eigenvectors of $H_0 + H$ by the eigenvalues q of Q if we require that these eigenvectors be of the form

$$\Omega_q = \omega_q + \chi_q. \tag{7-22}$$

This form of the state vector is the one which corresponds to the physical situations which we intend to describe with the Ω_q. The state vector is decomposed into an *incident* wave, ω_q, and a *scattered* wave, χ_q.

In ordinary wave mechanics the scattered wave is assumed to be an outgoing wave. The square of the amplitude of this wave for large separation of the particles involved determines the scattering cross section. It is well known that there also exist in this case solutions with an ingoing scattered wave, which are usually discarded as not suitable for the description of the physical situation.*

In the general theory of radiative reactions the simple characterization of the scattered wave as an outgoing spherical wave in a three-dimensional configuration space is not sufficient. We wish to keep the formalism sufficiently general so that the state vector Ω_q describes not only the ordinary wave-mechanical scattering processes, but also general reaction processes involving the creation or destruction of an unspecified number of particles.

We shall see, furthermore, that it is of **great** advantage not to consider separately those solutions which contain an outgoing scattered wave. We obtain a much more symmetrical treatment if we include on an equal footing the solutions which contain ingoing or standing scattered waves.

Loosely speaking, the form (7-22) of the state vector means that the χ_q which describes the scattering part does not contain any eigenfunction of the operator H_0. We can make this statement more precise by writing the state vector Ω_q in the complete system furnished by the ω_q. With each Ω_q we associate, then, a matrix $\Omega(q'q)$:

$$\Omega_q = \int \Omega(q'q)\omega_{q'}dq'. \tag{7-23}$$

---

* See N. F. Mott and H. S. W. Massey, *The Theory of Atomic Collisions*, Clarendon Press, Oxford, 2nd edition, 1949, especially Chapter II.

The transformation coefficients

$$\Omega(q'q) = (\omega_{q'}, \Omega_q) \tag{7-24}$$

are the elements of the *wave matrix* in the q-representation. We write for this matrix

$$\Omega(q'q) = \delta(q'q) + T(q'q), \tag{7-25}$$

where

$$T(q'q) = (\omega_{q'}, \chi_q). \tag{7-26}$$

We may now characterize χ_q as a scattered wave if $T(q'q)$ does not contain a singularity of the δ-type with respect to *all* the variables q. This may be formulated more precisely by stating that for any set of functions $F_n(q) > 0$ for which

$$\lim_{n \to \infty} \int F_n(q)dq = 0, \tag{7-27}$$

the integrals

$$\phi_n(q') \equiv \int \Omega(q'q)F_n(q)dq$$

converge to

$$\lim_{n \to \infty} \phi_n(q') = \lim_{n \to \infty} F_n(q'). \tag{7-28}$$

It is easily seen that this definition of the "scattered wave" is equivalent to the usual definition in ordinary wave mechnics. But it is more general, since it does not refer to a particular number of particles.

The state vectors Ω_q of the form (7–22) subject to the condition (7–27), (7–28), represent the *scattering states*. It is important to realize that these conditions are not sufficient to determine the scattering states completely. In order to see this explicitly, we write this equation in the representation (7–23). To this end we multiply the equation

$$(H_0 + H)\Omega_q = E\Omega_q$$

from the left with ω_q, use Eqs. (7–23) and (7–20), and obtain

$$(E - E')\Omega(q'q) = \int H(q'q'')\Omega(q''q)dq'', \tag{7-29}$$

where

$$H(q'q) = (\omega_{q'}, H\omega_q).$$

If we use Eq. (7–25), this becomes

$$(E-E')T(q'q) = G(q'q). \tag{7-30}$$

We have introduced the abbreviation

$$G(q'q) = H(q'q) + \int H(q'q'')T(q''q)dq''. \tag{7-31}$$

In order to obtain an integral equation for $G(q'q)$, we solve (7–30) for $T(q'q)$ and substitute in (7–31). From Eq. (7–30) we find the general solution*

$$T(q'q) = P\frac{G(q'q)}{E - E'} + F(q'q)\delta(E - E'). \tag{7-32}$$

The symbol P indicates that on integration over a region E which contains the point E' the Cauchy principal value is to be understood. The function $F(q'q)$ is so far arbitrary but can be uniquely determined by the requirement that asymptotically $T(q'q)$ corresponds to outgoing or ingoing waves only. Let us denote these two cases by $T_+(q'q)$ and $T_-(q'q)$, respectively. In a well-known manner one finds†

$$F_\pm(q'q) = \mp i\pi G_\pm(q'q) \tag{7-33}$$

and therefore

$$T_\pm(q'q) = \left(P\frac{1}{E - E'} \mp i\pi\delta(E - E')\right)G_\pm(q'q)$$

$$= \lim_{\epsilon \to +0} \frac{G_\pm(q'q)}{E - E' \pm i\epsilon}. \tag{7-34}_\pm$$

The integral equation for $G_\pm(q',q)$ follows from (7–31) and (7–34),

$$G_\pm(q'q) = H(q'q) + \lim_{\epsilon \to +0} \int \frac{H(q'q'')G_\pm(q''q)}{E - E'' \pm i\epsilon} dq''. \tag{7-35}_\pm$$

With the two forms of the matrix $T(q'q)$ we can now define two wave matrices

$$\Omega_\pm(q'q) = \delta(q'q) + T_\pm(q'q). \tag{7-36}_\pm$$

---

* $G(q'q)$ cannot vanish for all values of q,q' for which $E = E'$. If this were so, $G(q'q)$ could be written as

$$G(q'q) = (E - E')^\nu g(q'q)$$

with $g(q'q)$ regular for $E = E'$. From (7–29) we then obtain

$$\Omega(q'q) = (E - E')^{\nu-1} g(q'q),$$

and consequently

$$g(q'q) = \frac{1}{(E - E')^\nu} \int (E - E'')^{\nu-1} H(q'q'')g(q''q) dq''.$$

This shows that $g(q'q)$ cannot be regular at $E = E'$, except for $\nu = 0$, as is implied in the text.

† See P. A. M. Dirac, *loc. cit.* § 50. The generalization to reaction processes with an unlimited number of particles was given by W. Heisenberg, *Z. Physik* **120**, 513 (1943). An alternative proof, based on the interpretation of the wave operator in the interaction picture, will be given in Section 7–3.

A third solution of Eq. (7–30) is obtained if we put $F(q'q) = 0$ in Eq. (7–32). In place of Eq. (7–34), we then obtain

$$T_0(q'q) = P\frac{1}{E - E'}G_0(q'q). \qquad (7\text{–}34)_0$$

The $G_0(q'q)$ now satisfies the integral equation

$$G_0(q'q) = H(q'q) + P\int \frac{H(q'q'')G_0(q''q)}{E - E''}\,dq''. \qquad (7\text{–}35)_0$$

In this last equation the integration over the pole $E'' = E$ is to be understood as the Cauchy principal value.

7–3 The wave operator. In the development so far, which is more or less the traditional approach, we have used a special reference system Ω_q and have written matrix equations for what are really operator relations. These relations are clearly independent of any special choice of the base vector system and it is therefore desirable to develop the formalism with general operators.

Thus, we would consider the matrix $\Omega(q'q)$ as the matrix element in the coordinate system ω_q of an *operator* Ω, defined by

$$\Omega(q'q) = (\omega_{q'}, \Omega\omega_q). \qquad (7\text{–}37)$$

The operator Ω therefore has the property that when it operates on ω_q it changes it into Ω_q:

$$\Omega_q = \Omega\omega_q. \qquad (7\text{–}38)$$

We shall call Ω the *wave operator*.

The Schroedinger equation

$$(H_0 + H)\Omega_q = E(q)\Omega_q, \qquad (7\text{–}39)$$

which is satisfied by all Ω_q, can be transformed into an operator relation for Ω. Indeed, if we substitute (7–38) into (7–39) and multiply from the left with $\omega_{q'}^*$, we obtain the general matrix element of the operator relation

$$(H_0 + H)\Omega = \Omega H_0. \qquad (7\text{–}40)$$

We need next a convenient way of characterizing the solutions $\Omega_\pm$ which correspond to outgoing and ingoing scattered waves. To this end we introduce a τ-dependent operator family $\Omega(\tau)$ defined by

$$\Omega(\tau) = e^{iH_0\tau}\Omega e^{-iH_0\tau}. \qquad (7\text{–}41)$$

According to Eq. (7–40), the $\Omega(\tau)$ satisfy the differential equation

$$i\dot{\Omega}(\tau) = H(\tau)\Omega(\tau) \qquad (7\text{–}42)$$

with
$$H(\tau) = e^{iH_0\tau} H e^{-iH_0\tau}. \qquad (7\text{-}43)$$

The transformation (7-41) defines the operator Ω in the interaction picture. A comparison of (7-42) with (7-2) shows that the operators $\Omega(\tau)$ satisfy the same differential equation as the transformation matrices $V(\tau)$.

We shall now show that the particular wave operators $\Omega_\pm$ are those which satisfy the following boundary conditions at infinity:

$$\Omega_+(-\infty) = 1, \quad \Omega_-(+\infty) = 1. \qquad (7\text{-}44)_\pm$$

Before proving these limiting properties rigorously, it may be worth while to convince oneself that these properties correspond exactly to the physical interpretation of the wave operators.

In the interaction picture the wave operator $\Omega_+(\tau)$ is that particular solution of Eq. (7-42) which for $\tau \to -\infty$ becomes the incident wave. The free particle state at $\tau = -\infty$ changes because of the interaction into superpositions of other free particle states which at $\tau = 0$ (or any other finite τ-value) have the form of the original incident wave together with an outgoing scattered wave. If the τ-direction is reversed, we obtain a similar interpretation for the solution $\Omega_-(\tau)$.

The equations $(7\text{-}44)_\pm$ are equivalent to

$$T_+(-\infty) = 0, \quad T_-(+\infty) = 0. \qquad (7\text{-}45)_\pm$$

We shall now prove Eq. $(7\text{-}45)_+$. From the definition of the T_+-solution given in $(7\text{-}34)_+$ we see that $(7\text{-}45)_+$ is equivalent to the statement

$$0 = \lim_{\tau \to -\infty} e^{i(E'-E)\tau} \lim_{\epsilon \to +0} \frac{1}{E - E' + i\epsilon}.$$

This relation is proved in the mathematical appendix, Eq. (A6-1). In a similar manner, using Eq. (A6-2), one finds Eq. $(7\text{-}45)_-$.

The boundary conditions for $\Omega_0(\tau)$ can be established by using the limiting properties

$$\lim_{\tau \to \mp\infty} e^{i(E'-E)\tau} P \frac{1}{E - E'} = \pm \pi i \delta(E' - E), \qquad (7\text{-}46)$$

which are a simple consequence of Eqs. (A6-1) and (A6-2). In order to express these limiting properties in operator form, we define the operator K with the matrix elements $K(q',q)$ in the q-representation,

$$K(q'q) = 2\pi \delta(E'-E) G_0(q'q). \qquad (7\text{-}47)$$

It follows from (7-46), (7-34), and the definition (7-41) of $\Omega(\tau)$ that

$$\Omega_0(-\infty) = 1 + \frac{i}{2} K, \quad \Omega_0(+\infty) = 1 - \frac{i}{2} K. \qquad (7\text{-}44)_0$$

Since the state vectors $\Omega_{\pm q}$ and Ω_{0q} are eigenstates of the Hermitian operator $H_0 + H$ with eigenvalues $E(q)$, we know that two such vectors which belong to different eigenvalues E and E' must be orthogonal. When this is expressed in operator form, it means that for any of the three solutions the operator $\Omega^*\Omega$ must be proportional to $\delta(E-E')$.

We shall prove the stronger relations

$$\Omega_+^*\Omega_+ = 1, \quad \Omega_-^*\Omega_- = 1, \quad \Omega_0^*\Omega_0 = 1 + \tfrac{1}{4}K^2. \tag{7-48}$$

The proof is quite simple once we establish the fact that the operator $\Omega^*(\tau)\Omega(\tau)$ is independent of τ, that is,

$$[\Omega^*\Omega, H_0] = 0. \tag{7-49}$$

In order to see this, we recall Eq. (7-40),

$$(H_0 + H)\Omega = \Omega H_0,$$

and its Hermitian conjugate

$$\Omega^*(H_0 + H) = H_0 \Omega^*.$$

(We have used the fact that H_0 and H are Hermitian.) Multiplying the first of these equations from the left by Ω^* and the second from the right by Ω, and taking the difference, we find Eq. (7-49). The relations (7-48) are now obtained immediately by evaluating with Eqs. $(7-44)_\pm$ and $(7-44)_0$ the τ-independent operators $\Omega^*\Omega$ in the limits $\tau = \pm\infty$.

Thus, we find that the scattering states Ω_+ and Ω_- are indeed an *orthogonal normal* system of state vectors. But these orthogonality relations do not imply that they are also a complete system. Indeed, there are also the bound states Ω_α. The two sets of states taken together form a complete system by assumption. Moreover, because of condition (3), all bound states are orthogonal to the scattering states. We can formulate this situation compactly by defining the *unitary deficiency* Λ, as the operator

$$\Lambda = \sum_\alpha \Omega_\alpha \Omega_\alpha^*.$$

It is a projection operator satisfying

$$\Lambda^2 = \Lambda, \quad \Lambda^* = \Lambda.$$

The completeness relation is expressible in the form

$$\Omega_+\Omega_+^* = \Omega_-\Omega_-^* = 1 - \Lambda, \tag{7-50}$$

and the orthogonality of the state vectors Ω_α and $\Omega_{\pm q}$ is expressed by the relations

$$\Lambda \Omega_\pm = \Omega_\pm^* \Lambda = 0. \tag{7-51}_\pm$$

7-4 Integral representation of the wave operator.

We shall now derive an explicit expression for $\Omega_\pm$ in terms of H_0 and H in the form of an integral. The following basic integral formula is very useful in this connection.

Let A, B, and C be any three Hermitian operators; then the identity

$$-i\int_{-\infty}^{0} e^{iA\tau}(CB - AC)e^{-iB\tau+\epsilon\tau}d\tau = C - \epsilon\int_{-\infty}^{0} e^{iA\tau}Ce^{-iB\tau+\epsilon\tau}d\tau \qquad (7\text{--}52)$$

holds for every $\epsilon > 0$. This may be proved by integrating the identity

$$\frac{d}{d\tau}(e^{iA\tau}Ce^{-iB\tau+\epsilon\tau}) = -ie^{iA\tau}(CB - AC)e^{-iB\tau+\epsilon\tau} + \epsilon e^{iA\tau}Ce^{-iB\tau+\epsilon\tau} \qquad (7\text{--}53)$$

from $-\infty$ to 0.

We shall use this identity in the limit $\epsilon \to +0$. The term on the right side of Eq. (7–52) which is proportional to ϵ goes to zero if the integral is finite in the limit $\epsilon \to +0$. In that case it can be dropped. There are important cases, as we shall see, where the integral becomes infinite in the limit $\epsilon \to +0$ and the term contributes a finite operator. A simple example of this situation is obtained if we choose $A = B$, $C = 1$. Then the relation (7–52) becomes

$$\lim_{\epsilon \to +0} \epsilon \int_{-\infty}^{0} e^{\epsilon\tau}d\tau = 1.$$

Let us assume that the integral*

$$X = -i\int_{-\infty}^{0} e^{iA\tau}Ce^{-iB\tau+\epsilon\tau}d\tau \qquad (7\text{--}54)$$

exists and is finite in the limit $\epsilon \to +0$. It follows then from (7–52) that X is a solution of the operator relation

$$XB - AX = C. \qquad (7\text{--}55)$$

Now, because of Eq. (7–40), the operator $T = \Omega - 1$ satisfies the operator relation

$$TH_0 - (H_0 + H)T = H. \qquad (7\text{--}56)$$

It follows, therefore, from the last three equations that an expression for T can be obtained in the form

$$T = -i\int_{-\infty}^{0} e^{i(H_0+H)\tau}He^{-iH_0\tau+\epsilon\tau}d\tau. \qquad (7\text{--}57)$$

---

* Here and in all subsequent formulas the limit $\epsilon \to +0$ is always understood and will not be indicated explicitly.

This expression can be transformed further by using the identity (7–52) once more (substituting in it $A = H_0 + H, B = H_0, C = 1$). The result is

$$-T = 1 - \epsilon \int_{-\infty}^{0} e^{i(H_0+H)\tau} e^{-iH_0\tau + \epsilon\tau}. \tag{7-58}$$

Since $\Omega = 1 + T$, it follows that

$$\Omega = \epsilon \int_{-\infty}^{0} e^{i(H_0+H)\tau} e^{-iH_0\tau + \epsilon\tau} d\tau \tag{7-59}$$

is a solution of (7–40).

We shall now prove that this solution is equal to Ω_+. The first step is to show that

$$\Omega_+^* \Omega = 1. \tag{7-60}$$

We begin with the observation that for any function $f(x)$ and any Ω,

$$f(H_0 + H)\Omega = \Omega f(H_0). \tag{7-61}$$

This is a consequence of Eq. (7–40). In particular, for $f(x) = e^{ix\tau}$,

$$\Omega_+(\tau) = e^{iH_0\tau} \Omega_+ e^{-iH_0\tau} = e^{iH_0\tau} e^{-i(H_0+H)\tau} \Omega_+. \tag{7-62}$$

We take the Hermitian conjugate of this, multiply by $\eta e^{\eta\tau}$ ($\eta > 0$), and integrate from $-\infty$ to 0. The result is

$$\Omega_+^* \Omega = \eta \int_{-\infty}^{0} \Omega_+^*(\tau) e^{\eta\tau} d\tau. \tag{7-63}$$

The integral on the right side is understood in the limit $\eta \to +0$. It can be evaluated† by substituting $\eta\tau = x$:

$$\lim_{\eta \to +0} \eta \int_{-\infty}^{0} \Omega_+^*(\tau) e^{\eta\tau} d\tau = \lim_{\eta \to +0} \int_{-\infty}^{0} \Omega_+^*\left(\frac{x}{\eta}\right) e^x dx$$

$$= \Omega_+^*(-\infty) \int_{-\infty}^{0} e^x dx = 1.$$

---

†The limiting relation

$$\lim_{\tau \to -\infty} \Omega(\tau) = \lim_{\eta \to +0} \eta \int_{-\infty}^{0} e^{\eta\tau} \Omega(\tau) d\tau$$

which is implied in the text needs more careful justification. If it is written for a matrix element in the q-representation of the operators, it is equivalent to the relation

$$\lim_{\eta \to +0} \eta \int_{-\infty}^{0} e^{+ix\tau + \eta\tau} \lim_{\epsilon \to +0} \frac{1}{x + i\epsilon} d\tau = -2\pi i \delta(x),$$

which can be proved with usual methods of analysis. In discussing these kinds of relations, extreme care must be taken to observe the correct order of limits. Thus, in this example, for instance, the limit $\epsilon \to +0$ must be taken first, followed by the limit $\eta \to +0$.

Therefore
$$\Omega_+^*\Omega = 1. \tag{7-60}$$

We now use the fact that any solution of Eq. (7-40) is orthogonal to the bound states, or
$$\Lambda\Omega = \Omega^*\Lambda = 0. \tag{7-64}$$

If we multiply Eq. (7-60) from the left by Ω_+ and use Eq. (7-64), we obtain
$$\Omega = \Omega_+. \tag{7-65}$$

This completes the proof that Ω_+ is given by Eq. (7-59):
$$\Omega_+ = \epsilon \int_{-\infty}^{0} e^{i(H_0+H)\tau} e^{-iH_0\tau + \epsilon\tau} d\tau. \tag{7-66}_+$$

It is fairly clear how this result can be extended to show in a similar manner that
$$\Omega_- = \epsilon \int_{0}^{\infty} e^{i(H_0+H)\tau} e^{-iH_0\tau - \epsilon\tau} d\tau. \tag{7-66}_-$$

7-5 Definition of the S-matrix. The S-matrix is usually defined by giving its general matrix element in the q-representation.† We can define two S-matrices, corresponding to the two solutions $G_\pm(q'q)$ of the integral equation (7-35). Thus, we define with Møller,
$$S_\pm(q'q) = \delta(q'q) \mp 2\pi i \delta(E - E') G_\pm(q'q). \tag{7-67}_\pm$$

The S-matrix proper (without $\pm$ signs) is $S(q'q) \equiv S_+(q'q)$.

This traditional approach has the disadvantage that the physical significance of this matrix is not immediately obvious. In the following we shall use a formally different, though essentially equivalent, definition which is mathematically much simpler and admits a more direct physical interpretation.

Instead of the S-matrix, we define a general *S-operator* such that the S-matrix is simply the general matrix element of this operator in the base vector system ω_q.
$$S_\pm(q'q) = (\omega_{q'}, S_\pm \omega_q). \tag{7-68}_\pm$$

The S-operators are directly connected with the wave operators $\Omega_\pm(\tau)$ defined in the preceding section. Indeed, we define
$$S_+ = \lim_{\tau \to +\infty} \Omega_+(\tau), \tag{7-69}_+$$
$$S_- = \lim_{\tau \to -\infty} \Omega_-(\tau). \tag{7-69}_-$$

---

† See, for instance, C. Møller, *loc. cit.*

The equivalence of this definition of $S_\pm$ with (7–67)$_\pm$ is a direct consequence of the limiting relations (A6–1) and (A6–2) which are proved in the mathematical appendix.

The greater flexibility of the definition (7–69)$_\pm$ becomes evident, for instance, if we desire to prove the unitary property of the operators $S_\pm$ and establish the relationship of S_+ to S_-.

To this end we study the operator $\Omega_-^*(\tau)\Omega_+(\tau)$. We have previously shown that an operator of the form $\Omega^*(\tau)\Omega(\tau)$ is independent of τ [cf. Eq. (7–49), Section 7–3]. A glance at the proof shows that this remains true if we take two different Ω's in this expression. Consequently, taking in this expression for τ the values $\tau = -\infty,\ 0,\ +\infty$ and using Eqs. (7–69)$_\pm$ and (7–44)$_\pm$, we obtain the two relations

$$S_-^* = \Omega_-^*\Omega_+ = S_+ \equiv S. \tag{7-70}$$

It follows that

$$S^*S = \Omega_+^*\Omega_-\Omega_-^*\Omega_+ = \Omega_+^*(1-\Lambda)\Omega_+ = \Omega_+^*\Omega_+ = 1, \tag{7-71}$$

and similarly,

$$SS^* = \Omega_-^*\Omega_+\Omega_+^*\Omega_- = \Omega_-^*(1-\Lambda)\Omega_- = \Omega_-^*\Omega_- = 1. \tag{7-72}$$

The last two equations show that S is unitary.

The operator family $\Omega_+(\tau)$ with variable τ consists of operators which, in general, are not unitary. The only exceptions occur for $\tau = -\infty$ (where Ω_+ is 1) and $\tau = +\infty$ (where Ω_+ is S). A similar situation prevails for the operators $\Omega_-(\tau)$.

The definition (7–70) of the S-operator allows a simple physical interpretation. Let us consider the solutions Ω_{+q} which contain the outgoing scattered waves. These solutions can be written as linear superpositions of the solutions Ω_{-q} which contain ingoing scattered waves.

$$\Omega_{+q} = \int S(q'q)\Omega_{-q'}dq'.$$

The elements $S(q'q)$ of this transformation matrix are

$$S(q'q) = (\Omega_{-q'}, \Omega_{+q}).$$

Now, since $\Omega_{\pm q} = \Omega_\pm \omega_q$, this is also given by

$$S(q'q) = (\omega_{q'}, \Omega_-^*\Omega_+ \omega_q) = (\omega_{q'}, S\omega_q).$$

Thus we see that this matrix is the S-operator in the representation ω_q.

The S-matrix is thus the matrix which transforms the ingoing into the outgoing scattering states.

Let us now compare this definition of the S-operator with the preliminary definition (7–14). We assume first that the limit $\tau \to \pm\infty$ exists for all

operators involved. For the τ-dependent interaction operator $H(\tau)$, we choose
$$H(\tau) = H(\tau, -\infty). \tag{7-73}$$

It follows from Eq. (7-9) that
$$H(\tau, \infty) = U(\infty, -\infty) H(\tau, -\infty) U^{-1}(\infty, -\infty). \tag{7-74}$$

The solution of the operator relation
$$i\dot{V}(\tau) = H(\tau) V(\tau) \tag{7-75}$$
with the initial condition
$$V(-\infty) = 1 \tag{7-76}$$
is, therefore,
$$V(\tau, -\infty) = \Omega_+(\tau) = U(-\infty, \tau). \tag{7-77}$$

Because of Eqs. (7-7) and (7-74) the solution of Eq. (7-75) which satisfies $V(\infty) = 1$ is given by
$$\Omega_-(\tau) = U^{-1}(\infty, -\infty) V(\tau, +\infty) U(\infty, -\infty), \tag{7-78}$$
or, using Eq. (7-11),
$$\Omega_-(\tau) = U(-\infty, \tau) U(\infty, -\infty). \tag{7-79}$$
It follows that
$$S = \Omega_-^* \Omega_+ = U(-\infty, +\infty), \tag{7-80}$$

as in Eq. (7-14). We see that the preliminary definition of the S-operator agrees with the definition of this section under the assumption that the limits exist.

We can now see also that the existence of this limit $\tau \to \pm\infty$ in Eq. (7-14) implies a strong restriction on the operators H_0, H. Indeed, if this limit exists, we deduce from Eqs. (7-77) or (7-79) that both operators $\Omega_\pm(\tau)$ are unitary for all τ. This is so because the operators $U(\tau, \tau_0)$ are unitary for all τ, τ_0 and therefore any product of such operators is also unitary. We have seen that the operators $\Omega_\pm$ are not unitary whenever there exist bound states [see Eq. (7-50)]. More generally, the operators $\Omega_\pm$ are not unitary whenever the spectra of eigenvalues of the H_0 and $H_0 + H$ are not identical. For if Ω is a unitary operator, it has an inverse (namely, Ω^*) and the relation
$$(H_0 + H)\Omega = \Omega H_0$$
implies
$$H_0 + H = \Omega H_0 \Omega^{-1}, \tag{7-81}$$

which can hold only if the eigenvalue spectrum of the two operators H_0, $H_0 + H$ are identical. Thus, we conclude that the limits $U(-\infty, +\infty)$ cannot exist, if the eigenvalue spectra of H_0 and $H_0 + H$ are not identical.

In this case, the preliminary definition (7–14) must be replaced by the more general definition of this section.*

One could try a more general time-dependent definition of the S-operator by introducing an adiabatic "switching on" and "switching off" of the interaction at infinity. In this procedure, one defines a family of auxiliary systems described by an interaction operator $H(\epsilon,\tau) \equiv e^{-\epsilon|\tau|} H(\tau) (\epsilon > 0)$. Instead of the Schroedinger equation (7–75), one solves the equation

$$i\dot{V}(\epsilon,\tau) = H(\epsilon,\tau) V(\epsilon,\tau) \qquad (7\text{–}75)_\epsilon$$

with the initial condition $V(\epsilon,-\infty) = 1$. This defines a unitary operator $V(\epsilon,\tau)$. In a similar way, one defines a generalized transformation operator $U(\epsilon;\tau,\tau_0)$. It is then possible to show that for finite ϵ the double limit

$$\lim_{\substack{\tau \to -\infty \\ \tau_0 \to +\infty}} U(\epsilon;\tau,\tau_0) = S_\epsilon$$

always exists. The S-operator is then defined by the limit

$$\lim_{\epsilon \to +0} S_\epsilon = S.$$

The S-operator thus defined can be shown to be identical with the S-operator defined by Eq. (7–70).† For a detailed discussion of related questions we refer to the paper by F. J. Belinfante and C. Møller, *Kgl. Danske Videnskab. Selskab* **28**, No. 6 (1954).

There are simple examples for both cases in wave-mechanical scattering theory. For instance, the scattering in a Coulomb field with repulsive potential is described by an operator $H_0 + H$ which has the same spectrum as H_0. On the other hand, if the Coulomb force is attractive, then there exist bound states for the system $H_0 + H$ which are not present in H_0.

---

* In the past the equivalence of the two definitions (7–14) and (7–70) has almost always been assumed [for instance, E. C. G. Stueckelberg, *Helv. Phys. Acta* **17**, 13 (1943)], and (7–14) has been the starting point of nearly all calculations on the S-matrix in field theory [see, for instance, F. J. Dyson, *Phys. Rev.* **75**, 486 (1949), especially the end of Section III; B. A. Lippman and J. Schwinger, *Phys. Rev.* **79**, 469 (1950)]. The inconsistencies in this assumption have been noted by S. T. Ma, *Phys. Rev.* **87**, 652 (1952) and B. Ferreti, *Nuovo Cimento* **8**, 108 (1951). The definition (7–70) and the integral representations of the wave operators (7–66)$_\pm$ were given by M. Gell-Mann and M. L. Goldberger, *Phys. Rev.* **91**, 398 (1953). The equivalence proof given by S. Fubini, *Atti della Accad. Nat. dei Lincei* **12**, 298 (1952) ignores the complication due to bound states.

† This was recently proved by F. Coester, M. Hamermesh, and K. Tanaka, *Phys. Rev.* **96**, 1142 (1954).

It follows therefore from the general result of the preceding paragraph that in this case Ω cannot be unitary and the limit $U(-\infty,+\infty)$ cannot exist.

The definition (7–70) expresses the scattering operator in terms of the particular wave operators Ω_+ and Ω_-. It is possible to give a more general definition of the S-operator in terms of any solution Ω of the Eq. (7–40). The corresponding time-dependent operator $\Omega(\tau)$ is defined by

$$\Omega(\tau) = e^{iH_0\tau}\Omega e^{-iH_0\tau} = e^{iH_0\tau}e^{-i(H_0+H)\tau}\Omega.$$

The same train of reasoning which led to Eq. (7–64) results in the relation

$$\Omega(-\infty) = \Omega_+^*(\tau)\Omega(\tau)$$

for any Ω. If we substitute on the right side $\tau \to \infty$, we obtain, after rearranging,

$$\Omega(+\infty) = S\Omega(-\infty). \tag{7–82}$$

Since this relation holds for any solution Ω of Eq. (7–40), it holds in particular for $\Omega_\pm$ and Ω_0. For $\Omega_\pm$, we again obtain the relations (7–69)$_\pm$. But for Ω_0 we find a new relation. With the boundary conditions Eq. (7–44)$_0$, we find

$$1 - \frac{i}{2}K = S\left(1 + \frac{i}{2}K\right),$$

or

$$S = \frac{1 - \dfrac{i}{2}K}{1 + \dfrac{i}{2}K}. \tag{7–83}$$

Since we have proved earlier that S is unitary, it follows from Eq. (7–83) that K is Hermitian,

$$K^* = K. \tag{7–84}$$

Equation (7–83) is often put into another form by writing it in terms of $R = S - 1$. It then becomes

$$R = -iK(1 + \tfrac{1}{2}R). \tag{7–85}$$

In this form it is known as *Heitler's integral equation*.† The use of this equation for the problems of radiation damping in collision problems will be explained in Section 16–2.

---

† W. Heitler, *Proc. Cambridge Phil. Soc.* **37**, 291 (1941). Also W. Pauli, *Meson Theory*, Interscience Publications, New York, 1948. The derivation of this equation given in the text goes beyond the older derivations, insofar as no use of perturbation theory is made.

We shall finally identify the operators H_0 and H for the case of a field such as the radiation and matter field under discussion in this book. We denote by p^μ the four-momentum operator of the free fields, i.e., the sum of the expressions (2–38) and (3–69). We recall its defining property:

$$i[p_\mu,\varphi] = -\partial_\mu \varphi \tag{7-86}$$

for any field variable φ.

The operator H_0 is defined as

$$H_0 = -n^\mu p_\mu. \tag{7-87}$$

Because of the conservation laws for the free fields, H_0 is independent of τ.

The defining property (7–86) implies that for any functional $K(\tau)$ of the free fields φ on the surface τ,

$$i[H_0,K(\tau)] = \dot{K}(\tau). \tag{7-88}$$

In other words, the operator H_0 generates the infinitesimal displacement in the τ-direction for any functional of the fields. In particular, for $H(\tau)$ given by (7–6)

$$i[H_0,H(\tau)] = \dot{H}(\tau).$$

From the last relation follows that the operator

$$H = e^{-iH_0\tau}H(\tau)e^{+iH_0\tau}$$

is also independent of τ. We have thereby obtained the operators H_0, H which are used in the definition of the S-matrix.

7–6 Invariance properties of the S-matrix. The S-matrix has certain invariance properties which have their origin in the symmetry properties of the dynamical laws discussed in Chapter 5.

In order to illustrate this, we consider briefly the invariance properties resulting from the principle of relativity. Since this example is merely illustrative, we shall assume a scalar field for simplicity.

Let us consider an observer O who describes in his frame of reference the system with a field $\varphi(x)$. Let O' be another observer in a state of uniform motion with respect to O. The observer O' describes the same system by a new field φ' which is related to the field φ by a transformation law of the form (1–14):

$$\varphi'(x') = \varphi(x).$$

The new field φ' has the same dynamical properties as the original field φ, and it is therefore connected to the original field by a unitary transformation U:

$$\varphi'(x) = U\varphi(x)U^{-1}.$$

Each of the two observers calculates an S-operator, but O obtains an operator S which is a functional of the field φ, while O' obtains an operator

S' which is the same functional of the field φ'. The two S-operators are therefore connected by the same unitary operator U which connects the two fields φ, φ':

$$S' = USU^{-1}. \tag{7-89}$$

The observable effects are calculated from the matrix elements of the S-operators. The matrix element which corresponds to the transition from a state ω_a to a state ω_b is given by

$$S_{ba} = (\omega_b, S\omega_a) \text{ for the observer } O,$$

and

$$S_{ba}' = (\omega_b, S'\omega_a) \text{ for the observer } O'.$$

In the description of symmetry operations which we have used throughout this book we always assume that the transformations affect the operators but not the state vectors. This is an arbitrary convention. It follows then that the *same* state for the two observers is given by the same state vector ω in Hilbert space.

The squares of the magnitude of these two matrix elements are then proportional to the transition probability from the state ω_a to the state ω_b for the two observers. According to the principle of relativity, these two probabilities must be the same for all observers and all transitions; hence

$$|S_{ba}|^2 = |S_{ba}'|^2 \tag{7-90}$$

for all ω_a, ω_b.

Since the phase factors of the state vectors are arbitrary and have no physical significance, we can relate them to the unitary transformations U in such a manner that even

$$S_{ba} = S_{ba}'. \tag{7-91}$$

This relation must hold for all state vectors ω_a, ω_b. It follows that the two operators are identical:

$$S = S', \tag{7-92}$$

or, because of (7-89), that

$$[U, S] = 0. \tag{7-93}$$

Thus, S must commute with the unitary operator U which generates the symmetry transformation.

In this example we have described the situation encountered for Lorentz transformations. But it is clear that a commutator relation (7-93) holds for all types of symmetry transformations. The known symmetry transformations of the coupled radiation and matter field were discussed in Chapter 5. We shall now proceed to express the invariance properties of the S-operator arising from all these transformations.

(a) *Proper Lorentz transformations.* For the discussion of the invariance properties related to the proper Lorentz transformations we need to consider the infinitesimal transformations only. The general infinitesimal transformation for the free fields is given by the operator

$$U = 1 + \alpha_\mu p^\mu + \tfrac{1}{2}\alpha_{\mu\nu} m^{\mu\nu}, \tag{7-94}$$

where p^μ, $m^{\mu\nu}$ are the momentum operators for the free fields. The coefficients α_μ, $\alpha_{\mu\nu}$ are regarded as infinitesimals of the first order. The general invariance property (7-93) of the S-operator is expressed by the equations

$$[p_\mu, S] = 0, \quad [m_{\mu\nu}, S] = 0. \tag{7-95}$$

These equations express the conservation of momentum, energy, and angular momentum in general scattering processes. Indeed, if the initial state ω_0 is an eigenstate of the momentum operators p_μ with eigenvalues p_μ':

$$p_\mu \omega_0 = p_\mu' \omega_0,$$

then the final state $S\omega_0$ is also an eigenstate of p_μ with the same eigenvalues:

$$p_\mu S \omega_0 = p_\mu' S \omega_0.$$

Conversely, if the final state ω_1 is an eigenstate of p_μ with different eigenvalues $p_\mu'' \neq p_\mu'$, then the probability for the transition $\omega_0 \to \omega_1$ is zero:

$$(\omega_1, S\omega_0) = 0.$$

This is so because ω_1 and ω_0 are then two eigenstates of the Hermitian operators p_μ with different eigenvalues, and are therefore orthogonal (*cf.* Section 1-3).

Similar conservation laws hold for the angular momentum operators $m_{\mu\nu}$.

(b) *Gauge transformations.* The gauge transformation (2-45) for the free radiation field

$$a_\mu(x) \to a_\mu(x) + \partial_\mu \varphi(x) \tag{7-96}$$

is generated by the operator (2-47):

$$F = \int_\sigma \{\partial^\mu \varphi(x) \partial^\lambda a_\lambda(x) - \varphi(x) \partial^\mu \partial^\lambda a_\lambda(x)\} d\sigma_\mu. \tag{7-97}$$

Since the transformation is a symmetry transformation, it follows that

$$[F, S] = 0. \tag{7-98}$$

An important consequence of this relation is the *conservation of the subsidiary condition*. If the initial state vector ω_0 satisfies the subsidiary condition

$$\partial^\lambda a_\lambda(x) \omega_0 = 0 \quad \text{for all } x, \tag{7-99}$$

then the final state vector $\omega \equiv S\omega_0$ also satisfies this same condition. Indeed, all the state vectors which satisfy (7–99) are eigenstates of the operator F with the eigenvalue 0 and, conversely, all these eigenstates of F with eigenvalue 0 satisfy (7–99). Since F commutes with S, the final state, $S\omega_0$, also satisfies

$$\partial^\lambda a_\lambda(x) S\omega_0 = 0. \qquad (7\text{–}100)$$

(c) *Phase transformations.** The infinitesimal phase transformations [cf. Eq. (3–53)]

$$\psi(x) \to \psi'(x) = \psi(x) + i\lambda\psi(x) \qquad (7\text{–}101)$$

are generated by the unitary operator

$$U = 1 + i\lambda Q, \qquad (7\text{–}102)$$

where

$$Q = i \int \bar\psi(x) \gamma^\mu \psi(x) d\sigma_\mu. \qquad (7\text{–}103)$$

It follows that

$$[Q, S] = 0. \qquad (7\text{–}104)$$

According to Eqs. (3–103) and (3–106) Q is the total charge operator. The relation (7–104) expresses thus the conservation of charge.

(d) *Space inversion.* The transformation law of the field variables under space inversion, σ, was defined in Eqs. (5–17) and (5–19). The transformed radiation field $a_\mu^\sigma(x)$ is

$$a_i^\sigma(x') = -a_i(x), \quad a_0^\sigma(x') = a_0(x), \quad (i = 1, 2, 3). \qquad (7\text{–}105)$$

The transformed matter field $\psi^\sigma(x)$ is

$$\psi^\sigma(x') = i\gamma_0 \psi(x), \qquad (7\text{–}106)$$

where

$$x_i' = -x_i, \quad x_0' = x_0, \quad (i = 1, 2, 3).$$

The operator U for this transformation is denoted by π and is called the *space-parity operator*. Its eigenvalues are ± 1 and are called the *space parity*. It satisfies

$$a_\mu^\sigma(x) = \pi a_\mu(x) \pi^{-1}, \quad \psi^\sigma(x) = \pi \psi(x) \pi^{-1}. \qquad (7\text{–}107)$$

Since the space inversion is a symmetry operation, S satisfies

$$[\pi, S] = 0. \qquad (7\text{–}108)$$

The last relation is the mathematical expression for the conservation of parity in general transition processes.

---

* These transformations are often called "gauge transformations of the first kind," whereas the gauge transformations discussed under (b) are often referred to as "gauge transformations of the second kind."

(e) *Time inversion.* The invariance properties of the S-operator which result from the invariance of the dynamical laws under time reversal need some special consideration. Here the situation is different from that in other symmetry transformations and gives rise to a different transformation law for the time-inversion operator T. This transformation law was given in Eq. (5–50). We need a slight extension of this result to the case where the transformation operator U is itself τ-dependent.

In order to emphasize the characteristic difficulty of the time-reversal transformation, we discuss the simpler case where the S-operator can be defined with the preliminary definition (7–14):

$$S = U(-\infty, +\infty) = V(+\infty, -\infty).$$

We set $V(\tau, -\infty) = V(\tau)$. The unitary operator family $V(\tau)$ satisfies the relations

$$V(-\infty) = 1, \quad V(+\infty) = S, \tag{7-109}$$

and transforms the interacting Heisenberg fields $\Phi(x)$ into the free fields because of the relation (7–1):

$$\varphi(x) = V(\tau)\Phi(x)V^{-1}(\tau), \quad x \in \sigma(\tau). \tag{7-110}$$

The time-reversal transformation was defined in Section 5–4. It associates the new fields $\Phi^\tau(x)$ with the fields $\Phi(x)$ [*cf.* Eq. (5–42)]. We have shown that these new fields are connected to the old ones by a unitary transformation T:

$$\Phi^\tau(x) = T\Phi(x)T^{-1}. \tag{7-111}$$

Let us now determine the corresponding operator for the time-reversal transformation of the free fields $\varphi(x) \rightarrow \varphi^\tau(x)$:

$$\varphi^\tau(x) = t\varphi(x)t^{-1}. \tag{7-112}$$

We are particularly interested in finding the relation between T and t.

There are two features which make time inversion different from any other symmetry transformation. The first is that the time inversion is a nonlinear transformation and that the operator transforms according to the somewhat unfamiliar transformation law (5–50). The second is the fact that the reversal of the time axis also reverses the direction of the instant parameter τ in (7–110). Let us now see in detail how these features affect the relation between T and t.

We simplify the formulas by ignoring the fact that under time reversal the various components of the field variables Φ in general undergo a linear transformation [*cf.* Eq. (5–42)]. Thus, we are essentially discussing the situation for a scalar field. There is no difficulty in generalizing the subsequent formulas for the case of fields with more components. The essential features of the problem are all present in the simpler case which we are selecting for discussion.

The time-reversal transformation for the Heisenberg fields is then

$$\Phi^\tau(x) = \Phi^\times(x'), \tag{7-113}$$

and for the free fields, we find

$$\varphi^\tau(x) = \varphi^\times(x'). \tag{7-114}$$

Here $\Phi^\times$ is the complex conjugate of Φ (not Hermitian conjugate! See the discussion in Section 5–4) and

$$x_i' = x_i, \quad x_0' = -x_0, \quad (i = 1, 2, 3).$$

The last five equations are sufficient to establish the connection between T and t; for instance, we may employ the following sequence [resembling closely the corresponding sequence (5–49)]:

$$\varphi^\tau(x) = (V(\tau)\Phi(x)V^{-1}(\tau))^\tau = V^\times(\tau')\Phi^\times(x')(V^\times)^{-1}(\tau')$$
$$= V^\times(\tau')\Phi^\tau(x)(V^\times)^{-1}(\tau') = V^\times(\tau')T\Phi(x)T^{-1}(V^\times)^{-1}(\tau')$$
$$= V^\times(\tau')TV^{-1}(\tau)\varphi(x)V(\tau)T^{-1}(V^\times)^{-1}(\tau')$$
$$= t\varphi(x)t^{-1}.$$

The last equation says that

$$t = V^\times(\tau')TV^{-1}(\tau), \tag{7-115}$$

where τ' is the instant parameter of the plane σ' which is obtained from σ by the reversal of the time axis ($x_0' = -x_0$). It follows, therefore, that $\tau = +\infty$ for $\tau' = -\infty$ and that $\tau = -\infty$ for $\tau' = +\infty$. Substituting these values for τ, τ' in Eq. (7–115) and using Eq. (7–109), we obtain the two relations

$$t = TS^{-1} = S^\times T. \tag{7-116}$$

This leads to the important result (using $S^{\times\sim} \equiv S^* = S^{-1}$) that

$$S^\sim = tSt^{-1}. \tag{7-117}$$

The last relation is the mathematical expression of the *principle of reciprocity*. This principle may be formulated in the following manner.

Let ω_a, ω_b be two state vectors representing two arbitrary states a and b of the system. Denote by ω_{-a}, ω_{-b} the state vectors for the *time-reversed* states $-a, -b$:

$$\omega_{-a} = t^{-1}\omega_a^\times,$$
$$\omega_{-b} = t^{-1}\omega_b^\times \tag{7-118}$$

[*cf.* Eq. (5–48)]. The principle of reciprocity states that the transition probability P_{ab} for the transition from a into b is equal to the transition probability P_{-b-a} from $-b$ into $-a$.†

Since the transition probability is proportional to the square of the matrix element of the S-operator

$$P_{ab} \sim |S_{ba}|^2,$$

it is sufficient to prove that

$$S_{ba} = S_{-a-b}{}^*. \tag{7–119}$$

The proof is based on Eq. (7–117). With (7–118) we obtain

$$S_{-a-b}{}^* = (t^{-1}\omega_a{}^\times, St^{-1}\omega_b{}^\times)^* = (\omega_a{}^\times, tSt^{-1}\omega_b{}^\times)$$
$$= (\omega_a{}^\times, S^\sim \omega_b{}^\times) = (\omega_b, S\omega_a) = S_{ba}.$$

Thus, Eq. (7–119) and the general reciprocity principle are verified.

From the relation (7–117) follows further that the unitary operator $t^\times t$ commutes with S:

$$[t^\times t, S] = 0. \tag{7–120}$$

As in Section 5–4, we can show that the operator $t^\times t$ has eigenvalues ± 1. Therefore we obtain from the invariance of the theory under time reversal a new kind of selection rule. In Section 5–4 we have further shown that this is even a *superselection* rule. This means that an observable cannot have any matrix elements connecting the two subspaces which belong to the eigenvalues ± 1 of $t^\times t$, respectively.

(f) *Charge conjugation.* Finally, we derive a commutator relation which stems from the invariance of the theory under the transformation of charge conjugation. This transformation is defined for the free fields by

$$\psi^c(x) = C^* \psi^*(x), \quad a_\mu{}^c(x) = -a_\mu(x). \tag{7–121}$$

We have shown in Section 5–5 that this transformation leaves the field equations and commutator relations invariant. Therefore it is a symmetry transformation. There exists a unitary operator γ with the properties

$$\psi^c(x) = \gamma \psi(x) \gamma^{-1}, \quad a_\mu{}^c(x) = \gamma a_\mu(x) \gamma^{-1}, \tag{7–122}$$

$$[\gamma, S] = 0. \tag{7–123}$$

---

† This principle is more general than the *principle of detailed balance*, which may be stated as $P_{ab} = P_{ba}$. The principle of detailed balance is valid only for special cases, for instance if the interaction is sufficiently weak, or for particles without spin. J. Hamilton and H. W. Peng, *Proc. Roy. Soc. Ir. Acad.* **A49**, 197 (1944) have given an example for a vector meson field where this principle does not hold. The recent statement to the contrary by K. Goro, *Progr. Theor. Phys.* **8**, 565 (1952) is incorrect. F. Coester [*Phys. Rev.* **84**, 1259 (1951)] has shown that the principle of detailed balance holds generally if the summation over the spins of initial and final states is carried out.

Since $\gamma^2 = 1$, it follows that γ has the eigenvalues ± 1. We shall call the operator γ the *charge-parity operator* and its eigenvalues the *charge parity*. Equation (7–123) is then the mathematical expression for the conservation of charge parity.

These are all the known invariance properties of the S-matrix which stem from the general symmetry properties of the dynamical laws. These properties of the S-operator are of great importance because they are of general validity. This means that we expect them to hold even if in a future theory the detailed form of the interaction law will be modified. They are also independent of any approximation technique which may be needed for the actual evaluation of the S-operator. They are thus the only truly general laws of the theory which we can obtain without actually solving the field equations.

It is an interesting question whether there are perhaps hidden symmetries in the theory which are not yet discovered. This is very well possible, since we have not sufficient insight into the general symmetry properties of dynamical systems such as the coupled radiation and matter field. The time-reversal and charge conjugation transformations are instances of relatively recent discoveries of symmetry properties of this system.

CHAPTER 8

EVALUATION OF THE S-MATRIX

8–1 The iteration solution. Up to the present the explicit evaluation of the S-operator in quantum electrodynamics has not been possible. It is therefore necessary to develop an approximation method. In the past this approximation has always been based on Eq. (7–8) together with the definition (7–14). As we have seen, this definition is not adequate when the operators $\Omega_\pm$ are not unitary.* The approximation method for the S-operator based on this definition is therefore of only limited validity. In spite of this limitation, it was found that for the lowest orders of approximation the S-matrix elements thus calculated give the correct values of the transition probabilities (*cf.* Section 7–5).

We obtain first a formal solution of the Eq. (7–8) for the operator $V(\tau)$:

$$V(\tau) = 1 - i\int_{-\infty}^{\tau} H(\tau')V(\tau')d\tau' \tag{7-8}$$

in the form of a series

$$V(\tau) = \sum_{n=0}^{\infty} V_n(\tau) \tag{8-1}$$

with

$$V_0 = 1, \tag{8-2}$$

$$V_n(\tau) = -i\int_{-\infty}^{\tau} H(\tau')V_{n-1}(\tau')d\tau', \quad (n \geq 1). \tag{8-3}$$

The recursion relation for the operators $V_n(\tau)$ can be solved. The result is

$$V_n(\tau) = (-i)^n \int_{-\infty}^{\tau} d\tau_n \cdots \int_{-\infty}^{\tau_3} d\tau_2 \int_{-\infty}^{\tau_2} d\tau_1 H(\tau_n) \cdots H(\tau_1). \tag{8-4}$$

We evaluate this operator for $\tau = \infty$ and define

$$S^{(n)} = V_n(\infty). \tag{8-5}$$

A more symmetrical form of $S^{(n)}$ is obtained if we introduce the *time-ordered* or *chronological product*. It is defined as follows:

$$P(H(\tau_n) \cdots H(\tau_1)) = H(\tau_{i_n}) \cdots H(\tau_{i_1}), \quad \tau_{i_n} \geq \cdots \geq \tau_{i_1}. \tag{8-6}$$

---

* This is always the case when the system has bound state solutions. Since the series given below Eq. (8–1) is formally unitary, this series cannot converge in the presence of bound states.

This product is, by definition, a symmetrical function of the n operators $H(\tau_i)$ ($i = 1, \cdots, n$). If we replace the integrand in Eq. (8–4) by the chronological product and let all the integration variables τ_i range over the entire set, we obtain $n!$ times the original integral. Therefore, the S-operator is given by

$$S = V(\infty) = \sum_{n=0}^{\infty} S^{(n)} \tag{8-7}$$

with

$$S^{(0)} = 1 \tag{8-8}$$

and

$$S^{(n)} = \frac{(-i)^n}{n!} \int_{-\infty}^{\infty} \cdots \int_{-\infty}^{\infty} d\tau_n \cdots d\tau_1 P(H(\tau_n) \cdots H(\tau_1)), \quad (n \geq 1).$$

Recalling the definition of $H(\tau)$ [Eq. (7–3)] in terms of the free fields,

$$H(\tau) = ie \int_{\sigma(\tau)} \bar{\psi}(x)\gamma^\mu \psi(x) a_\mu(x) d\sigma, \tag{7-3}$$

we can rewrite the expression for $S^{(n)}$:

$$S^{(n)} = \frac{e^n}{n!} \int_{-\infty}^{\infty} \cdots \int_{-\infty}^{\infty} dx_n \cdots dx_1 P(\bar{\psi}(x_n)\gamma^{\mu_n}\psi(x_n) \cdots$$
$$\cdots \bar{\psi}(x_1)\gamma^{\mu_1}\psi(x_1)) P(a_{\mu_n}(x_n) \cdots a_{\mu_1}(x_1)). \tag{8-9}$$

We remind the reader that according to our convention each pair of operators $\bar{\psi}(x_i)\psi(x_i)$ in (8–9) is understood to be ordered in creation and annihilation operators.

In the description of physical processes we are interested in the evaluation of the matrix elements of the S-operator (8–7). To this end one assumes that the first terms of this series will give a sufficiently accurate approximation. Since each operator $S^{(n)}$ contributes matrix elements to many different processes, we are confronted with the problem of extracting a particular matrix element $S_{fi}^{(n)}$ which refers to a specific initial state ω_i and a specific final state ω_f.

We consider, for instance, the transition from an initial state with r negatons, s positons, and t photons into a final state with r' negatons, s' positons, and t' photons. The matrix element for such a process receives contributions only from the operator which contains r absorption operators for negatons, s absorption operators for positons, and t absorption operators for photons multiplied on the left by r' emission operators for negatons, s' emission operators for positons, and t' emission operators for photons. This operator occurs in the ordered product containing $r + s' = r' + s$ factors $\bar{\psi}\psi$ and $t + t'$ factors a_μ. It is therefore necessary to write the general expression $S^{(n)}$ as a sum of ordered products. This is precisely the problem which is solved in Appendix 4.

8–2 The Feynman-Dyson diagrams.

We shall present the result of the ordering process in graphical form.* The graphic representation of S_{fi} associates a set of *diagrams* to every $S_{fi}^{(n)}$. This set will be called the nth order diagrams of the process $\omega_i \to \omega_f$. The importance of this association lies in its uniqueness; i.e., there is a one-to-one correspondence between the diagrams in the set of nth order and certain groups of terms in the sum $S_{fi}^{(n)}$.

In order to construct the diagram corresponding to a given term in $S_{fi}^{(n)}$, we associate n distinct points with the n variables $x_1, \cdots, x_n$, each representing a point in a four-dimensional space-time continuum. These points constitute the n *corners* of the diagram. Each x will occur in exactly three factors in the product; this is assured by the structure of the interaction operator density $\bar{\psi}(x)\mathbf{a}(x)\psi(x)$. These three factors will be associated with three lines which join at the point x. The lines are defined in Table 8–1.

TABLE 8–1

The correspondence between diagrams and S-matrix in position space

Component of Diagram		Factor in S-Matrix	
Internal photon line	$\begin{array}{c} 1,\mu \quad\quad\quad 2,\nu \\ 2,\nu \text{ - - - - - } 1,\mu \end{array}$	$-ig_{\mu\nu}D_c(x_2 - x_1)$	photon propagation function
Internal electron line	$\xrightarrow{}$ $\;$ 1 $\quad\quad$ 2	$iS_c(x_2 - x_1)$	electron propagation function
Corner	$x\!\!\diagdown\!\!-\,-$ $\;\;\;\;\;\mu$	γ^μ	
External photon lines	$x,\mu\uparrow \quad\quad \uparrow$ $\quad\quad\quad\; \cdot x,\mu$	$a_\mu^{(-)}(x), a_\mu^{(+)}(x)$	ingoing and outgoing photon field operators
External negaton lines	$\cdot x \quad\quad \downarrow$ $\uparrow \quad\quad\quad \uparrow\! x$	$\psi^{(-)}(x), \overline{\psi^{(-)}}(x)$	ingoing and outgoing negaton field operators
External positon lines	$\cdot x \quad\quad \downarrow$ $\downarrow \quad\quad\quad \cdot x$	$\overline{\psi^{(+)}}(x), \psi^{(+)}(x)$	ingoing and outgoing positon field operators

---

* Diagrams for the representation of physical processes were first introduced by R. P. Feynman, *Phys. Rev.* **76**, 769 (1949); **80**, 440 (1950). The connection of Feynman's approach with conventional quantum electrodynamics was established by F. J. Dyson, *Phys. Rev.* **75**, 486, 1736 (1949).

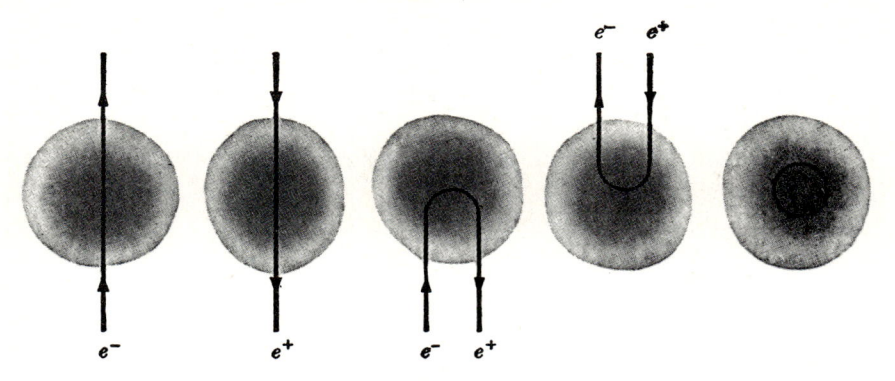

FIG. 8–1. The five types of basic electron paths.

All *photon lines* are dashed and all *electron lines* are solid. They form a connected array of lines for every product, provided the associated integral cannot be written as the product of two integrals. We note that all propagation functions, S_c and D_c, are associated with lines joining two points, i.e., *internal lines*, whereas all free operators are associated with lines which are attached to the diagram at only one corner, i.e., *external lines*. These lines suggest the obvious meaning of world lines for particles entering and leaving the reaction. For definiteness, we shall assume that the positive time direction is upward, such that the reaction products will enter at the bottom and leave at the top of the diagram.

In order to distinguish between positons and negatons, we follow Feynman and associate an arrow with the solid lines: solid lines directed towards increasing and decreasing time shall designate negatons and positons, respectively. The implication of this notation is shown in Fig. 8–1, where the five basic cases of *electron paths* are indicated schematically. By an electron path we mean the *whole* electron line, which will, in general, be a sum of two external and many internal lines. The structure of $H(\tau)$ makes it impossible that two electron paths cross each other.

In Fig. 8–1 the shaded areas indicate diagrams with various external and internal lines. Only one electron path is shown for each diagram. In the last case the electron path consists of internal lines only (closed loop). One often refers to processes whose diagrams contain only one such path as *one-particle processes*, although closed-loop diagrams are usually not included in this notion.

The resultant topology of the diagrams is such that of two adjacent arrows one and only one will point toward their common corner. This implies that by following the arrows one can trace every solid line through the whole diagram from its entrance to its exit. Therefore, every electron path has a definite direction, and we see that the notation of arrows in negative time directions for positons assures that there is a single arrow-direction along the whole electron path. Similarly, it follows that every electron path which does not leave the diagram necessarily forms a closed loop with a uniform sense of direction. These closed loops will be discussed further in Section 8–4.

It is important to keep in mind that in this figure negatons move in the direction in which the arrows point, while positons move *opposite* to the arrow direction, i.e., toward increasing time.

There is no need to associate a direction to the internal photon lines, since $D_c(-x) = D_c(x)$. The external photon lines also do not need an arrow if one distinguishes clearly between the past and future light cone on which the ingoing and outgoing photon lines must lie. However, we shall use the arrow at times for emphasis.

The factor $g_{\mu\nu}$ in the photon's propagation function contracts the product of γ^μ and γ^ν associated with the two corners which form the end points of that photon line. This is indicated in Table 8–1.

We now remind the reader that according to the ordering theorem, $S_{fi}^{(n)}$ is a sum of integrals over $x_1, \cdots, x_n$, which contains for each integrand also all integrands which are obtained by all possible permutations of the variables $x_1, \cdots, x_n$. There will thus be diagrams which differ only in the labeling of the corners, but which are topologically identical. Such diagrams we call *equivalent diagrams*. Since the corresponding integrals are identical after integration over the dummy variables, and can therefore be combined easily, it is customary to make no distinction between equivalent diagrams, and to draw one typical diagram only.

If two diagrams differ by the interchange of two ingoing or outgoing particles of the same kind they will not be equivalent. The sum over these interchanges will assure the proper symmetrization of the initial and final states. The associated signs in the electron case all arise from the matrix elements of the ordered creation and annihilation operators. A typical example of this kind is electron-electron scattering, which is discussed below.

It should be noted that in Table 8–1 a correspondence is made between the external lines and the ingoing and outgoing field *operators*. In position space this is easier to do. Actually, however, a diagram is meant to correspond to the *matrix element* of the operator whose factors are indicated on the right side of Table 8–1. This correspondence will become clearer in notation and also more meaningful in momentum space.

The one-to-one correspondence between the S-matrix elements and the Feynman-Dyson diagrams follows from the above discussion of these diagrams. It is possible to use this correspondence to great advantage: one can *first* draw the diagrams for the process of interest and *then* construct the associated ordered product. Table 8–1 will thereby be useful. Of course, this procedure does not at all mean that one discards the whole theory which leads to the ordered product representation of the S-matrix;

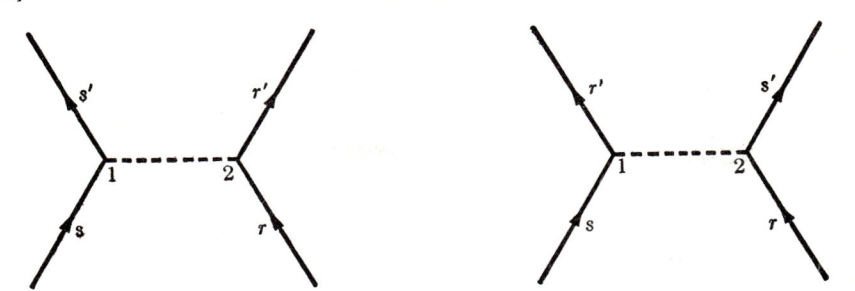

FIG. 8–2. Negaton-negaton scattering diagrams for position space in lowest order.

it is simply a convenient way of obtaining the ordered products without going through the cumbersome ordering process explicitly. This short-cut is used extensively in the applications of the theory.

As illustrations of the construction of diagrams for a given physical process we consider two simple examples, *Møller scattering* and *Compton scattering*, in lowest order. The ordered product in $S^{(2)}$ which describes the scattering of two negatons by each other is

$$S^{(2)}(e^--e^-)$$
$$= -ie^2 \int \overline{\psi_{r'}^{(-)}}(x_2)\gamma^\mu\psi_r^{(-)}(x_2)d^4x_2 D_c(x_2 - x_1)d^4x_1 \overline{\psi_{s'}^{(-)}}(x_1)\gamma_\mu\psi_s^{(-)}(x_1)$$
$$- ie^2 \int \overline{\psi_{s'}^{(-)}}(x_2)\gamma^\mu\psi_r^{(-)}(x_2)d^4x_2 D_c(x_2 - x_1)d^4x_1 \overline{\psi_{r'}^{(-)}}(x_1)\gamma_\mu\psi_s^{(-)}(x_1).$$
(8–10)

In this expression the initial states r, s, and the final states r', s', are explicitly indicated. The two terms correspond to a different association of final states to the two particles. (It is important for the representation by diagrams to interchange the labeling and keep the particles fixed rather than vice versa.) The correct antisymmetrization is assured in Eq. (8–10) because the ordered product of creation and annihilation operators in the first term will give $+1$, whereas in the second term it will give -1. The two diagrams which correspond to the two terms in Eq. (8–10) are shown in Fig. 8–2.*

---

* In this and the following figures the diagrams are stylized. Although an interpretation in terms of world lines is sometimes given, this fact is irrelevant here and will be ignored. Consequently, external photon lines are not always drawn at 45° to the time axis (which is upward), and external electron lines are drawn parallel to the time axis even when these electrons are not meant to be at rest. Similarly, internal photon lines are often drawn perpendicular to the time axis, even though instantaneous interaction is not necessarily implied. The important distinction between ingoing and outgoing particles will always be clearly shown.

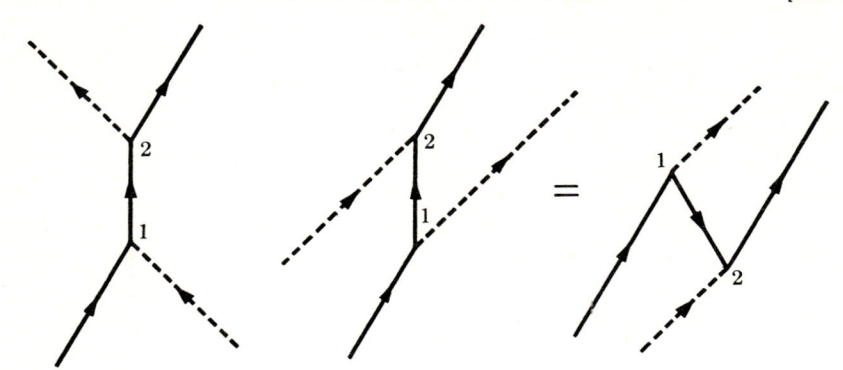

FIG. 8-3. The two second order diagrams for Compton scattering.

We note that each of the two integrals in (8-10) comes from a pair of integrals which differ from each other only in the permutation of x_1 and x_2. These latter two integrals are thus equivalent, and each pair has been combined into one term, thereby canceling a factor $1/2!$ which arises from the expression (8-8). Correspondingly, each diagram in Fig. 8-2 is representative of two equivalent diagrams which differ in the interchange of the labels 1 and 2.

In a similar way, one can find the diagrams for positon-positon scattering. They are identical with Fig. 8-2 except that all arrows are reversed.

The scattering of a photon by a free electron, e.g., a negaton (*Compton scattering*), is, in lowest order,

$$S^{(2)}(\gamma\text{-}e^-) = i\frac{e^2}{2!} \int \overline{\psi^{(-)}}(x_2) a^{(+)}(x_2) d^4x_2 S_c(x_2 - x_1) d^4x_1 a^{(-)}(x_1) \psi^{(-)}(x_1)$$

$$+ i\frac{e^2}{2!} \int \overline{\psi^{(-)}}(x_2) a^{(-)}(x_2) d^4x_2 S_c(x_2 - x_1) d^4x_1 a^{(+)}(x_1) \psi^{(-)}(x_1). \quad (8\text{-}11)$$

There are again two diagrams, one for each product. They differ in the succession of photon absorption and emission when one follows the electron path. In the first diagram the photon is absorbed at the same point at which the electron enters, but in the second diagram it is absorbed where the electron leaves. It is thereby irrelevant whether or not point 2 is later in time than point 1. The relative position of the corners in a diagram is not significant, since the diagram stands for a typical integrand, and x_1 and x_2 are dummy variables. This fact is indicated in Fig. 8-3, where the diagram corresponding to the second term in Eq. (8-11) is drawn in two different ways which are topologically identical.

It is important to realize that only the *sum* of the two Compton diagrams describes photon-electron scattering. The separation of a matrix element into terms corresponding to the individual diagrams, though extremely useful, has in general no physical meaning. Only the sum is observable.

8-3 Diagrams in momentum space.

When compared with position space, momentum space offers a much more satisfactory description of a physical process which takes place between initially free particles. Therefore, momentum space is much more important for applications.

A physical process is completely specified if we indicate what kind of particles enter and leave the reaction and in what state these particles are. A "state" can be specified by giving the momentum and the internal degrees of freedom (i.e., state of polarization for photons and spins for the electrons). This is, of course, not the only way of describing free-particle states. One could choose, for instance, the magnitude and 3-component of the total angular momentum and the parity instead of the three components of the momentum. This is done in the theory of multipole radiation. We shall choose here plane waves for the initial and final states, which is sufficient for most of the applications.

The initial state, for instance, would be characterized by a set of g momentum vectors,

$$p_1, \cdots, p_r; q_1, \cdots, q_s; k_1, \cdots, k_t \qquad (8\text{-}12)$$

corresponding to the momentum values of the r initial negatons, s initial positons, and t initial photons.

$$g = r + s + t \qquad (8\text{-}13)$$

is the total number of particles in the initial state.

Similarly, the final state would be characterized by a set of g' momentum vectors,

$$p_1', \cdots, p_{r'}'; q_1', \cdots, q_{s'}'; k_1', \cdots, k_{t'}' \qquad (8\text{-}14)$$

corresponding to the momentum values of the r' final negatons, s' final positons, and t' final photons.

$$g' = r' + s' + t' \qquad (8\text{-}15)$$

is the total number of particles in the final state.

We shall agree to include the spin and polarization as degrees of freedom in this specification without so indicating explicitly in the notation. For instance, the momentum vector p_1 for the first negaton really stands for a symbol (p_1, ρ_1), where ρ_1 is a discrete double-valued index which labels the spin of the first negaton. For the sake of a simpler notation, we shall omit ρ_1. This is done for all initial and final momentum variables.

The contribution to the matrix element for the specified process is now obtained by substituting the appropriate Fourier transforms. These are given for photons in Eqs. (2–63) and (2–64),

$$a_\mu^{(+)}(x) = \frac{1}{(2\pi)^{3/2}} \int \frac{d^3k}{\sqrt{2\omega}} a^*(\mathbf{k}) e_\mu(\mathbf{k}) e^{-ikx}$$

$$a_\mu^{(-)}(x) = \frac{1}{(2\pi)^{3/2}} \int \frac{d^3k}{\sqrt{2\omega}} a(\mathbf{k}) e_\mu(\mathbf{k}) e^{ikx}.$$

(8–16)

Here $e_\mu(\mathbf{k})$ stands for the transverse polarization vector of a free photon of momentum k ($e \cdot k = 0$). In agreement with our convention, we have omitted an explicit label for the two states of polarization.

The Fourier transforms for the electron field operators are given in Eqs. (3–87):

$$\overline{\psi^{(-)}}(x) = \frac{1}{(2\pi)^{3/2}} \int d^3p \sqrt{\frac{m}{\epsilon}} a^*(\mathbf{p}) \bar{u}(\mathbf{p}) e^{-ipx}, \quad \text{outgoing negaton,}$$

$$\psi^{(-)}(x) = \frac{1}{(2\pi)^{3/2}} \int d^3p \sqrt{\frac{m}{\epsilon}} a(\mathbf{p}) u(\mathbf{p}) e^{ipx}, \quad \text{ingoing negaton,}$$

$$\overline{\psi^{(+)}}(x) = \frac{1}{(2\pi)^{3/2}} \int d^3p \sqrt{\frac{m}{\epsilon}} b(\mathbf{p}) \bar{v}(\mathbf{p}) e^{ipx}, \quad \text{ingoing positon,}$$

$$\psi^{(+)}(x) = \frac{1}{(2\pi)^{3/2}} \int d^3p \sqrt{\frac{m}{\epsilon}} b^*(\mathbf{p}) v(\mathbf{p}) e^{-ipx}, \quad \text{outgoing positon.}$$

(8–17)

The spin indices are again omitted.

The only other functions which occur in the integrals over ordered products are $D_c(x)$ and $S_c(x)$. Their Fourier transforms follow from Eq. (A1–1):

$$D_c(x) = \frac{1}{(2\pi)^4} \int \frac{e^{ikx}}{k^2 - i\mu} d^4k, \qquad (8\text{–}18)$$

$$S_c(x) = \frac{1}{(2\pi)^4} \int \frac{i\not{p} - m}{p^2 + m^2 - i\mu} e^{ipx} d^4p. \qquad (8\text{–}19)$$

In these integrals we took the path of integration in the k^0 and p^0 planes along the real axis and displaced the poles slightly off the real axis by approximately $\mu/2$ ($\mu > 0$). The integrals are to be taken in the limit $\mu \to 0$. In this limit they become identical with the integrals over the path C_{1R} and the undisplaced poles, as shown in Fig. (A1–5).

After all these substitutions are performed, the resultant expression contains the integration variables x only in exponential functions. All x integrations can therefore be carried out. One obtains $(2\pi)^4 \delta(p \pm q \pm k)$ for each corner, where p and q are the momenta of the ingoing and outgoing negatons and positons at that corner, and k is the momentum of the photon. Any or all of these momenta may be associated with the initial or final particle states. The two signs of k correspond to opposite ends of an internal photon line. If the lines are all internal, one of the integrals, for instance that over k, can be carried out, replacing, in effect, $\pm k$ by $\pm q - p$.

There remains an over-all δ-function, $\delta(P - P')$, depending on the initial and final total momentum.

$$P = p_1 + \cdots + p_r + q_1 + \cdots + q_s + k_1 + \cdots + k_t,$$

$$P' = p_1' + \cdots + p_{r'}' + q_1' + \cdots + q_{s'}' + k_1' + \cdots + k_{t'}'.$$

This δ-function expresses the conservation of energy and momentum of the whole system.

We next dispose of the ordered products of emission and absorption operators by observing that the matrix elements of such products are ± 1, depending on the order of the anticommuting operators. In simple cases the sign is most easily determined by direct inspection.

From these remarks it becomes clear that it is possible to associate a diagram to each ordered product in p-space which is of exactly the same topological structure as the corresponding diagram in x-space. But, whereas in x-space the *corners* are labeled with the variables $x_1, x_2, \cdots, x_n$, in p-space the *lines* are labeled with the variables p_i, q_j, and k_l, depending on the type of particle. Because of the δ-functions which occur at every corner, the four-momenta of the three lines associated with the same corner will always be such that the sum of the ingoing momenta equals the sum of the outgoing momenta.

The p-space diagrams can be constructed in much the same way as the diagrams in x-space. The similarity is evident from a comparison of Table 8–2 for the momentum-space diagrams with Table 8–1. In particular, the one-to-one correspondence is unaltered by the Fourier transform. A number of essential differences between position space and momentum space should be noted, however, and are apparent from comparison of the two tables. The association of four-vectors to the lines of the diagram requires that we give a direction to *every* line. The electron lines are already directed, and so are the external photon lines; but we must also give an arrow to the internal photon lines. The direction of this arrow is arbitrary, and its sole purpose is to assure the correct signs of k in the two δ-functions associated with the corners at the two end points of that internal photon line.

TABLE 8–2

The correspondence between diagrams and S-matrix elements in momentum space

Component of Diagram		Factor in S-Matrix Element	
Internal photon line	$\nu\ \text{-----}\ \lambda$	$g_{\nu\lambda} \dfrac{1}{k^2 - i\mu}$	photon propagation function
Internal electron line	$\longrightarrow$	$\dfrac{ip - m}{p^2 + m^2 - i\mu}$	electron propagation function
Corner	$\begin{array}{c} p' \\ \nu\ k \\ p \end{array}$	$\gamma^\nu \delta(p - p' - k)$	
External photon lines	$\begin{array}{cc} \mu & k \\ k & \mu \end{array}$	$\dfrac{1}{\sqrt{2\omega}} e_\mu(\mathbf{k}),\ \dfrac{1}{\sqrt{2\omega}} e_\mu(\mathbf{k})$	ingoing and outgoing photons
External negaton lines	$p \quad p$	$\sqrt{\dfrac{m}{\epsilon}}\, u(\mathbf{p}),\ \sqrt{\dfrac{m}{\epsilon}}\, \bar{u}(\mathbf{p})$	ingoing and outgoing negatons
External positon lines	$p \quad p$	$\sqrt{\dfrac{m}{\epsilon}}\, \bar{v}(\mathbf{p}),\ \sqrt{\dfrac{m}{\epsilon}}\, v(\mathbf{p})$	ingoing and outgoing positons

As mentioned in the previous section, positons move opposite to the arrows associated with positon lines. Therefore, the momentum associated with a positon line is $-q$ when we follow the arrow, but $+q$ when we follow the positon. This sign is important when we want to use the above rule that the sum of the momenta which enter a corner equals the sum of the momenta which leave this corner. This rule enables one to write down the matrix element directly in p-space and eliminates the awkward Fourier transform from x-space.

Another difference between Tables 8–1 and 8–2 is in the external lines. As was pointed out previously, in position space it is simpler to associate the *operators* to the external lines, but for the momentum space the corresponding *functions* are tabulated. The diagrams are drawn *after* the product of all creation and annihilation operators has been replaced by its matrix element ± 1. Therefore, each diagram is associated with a definite over-all sign which is to be determined from the ordered operator product. This sign expresses exactly the Pauli exclusion principle for electrons and assures the correct antisymmetrization of the initial and final states. Electron-

electron scattering, which is discussed below, furnishes a detailed example of this situation.

As before, Table 8–2 can be used to construct the S-matrix element for a given diagram. Since it is quite easy to enumerate all Feynman diagrams associated with a given process and a given order n of approximation (at least for those small n for which an evaluation is feasible), the construction of the matrix elements from the diagrams is the method actually adopted in applications. The p-space diagrams are usually preferred because of the simpler structure of the integrand, and because of the fact that the initial and final particle states are usually known in terms of their momenta rather than their positions.

However, the actual construction of the matrix elements from the diagrams, following Tables 8–1 or 8–2, is incomplete until we determine the correct numerical factor which multiplies the ordered product of the same n and the same process. This factor can be obtained as follows.

We first note that from the above discussion of the Fourier transform and the general structure of the ordered product it follows that the factor must be of the form

$$(-1)^l \delta_P i^\alpha (2\pi)^\beta e^n. \qquad (8\text{–}20)$$

In this numerical factor $(-1)^l$ is the sign associated with the closed loops and will be discussed in Section 8–4. δ_P is the sign which arises from the matrix elements of the ordered product, as explained earlier. If we want to construct $S_{fi}^{(n)}$ from the diagrams, we must determine this sign by other means. To this end we remember the physical meaning of this factor as explained above and conclude that it must be possible to determine this sign by proper antisymmetrization of the initial and final states of the electrons. We do this by keeping the particles fixed and permuting the states which we associate to these particles. It is sufficient to permute the final states only because any additional permutation of the initial states would be equivalent to a relabeling of the particles. We conclude that δ_P is the signature of the permutation P of final electron states.

The remaining factors in the expression (8–20) arise from the expansion (8–7) and the Fourier transforms. A counting of the internal and external lines yields the result

$$\alpha = E_i - P_i,$$

$$\beta = 4n - \tfrac{3}{2}(E_e + P_e) - 4(E_i + P_i).$$

The number of internal lines can be expressed in terms of the number of external lines and the order n (Section 10–1). The exponents α and β can therefore be written

$$\alpha = \tfrac{1}{2}(n + P_e - E_e), \qquad (8\text{–}21)$$

$$\beta = \tfrac{1}{2}(P_e + E_e) - 2n. \qquad (8\text{–}22)$$

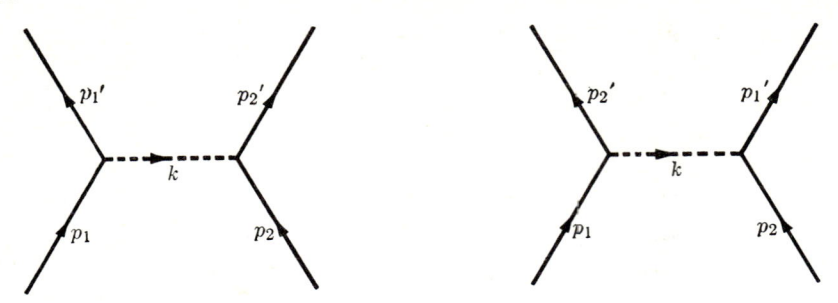

FIG. 8–4. Negaton-negaton scattering diagrams for momentum space in lowest order.

The factor $1/n!$ always cancels exactly. This statement is meaningful only where it is made concomitant with a definition which specifies when two diagrams are considered to be different. Since each photon is associated with a different momentum four-vector and each electron path can be labeled as referring to the first, second, etc., electron, we can state the following definition: *two diagrams are different when they differ in the sequence of the emitted and absorbed photons as one follows the arrows in the electron paths.* When we associate exactly one term to each of the different diagrams of a process, the factor $1/n!$ will always cancel. To see this, we need only to recognize that in the Fourier transform each dummy coordinate x can be associated to the n "definite" corners in exactly $n!$ ways. The corners are made definite by the connected momenta. For an open electron path this is quite simple to verify. For a closed loop we note that it does not matter where on the loop one starts going around. Therefore, external photon lines can be attached to a closed loop of p corners in $p!/p = (p-1)!$ different ways. This means that for a diagram that contains a closed loop of p corners there are $(p-1)!$ ways to insert this closed loop, and there are, therefore, $(p-1)!$ diagrams which differ only in the closed loop. A typical example will be discussed in Sections 13–1 and 13–2. The notation used in the expressions (8–20) to (8–22) is summarized below:

$$\left.\begin{array}{l} n = \text{number of corners} = \text{order of term in Eq. (8–7)} \\ l = \text{number of closed loops} \\ \delta_P = \text{signature of the permutations of final electron states} \\ E_i = \text{number of internal electron lines} \\ P_i = \text{number of internal photon lines} \\ E_e = \text{number of external negaton and positon lines} \\ P_e = \text{number of external photon lines} \end{array}\right\} \quad (8\text{–}23)$$

We are now ready to construct an S-matrix element from a diagram in momentum space. Let us consider the second order processes discussed in

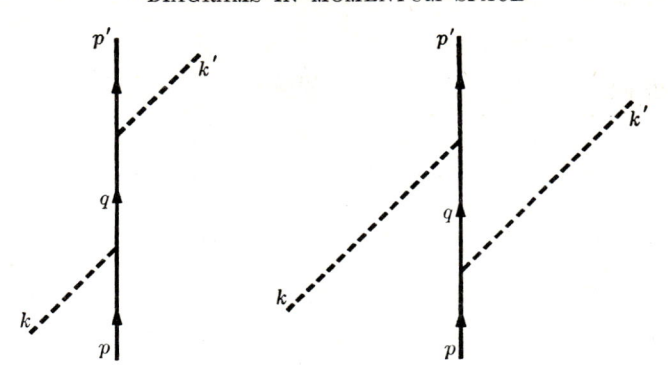

FIG. 8-5. Photon-negaton scattering diagrams for momentum space in lowest order.

Section 8-2. The diagrams for $S^{(2)}(e^- \text{-} e^-)$ in p-space are shown in Fig. 8-4. They differ only in the permutation of the final states.

Following Table 8-2 and the expressions (8-20) to (8-23), we obtain for the first diagram the values

$$n = 2, \quad l = 0, \quad \delta_P = 1, \quad E_i = 0, \quad P_i = 1, \quad E_e = 4, \quad P_e = 0,$$

and the matrix element

$$S^{(2)}(e^- \text{-} e^-)_1 = i^{-1}(2\pi)^{-2} e^2 \frac{m}{\sqrt{\epsilon_1' \epsilon_2'}} \frac{m}{\sqrt{\epsilon_1 \epsilon_2}}$$

$$\times \int \bar{u}(\mathbf{p}_1')\gamma^\nu \delta(p_1 - p_1' - k)u(\mathbf{p}_1) \frac{g_{\nu\lambda} d^4k}{k^2 - i\mu} \bar{u}(\mathbf{p}_2')\gamma^\lambda \delta(p_2 - p_2' + k)u(\mathbf{p}_2).$$

The second diagram will give exactly the same result, but p_1' and p_2' will be interchanged, and $\delta_P = -1$. After integration over k, we find for the sum of both diagrams

$$S^{(2)}(e^- \text{-} e^-) = \frac{e^2}{4\pi^2 i} \delta(p_1' + p_2' - p_1 - p_2) \frac{m^2}{\sqrt{\epsilon_1' \epsilon_2' \epsilon_1 \epsilon_2}}$$

$$\times \left[\frac{\bar{u}(\mathbf{p}_1')\gamma^\nu u(\mathbf{p}_1)\bar{u}(\mathbf{p}_2')\gamma_\nu u(\mathbf{p}_2)}{(p_1' - p_1)^2 - i\mu} - \frac{\bar{u}(\mathbf{p}_2')\gamma^\nu u(\mathbf{p}_1)\bar{u}(\mathbf{p}_1')\gamma_\nu u(\mathbf{p}_2)}{(p_2' - p_1)^2 - i\mu} \right]. \quad (8\text{-}24)$$

In a similar way, we can find the matrix element for Compton scattering from the diagrams of Fig. 8-5. With the values

$$n = 2, \quad l = 0, \quad \delta_P = 1, \quad E_i = 1, \quad P_i = 0, \quad E_e = 2, \quad P_e = 2,$$

we obtain for both diagrams

$$S^{(2)}(\gamma\text{-}e^-) = i(2\pi)^{-2}e^2 \frac{m}{\sqrt{\epsilon'\epsilon}} \int \bar{u}(\mathbf{p}') \left[\frac{\delta(q-p'-k')}{\sqrt{2\omega'}} e' \frac{i q - m}{q^2 + m^2 - i\mu} \right.$$

$$\times e \frac{\delta(p+k-q)}{\sqrt{2\omega}} + \frac{\delta(q-p'+k)}{\sqrt{2\omega}} e \frac{i q - m}{q^2 + m^2 - i\mu}$$

$$\left. \times e' \frac{\delta(p-q-k')}{\sqrt{2\omega'}} \right] u(\mathbf{p}) d^4q$$

$$= \frac{ie^2}{8\pi^2} \frac{m}{\sqrt{\epsilon'\epsilon\omega'\omega}} \delta(p'+k'-p-k)$$

$$\times \bar{u}(\mathbf{p}') \left[e' \frac{i(\not p'+\not k')-m}{(p'+k')^2+m^2-i\mu} e \right.$$

$$\left. + e \frac{i(\not p-\not k')-m}{(p-k')^2+m^2-i\mu} e' \right] u(\mathbf{p}). \quad (8\text{--}25)$$

These simple examples illustrate the effectiveness of the Feynman-Dyson diagrams. The evaluation of matrix elements by this method constitutes a significant advance in mathematical technique for treating problems in field theory. In the older form of the perturbation theory the time variable played a role distinguished from the space variables. The resulting unsymmetrical formulas were much more complicated than the corresponding formulas in the invariant perturbation treatment. Corresponding to this lack of symmetry, one had in the older formulation conservation of momentum but not conservation of energy for the intermediate states. In the present formulation one has conservation of energy and momentum at each corner of the diagram. On the other hand, the energy-momentum relation which is valid for free particles is not true for the particles in the intermediate states. The energy denominators of the old-fashioned perturbation theory are here replaced by the invariant factors k^2 and $p^2 + m^2$ in the denominators of the propagation factors.

This symmetrical treatment of the space-time variables has also enabled us to simplify the treatment of virtual negatons and positons in the intermediate states. The contributions from the two types of particles are contained in the one propagation function for the matter field. In the older formalism the two terms referring to negatons and positons had to be written separately and could be united later only by using certain identities involving the spinor wave functions.

These remarks show why the covariant formulation is actually simpler and quicker to use than the older noncovariant formulation which was based

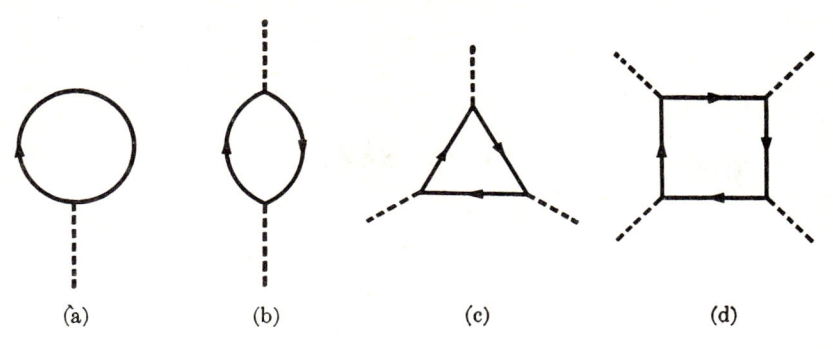

FIG. 8-6. Closed-loop diagrams.

on the classical Hamilton-Jacobi theory and its canonically conjugate variables. A second, and much more relevant, advantage of the new formulation will be discussed in connection with the problem of divergences in Chapters 9 and 10.

The graphical representation of the covariant perturbation calculation has proved not only an aid in computation but also a powerful tool in the investigation of general problems. An example will be found in the next section, where Furry's theorem is proved. This theorem is much more difficult to prove in the older formalism.

8-4 Closed loops. There are a number of important features connected with the occurrence of closed loops in diagrams. We want to discuss them in this section. A typical closed loop was shown in the last drawing of Fig. 8-1. Closed loop diagrams with one to four corners in the loop are shown in Fig. 8-6.

A closed loop is an electron path which is closed in itself and therefore does not contain free electron functions, but only electron propagation functions. It follows that a closed loop must necessarily correspond to a double sum over spinor indices, i.e., to a trace.

An important property of diagrams containing closed loops was found in 1937 by Furry.* He proved a special case of a theorem which may be stated as follows.

FURRY'S THEOREM. *Diagrams which contain a closed loop with an odd number of corners contribute nothing to the matrix elements.*

The proof of this theorem is based on the charge symmetry of the theory. Consider a diagram, D, which contains a closed loop, C. This loop will have a uniform sense of direction determined by the arrows of this closed electron line. It then follows that in the set of diagrams of the same order

---

* W. H. Furry, *Phys. Rev.* **51**, 125 (1937).

which belongs to the same physical process, there is another diagram, D', which differs from D only in the sense of direction of its closed loop, C'. This loop has the same number of corners, m, as C has. Since the only difference between C and C' is the arrow direction, the loop C' corresponds to a factor in the integrand associated with D' which is exactly the charge conjugate to that factor in the integral for D which corresponds to C. This factor is a trace, and arose from the complete contraction of the ordered product of m factors $\bar{\psi}\gamma^\mu\psi$. Since $\bar{\psi}^c\gamma^\mu\psi^c = -\bar{\psi}\gamma^\mu\psi$, the charge conjugate to this trace is exactly $(-1)^m$ times that trace. Therefore, the integrals for D and D' differ only by the factor $(-1)^m$. The sum of these two diagrams (which is part of the corresponding matrix element $S_{fi}^{(n)}$) is $2D$ for even m, and 0 for odd m. This concludes the proof of Furry's theorem. We can generalize this theorem as follows.

GENERALIZED FURRY THEOREM. *Any diagram or part of a diagram from which only photon lines emerge does not contribute to the matrix element if the number of these photon lines is odd.*

In order to prove this generalization, we first consider a diagram with more than one closed loop. Since the above argument can be applied to each closed loop separately, it follows that a diagram will not contribute whenever it contains one or more closed loops with an odd number of corners. As a second step, we can verify that any diagram or part of a diagram from which only an odd number of photon lines emerges necessarily contains at least one closed loop with an odd number of corners. The above generalization is an immediate consequence of these remarks.

As an example, we consider the diagrams in Fig. 8–6. According to the above theorems, diagrams (a) and (c) do not contribute to the matrix elements, whereas the contributions from diagrams (b) and (d) are exactly the same as those of their charge conjugate diagrams.

We now want to establish the factor $(-1)^l$ in the expression (8–20). To this end, we consider two diagrams which describe the same process and differ only in the number of closed loops which they contain. If they differ by one closed loop their difference must be due to the contraction of an *even* number of products $\bar{\psi}\gamma^\mu\psi$, since by Furry's theorem that closed loop must contain an even number of corners. But the signature of the cyclic permutation of an even number of elements is odd, such that the two diagrams must differ in their signature. It follows that if two diagrams differ by l closed loops their signatures will differ by $(-1)^l$.

Another important theorem about closed loops concerns their non-relativistic limit:

The S-matrix elements of any process described by diagrams in which all external lines are photon lines vanishes in every order of approximation when any one of the external photon momenta vanishes.

This theorem is a direct consequence of gauge invariance. It is a theorem about closed loops, because the diagrams in question necessarily consist of closed loops only.

In order to prove this theorem, we write the matrix element in its most general form. Let $e_\mu^{(1)}, e_\nu^{(2)}, \cdots, e_\rho^{(m)}$ be the polarization vectors associated with the external photon lines, and let $k^{(1)}, \cdots, k^{(m)}$ be their momenta. The matrix element can be written

$$M = e_\mu^{(1)} e_\nu^{(2)} \cdots e_\rho^{(m)} M^{\mu\nu\cdots\rho}(k^{(1)}, k^{(2)}, \cdots, k^{(m)}).$$

The γ-matrices will occur in $M^{\mu\nu\cdots\rho}$ only in the form of traces. The tensor $M^{\mu\nu\cdots\rho}$ must therefore be constructed from the m vectors $k^{(1)} \cdots k^{(m)}$. Since M must be invariant under the gauge transformations

$$e_\mu^{(1)} \to e_\mu^{(1)} + k_\mu^{(1)} \chi,$$

$$e_\nu^{(2)} \to e_\nu^{(2)} + k_\nu^{(2)} \chi,$$

etc.,

it follows that

$$k_\mu^{(1)} M^{\mu\nu\cdots\rho} = 0,$$

$$k_\nu^{(2)} M^{\mu\nu\cdots\rho} = 0,$$

etc.,

provided $k_\mu^{(1)} k^{\mu(1)} = 0$, $k_\nu^{(2)} k^{\nu(2)} = 0$, etc. These equations are satisfied if and only if for small $k^{(i)}$

$$M^{\mu\nu\cdots\rho} = \text{const.}\, k^{(1)\mu} k^{(2)\nu} \cdots k^{(m)\rho}(1 + O(k^{(1)} \cdots k^{(m)})).$$

From this result follows the above theorem.

A generalization of this theorem is obtained when some (but not all) of the external photon lines are attached to another diagram which is not necessarily a closed-loop diagram. We can conclude from the above arguments that in this case *the contributions from any arbitrary diagram vanish when the external photon momentum vanishes, provided this photon is attached to a closed loop in this diagram.* This conclusion is based on the fact that the structure of $M^{\mu\nu\cdots\rho}$ remains essentially unchanged when the closed loop diagram is part of a larger diagram. The fact that certain $k_\mu^{(i)} k^{(i)\mu} \neq 0$ will not affect the argument, so long as the momentum of the *external* photon line is a null vector.

8–5 The substitution law. One of the most powerful consequences of the structure of the S-matrix in the iteration solution is the substitution law. It is most easily expressed in terms of the diagrams in momentum space, as follows.

If two diagrams M' and M differ only in one external line, such that this line is an *outgoing* photon, negaton, or positon line in M' and an *ingoing* photon, positon, or negaton line, respectively, in M, then the S-matrix elements associated with M' and M are related as follows:

M'	M		
k' out	k in	$k' \leftrightarrow -k$	$e' \leftrightarrow e$
p' out	q in	$p' \leftrightarrow -q$	$u(p') \leftrightarrow v(q)$
q' out	p in	$q' \leftrightarrow -p$	$v(q') \leftrightarrow u(p)$

In case of circular polarization: right-circular $\leftrightarrow$ left-circular polarization.

The double arrow indicates that the substitution necessary to obtain M from M' is reversible, so that one can also obtain M' from M. The relation is therefore a one-to-one correspondence.

The proof of this law follows by inspection from the general structure of the S-matrix elements.

For the purpose of clarification, we remind the reader that the negaton momenta p, p' and the positon momenta q, q' are the actual physical four-vectors, so that q and q' are directed opposite to their diagram arrows. The substitutions mean simply that the corresponding outgoing electron line of the diagram was "bent" into an ingoing electron line; the arrow thereby changes its time-direction, so that every complete electron path remains associated to a single arrow-direction (but not necessarily time-direction).

This substitution law can also be applied to the square of the matrix elements after a spin summation has been carried out. In that case, the wave functions have disappeared and projection operators take their place. The substitution can then be carried out directly in these, as follows (*cf.* Section A2–5):

$$\text{for } p' \leftrightarrow q: \quad \Lambda_-(p') \leftrightarrow -\Lambda_+(q) = -\Lambda_-(-q),$$
$$\text{for } q' \leftrightarrow p: \quad \Lambda_+(q) \leftrightarrow -\Lambda_-(p) = -\Lambda_+(-p). \tag{8-26}$$

This simply means that there is also an over-all sign change of the trace in addition to the momentum substitution.

It should be clear that repeated application of the substitution law permits "bendings" of the external lines of a diagram in all possible ways. Some of the resulting diagrams will be excluded, of course, by conservation of energy and momentum.

The importance of the substitution law for the evaluation of matrix elements and cross sections is rather plain from these remarks. As an example, we cite the three processes of Compton scattering, two-photon free-pair annihilation, and pair production by two photons. When any one of these processes is computed the others follow by substitution (*cf.* Sections

11–1, 12–4, and 13–3). However, the original matrix elements (or trace) must be known in complete generality (without restriction to a special coordinate system) in order to make this procedure useful.

Special cases of the substitution law have been known for some time. In particular, the relation between bremsstrahlung and pair production was recognized very early.* Later work includes the relation of Møller scattering and *Bhabha scattering*,† and a complete list of all possible processes associated with lowest-order Compton and double Compton scattering‡. But a complete and general discussion has never been given.

8–6 Lifetimes and cross sections. The S-matrix permits the calculation of two types of observable quantities, lifetimes and cross sections. Both can be calculated from the transition probability per unit space-time volume.

Consider the transition of a system from an initial state ω_i to a final state $\omega_f \neq \omega_i$. The matrix element of S may then be replaced by that of

$$R = S - 1, \tag{8-27}$$

$$S_{fi} \equiv (\omega_f, S\omega_i) \equiv (f|S|i) = (f|R|i), \quad (\omega_f \neq \omega_i). \tag{8-28}$$

We conclude from Section 8–3 that the matrix elements of the operator R are of the form

$$(f|R|i) = \delta(P_f - P_i)(f|M|i), \tag{8-29}$$

where P_i and P_f are the initial and final four-momenta of the system. Apparently, $(f|R|i)$ is the *transition probability amplitude* for a transition taking place over all space and all time from the infinite past to the infinite future. The corresponding *transition probability*, $|(f|R|i)|^2$, is not a meaningful quantity (and not at all a probability), since observations are carried out over finite times and only the *transition probability per unit time* is essentially measurable. Indeed, Eq. (8–29) gives infinity for $|(f|R|i)|^2$ and simply expresses the fact that, during an infinite time, a nonzero incident flux of particles will cause an infinite number of repetitions of the elementary process under consideration.

Hence the interesting quantity is the transition probability per unit time, Γ, or, for convenience in our covariant formalism, the *transition probability per unit space-time volume* Γ_1. The latter can be obtained as a

---

* H. A. Bethe and W. Heitler, *Proc. Roy. Soc.* **A146**, 83 (1934).
† K. Baumann, *Acta Phys. Austriaca* **7**, 96 (1953).
‡ B. Kockel, *Ann. Phys.* **4**, 279 (1949).

limit from finite space-time volumes,

$$\Gamma_1 = \lim_{\substack{V \to \infty \\ T \to \infty}} \frac{|(f|R|i)|_{VT}{}^2}{VT}. \tag{8-30}$$

Here, $|(f|R|i)|_{VT}{}^2$ is $|(f|R|i)|^2$ calculated for a finite space-time volume VT. We remember that the expression (8–29) resulted from an integration over *infinite* space time,

$$(f|R|i) = (2\pi)^{-4}(f|M|i)\int e^{i(P_f - P_i)x}d^4x.$$

Therefore,

$$(f|R|i)_{VT} = (2\pi)^{-4}(f|M|i)\int_{VT} e^{i(P_f - P_i)x}d^4x. \tag{8-31}$$

The evaluation of Γ_1 involves the limit

$$\lim_{\substack{V \to \infty \\ T \to \infty}} \frac{1}{VT}\left|\int_{VT} e^{i(P_f - P_i)x}d^4x\right|^2 = (2\pi)^4\delta(P_f - P_i). \tag{8-32}$$

The result (8–32) is easily obtained as the four-dimensional generalization of the limit

$$\lim_{T \to \infty} \frac{1}{T}\left|\int_{-T/2}^{T/2} e^{iEt}dt\right|^2 = \lim_{T \to \infty} \frac{1}{T}\left[\frac{\sin(ET/2)}{E/2}\right]^2 = 2\pi\delta(E). \tag{8-33}$$

We combine Eqs. (8–30) to (8–32) and find for the transition probability per unit space-time volume,*

$$\Gamma_1 = (2\pi)^{-4}\delta(P_f - P_i)|(f|M|i)|^2. \tag{8-34}$$

In this transition probability the normalization of the particle flux is determined by the normalization of the wave functions. These were so chosen that we have a density of one electron per invariant space volume $V(m/\epsilon)$ with $V = 1$. Therefore, the number of states available for a particle of momentum between p^μ and $p^\mu + dp^\mu$, in a unit volume and of definite spin, is

$$\frac{m}{\epsilon}\frac{d^3p}{(2\pi)^3}. \tag{8-35}$$

A comparison with Eqs. (8–17) tells us that this is exactly the factor which will occur in our momentum-space integrals for every electron line which is not closed. Thus, in these integrals electrons are normalized with respect to the volume $(2\pi)^3$ of the unit cell in phase space.

We further conclude that in position space all particle wave functions are normalized to a spatial volume $(2\pi)^3$. It will be more convenient in

---

* A more satisfactory derivation was given by J. M. Jauch, *Helv. Phys. Acta* **31**, 127 and 661 (1958).

what follows to normalize in a volume V. This means that we must multiply the integrals by $(2\pi)^{3/2}/V^{1/2}$ for every external line. We want to show explicitly how the physically meaningful quantities emerge and that they are independent of V. First we note that the *density of final states* will now be different from (8–35),

$$\frac{m}{\epsilon} \frac{V d^3 p}{(2\pi)^3} \quad \text{for electrons,} \tag{8-36}$$

and

$$\frac{V d^3 k}{(2\pi)^3} \quad \text{for photons.} \tag{8-37}$$

The factors $1/V$ which occur for each outgoing line in the transition probability will therefore exactly cancel the factors V in Eqs. (8–36) and (8–37). There remains a factor

$$\frac{(2\pi)^{3g}}{V^g} \tag{8-38}$$

if there are g ingoing particles.

With this choice, Γ_1 still depends on the details observed in the experiments. For example, if the polarization of the outgoing electron is not observed, a summation over the final polarization states must be carried out. For the initial states, a suitable average must be found. Thus, with (8–38),

$$\Gamma_1 = \frac{(2\pi)^{3g-4}}{V^g} \mathsf{S}_f \overline{\mathsf{S}_i} \delta(P_f - P_i) |(f|M|i)|^2, \tag{8-39}$$

where the generalized summation symbols S indicate integrations over momenta and summations over spins and polarizations, depending on the type of process. The bar over S_i indicates the averaging process.

In the expression (8–39) it is not necessary to restrict the considerations to initial negatons, positons, and photons. Indeed, it is, in general, possible to have the interaction between g *systems*, each of which consists of these three types of particles. Equation (8–39) will be formally the same. Let us then consider one single initial system ($g = 1$). The transition probability per unit time into *all* other possible states will be

$$\Gamma_{\text{tot}} = V \Gamma_1 = V \frac{1}{2\pi} \frac{1}{V} \mathsf{S}_f \overline{\mathsf{S}_i} \delta(P_f - P_i) |(f|M|i)|^2. \tag{8-40}$$

We see that Γ_{tot} is independent of V. Its inverse,

$$\tau = \frac{1}{\Gamma_{\text{tot}}} \tag{8-41}$$

is called the *lifetime* of the system.

Consider the case of two systems or particles ($g = 2$). Let p_1 and p_2 be their four-vectors of momentum and energy. Let $q_1 = |\mathbf{p}_1|$ and $q_2 = |\mathbf{p}_2|$ be the absolute values of their momenta, ϵ_1 and ϵ_2 the values of their energies,

$$p_i^2 + m_i^2 = q_i^2 - \epsilon_i^2 + m_i^2 = 0.$$

In the center-of-momentum system, $\mathbf{p}_1 = -\mathbf{p}_2$ and $q_1 = q_2 = q$. The flux density I is determined by the velocities $\beta_i = q_i/\epsilon_i$,

$$I = \frac{1}{V}(\beta_1 + \beta_2) = \frac{q}{V}\left(\frac{1}{\epsilon_1} + \frac{1}{\epsilon_2}\right). \tag{8-42}$$

This last expression can be written in such a way that it is valid in any reference system,*

$$I = \frac{1}{V}\frac{1}{\epsilon_1 \epsilon_2}|F|, \tag{8-43}$$

$$F^2 = (p_{1\mu}p_2^\mu)^2 - m_1^2 m_2^2 = -\tfrac{1}{2}(p_{1\mu}p_{2\nu} - p_{1\nu}p_{2\mu})(p_1^\mu p_2^\nu - p_1^\nu p_2^\mu).$$

One verifies easily that Eq. (8–43) is equivalent to Eq. (8–42) in the center-of-momentum system. On the other hand, in the rest-frame of system 1 ($m_1 \neq 0$), we obtain

$$F = m_1 p_2 \tag{8-44}$$

and

$$I = \frac{1}{V}\frac{p_2}{\epsilon_2} = \frac{\beta_2}{V}. \tag{8-45}$$

If one of the systems (system 1, say) is a set of photons,

$$F = |p_{1\mu}p_2^\mu| \tag{8-46}$$

and

$$I = \frac{1}{V}\frac{|p_{1\mu}p_2^\mu|}{\omega_1 \epsilon_2}. \tag{8-47}$$

If both systems are photons then, in the center-of-momentum system,

$$I = \frac{2}{V}. \tag{8-48}$$

We now notice from Eq. (8–39) that for $g = 2$, $\Gamma_1 \to 0$ for $V \to \infty$, and also $\Gamma = V\Gamma_1 \to 0$ in this limit. But the ratio of the transition probability per unit time, Γ, and the flux density in the initial state, I, is finite. This ratio for a transition into a given set of final states is called the cross section,

---

* C. Møller, *Kgl. Danske Videnskab. Selskab* **23**, No. 1, Section 2, p. 18 ff.

σ, of that process:

$$\sigma = (2\pi)^2 \frac{\epsilon_1 \epsilon_2}{F} \overline{S_f S_i} \delta(P_f - P_i) |\langle f|M|i\rangle|^2, \qquad (8\text{--}49)$$

$$F = \sqrt{(p_{1\mu} p_2{}^\mu)^2 - m_1{}^2 m_2{}^2}. \qquad (8\text{--}50)$$

According to the summations in S_f there exist various *partial cross sections*. If the momentum vectors of the final state fall within certain differential ranges we use the term *differential cross sections*.

8–7 Evaluation of the S-matrix in the Heisenberg picture.* We conclude this chapter with an approximate evaluation of the S-matrix in the Heisenberg picture. This evaluation will be subject to the same restrictions as the iteration method discussed in Section 8–1. The validity of Eq. (7–80) is assumed.

We start with the inhomogeneous field equations (4–6), (4–15) for the Heisenberg fields. The general form of these equations is

$$L(\Phi(x)) = F(x), \qquad (8\text{--}51)$$

where L is a linear differential operator and $F(x)$ is the interaction term which generally involves the product of several field components. For instance, for Eq. (4–6), $L = \partial^\nu \partial_\nu$ and $F(x) = ie\,\overline{\Psi}(x)\gamma^\mu \Psi(x)$.

We can replace Eq. (8–51) by an integral equation. For this purpose, we introduce the Green's function $G(x_1, x_2)$ with the property

$$L_1(G(x_1 x_2)) = -\delta(x_1 - x_2). \qquad (8\text{--}52)$$

L_1 is the differential operator with respect to the first variable x_1, and $\delta(x_1 - x_2)$ is the four-dimensional δ-function. It follows from (8–51) and (8–52) that

$$\varphi(x) = \Phi(x) + \int G(x, x') F(x') dx' \qquad (8\text{--}53)$$

is a solution of the homogeneous equation

$$L(\varphi(x)) = 0. \qquad (8\text{--}54)$$

Equation (8–52) does not determine the Green's function uniquely. It is necessary to specify the boundary conditions. But, since L is linear, two solutions of (8–52) differ by a solution of the homogeneous equation $L(g(x)) = 0$.

---

* This section is based on the work of C. N. Yang and D. Feldman, *Phys. Rev.* **79**, 972 (1950) and G. Källén, *Ark. Fys.* **2**, No. 37, 371 (1950).

To each solution of (8–52) corresponds a solution $\varphi(x)$ of (8–54), according to (8–53). We determine the free field $\varphi(x)$ and with it the Green's function by imposing certain boundary conditions on these fields. For instance, if we put

$$\varphi(x,-\infty) \equiv \varphi(x), \quad \varphi(x,+\infty) \equiv \varphi'(x),$$

where $\varphi(x,\tau)$ are the τ-dependent free fields which were defined in (4–21), we obtain two equations from (8–53):

$$\varphi(x) = \Phi(x) + \int G_R(x,x')F(x')dx', \qquad (8\text{–}53)_R$$

$$\varphi'(x) = \Phi(x) + \int G_A(x,x')F(x')dx'. \qquad (8\text{–}53)_A$$

The functions G_R, G_A are special solutions of Eq. (8–52) defined by the properties

$$G_R(x_1,x_2) = 0 \quad \text{for } \tau(x_1) = -\infty, \qquad (8\text{–}55)_R$$

$$G_A(x_1,x_2) = 0 \quad \text{for } \tau(x_1) = +\infty. \qquad (8\text{–}55)_A$$

They correspond to the retarded and advanced solutions of (8–52). For the operator $L = \partial^\nu \partial_\nu$ these functions are simply D_R and D_A (*cf.* Appendix Section A1–1).

From $(8\text{–}53)_R$ and $(8\text{–}53)_A$, we obtain

$$\varphi(x) = \varphi'(x) + \int g(x,x')F(x')dx', \qquad (8\text{–}56)$$

where $g(x,x') = G_R(x,x') - G_A(x,x')$ is a solution of the homogeneous equation

$$L(g(x,x')) = 0. \qquad (8\text{–}57)$$

We recall the relation

$$\varphi'(x) = S^{-1}\varphi(x)S, \qquad (8\text{–}58)$$

which is a special case of (7–9). Substituting this into (8–56) and multiplying through with S, we obtain the basic equation of this method:

$$[S,\varphi(x)] = S \int g(x,x')F(x')dx'. \qquad (8\text{–}59)$$

This relation is still rigorous. No approximations have been made up to this point. But the equation can only be solved with approximations. We obtain first a solution for the Heisenberg fields in powers of e.

$$\Phi = \varphi_0 + \varphi_1 + \cdots + \varphi_n + \cdots, \qquad (8\text{–}60)$$

where φ_n is of nth order in e, and $\varphi_0 = \varphi$. The φ_n then satisfy a recursion relation

$$\varphi_n(x) = -\int G_R(x,x')F_{n-1}(x')dx' \qquad (8\text{-}61)$$

which is obtained from $(8\text{-}53)_R$. F_{n-1} is the term of order $n-1$ in the inhomogeneous part of the field equations. It involves only the fields up to the $(n-1)$th approximation. Thus, Eq. $(8\text{-}61)$ gives a recursion relation of the φ_n in terms of the φ_r $(r < n)$.

We solve $(8\text{-}59)$ with a similar procedure:

$$S = S_0 + S_1 + \cdots + S_n + \cdots . \qquad (8\text{-}62)$$

If we denote the right side of $(8\text{-}59)$ by T, and call T_n its nth order term in powers of e, we may write

$$[S_n,\varphi(x)] = T_n. \qquad (8\text{-}63)$$

The term T_n involves only S_r with $r < n$, and $(8\text{-}63)$ is therefore a recursion relation for S_n. It is thus possible to solve S_n in terms of the S_r with $r < n$.

The method sketched here can be applied to the coupled radiation and matter fields. It is not difficult to show that for the lowest order terms one obtains the same results as with the diagrams. The direct proof of the equivalence of the general terms to the result obtained with diagrams is exceedingly complicated* and will not be given here.

It is seen that this method is much less convenient than the diagram method. It is not of much use where the diagram method can be applied.

In the field theories with nonlocal interactions† it has so far been impossible to construct solutions with diagrams. In such a case the method of this section is the only one available for the evaluation of the S-matrix.

---

* *Cf.* G. Källén, *loc. cit.*

† Field theories with nonlocal interaction were studied by G. Wataghin, *Z. Physik* **86**, 92 (1934), **92**, 547 (1935); H. McManus, *Proc. Roy. Soc.* **A195**, 323 (1949); J. Rayski, *Phil. Mag.* **42**, 1289 (1951); P. Kristensen and C. Møller, *Kgl. Danske Videnskab. Selskab* **27**, No. 7 (1952).

CHAPTER 9

THE DIVERGENCES IN THE ITERATION SOLUTION

If quantum electrodynamics were a complete and satisfactory theory, it would be a mathematically rigorous and logically consistent structure which allows—at least in principle—the calculation of all radiative processes. This cannot be said without qualifications of the theory in its present state of development. The equations of this theory seem to lead to mathematically inconsistent consequences. In spite of this defect, it has been possible to extract from this incomplete theory results which are in perfect agreement with empirical data. If we bear in mind that some of these results are far from simple in their analytical form, and that they can be derived from the basic equations of the theory so far developed without any further *ad hoc* assumptions, we are driven to the conclusion that the theory in its present form is essentially correct. There is not one single experimental fact known today concerning radiative processes which could not be quantitatively explained by this theory.

Before we elaborate on these satisfactory aspects of the theory we have to dwell on its pathological features. These appear always in the form of infinities or ambiguities at certain stages of the calculations. The handling of these infinities involves delicate considerations which are neither unambiguous nor final. No pretense is made to cover up the inherently heuristic and tentative character of some of these manipulations. We rather hope that the attentive reader will feel the challenge offered by this state of affairs.

We note here that all these difficulties appear when we try to obtain solutions of the field equations with the approximation method which we previously described as the *iteration solution*. An important problem is therefore this: Are these divergences a consequence of this particular approximation method or are they difficulties inherent in the structure of the theory? G. Källén has recently been able to show that the second alternative is the correct one.* This was accomplished by demonstrating that the renormalized field equations lead to an inconsistency unless the renormalization constants are infinite, independent of the use of the iteration solution. In this and the following chapter we shall discuss the renormalization theory in the iteration solution. A further study of Källén's interesting work must be left to the reader.

---

* G. Källén, *Kgl. Danske Videnskab. Selskab* **27**, 12 (1953); *Helv. Phys. Acta* **25**, 417 (1952).

We shall begin in Section 9–1 with a brief historical sketch of the problem. In Section 9–2 we give a preliminary classification of the different types of divergent expressions encountered in this theory. In the subsequent four sections, 9–3, 9–4, 9–5, and 9–6, we discuss in detail the vacuum fluctuations, the self-energy of the electron and of the photon, and, finally, the so-called vertex parts. The last three of these are evaluated to second order in e and the removal of the divergences by the method of renormalization of mass and charge is indicated.

9–1 Historical background. The divergences which we encounter in quantum electrodynamics have different origins. Some of them can be avoided with a relatively minor formal modification of the theory, some of them have their roots almost certainly in the inadequate mathematical methods employed, still others are probably introduced by the very concept of the localizable field and the particular form of the interaction. All of them are intimately connected with the fact that a field has infinitely many degrees of freedom.

We gain a better insight into the origin of the various difficulties if we start with a brief historical sketch.* Before the advent of quantum mechanics there were two outstanding problems which marred the success of the classical theory of the electromagnetic field. They were the blackbody radiation and the theory of the electron. It is well known now how Planck's postulate of the finite quantum of action successfully resolved the first of these difficulties. The difficulties connected with the existence of electrons and other point particles are much more serious. They lie essentially in the fact that the existence of the electron leads in the classical theory to a dilemma. Either one assumes the electron as a point without structure; then the total energy of the electromagnetic field becomes infinite and, therefore, the mass of the electron is infinitely large, too. Or, one assumes a finite extension of the electron; then the stress of the electromagnetic field will tend to explode this charge distribution. In the latter case, the electron may also have properties which depend on the actual distribution of the charge. Such structure-dependent properties are characteristic of the model, but have nothing to do with reality.

It has become possible, by a more sophisticated approach, to overcome some of these defects in the classical models. However, we shall not discuss these developments. They have shed no light on the corresponding problem of the self-energy in the quantum theory.

Unlike the case of the blackbody radiation, the quantum theory of the electron has brought no relief from the above-mentioned problems of the

---

* For a more detailed account of the history of some of these problems we refer to A. Pais, *Developments in the Theory of the Electron*, Institute for Advanced Study and Princeton University, 1948, which covers the period up to December 1947.

self-energy. In fact, the difficulty becomes worse. In addition to the electrostatic self-energy which already occurs in the classical problem, there is electromagnetic self-energy caused by the interaction of the electron with the radiation field. This effect can be calculated only with perturbation theory, but we know that it is divergent in every approximation.† A similar divergence occurs for electrons which are bound by an external field.‡

Self-energy is not the only problem relating to the structure of the electron. In fact, closely related to it is the problem of *self-stress*, which needs satisfactory treatment in a consistent quantum electrodynamics.§

Although the presence of virtual pairs seems to ameliorate the divergence of the self-energy, it is itself the cause of new complications. For in pair theory the vacuum can be polarized by an external field and the polarizability is infinite, too. This problem does not exist in the classical case. Therefore, the quantum theory of the electron, which is necessarily a many-particle theory, introduces new complications. This fact makes it very improbable that a solution of the divergence problem can be found by searching for a more suitable *classical* model. We believe the divergence problem must eventually be solved within the framework of relativistic *quantum* theory.

We have actually seen an example of a divergence of purely quantum theoretical origin in connection with the zero-point energy of the radiation field discussed in Section 2–9. The energy for each radiation field oscillator, according to Eqs. (2–77) and (2–78), is of the form

$$\mathsf{E} = \frac{\omega}{2}(aa^* + a^*a).$$

For the vacuum state this operator assumes its minimum value $\frac{1}{2}\omega$. The total contribution of all degrees of freedom to the energy of the vacuum state is therefore infinite. In Section 2–9 we circumvented this difficulty by the simple requirement that quadratic expressions of the field operators should always be replaced by their ordered products [for the definition see Eq. (2–118)]. With this postulate we obtained the energy operator for each degree of freedom,

$$\mathsf{E}' = \omega a^* a = \mathsf{E} - \tfrac{1}{2}\omega.$$

---

† The first calculation is due to I. Waller, *Z. Physik* **62**, 673 (1930), who found that the self-energy is quadratically divergent. However, the inclusion of the contribution due to virtual pair creation reduces it to a logarithmic divergence; V. Weisskopf, *Phys. Rev.* **56**, 72 (1939).

‡ J. R. Oppenheimer, *Phys. Rev.* **35**, 461 (1930).

§ This problem is discussed in Section 16–4.

The total energy as well as the total momentum for the vacuum is then zero.

A similar situation was encountered in connection with the matter field. In Sections 3–8 and 3–10 we showed that a satisfactory definition of the vacuum state, as in the previous case, can only be given by defining quadratic expressions as ordered products. We have seen also that this definition is in accord with a basic requirement of the theory, namely, the charge symmetry. This is a simple example illustrating the fact that general requirements of invariance may be useful in deciding the value of divergent or ambiguous expressions. While these infinities are relatively harmless and could easily be disposed of in this manner, the divergences connected with the self-energy and the vacuum polarization remained an unsolved problem for a long time.

In 1947 a new development started with the discovery of the electromagnetic level shifts. The theoretical problems raised by these experimental results were successfully solved by the introduction of the concept of *renormalization*. The subsequent theoretical developments led to a reformulation of the theory in which the covariant aspects were more strongly emphasized. This made it possible to extend the idea of renormalization to all orders of approximation, and, consequently, many effects which formerly were inaccessible to calculations could be described quantitatively.

The fundamental problems which remain at the present stage of development center around the correct description of the "free particles" and their interaction. If a description could be found so that it incorporated the properties due to all the virtual processes, we would presumably obtain a satisfactory theory.

9–2 Classification of divergences. In this chapter we shall be concerned with the divergences which occur in the S-matrix. In the preceding chapter we have developed a method for evaluating any element of the S-matrix to an arbitrary order of approximation. Any given process is characterized by a family of diagrams with a certain number of incoming and outgoing external lines. To each such diagram there is associated a definite analytical expression which can be written down with the help of Tables 8–1 or 8–2.

The matrix elements which are obtained in this manner for the higher order approximations are, however, useless in the present form. In fact, only the lowest order diagrams for a given process are usually finite. The higher order corrections lead to divergent integrals. Thus, it may seem impossible to include in a consistent manner the radiative corrections of higher orders in e. Nevertheless, as was first shown by Dyson, with certain assumptions the S-matrix elements can be consistently evaluated to arbitrarily high orders in e.

The divergent diagrams are dealt with by various methods. In order to characterize these methods briefly it is convenient to classify the divergent diagrams as follows:

(a) Divergences associated with the description of the vacuum.
(b) Infrared divergences.
(c) Divergences associated with certain closed loops.
(d) "Serious" divergences.

All of these classes except (b) occur for large momenta of the intermediate states.

The divergences of class (a) are discussed in Section 9–3, where it is shown that these terms simply have the effect of multiplying any element of the S-matrix by the same phase factor. Although the phase is infinite, this factor cannot lead to any observable effects, since such effects are always expressed by the absolute value of the matrix elements. This circumstance makes it possible to ignore these terms altogether.

The *infrared* divergences of class (b) are entirely caused by the particular mathematical method employed for constructing solutions. We shall give a detailed analysis of this problem in Section 16–1, where we shall also show that it is possible to avoid this difficulty by an improved mathematical treatment of the soft photon emission processes.

Class (c) contains the divergences associated with the photon self-energy and the scattering of light by light. The handling of these divergences is a much more delicate problem. A possible approach is to invoke the invariance of the theory under gauge transformations. We have formulated the free fields and the coupling of the fields in a gauge-invariant manner. The consistency of the theory then requires that all the results also be gauge-invariant. We must therefore choose the mathematically undetermined values of the divergences in class (c) so as to conform with the postulate of gauge invariance. This procedure has been widely used in the past and it is perhaps the easiest way out of a not entirely satisfactory situation.

However, it should not be overlooked that there are other methods available to handle the difficulties caused by these divergences which lead to the same final result. We mention here one which has important arguments in its favor. This method introduces gauge invariance only for the coupled and renormalized fields. The free photons are treated as a neutral vector meson field of undetermined mass κ. In addition to the ordinary interaction, one assumes a direct interaction of the form $\lambda(a_\mu a^\mu)^2$. The constants κ and λ are new parameters ultimately to be determined by the physical requirements of gauge invariance of the renormalized S-matrix. With this approach, the divergences of class (c) can be treated in exactly the same spirit as those of class (d) (see below).

The main advantage of this method is that it is not necessary to invoke gauge invariance in order to obtain definite values for the undetermined integrals. In fact, the values of these integrals never appear explicitly in any final result. An additional advantage is the fact that it is easier to discuss the subsidiary condition for a vector meson field with finite mass. These advantages are partly offset by the considerable complexity of the mathematical manipulations. We shall treat the second order photon self-energy with this method for illustrative purposes. Since the final results obtained with the two methods are identical, we shall use the first method, which is mathematically simpler, for the rest of the discussion on these divergences.

The "serious" divergences (class d) are associated with vacuum polarization, electron self-energy, and the so-called vertex part. They are called "serious" because there seems to be no simple, consistent way to eliminate them. In particular, no appeal to invariance properties affects them. One proceeds, instead, in two steps. In the first step we separate the infinite from the finite, observable parts of the matrix elements. This can be accomplished relatively easily. The second step consists in a proof that the infinite parts can all be combined with the two phenomenological constants m and e. These divergences are then eliminated by a redefinition of mass and charge. This procedure is called *mass and charge renormalization*. In Sections 9–4, 9–5, and 9–6 we show how this is carried out for the second order divergences of this class. In Chapter 10 we complete this discussion by the proof that the renormalization procedure can be consistently extended to all orders in the charge e. Thereby all divergences are removed from the theory and it then becomes possible to calculate radiative processes to any desired accuracy.

This daring manipulation of infinite quantities raises many questions which are not yet fully understood. From a purely pragmatic point of view this procedure is perhaps justifiable. But there is no doubt that the necessity for such drastic steps is a clear indication of the need for further revision of the fundamental concepts on which this theory is built.

We might mention one of these questions here. In the preceding chapter we have proved that the S-matrix is unitary. Consequently, each matrix element must be bounded and therefore cannot be infinite. It appears, then, that either the proof of the unitary property is incorrect or we are not calculating the S-matrix for which we have proved this property.

The following points must be kept in mind when this question is considered. The proof of the unitary property of S was based on the assumption that the total energy operator $H_0 + H$ in the Schroedinger picture is hypermaximal. We do not know whether this is the case in quantum electrodynamics.

Furthermore, the diagram method can only be derived from the field equations with the iterated solution of Dyson (8–7). But we have shown that the correct definition of the S-operator is (7–70) and that this definition coincides with the definition (7–80) only under very special assumptions.

Finally, the expression of the general S-matrix element in the form of a power series implies that these matrix elements, when regarded as functions of e, are analytic in the neighborhood of $e = 0$. This is not necessarily the case* and the formal evaluation of such matrix elements as power series in e could then lead to divergent results. The amazingly accurate experimental verification of the predictions of this theory indicates, however, that the expansion in e after renormalization is at least semiconvergent; i.e., it is an asymptotic expansion.

9–3 The vacuum fluctuations. The first example of diagrams associated with divergent matrix elements are the *vacuum diagrams*. A vacuum diagram is one with no external lines of any kind. It follows from this definition that these diagrams when considered by themselves contribute only to the diagonal matrix element of the vacuum state. It is clear, then, that these diagrams do not give rise to any new transitions at all. Some typical examples of such vacuum diagrams are given in Fig. 9–1. There exists an infinite set of such diagrams.

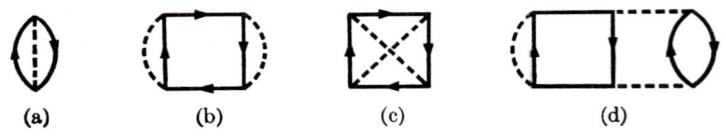

Fig. 9–1. Examples of vacuum diagrams.

Every one of them when written in analytical form according to Table 8–2 leads to an infinite expression. For instance, the expression associated with diagram 9–1(a) would be proportional to

$$M = \iiint \delta(k + p - p')\delta(k + p - p') \frac{1}{k^2} \text{Tr}\left(\frac{i\mathbf{p} - m}{p^2 + m^2} \gamma^\mu \frac{i\mathbf{p}' - m}{p'^2 + m^2} \gamma_\mu\right) \\ \times d^4p\, d^4p'\, d^4k.$$

When we carry out the integration over k, we find

$$M = \delta(0) \iint \frac{1}{(p - p')^2} \text{Tr}\left(\frac{i\mathbf{p} - m}{p^2 + m^2} \gamma^\mu \frac{i\mathbf{p}' - m}{p'^2 + m^2} \gamma_\mu\right) d^4p\, d^4p'. \quad (9\text{–}1)$$

There are two reasons for the divergence of this expression. First, there is the factor $\delta(0)$ in front, which stems from the fact that the vacuum diagrams have no external lines and therefore there are in these expressions as many δ-functions as there are corners in the diagram. This factor $\delta(0)$ is, of course, infinite, and shows, in fact, that all these matrix elements are pro-

---

* F. J. Dyson, *Phys. Rev.* **85**, 631 (1952).

portional to the four-dimensional volume in x-space in which the field is considered.

Secondly, this infinite factor $\delta(0)$ is multiplied in the above example with a double integral over p, p', which is also divergent. All of the vacuum diagrams lead to divergent integrals of this type.

According to our interpretation, the S-matrix elements describe the transformation of state vectors in the interaction picture from the remote past to the remote future. The vacuum state must remain unchanged under this transformation. It is shown below that the vacuum fluctuation diagram can contribute merely a constant multiplicative factor of absolute value one to the state vector.

The vacuum diagrams can also occur in conjunction with other diagrams describing real physical processes. For instance, the diagram shown in Fig. 9–2 is a contribution of the fourth order in e to the Compton effect. This diagram consists of two disconnected portions, one of which is a Compton diagram of second order, the other a vacuum diagram of second order. Such disconnected diagrams are associated with expressions which have the form of a product, each factor corresponding to one of the disconnected portions.

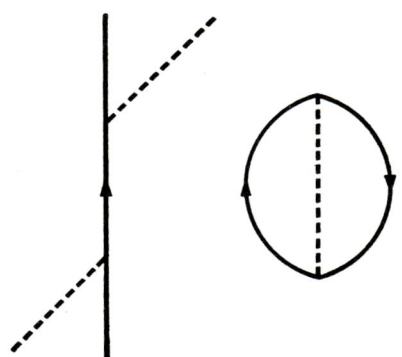

FIG. 9–2. A fourth order contribution to the Compton effect, involving a vacuum diagram.

Since this situation can occur for any diagram, it follows that each matrix element for any type of real process is multiplied by a numerical factor, which is the sum of all the terms associated with vacuum diagrams. This factor is common to all the matrix elements. If we denote this sum by

$$C = \sum M, \qquad (9\text{--}2)$$

where the summation is extended over all vacuum diagrams, we can write the S-matrix in the form

$$S = CS', \qquad (9\text{--}3)$$

where S' is the S-matrix constructed with omission of *all* the vacuum diagrams.

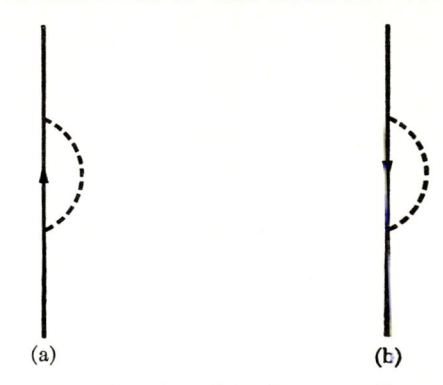

FIG. 9–3. Negaton (a) and positon (b) self-energy diagram of second order.

We observe that, because of conservation of momentum, there exists no matrix element which connects the vacuum state with any other state. It follows that the vacuum state is an eigenstate of S and its eigenvalue is just the number C:

$$S\omega_0 = C\omega_0. \tag{9-4}$$

If S is unitary it follows that C must be a factor of magnitude 1:

$$|C|^2 = 1 \tag{9-5}$$

or

$$C = e^{i\alpha}, \quad (\alpha \text{ real}). \tag{9-6}$$

The above-mentioned fact that each vacuum diagram contributes an infinite number to C must be interpreted to mean that the phase α is infinite and that therefore the power series development in e is not valid.

We can formally justify the omission of all vacuum diagrams by transforming all state vectors in accordance with

$$\omega \rightarrow \omega' = e^{-i\alpha}\omega. \tag{9-7}$$

Since phase transformations of the state vector have no effect on any calculation of physical processes, we can calculate all these processes with the modified S-matrix S' which does not contain any vacuum diagrams. This matrix is also unitary if S is unitary.

Thus, we shall henceforth drop all vacuum diagrams and work with the matrix S' which we again denote by S in the future work.

9–4 The self-energy of the electron. The difficulties connected with the structure of the electron outlined in Section 9–1 reappear in the theory of the S-matrix. They reveal themselves as divergent terms corresponding to diagrams with two external electron lines and no external photon lines. We call these *electron self-energy diagrams*. The simplest examples are the negaton and positon self-energy diagrams of second order shown in Fig. 9–3.

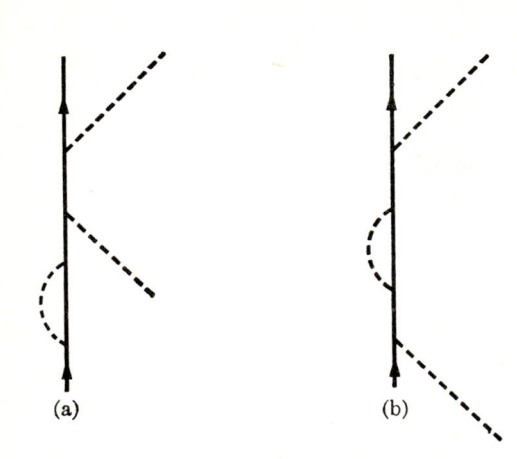

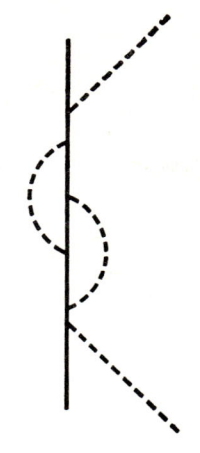

FIG. 9-4. Examples of second order self-energy parts inserted into (a) external and (b) internal electron lines of a Compton diagram.

FIG. 9-5. A fourth order self-energy part inserted into an internal electron line of a Compton diagram.

Diagrams of this structure may appear as parts of more complex diagrams. We call a part of a diagram an *electron self-energy part* if, upon separation from the rest of the diagram, it becomes an electron self-energy diagram. Such a self-energy part may be connected to the rest of the diagram by one or two electron lines. In the former case it is inserted into an external electron line and in the latter case it is inserted into an internal electron line. Examples of second order self-energy parts are shown in Fig. 9-4. A fourth order example is shown in Fig. 9-5. The insertion of a self-energy part into an internal electron line has the effect of replacing the electron propagation function* $S(p)$ corresponding to this line by another function $S'(p)$. For the second order self-energy parts, for instance, we find

$$S'(p) = S(p) + S(p)\Sigma(p)S(p), \tag{9-8}$$

where $\Sigma(p)$ can be readily calculated with the help of Table 8-2:

$$\Sigma(p) = \frac{ie^2}{(2\pi)^4} \int \gamma_\mu \frac{i(\boldsymbol{p} - \boldsymbol{k}) - m}{(p-k)^2 + m^2} \gamma^\mu \frac{1}{k^2} d^4k. \tag{9-9}$$

This second order radiative correction to the electron propagation function is independent of the rest of the diagram. It can therefore be calculated once and for all and applied to any diagram.

---

* In this and the following chapter we omit the index c which was previously used to indicate the special path of integration to be followed in the neighborhoods of the poles. The rule $S(p) = -\lim_{\mu \to +0} (i\boldsymbol{p} + m - i\mu)^{-1}$ is always understood.

The expression $\Sigma(p)$ also enters into the S-matrix elements corresponding to the self-energy diagrams with two external electron lines (Fig. 9-3). We can write this part of the S-matrix compactly in operator form by introducing the operators $\varphi(p)$:

$$\psi(x) = \frac{1}{(2\pi)^2} \int \varphi(p) e^{ipx} d^4p, \tag{9-10}$$

without separating them into emission and absorption operators.

The second order S-operator corresponding to the self-energy diagrams of Fig. 9-3 is then obtained, with the help of Table 8-2, in the form

$$S^{(2)} = -i \int \bar{\varphi}(p) \Sigma(p) \varphi(p) d^4p. \tag{9-11}$$

When Eq. (9-9) is inserted for $\Sigma(p)$ this expression can be simplified because the operators $\varphi(p)$ satisfy the free-particle equations

$$(i\slashed{p} + m)\varphi(p) = 0, \quad \bar{\varphi}(p)(i\slashed{p} + m) = 0, \tag{9-12}$$

which imply $p^2 + m^2 = 0$. In the following, we shall first evaluate $\Sigma(p)$ in Eq. (9-9), i.e., without any restrictions regarding the values of p, and then apply (9-12) to find $S^{(2)}$ of Eq. (9-11).

Inspection of Eq. (9-9) shows that the integral is linearly divergent. Following the program outlined in Section 9-1, we proceed in a purely formal way to evaluate (9-9) with the method described in Appendix 5.

Combining the denominators according to Eq. (A5-2), we obtain

$$\Sigma(p) = \frac{ie^2}{(2\pi)^4} \int d^4k \int_0^1 dx \gamma_\mu \frac{i(\slashed{p} - \slashed{k}) - m}{[(k - px)^2 + a^2]^2} \gamma^\mu, \tag{9-13}$$

where $a^2 = (p^2 + m^2)x(1 - x) + m^2 x^2$. The numerator can be simplified, since $\gamma_\mu \slashed{p} \gamma^\mu = -2\slashed{p}$, $\gamma_\mu \gamma^\mu = 4$. Thus,

$$\Sigma(p) = -\frac{2ie^2}{(2\pi)^4} \int d^4k \int_0^1 dx \frac{i\slashed{p} + 2m - i\slashed{k}}{[(k - px)^2 + a^2]^2}. \tag{9-14}$$

The first two terms in the numerator multiply a logarithmically divergent integral and the last term is linearly divergent.

As is shown in the Appendix, a careful manipulation of these integrals reveals that we can carry out a transformation corresponding to a shift of $k, k \to k + px$ in the whole expression if we follow Eqs. (A5-14) and (A5-16). Using $\alpha = e^2/4\pi$ for the fine-structure constant, we find

$$\Sigma(p) = \frac{\alpha}{2\pi} \left(\frac{1}{4} i\slashed{p} + \frac{1}{i\pi^2} \int d^4k \int_0^1 dx \frac{i\slashed{p}(1 - x) + 2m - i\slashed{k}}{(k^2 + a^2)^2} \right). \tag{9-15}$$

The last term in the numerator of the integrand will vanish upon symmetrical integration (*cf.* Appendix, end of Section A5–1 for the definition of symmetrical integration). The integral reduces then from a linearly to a logarithmically divergent one.

The physical interpretation of this expression is facilitated if we write it in the form

$$\Sigma(p) = A + (i\not{p} + m)B + (i\not{p} + m)^2 \Sigma_f(p), \qquad (9\text{–}16)$$

where A and B are determined by the condition that they be constants (independent of p). For instance, A is readily obtained from Eq. (9–15) by replacing everywhere $i\not{p}$ by $-m$, its free-particle value, giving

$$A = \frac{\alpha m}{2\pi} \left(\frac{1}{i\pi^2} \int d^4k \int_0^1 dx \, \frac{1+x}{(k^2 + m^2 x^2)^2} - \frac{1}{4} \right). \qquad (9\text{–}17)$$

The integral may be expressed in terms of the standard divergent integral

$$D = \frac{1}{i\pi^2} \int \frac{d^4k}{(k^2 + m^2)^2}. \qquad (9\text{–}18)$$

If we restrict the domain of integration to values of k for which $k^2 \leq M^2$, then D can be evaluated by using the method explained in Section A5–1 of the Appendix:

$$D = \ln \frac{M^2 + m^2}{m^2} + \frac{m^2}{M^2 + m^2} - 1.$$

Retaining only terms of the highest order in m/M, we write for this

$$D \simeq 2 \ln \frac{M}{m} - 1. \qquad (9\text{–}19)$$

Returning now to the expression (9–17) for A, we transform the integrand by partial integration:

$$\int_0^1 dx \, \frac{1+x}{(k^2 + m^2 x^2)^2} = \frac{3}{2} \frac{1}{(k^2 + m^2)^2} + 4m^2 \int_0^1 \frac{x^2(1 + \frac{1}{2}x)}{(k^2 + m^2 x^2)^3} \, dx. \qquad (9\text{–}20)$$

If this is substituted in (9–17) and then integrated over k, the first term gives $\frac{3}{2}D$ and the second term is absolutely convergent. According to Eq. (A5–12), its value is

$$2 \int_0^1 (1 + \tfrac{1}{2}x) \, dx = \tfrac{5}{2}.$$

Thus,

$$A = \frac{\alpha}{2\pi} m \left(\frac{3}{2} D + \frac{9}{4} \right) = \frac{3\alpha}{2\pi} m \left(\ln \frac{M}{m} + \frac{1}{4} \right). \qquad (9\text{–}21)$$

We next proceed to the evaluation of B and Σ_f in Eq. (9–16). By subtracting A [Eq. (9–17)] from Eq. (9–15), we obtain first

$$(i\mathbf{p} + m)B + (i\mathbf{p} + m)^2 \Sigma_f(p)$$

$$= \frac{\alpha}{2\pi} \left\{ \frac{1}{4}(i\mathbf{p} + m) \right.$$

$$+ \frac{1}{i\pi^2} \int d^4k \int_0^1 dx[i\mathbf{p}(1-x) + 2m] \left[\frac{1}{(k^2+a^2)^2} - \frac{1}{(k^2+m^2x^2)^2} \right]$$

$$\left. + (i\mathbf{p} + m) \frac{1}{i\pi^2} \int d^4k \int_0^1 dx \frac{1-x}{(k^2+m^2x^2)^2} \right\}. \quad (9\text{–}22)$$

The divergent part is contained entirely in the last term. The integral in the second term is absolutely convergent and can be evaluated in closed form. The first and last term are constants multiplied by $(i\mathbf{p} + m)$. They contribute therefore only to the term B. In order to extract from the convergent integral the contribution to B, we proceed as follows.

By use of the identity (A5–13) and the formula (A5–12), and by substitution of the value $a^2 = m^2x^2 + (p^2 + m^2)x(1-x)$, we obtain

$$\frac{1}{i\pi^2} \int d^4k \left[\frac{1}{(k^2+a^2)^2} - \frac{1}{(k^2+m^2x^2)^2} \right]$$

$$= -\int_0^1 dz \frac{(p^2+m^2)x(1-x)}{m^2x^2 + (p^2+m^2)x(1-x)z}.$$

The convergent integral in (9–22) becomes

$$-\int_0^1 dx \int_0^1 dz [i\mathbf{p}(1-x) + 2m] \frac{(p^2+m^2)x(1-x)}{m^2x^2 + (p^2+m^2)x(1-x)z}$$

$$= +\int_0^1 dx \int_0^1 dz \left\{ (i\mathbf{p}+m)^2 x(1-x) \right.$$

$$\times \frac{m(1+x) + (i\mathbf{p}-m)(1-x)\left[1 - \frac{2z}{x}(1+x)\right]}{m^2x^2 + (p^2+m^2)x(1-x)z}$$

$$\left. - 2(i\mathbf{p}+m)\left(\frac{1}{x} - x\right) \right\}.$$

The transformation carried out in this last equation is based on the identity

$$(i\mathbf{p}+m)^2 = 2m(i\mathbf{p}+m) - (p^2+m^2).$$

The purpose of this transformation is to cast the integral into the required form of $(i\mathbf{p}+m)^2 \Sigma_f(p) + (i\mathbf{p}+m)B'$ with B' independent of p. This separation is unique. Substituting this result into (9–22) and identifying the coefficients of $i p + m$, we find*

$$B = \frac{\alpha}{8\pi} - \frac{\alpha}{\pi} \int_0^1 dx \left(\frac{1}{x} - x\right) + \frac{\alpha}{2\pi^3 i} \int d^4k \int_0^1 dx \, \frac{1-x}{(k^2 + m^2 x^2)^2}, \quad (9\text{–}23)$$

$$\Sigma_f(p) = \frac{\alpha}{2\pi} \int_0^1 dx\, x(1-x) \int_0^1 dz \, \frac{m(1+x) + (ip-m)\left[1 - x + 2z\left(x - \frac{1}{x}\right)\right]}{m^2 x^2 + (p^2 + m^2)x(1-x)z}. \quad (9\text{–}24)$$

The expression (9–23) for B thus obtained still contains a divergent k-integral. But a glance at the second term shows that a new divergence has appeared. It is of the form $\int_0^1 dx/x$. A similar term occurs in (9–24), since the two terms together are an absolutely convergent integral. The origin of this new divergence can be traced back to the D-function and the fact that this function is not well defined at $k^2 \to 0$. These divergences would be absent if the photon had a finite rest-mass. In fact, one can easily show that replacement of k^2 in D by $k^2 + \lambda^2 m^2$, with $\lambda \ll 1$, has the effect of replacing the integrals $\int_0^1 dx/x$ by $\int_\lambda^1 dx/x$. This divergence is of class (b) (*cf.* Section 9–2) and does not cause difficulties.

The expression (9–23) can be further reduced by the same method which was used for the evaluation of (9–20),

$$B = \frac{\alpha}{4\pi}\left(D - 4\int_0^1 \frac{dx}{x} + \frac{11}{2}\right). \quad (9\text{–}25)$$

A similar procedure which is given in more detail in Appendix A5–4 gives for $\Sigma_f(p)$

$$\Sigma_f(p) = \frac{\alpha}{2\pi m} \left\{ \frac{1}{2(1-\rho)}\left(1 - \frac{2 - 3\rho}{1-\rho} \ln \rho\right) \right.$$

$$+ \frac{ip - m}{m}\left[\frac{1}{2\rho(1-\rho)}\left(+2 - \rho + \frac{-4 + 4\rho + \rho^2}{1-\rho} \ln \rho\right)\right.$$

$$\left.\left. - \frac{2}{\rho}\int_0^1 dx \left(\frac{1}{x} - x\right)\right]\right\}, \quad (9\text{–}26)$$

---

* R. Karplus and M. Kroll, *Phys. Rev.* **77**, 536 (1950), esp. Eq. (23). Unfortunately, this paper is marred by many misprints.

with $\rho = (p^2 + m^2)/m^2$. The explicit evaluation of A, B, and $\Sigma_f(p)$ in the expression (9–16) for $\Sigma(p)$ is thereby accomplished.

We proceed now to give the physical interpretation of this result. Let us first consider the self-energy diagram of Fig. 9–3. The part of the S-matrix associated with this diagram was given in Eq. (9–11). In this case, both electron lines are external. We can use the relations (9–12), and the expression simplifies if we substitute (9–16) for $\Sigma(p)$. Of the three terms in $\Sigma(p)$ only the first one contributes to $S^{(2)}$. It is

$$S^{(2)} = -iA \int \bar{\varphi}(p)\varphi(p)d^4p. \tag{9–27}$$

This particular form of $S^{(2)}$ is exactly the same that would have been obtained for $S^{(2)}$ from a term in the interaction operator of the form

$$\delta m \int_\sigma \bar{\psi}(x)\psi(x)d\sigma.$$

According to Eq. (8–7), we would have calculated with this interaction operator a contribution of the form (9–27) with

$$A = \delta m. \tag{9–28}$$

This observation permits the elimination of the term (9–22) from the S-matrix. For, let us assume that the observed electron mass m_0 consists of two parts: the mass m which enters into the Dirac equation, and the mass δm of electromagnetic origin, so that

$$m_0 = m + \delta m. \tag{9–29}$$

The free-particle operator $H_0(\tau)$ contains the mass in the term $m\bar{\psi}\psi$ [cf. Eq. (7–87)]. We replace m by $m_0 - \delta m$ in this operator and keep the observable mass m_0 in the free-particle operator H_0. The "electromagnetic" mass δm is combined with the interaction operator $H(\tau)$:

$$H'(\tau) = H(\tau) - \delta m \int_{\sigma(\tau)} \bar{\psi}(x)\psi(x)d\sigma. \tag{9–30}$$

The new S-operator,

$$S' = 1 - i\int H'(\tau)d\tau + \cdots = 1 - i\int H(\tau)d\tau + i\delta m \int \bar{\psi}(x)\psi(x)d^4x + \cdots, \tag{9–31}$$

will contain a new term of order α. When we calculate the second order electron self-energy from this new S-matrix, we obtain

$$S^{(2)} + i\delta m \int \bar{\psi}\psi d^4x = 0. \tag{9–32}$$

Therefore, there is (at least to second order) no electron self-energy other than the rest energy m_0. It is clear that the separation of the mass into two parts, a "mechanical" mass m and an "electromagnetic" mass δm, cannot be observed and is therefore physically meaningless. It is a particular consequence of the unphysical separation of the total energy operator $H_0 + H$ into two parts such that H_0 describes a "bare" electron, i.e., an electron without its surrounding field. It is therefore extremely satisfactory that it is possible to eliminate the electron self-energy from the S-matrix by the procedure indicated above. This procedure is called *mass renormalization*.

Further inspection of this method raises many questions. First of all, it will be necessary to show that mass renormalization can be carried out consistently to all orders in the expression (8-7), and that this procedure is consistent with the unitary and invariance properties of the S-matrix. The former was discussed at the end of Section 9-2 and we shall return to this question in Section 10-8.

Let us now examine the physical significance of the self-energy part, $\Sigma(p)$, when it is inserted into an electron line of a diagram. This electron line may be either an internal or an external line. For an internal line this insertion, when combined with the original line, corresponds, according to Eq. (9-8), to a replacement of S by

$$S'(p) = S(p) + S(p)\Sigma(p)S(p). \tag{9-33}$$

When $\Sigma(p)$ is inserted into an external electron line, the sum of the original external line and its radiative correction is obtained by replacing φ and $\bar{\varphi}$ by

$$\varphi'(p) = \varphi(p) + S(p)\Sigma(p)\varphi(p), \quad \bar{\varphi}'(p) = \bar{\varphi}(p) + \bar{\varphi}(p)\Sigma(p)S(p). \tag{9-34}$$

We shall assume in the following that mass renormalization has been carried out, so that by Eq. (9-28) the last term in Eq. (9-30) and the first term in Eq. (9-16) have been eliminated.

Equation (9-33) now becomes, with (9-16) and $S(p) = -(i\mathbf{p} + m)^{-1}$,

$$S'(p) = (1 - B)S(p) + \Sigma_f(p). \tag{9-35}$$

Similarly, using Eq. (9-12), Eqs. (9-34) can be written

$$\varphi'(p) = \varphi(p) - B(i\mathbf{p} + m)^{-1}(i\mathbf{p} + m)\varphi(p),$$
$$\bar{\varphi}'(p) = \bar{\varphi}(p) - \bar{\varphi}(p)(i\mathbf{p} + m)(i\mathbf{p} + m)^{-1}B. \tag{9-36}$$

The second term on the right side is undetermined, since

$$[(i\mathbf{p} + m)^{-1}(i\mathbf{p} + m)]\varphi(p) = \varphi(p),$$

but $(i\mathbf{p} + m)^{-1}[(i\mathbf{p} + m)\varphi(p)] = 0$ by Eqs. (9-12). Actually, Eqs. (9-36) are uniquely determined by the requirement of internal consistency, as we shall see below.

The unsatisfactory feature in Eqs. (9–35) and (9–36) is, of course, the appearance of the constant B, which is infinite although formally of order α. At the same time, we notice that the radiative corrections to $S(p)$ reproduce this same function apart from a constant multiplicative factor, $-B$. This feature is in many ways similar to the reproduction of the term $m\bar{\psi}\psi$ of $H_0(\tau)$ in the form $A\bar{\psi}\psi$.

It is therefore completely within the spirit of the previous arguments to remove the appearance of B by an extension of the idea of renormalization. Since the considerations of this section are correct only to second order, we can write Eq. (9–35):

$$S'(p) = (1 - B)S_0(p),$$
$$S_0(p) = S(p) + \Sigma_f(p), \tag{9–37}$$

where $S_0(p)$ is the renormalized electron propagation function including its radiative corrections (to second order).

These equations mean that *the inclusion of the radiative corrections due to electron self-energy parts modifies $S(p)$ into the finite function $S_0(p)$ and multiplies it by an infinite constant, $1-B$.*

The question immediately arises whether the renormalization (9–37) can be carried out consistently. The following arguments will serve to give an affirmative answer to second order. The higher orders will be discussed in Chapter 10.

The function $S(p)$ had its origin in the contraction symbol. Radiative corrections to $S(p)$ therefore correspond to radiative corrections of operators $\varphi(p)$, as is also indicated by Eqs. (9–34). Similarly, a renormalization of $S(p)$ implies a corresponding renormalization of $\varphi(p)$ and $\bar{\varphi}(p)$. Indeed, the structure of the contraction symbol requires the renormalization

$$\varphi'(p) = (1 - B)^{1/2}\varphi_0(p) \approx (1 - \tfrac{1}{2}B)\varphi_0(p),$$
$$\bar{\varphi}'(p) = (1 - B)^{1/2}\bar{\varphi}_0(p) \approx (1 - \tfrac{1}{2}B)\bar{\varphi}_0(p), \tag{9–38}$$

where φ_0 and $\bar{\varphi}_0$ are the renormalized "wave functions" including the radiative corrections.* However, the finite terms of the radiative corrections will not contribute, since Eq. (9–12) must be satisfied, so that we can identify $\varphi_0(p)$ and $\varphi(p)$, and therefore also Eqs. (9–38) and 9–36). The undetermined last term in Eqs. (9–36) is now found to be $-\tfrac{1}{2}B\varphi(p)$.

We conclude that *the inclusion of the radiative corrections due to electron self-energy parts has no effect on the operators φ and $\bar{\varphi}$ other than to multiply*

---

* In these equations we have used the *formal* expansion in powers of B and retained the first order term only. Since these manipulations are purely algebraic, the question of convergence is not involved. The justification for this procedure will be given in Chapter 10.

each by *a factor** $\sqrt{1-B} \approx 1 - \frac{1}{2}B$. It is to be noted that the identification of the renormalized operators $\varphi_0(p)$ with the original operators $\varphi(p)$ depends entirely on Eq. (9–12). When these equations do not hold, as in the case of the operators which are contracted into $S(p)$, $\varphi_0(p)$ will differ from $\varphi(p)$ by finite terms which exactly account for the difference between $S_0(p)$ and $S(p)$.

The *"wave function" renormalization*† (9–38) and the corresponding renormalization of $S'(p)$ in Eq. (9–37) enable one to eliminate B completely from the matrix element of any arbitrary diagram by a subsequent charge renormalization. Consider a diagram M of nth order which contains E_e external electron lines and possibly a certain number of external photon lines. This diagram will consist of $E_e/2$ open electron paths and possibly a certain number of closed electron paths, i.e., closed loops. The total number of corners will be n. Therefore, the matrix element will contain $E_e/2$ operators φ (i.e., u or v, or both) and $E_e/2$ operators $\bar{\varphi}$ (i.e., $\bar{u}$ or $\bar{v}$, or both). It will also contain $n - E_e/2$ propagation functions S. The matrix element M' which is the sum of M and its second order radiative corrections is obtained by the replacements $S \to S'$, $\varphi \to \varphi'$, $\bar{\varphi} \to \bar{\varphi}'$, according to Eqs. (9–33) and (9–34). When these factors are renormalized following Eqs. (9–37) and (9–38), we obtain $n - E_e/2$ factors $(1-B)S_0$, $E_e/2$ factors $\sqrt{1-B}\,\varphi$, and $E_e/2$ factors $\sqrt{1-B}\,\bar{\varphi}$. The whole matrix element is therefore modified by *finite* terms only, but multiplied by $(1-B)^n = 1 - nB$, to second order. This important result enables us to eliminate the infinite factor by incorporating it into the factor e^n of our nth order matrix element. This amounts to a *renormalization of charge*,

$$e_0 = (1 - B)e,$$

or, in the form

$$e_0 = e + \delta e, \quad \frac{\delta e}{e} = -B, \tag{9–39}$$

it may be regarded as a radiative correction to the charge which enters the original equations of motion. As in the case of mass renormalization, e and δe are unobservable, and e_0 is to be identified with the physical and observable charge of the electron.

---

* Dyson has introduced the *intermediate* representation [F. J. Dyson, *Phys. Rev.* **83**, 608 (1951)] which enables one to define the "incident and outgoing particle states" by a limiting procedure. In this method, one can prove *by direct and unambiguous calculation* that the electron operators are multiplied by $\sqrt{1-B}$, since the complete binomial expansion of this expression can be obtained. Similarly, he shows that the photon operators are multiplied by $\sqrt{1-C}$ [*cf.* Eq. (9–72)].

† The renormalization of the operator $\varphi(p)$ is, of course, equivalent to a renormalization of the associated wave functions $u(\mathbf{p})$ and $v(\mathbf{p})$.

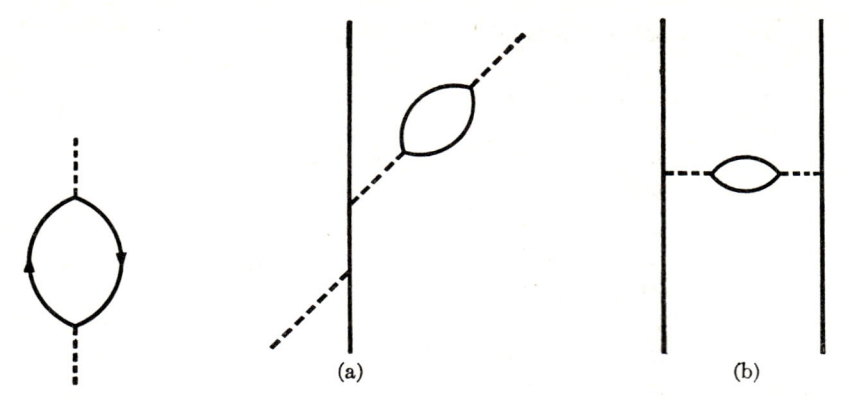

FIG. 9-6. Photon self-energy diagram of second order.

FIG. 9-7. Examples of second order photon self-energy parts inserted into (a) external and (b) internal photon lines.

The result (9-39) also tells us that the charge renormalization cannot affect the unitary property of the S-matrix, because this property is independent of the magnitude of the electronic charge. This argument, which is so far valid to second order only, will also be extended to all higher orders in Chapter 10 (Section 10-8).

After the renormalization process, only the finite observable parts, Σ_f, remain. The infinite constants A and B are both eliminated.

9-5 The self-energy of the photon. In addition to the self-energy of the electron, which has a classical counterpart, there also appears in quantum electrodynamics a self-energy of the photon which does not have a simple classical analogue. Such a self-energy could be caused only by a self-interaction of the electromagnetic field, which has no place in the classical Maxwell equations. The terms in the S-matrix which describe this effect are represented by diagrams which have only two external photon lines and no external electron lines. We call these *photon self-energy diagrams*. The simplest example is the photon self-energy diagram of second order shown in Fig. 9-6.

Diagrams of this structure may also appear as parts of larger diagrams. We call a part of a larger diagram a *photon self-energy part* if upon separation from the rest of the diagram it becomes a photon self-energy diagram. Such a part may be connected to the rest of the diagram by one or two photon lines. In the former case it is inserted into an external photon line and in the latter case it is inserted into an internal photon line. Examples are shown in Fig. 9-7.

The insertion of a photon self-energy part into an internal photon line has the effect of replacing the photon propagation function $g_{\mu\nu}D(k) = g_{\mu\nu}(1/k^2)$ corresponding to this line by another function, $D_{\mu\nu}'(k)$. For the

9-5] THE SELF-ENERGY OF THE PHOTON

second order self-energy parts, we find

$$D_{\mu\nu}'(k) = g_{\mu\nu}D(k) + D(k)\Pi_{\mu\nu}(k)D(k). \tag{9-40}$$

Table 8–2 gives

$$\Pi_{\mu\nu}(k) = \frac{ie^2}{(2\pi)^4} \text{Tr} \left\{ \int \gamma_\mu \frac{i(\not{p} - k) - m}{(p - k)^2 + m^2} \gamma_\nu \frac{i\not{p} - m}{p^2 + m^2} d^4p \right\}. \tag{9-41}$$

The integrand in this expression behaves like $1/p^2$ for large values of p. Thus, the integral is expected to be quadratically divergent.

Before we evaluate this integral, we note its formal properties, which follow from the relativistic and gauge invariance of the theory. Since $\Pi_{\mu\nu}(k)$ is a tensor of second rank which depends only on the four-vector k, it is of the form

$$\Pi_{\mu\nu}(k) = C(k^2)k_\mu k_\nu + D(k^2)g_{\mu\nu}. \tag{9-42}$$

Gauge invariance implies, however, that

$$k^\mu \Pi_{\mu\nu}(k) = 0, \quad \Pi_{\mu\nu}(k)k^\nu = 0. \tag{9-43}$$

In order to explain how these relations are connected with the gauge invariance of the theory, we recall that the S-matrix is invariant if the radiation field operators are subject to the gauge transformations

$$a_\mu(k) \to a_\mu(k) + k_\mu \chi(k),$$

where $\chi(k)$ is an arbitrary scalar function. This transformation produces additional terms of the form

$$a^\mu \Pi_{\mu\nu} k^\nu \chi + \chi k^\mu \Pi_{\mu\nu} a^\nu + \chi^2 k^\mu k^\nu \Pi_{\mu\nu}.$$

Since these terms must vanish for arbitrary χ, we must have the relations (9–43).

From Eq. (9–43) follows

$$k^2 C(k^2) + D(k^2) = 0, \tag{9-44}$$

and that $\Pi_{\mu\nu}(k)$ must be of the form

$$\Pi_{\mu\nu}(k) = (k_\mu k_\nu - k^2 g_{\mu\nu})C(k^2). \tag{9-45}$$

It is therefore sufficient to evaluate the scalar quantity $\Pi_\mu{}^\mu(k)$ which, according to Eq. (9–45), is

$$\Pi_\mu{}^\mu(k) = -3k^2 C(k^2). \tag{9-46}$$

The trace in Eq. (9–41) for $\Pi_\mu{}^\mu(k)$ can easily be evaluated and yields

$$8(p^2 + 2m^2 - p \cdot k). \tag{9-47}$$

We find therefore, from Eq. (9–42),

$$\Pi_\mu{}^\mu = k^2 C(k^2) + 4D(k^2) = -\frac{2\alpha}{\pi}\frac{1}{i\pi^2}\int \frac{p^2 + 2m^2 - p\cdot k}{[(p-k)^2 + m^2](p^2 + m^2)}\,d^4p.$$

(9–48)

If $\Pi_{\mu\nu}$ were gauge-invariant, i.e., if Eq. (9–44) were satisfied, $D(k^2)$ would have to vanish for $k^2 = 0$, and therefore

$$\Pi_\mu{}^\mu(0) = 0.$$

(9–49)

This equation is obviously not fulfilled, since Eq. (9–48) gives

$$\Pi_\mu{}^\mu(0) = 4D(0) = -\frac{2\alpha}{\pi}\frac{1}{i\pi^2}\int \frac{p^2 + 2m^2}{(p^2 + m^2)^2}\,d^4p,$$

(9–50)

which is quadratically divergent. One might argue that this result is due to our way of handling the undefined expression (9–41). Indeed, by suitable manipulation of the divergent integrals, Schwinger* was able to obtain a result in the gauge-invariant form (9–45); it is also possible to obtain results which differ from it by an additional term $A'g_{\mu\nu}$, where A' is a (finite or infinite) constant, as in (9–50). These lead to a nonvanishing photon self-energy and are, of course, not gauge-invariant.†

On the other hand, no invariance property can be lost in a consistent theory. From this point of view, we conclude that since the theory is originally gauge-invariant, this invariance property is to be considered as a guide to define the divergent and therefore undetermined integrals which occur here. Therefore, the expression $D(0)$ in Eq. (9–50) must vanish as a consistency requirement.

As mentioned in Section 9–2, the divergence (9–50) can also be treated by a very different method, in which the photon is first described by a neutral vector meson field of mass κ. The term (9–50) can then be interpreted as a mass term and can be eliminated by mass renormalization very similar to the electron mass. Gauge invariance can then be invoked *for the renormalized theory only* and requires that the renormalized mass $\kappa_0 = \kappa + \delta\kappa$ be zero. The following arguments exhibit these ideas in further detail.

We notice first that the limit $\kappa \to 0$ of a neutral vector meson theory exists and is uniform.‡ Secondly, consider the self-energy diagram Fig. 9–6 where the

---

* J. S. Schwinger, *Phys. Rev.* **74**, 1439 (1948), especially Section 2.
† G. Wentzel, *Phys. Rev.* **74**, 1070 (1948).
‡ This was proved by P. T. Matthews and by F. Coester, *loc. cit.*, Section 6–5. The renormalization of the neutral vector meson theory was first carried out correctly by R. J. N. Phillips, *Phys. Rev.* **96**, 1678 (1954).

photon lines are replaced by neutral vector-meson lines. The S-matrix element of second order associated with this diagram is

$$S^{(2)} = -i \int \varphi_\mu(k) \Pi^{\mu\nu}(k) \varphi_\nu(k) d^4k, \tag{9-51}$$

where $\varphi_\mu(k)$ is the meson field in momentum space. From Eq. (6-112) it follows that $\varphi_\mu(k)$ satisfies

$$(k^2 + \kappa^2)\varphi_\mu = 0, \tag{9-52}$$

$$k^\mu \varphi_\mu = 0. \tag{9-53}$$

According to the result (9-50) we cannot write $\Pi_{\mu\nu}(k)$ in the form (9-45), but we have instead

$$\Pi_{\mu\nu}(k) = D(0) g_{\mu\nu} + (k_\mu k_\nu - k^2 g_{\mu\nu}) C(k^2). \tag{9-54}$$

We shall prove later that $k^2 C(k^2) \to 0$ for $k^2 \to 0$ [cf. Eq. (9-64)]. Inserting (9-54) into (9-51) and observing (9-52) and (9-53), we find

$$S^{(2)} = -i(D(0) + \kappa^2 C(-\kappa^2)) \int \varphi_\mu(k) \varphi^\mu(k) d^4k \equiv -i\delta\kappa^2 \int \varphi_\mu(k) \varphi^\mu(k) d^4k. \tag{9-55}$$

On the other hand, the operator H_0 (Eq. 7-87), when suitably generalized for the vector meson field, will contain a term

$$\kappa^2 \int_{\sigma(\tau)} \varphi_\mu(x) \varphi^\mu(x) d\sigma.$$

We can therefore define a new interaction operator

$$H'(\tau) = H(\tau) - \delta\kappa^2 \int_{\sigma(\tau)} \varphi_\mu(x) \varphi^\mu(x) d\sigma \tag{9-56}$$

corresponding to a replacement of κ^2 by $\kappa_0^2 - \delta\kappa^2$, where

$$\kappa_0^2 = \kappa^2 + \delta\kappa^2. \tag{9-57}$$

The new S-operator will be

$$S' = 1 - i \int H(\tau) d\tau + i\delta\kappa^2 \int \varphi_\mu(x) \varphi^\mu(x) d^4k + \cdots,$$

so that the $\delta\kappa^2$ term arising from (9-56) will exactly cancel the self-energy term (9-55). These arguments can be extended to higher orders.

Once it is possible to renormalize the photon mass consistently, only κ_0 will occur, and we are free to identify this renormalized mass with the observed photon mass, $\kappa_0 = 0$. The stronger postulate of gauge invariance implies this value, too. If we work to a given order of approximation, κ_0 and κ can be identified to this order; i.e., one can regard the renormalization completed although it has been carried out consistently only to this order. We can then put $\kappa_0 = 0$ and also $k^2 = 0$. For example, in second order we can put $\kappa^2 = \kappa_0^2 = 0$ in Eq. (9-55), which yields, since $C(0)$ is a constant (see below),

$$\delta\kappa^2 = D(0) = -\frac{\alpha}{2\pi} m^2 (D' + D), \tag{9-58}$$

where
$$D' = \frac{1}{i\pi^2} \int \frac{d^4p}{p^2 + m^2}, \quad D = \frac{1}{i\pi^2} \int \frac{d^4p}{(p^2 + m^2)^2}.$$

Equation (9–54) can therefore be written

$$\Pi_{\mu\nu}(k) = \delta\kappa^2 g_{\mu\nu} + (k_\mu k_\nu - k^2 g_{\mu\nu}) C(k^2) \tag{9–59}$$

and shows that after mass renormalization $\Pi_{\mu\nu}(k)$ is indeed gauge-invariant.

We can now continue with the evaluation of Eq. (9–48). Using (9–46) and (9–50), we find

$$C(k^2) = \frac{1}{3k^2} \frac{2\alpha}{\pi} \frac{1}{i\pi^2} \int \left(\frac{p^2 + 2m^2 - p \cdot k}{[(p-k)^2 + m^2](p^2 + m^2)} - \frac{p^2 + 2m^2}{(p^2 + m^2)^2} \right) d^4p. \tag{9–60}$$

In order that this procedure be meaningful, we must have

$$C(k^2) = C(0) + 0(k^2).$$

We shall see below that this equation is indeed satisfied.

The methods of Appendix 5 can now be employed. Symmetrical integration allows us to add a term $-p \cdot k$ to the numerator of the second term in (9–60). The two factors in the first denominator can then be combined, and finally both terms can be written with only one denominator:

$$C(k^2) = \frac{1}{3k^2} \frac{2\alpha}{\pi} \frac{1}{i\pi^2} \int (p^2 + 2m^2 - p \cdot k) d^4p$$

$$\times \int_0^1 \left(\frac{dx}{[(p - kx)^2 + m^2 + k^2 x(1-x)]^2} - \frac{dx}{(p^2 + m^2)^2} \right)$$

$$= -\frac{1}{3k^2} \frac{2\alpha}{\pi} \frac{1}{i\pi^2} \int (p^2 + 2m^2 - p \cdot k) d^4p \int_0^1 dx \int_0^1 2 dz$$

$$\times \frac{k^2 x - 2p \cdot kx}{[p^2 + m^2 + (k^2 x - 2p \cdot kx)z]^3}$$

$$= -\frac{1}{k^2} \frac{4\alpha}{3\pi} \frac{1}{i\pi^2} \int (p^2 + 2m^2 - p \cdot k) d^4p \int_0^1 dx \int_0^1 x \, dz$$

$$\times \frac{k^2 - 2p \cdot k}{[(p - kxz)^2 + m^2 + k^2 xz(1 - xz)]^3}.$$

If we put $xz = y$, $x\,dz = dy$, and interchange the x and y integrals,

$$\int_0^1 dx \int_0^x dy f(y) = \int_0^1 dy \int_y^1 dx f(y) = \int_0^1 (1-y) dy f(y),$$

9-5] THE SELF-ENERGY OF THE PHOTON

and finally, writing x for y, we can bring $C(k^2)$ into the form

$$C(k^2) = -\frac{1}{k^2}\frac{4\alpha}{3\pi}\frac{1}{i\pi^2}\int d^4p \int_0^1 (1-x)dx \, \frac{(p^2+2m^2-p\cdot k)(k^2-2p\cdot k)}{[(p-kx)^2+m^2+k^2x(1-x)]^3}.$$

This is a linearly divergent integral. A shift of origin therefore produces the surface term

$$-\frac{i\pi^2}{2}(-2k^2x) = i\pi^2 k^2 x$$

since the linearly divergent term has the numerator $-2p^2 p\cdot k$ (cf. Section A5-2).

$$C(k^2) = -\frac{1}{k^2}\frac{4\alpha}{3\pi}\left\{k^2\int_0^1 x(1-x)dx \right.$$

$$\left. + \frac{1}{i\pi^2}\int d^4p \int_0^1 (1-x)dx \, \frac{k^2 N}{[p^2+m^2+k^2x(1-x)]^3}\right\}.$$

The numerator N, because of symmetrical integration, is

$$k^2 N = [(p+kx)^2 + 2m^2 - (p+kx)\cdot k][k^2 - 2(p+kx)\cdot k]$$

$$= k^2(1-2x)(\tfrac{3}{2}p^2 + 2m^2 - k^2 x(1-x))$$

$$= k^2(1-2x)[\tfrac{3}{2}(p^2 + m^2 + k^2 x(1-x))$$

$$+ \tfrac{1}{2}(m^2 + k^2 x(1-x)) - 3k^2 x(1-x)]. \tag{9-61}$$

A transformation $x \to 1-x$ yields

$$C(k^2) = -\frac{2\alpha}{9\pi} + \frac{4\alpha}{3\pi}\frac{1}{i\pi^2}\int d^4p \int_0^1 x dx \, \frac{N}{[p^2+m^2+k^2x(1-x)]^3},$$

where N is given by Eq. (9-61). The first term in N gives a logarithmically divergent integral. By partial integration, we find

$$\frac{3}{2}\frac{1}{i\pi^2}\int d^4p \int_0^1 \frac{x(1-2x)dx}{[p^2+m^2+k^2x(1-x)]^2}$$

$$= \frac{3}{2}\frac{1}{i\pi^2}\int d^4p\left[\frac{-1}{6(p^2+m^2)^2} + k^2\int_0^1 \frac{x^2(1-\tfrac{4}{3}x)(1-2x)dx}{[p^2+m^2+k^2x(1-x)]^3}\right]$$

$$= -\frac{D}{4} + \frac{3k^2}{4}\int_0^1 \frac{x^2(1-\tfrac{4}{3}x)(1-2x)dx}{m^2+k^2x(1-x)},$$

where D is given by Eq. (9-18).

The other terms in N are convergent and yield

$$\frac{1}{2}\int_0^1 \frac{x(1-2x)[\frac{1}{2}(m^2 + k^2 x(1-x)) - 3k^2 x(1-x)]}{m^2 + k^2 x(1-x)}\,dx$$

$$= -\frac{1}{24} - \frac{3}{2}k^2 \int_0^1 \frac{x^2(1-2x)(1-x)}{m^2 + k^2 x(1-x)}\,dx.$$

Combining these results, we find

$$C(k^2) = -\frac{\alpha}{3\pi}\left(D + \frac{2}{3} + \frac{1}{6}\right)$$

$$+ \frac{2\alpha}{\pi}k^2 \int_0^1 \frac{\frac{1}{2}x^2(1-2x)(1-\frac{4}{3}x) - x^2(1-2x)(1-x)}{m^2 + k^2 x(1-x)}\,dx$$

$$= -\frac{\alpha}{3\pi}\left(D + \frac{5}{6}\right) + \frac{2\alpha}{\pi}k^2 \int_0^1 \frac{(1-2x)\left(-\frac{x^2}{2} + \frac{x^3}{3}\right)dx}{m^2 + k^2 x(1-x)}$$

$$= -\frac{\alpha}{3\pi}\left(D + \frac{5}{6}\right) + \frac{2\alpha}{\pi}\left[\left(-\frac{x^2}{2} + \frac{x^3}{3}\right)\ln\left(1 + \frac{k^2}{m^2}x(1-x)\right)\right]_0^1$$

$$-\int_0^1 (-x + x^2) \ln\left(1 + \frac{k^2}{m^2}x(1-x)\right)dx$$

$$= -\frac{\alpha}{3\pi}\left(D + \frac{5}{6}\right) + \frac{2\alpha}{\pi}\int_0^1 x(1-x)\ln\left(1 + \frac{k^2}{m^2}x(1-x)\right)dx.$$

(9-62)

Equations (9-45) and (9-62) can be combined to

$$\Pi_{\mu\nu}(k) = [k_\mu k_\nu - k^2 g_{\mu\nu}][C - k^2 \Pi_f(k^2)], \quad (9\text{-}63)$$

where

$$C = C(0) = -\frac{\alpha}{3\pi}\left(D + \frac{5}{6}\right), \quad (9\text{-}64)$$

$$-k^2 \Pi_f = \frac{2\alpha}{\pi}\int_0^1 x(1-x)\ln\left[1 + \frac{k^2}{m^2}x(1-x)\right]dx. \quad (9\text{-}65)$$

This integral can be evaluated in closed form (cf. Section A5-3). The result is

$$+k^2 \Pi_f = \frac{\alpha}{3\pi}\left[\frac{5}{3} - \frac{1}{\rho} - \left(1 - \frac{1}{2\rho}\right)\sqrt{1 + \frac{1}{\rho}}\ln\frac{\sqrt{1 + (1/\rho)} + 1}{\sqrt{1 + (1/\rho)} - 1}\right], \quad (9\text{-}66)$$

where $\rho = k^2/4m^2$.

For small ρ we can expand $\Pi_f(k^2)$ in a power series,

$$-k^2 \Pi_f(k^2) = \frac{\alpha}{\pi} \left[\frac{1}{15} \frac{k^2}{m^2} - \frac{1}{140} \left(\frac{k^2}{m^2} \right)^2 + \cdots \right]. \tag{9-67}$$

We also note that*

$$\Pi_f(0) = -\frac{\alpha}{15\pi m^2}. \tag{9-68}$$

If we use the method of finite photon mass, Eq. (9-63) contains an additional term $\delta\kappa^2 g_{\mu\nu}$, where $\delta\kappa^2$ is given by Eq. (9-58).

Returning to the photon self-energy diagrams, we see that the second order diagram contributes at most a mass renormalization term, since the other terms in (9-63) are zero by $k^2 = 0$ to this order.

The contributions of photon self-energy parts depend on whether the photon line on which the insertion was made is an external line or an internal line. Either by postulate or after photon mass renormalization, and with use of $\kappa_0 = 0$, $k^2 = 0$, we have a gauge-invariant expression. It follows from this that terms in $k_\mu k_\nu$ of (9-63) do not contribute. In place of Eq. (9-63), we can substitute the expression

$$\Pi_{\mu\nu}(k) = g_{\mu\nu} \Pi(k^2),$$

$$\Pi(k^2) = k^2(-C + k^2 \Pi_f(k^2)). \tag{9-63a}$$

In order to show this, we consider all the diagrams of a fixed order n with a fixed number of internal and external photon lines. If the photon self-energy part $\Pi_{\mu\nu} = (k_\mu k_\nu - k^2 g_{\mu\nu}) C(k^2)$ is inserted into an external photon line of any of these diagrams, the term corresponding to $k_\mu k_\nu$ contributes a factor of the form $A k^\mu a_\mu$. Since this factor operates on a state vector ω which satisfies the subsidiary condition $k^\mu a_\mu \omega = 0$ (cf. Section 6-2), this term contributes nothing to the matrix element. We need only to prove the vanishing of the terms for the case of self-energy parts inserted into *internal* photon lines.

Let us consider then any one of the specified diagrams in which one internal photon line is replaced by a photon self-energy part. To each such diagram there corresponds a new diagram which is obtained by opening the photon line which contains $\Pi_{\mu\nu}$ and replacing $\Pi_{\mu\nu}$ by two external photon lines. The rest of the diagram is left unchanged. The resulting diagram describes a different physical process since, apart from irrelevant factors, $\Pi_{\mu\nu}$ is replaced by $a_\mu(\mathbf{k}) a_\nu(\mathbf{k}')$. The factor $a_\mu(\mathbf{k})$, for instance, appears multiplied by a term $A\gamma^\mu B$, where A and B are products of S-functions and γ-matrices which describe the electron line to which $a_\mu(\mathbf{k})$ is attached.

---

* This is the first term in an expansion for small k^2/m^2 of the observable function $\Pi_f(k^2)$. It was first calculated by E. A. Uehling, *Phys. Rev.* **48**, 55 (1935).

A gauge transformation will replace $a_\mu(\mathbf{k})$ by $a_\mu(\mathbf{k}) + k_\mu\chi(k)$, where $\chi(k)$ is an arbitrary function of $\mathbf{k}$. The sum $\sum A\mathbf{a}B$, which is obtained by summing over all possible places to which a_μ can be attached on the electron line, transforms under such a gauge transformation into $\sum A\mathbf{a}B + \sum A\mathbf{k}\chi B$. Gauge invariance implies that the second term must vanish for any χ; therefore,

$$\sum A\mathbf{k}B = 0.$$

Now we consider the original set of diagrams with the inserted photon self-energy part. They furnish an expression of the form

$$\sum A\gamma^\mu B \frac{1}{k^2}\Pi_{\mu\nu}\frac{1}{k^2}C\gamma^\nu D = \sum A\mathbf{k}B \frac{1}{k^2}C(k^2)\frac{1}{k^2}C\mathbf{k}D + \sum A\gamma^\mu B \frac{1}{k^2}\Pi\frac{1}{k^2}C\gamma_\mu D,$$

where $\Pi = -k^2 C(k^2)$ [Eq. (9–63)]. The first term on the right was just shown to vanish. This justifies the omission of the terms proportional to $k_\mu k_\nu$ in the photon self-energy part.*

The insertion of the photon self-energy part into an internal photon line modifies $D(k)$ into

$$\frac{g_{\mu\lambda}}{k^2}\Pi^{\lambda\sigma}(k)\frac{g_{\sigma\nu}}{k^2} = \frac{g_{\mu\nu}}{k^2}\Pi(k^2)\frac{1}{k^2}.$$

In order to include radiative corrections in an internal photon line, we therefore must replace $D(k)$ by

$$\begin{aligned}D'(k) &= D(k) + D(k)\Pi(k^2)D(k)\\ &= (1-C)D(k) + \Pi_f(k^2),\end{aligned} \qquad (9\text{–}69)$$

at least to second order.

Similarly, the insertion of $\Pi_{\mu\nu}$ into an external photon line yields a replacement of $a_\mu(\mathbf{k})$ by

$$a_\mu'(\mathbf{k}) = a_\mu(\mathbf{k}) + D(k)\Pi(k^2)a_\mu(\mathbf{k}). \qquad (9\text{–}70)$$

The analogy between Eqs. (9–69), (9–70), and the corresponding equations for the electron case, Eqs. (9–35) and (9–36), is evident. The last term in (9–70) is again undetermined and will be found later by a consistency argument.

Continuing the analogy, we renormalize the photon propagation function to second order,

$$D'(k) = (1-C)D_0(k), \quad D_0(k) = D(k) + \Pi_f(k^2), \qquad (9\text{–}71)$$

---

* R. P. Feynman, *Phys. Rev.* **76**, 769 (1949), especially Section 8, page 780, has verified this for open electron paths. His proof can be augmented to include the case of closed loops.

as is required by Eq. (9–69). The connection between the wave functions $a_\mu(\mathbf{k})$ and the contraction symbol then yields the wave function renormalization*

$$a_\mu'(\mathbf{k}) = \sqrt{1 - C}\, a_{\mu 0}(\mathbf{k}) \simeq (1 - \tfrac{1}{2}C) a_{\mu 0}(\mathbf{k}). \tag{9-72}$$

Again $a_{\mu 0}(\mathbf{k}) = a_\mu(\mathbf{k})$ as long as $k^2 = 0$ must be satisfied. The last term in (9–70) is thus found to be $\tfrac{1}{2}C a_\mu(\mathbf{k})$.

Consider a diagram of nth order containing P_e external photon lines. The topological structure of these diagrams then requires that there be exactly $P_i = (n - P_e)/2$ photon lines. The inclusion of all second-order corrections in the photon lines requires that we replace all D by D' and all a_μ by a_μ'. According to Eqs. (9–71) and (9–72), this will produce $(n - P_e)/2$ factors $(1 - C)D_0(k)$ and P_e factors $\sqrt{1 - C}\, a_{\mu 0}(\mathbf{k})$. The whole matrix element will therefore be modified by finite functions only, except that it will be multiplied by $(1 - C)^{n/2} \simeq 1 - (n/2)C$. Thus, we have again achieved that all unwanted infinite terms are lumped together into a multiplicative factor which can be removed by charge renormalization. The latter must apparently be of the form

$$e_0 = \sqrt{1 - C}\, e \simeq (1 - \tfrac{1}{2}C) e \tag{9-73}$$

in order to remove the infinite factor. All this is quite analogous to the electron case.

The two charge renormalizations, Eqs. (9–39) and (9–73), can be combined to give

$$e_0 = (1 - B)\sqrt{1 - C}\, e. \tag{9-74}$$

After completion of these renormalizations, the photon self-energy parts contribute only the finite, observable terms $\Pi_f(k^2)$.

9–6 The vertex part. A third type of divergence is associated with parts of diagrams which are connected to the rest of the diagram by exactly two electron lines and one photon line. Such parts are called *vertex parts*. The simplest vertex part is a corner. The next simplest is shown in Fig. 9–8, and will be studied in this section. A vertex part can never appear as a separate diagram of the S-matrix, since the momentum conservation between free-particle states would make this term vanish. But as a part of a larger diagram, it can occur at every corner. An example of vertex parts inserted into a Compton diagram is shown in Fig. 9–9.

The contribution of the vertex part, Fig. 9–8, to a diagram can easily be obtained from Table 8–2. It is a 4×4 matrix vector, $\Lambda_\mu(p',p)$, which de-

---

* *Cf.* the footnote on p. 187.

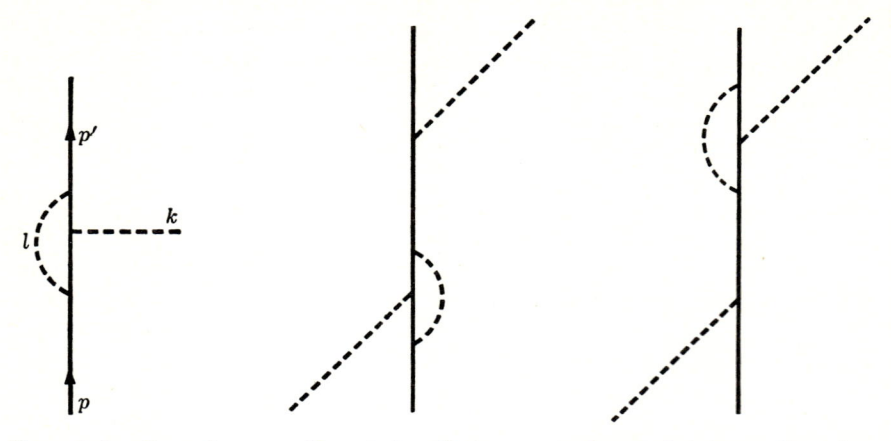

FIG. 9-8. Second order vertex part with attached photon line k and electron lines p and p'.

FIG. 9-9. Vertex parts inserted into a Compton diagram.

pends on the momentum vectors p' and p of the electron lines:

$$\Lambda_\mu(p',p) = \frac{ie^2}{(2\pi)^4} \int \gamma_\nu \frac{i(\not{p}' - \not{l}) - m}{(p' - l)^2 + m^2} \gamma_\mu \frac{i(\not{p} - \not{l}) - m}{(p - l)^2 + m^2} \gamma^\nu \frac{d^4l}{l^2}. \quad (9\text{-}75)$$

The effect on a diagram of this second order vertex part, which can be regarded as a second order correction to a corner, consists therefore simply in replacing the γ_μ of the corner by

$$\Gamma_\mu = \gamma_\mu + \Lambda_\mu(p',p). \quad (9\text{-}76)$$

The integral in (9-75) is logarithmically divergent. With Eq. (A5-2), we can bring it into the form

$$\Lambda_\mu(p',p) = \frac{2i\alpha}{4\pi^3} \int_0^1 dx \int_0^x dy \frac{N_\mu d^4l}{[(l - px + ky)^2 + a^2]^3}, \quad (9\text{-}77)$$

where

$$N_\mu = \gamma_\nu[i(\not{p}' - \not{l}) - m]\gamma_\mu[i(\not{p} - \not{l}) - m]\gamma^\nu, \quad (9\text{-}78)$$

$$a^2 = m^2x^2 + k^2y(x - y) + (p^2 + m^2)(1 - x)(x - y)$$
$$\qquad + (p'^2 + m^2)(1 - x)y, \quad (9\text{-}79)$$

and

$$k = p - p'.$$

The summation over ν can be carried out in Eq. (9-78), with the result

$$N_\mu = 2(\not{p} - \not{l})\gamma_\mu(\not{p}' - \not{l}) - 4im(P_\mu - 2l_\mu) - 2m^2\gamma_\mu, \quad (9\text{-}80)$$

where $P = p + p'$. Since the integral is only logarithmically divergent, it is permitted to shift the origin in l-space by substituting $l + px - ky$ for l (cf. Section A5–2). Symmetrical integration allows us to drop all terms of odd powers of l in the numerator. For the even powers, we can replace $l_\mu l_\nu$ by its average $\frac{1}{4} l^2 g_{\mu\nu}$. The resultant integral is of the form

$$\Lambda_\mu(p',p) = \frac{2i\alpha}{4\pi^3} \int_0^1 dx \int_0^x dy \, \frac{N_\mu d^4 l}{[l^2 + a^2]^3}. \tag{9-81}$$

The new expression N_μ for the numerator is

$$N_\mu = -l^2 \gamma_\mu + 2(\boldsymbol{p}(1-x) + \boldsymbol{k}y)\gamma_\mu(\boldsymbol{p}' - \boldsymbol{p}x + \boldsymbol{k}y)$$
$$- 4im\,(P_\mu - 2p_\mu x + 2k_\mu y) - 2m^2 \gamma_\mu. \tag{9-82}$$

We can bring this numerator into a form which is more suitable for the physical interpretation. This is achieved by separating the terms which vanish when p, p', and k assume the free-particle values. The result of this straightforward but rather lengthy algebraic calculation is

$$N_\mu = -\gamma_\mu(l^2 - 4m^2(1 - x - \tfrac{1}{2} x^2)) + 2K_\mu(p',p,x,y), \tag{9-83}$$

with

$$K_\mu(p',p,x,y) = (1-x)(i\boldsymbol{p}' + m)\gamma_\mu(i\boldsymbol{p} + m) + \gamma_\mu C - (i\boldsymbol{p}' + m)M_\mu'$$
$$- M_\mu(i\boldsymbol{p} + m) + ik_\mu m(1+x)(x-2y) - mx(1-x)\sigma_{\mu\nu}k^\nu, \tag{9-84}$$

and

$$C = k^2(1 - x + y)(1 - y) - (1 - x)[(p^2 + m^2)(1 - x + y)$$
$$+ (p'^2 + m^2)(1 - y)],$$

$$M_\mu = \gamma_\mu m(1 - x^2) + iP_\mu(1 - x)(1 - x + y)$$
$$- ik_\mu(1 + x - 2y)(1 - x + y),$$

$$M_\mu' = \gamma_\mu m(1 - x^2) + iP_\mu(1 - x)(1 - y)$$
$$+ ik_\mu(1 - x + 2y)(1 - y),$$

$$\sigma_{\mu\nu} = \frac{1}{2i}(\gamma_\mu \gamma_\nu - \gamma_\nu \gamma_\mu).$$

$$\tag{9-85}$$

All the terms in (9–84) except the last two vanish for the free-particle values of p, p', k. The next to the last one is proportional to k_μ and does not vanish. But if it is contracted on a free-photon operator, it contributes nothing to the matrix element, because of the subsidiary condition. The last term in Eq. (9–84) contains the second order radiative correction to the magnetic moment of the electron, as we shall see in Section 15–3.

Inserting this result into Eq. (9–81), we obtain

$$\Lambda_\mu(p',p) = \frac{i\alpha}{2\pi^3} \int_0^1 dx \int_0^x dy \left\{ \int \frac{-\gamma_\mu[l^2 - 4m^2(1 - x - \frac{1}{2}x^2)]}{(l^2 + a^2)^3} d^4l \right.$$

$$\left. + \int \frac{2K_\mu}{(l^2 + a^2)^3} d^4l \right\}. \quad (9\text{–}86)$$

Only the first integral in Eq. (9–86) is divergent. Since it is proportional to γ_μ, we can separate this divergent part by introducing a constant L such that

$$\Lambda_\mu(p',p) = L\gamma_\mu + \Lambda_{\mu f}(p',p), \quad (9\text{–}87)$$

where L is uniquely defined by the condition

$$\Lambda_{\mu f}(p',p) = 0, \quad \text{for } i p' = i p = -m. \quad (9\text{–}88)$$

Therefore,

$$L = -\frac{i\alpha}{2\pi^3} \int d^4l \int_0^1 x dx \frac{l^2 - 4m^2(1 - x - \frac{1}{2}x^2)}{(l^2 + m^2x^2)^3}. \quad (9\text{–}89)$$

The divergent part of this integral is brought into the standard form by partial integration with respect to x:

$$\int d^4l \int_0^1 x dx \frac{l^2}{(l^2 + m^2x^2)^3}$$

$$= \int d^4l\, l^2 \left\{ \frac{\frac{1}{2}x^2}{(l^2 + m^2x^2)^3} \bigg|_0^1 + 3m^2 \int_0^1 \frac{x^3 dx}{(l^2 + m^2x^2)^4} \right\}.$$

The second term in this last expression is absolutely convergent and is evaluated with Eq. (A5–12). The remaining term in Eq. (9–89) is

$$-4m^2 \int_0^1 x dx \int \frac{1 - x - \frac{1}{2}x^2}{(l^2 + m^2x^2)^3} d^4l = -2i\pi^2 \int_0^1 dx(1 - x - \frac{1}{2}x^2)\frac{1}{x}.$$

Inserting this into (9–89), we find, with (9–18),

$$L = \frac{\alpha}{4\pi}\left(D - 4\int_0^1 \frac{dx}{x} + \frac{11}{2}\right). \quad (9\text{–}90)$$

The expression for $\Lambda_{\mu f}(p',p)$ is absolutely convergent, so that the l-integrations can all be carried out with the aid of Eq. (A5–12). Using Eq. (9–89) and subtracting $L\gamma_\mu$ from $\Lambda_\mu(p',p)$, Eq. (9–86), we find, after a simple calculation,*

$$\Lambda_{\mu f}(p',p) = -\frac{\alpha}{2\pi}\int_0^1 dx \int_0^x dy \left\{ \frac{K_\mu}{a^2} + \gamma_\mu \int_0^1 dz \frac{a^2 - m^2x^2}{m^2x^2 + (a^2 - m^2x^2)z} \right.$$

$$\left. - 2m^2\gamma_\mu(1 - x - \tfrac{1}{2}x^2)\frac{a^2 - m^2x^2}{a^2 m^2 x^2} \right\}. \quad (9\text{-}91)$$

With regard to the physical interpretation of this result, we observe first that the separation of the constant L from the rest of the expression has introduced an infrared divergence, as in the similar case of Σ (*cf.* Section 9–4). The remarks made in that section regarding this point apply here too.

Comparison of Eq. (9–90) with Eq. (9–25) reveals the important relation

$$L = B. \quad (9\text{-}92)$$

If the vertex part is taken together with the γ_μ for the corner alone, we obtain for the total contribution (9–76) from this corner

$$\Gamma_\mu(p',p) = \gamma_\mu(1 + L) + \Lambda_{\mu f}(p',p). \quad (9\text{-}93)$$

The infinite constant L can be removed by renormalization in much the same way as B was removed in Σ. Just as in that case, the radiative corrections are in part a reproduction of the corrected term, which is γ_μ for the vertex part. Therefore we write

$$\Gamma_\mu = (1 + L)\Gamma_{\mu 0}, \quad \Gamma_{\mu 0} = \gamma_\mu + \Lambda_{\mu f}, \quad (9\text{-}94)$$

which is the same as Eq. (9–93) to second order. Since each corner of a diagram is associated with exactly one factor e, the factor $1 + L$ can be incorporated in the charge, so that we arrive at a charge renormalization

$$e_0 = (1 + L)e.$$

Combined with all the other charge renormalizations, Eq. (9–74), this renormalization yields

$$e_0 = (1 - B)\sqrt{1 - C}\,(1 + L)e.$$

Since the calculations of this and the previous two sections are valid only to second order, this can be written

$$e_0 = (1 - B + L)(1 - \tfrac{1}{2}C)e = (1 - \tfrac{1}{2}C)e. \quad (9\text{-}95)$$

---

* For further details see R. Karplus and N. Kroll, *Phys. Rev.* **77**, 536 (1950), especially Eqs. (31) and (32). We have corrected a sign error in the expression for K_μ and several printing errors in (32).

The terms B and L exactly cancel because of Eq. (9–92). We shall see that this cancellation is exact to all orders in the fine-structure constant α, so that the renormalizations arising from B in Σ and of L in Γ_μ always exactly compensate. For this reason they are referred to as *spurious charge renormalizations*.

The generalization of these arguments to higher orders for the removal of the infinite term in Γ_μ by renormalization and the proof of the internal consistency of this procedure is the subject of Chapter 10.

CHAPTER 10

RENORMALIZATION

The divergences which we found in Sections 9–4, 9–5, and 9–6 could all be disposed of by the method of mass and charge renormalization. However, the consistency of this procedure was demonstrated only up to second order in the coupling constant e. In this chapter we prove that it can be done to any order in e.*

The proof consists in the following steps. First we show that all divergences are obtained from a certain class called *primitive divergences* (Section 10–1). In the following Section 10–2 we define the irreducible and proper diagrams. The separation of divergent parts is carried out for the irreducible diagrams in Section 10–3 and for the reducible diagrams in Section 10–4. In Section 10–5 we show how the mass renormalization must be carried out. All the remaining divergences are canceled by charge renormalization. This is proved in Section 10–6. The wave function renormalization is defined in Section 10–7. The proof is finally completed by showing in Section 10–8 that the matrix element of the S-matrix is finite to any order of α. A final Section 10–9 gives a brief discussion of the method of regulators.

10–1 The primitive divergences. In the preceding chapter we discussed four different types of divergent diagrams: the vacuum diagram, the self-energy diagrams, and the vertex-diagram of second order. The question remains what other kind of divergent diagrams there are, and what their structure is. It is clear that there are many more divergent diagrams. Examples are very easily constructed, for instance, the one shown in Fig. 10–1. This example is a combination of an electron and a photon self-energy diagram. Divergences of this sort, which are obtained by combining any of the three divergent parts (excluding the vacuum diagrams), are, in a sense, not new divergences.

In order to survey the possible types of diagrams obtainable by insertion of divergent parts into larger diagrams, we introduce with Dyson the concept of the *primitive divergence*.† We call a diagram primitively divergent

---

* An incomplete proof was first given by F. J. Dyson, *Phys. Rev.* **75**, 1736 (1949). It was completed by A. Salam, *Phys. Rev.* **82**, 217 (1951). A much simpler proof was given by J. C. Ward, *Proc. Phys. Soc.* **A64**, 54 (1951). A modification of the latter is presented in this chapter.

† F. J. Dyson, *Phys. Rev.* **75**, 1736 (1949), especially Section V.

if upon the opening of any one of the internal lines, it becomes a convergent diagram. In other words, if any one of the integration variables, i.e., any one *four-vector* associated with an internal line of a primitively divergent diagram, is held fixed, the integral is convergent.

The diagrams discussed in the preceding sections are all primitively divergent. Figure 10-1 is an example of a divergent diagram not primitively divergent.

Let M be the analytical expression associated with a primitively divergent diagram. It will be a multiple integral of order 4ρ, say. The number of independent four-vectors, ρ, is determined by the number of corners n and the numbers E_i and P_i of internal electron and photon lines. There is a four-vector associated with each internal line, each of which corresponds to a fourfold integration. These vectors are not independent. For each corner there occurs a δ-function, establishing certain linear relations between the vectors. When these relations are utilized there remains the one δ-function which depends only on the external variables and which therefore remains in the final expression as a factor. The number of independent vector relations is therefore $(n - 1)$. Consequently, we find for the number of independent four-vectors:

$$\rho = E_i + P_i - n + 1.$$

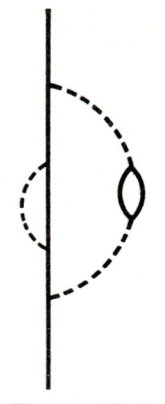

FIGURE 10-1

The integrand of M is a rational function. After combining the denominators according to the integral identity (A5-2), we obtain an integral of the form

$$M = \int_0^1 dx \int_0^x dy \cdots \int \frac{N}{D} d^4k_1 \cdots d^4k_\rho. \tag{10-1}$$

The integrals over the auxiliary variables $x, y, \cdots$ do not introduce any divergence. The convergence or divergence of the expression is entirely determined by the integral over the ρ four-vectors $k_1, \cdots, k_\rho$. To find the conditions under which this integral diverges, we change the path of integration of the variables $k_1{}^0, \cdots, k_\rho{}^0$ in the manner indicated in Section A5-1 such that the denominator becomes a positive definite expression for large values of the integration variables.

Both the numerator and the denominator are polynomials in the variables $k_1, \cdots, k_\rho$. The numerator is of degree E_i and the denominator is of degree $2E_i + 2P_i$. Since the integral converges when one of the k_i is held fixed (by definition), the integral is divergent if and only if

$$E_i + 4(E_i + P_i - n + 1) \geq 2(E_i + P_i).$$

We can define the degree of divergence

$$\kappa = 3E_i + 2P_i - 4n + 4.$$

When the degree of divergence is 0, 1, 2, etc., the integral is called logarithmically, linearly, quadratically, etc., divergent. The integral converges when $\kappa < 0$.

An important result is obtained if we express κ in terms of external variables only. The relations needed in this connection are

$$n = E_i + \tfrac{1}{2}E_e = 2P_i + P_e. \tag{10-2}$$

They are easily obtained from the topological structure of the diagrams by expressing the number of corners either in terms of the number of electron lines or in terms of the number of photon lines.

Substituting these relations into the right side of the definition for κ, we obtain

$$\kappa = 4 - (\tfrac{3}{2}E_e + P_e). \tag{10-3}$$

This expression for κ contains a remarkable result which is of the utmost importance for the renormalization theory. *The order n of the diagram does not occur in Eq. (10-3).* In other words, the coefficient κ is entirely determined by the number of *external* lines. If the right side of Eq. (10-3) increased with n then we would obtain primitively divergent diagrams for any given process (i.e., for any diagram with given numbers E_e and P_e) provided the order n is sufficiently great. This would lead to an infinite number of primitively divergent diagrams.*

The above equations also imply that a diagram must have at least as many corners as it has external lines in order to be divergent,

$$n \geq E_e + P_e.$$

We also note that $n - E_e - P_e$ is always an even number.

In Table 10-1 we enumerate the four primitive divergences which are not zero by Furry's theorem.

It is instructive to compare this list of primitive divergences with the classes of divergences discussed in Section 9-2. Since all these divergences are caused by the large values of the integration variables ("ultraviolet" divergence), class (b) is not represented in Table 10-1. Similarly, class (a) is not represented, since none of the vacuum diagrams is *primitively* divergent. But the other two classes, (c) and (d), are represented. The quadratic part of the divergence in the first line belongs to class (c), while the

---

* Certain meson theories and the quantum electrodynamics of particles with spin 1 are of this nature. These theories are apparently not renormalizable [cf. F. Rohrlich, *Phys. Rev.* **80**, 666 (1951)].

TABLE 10–1

Table of primitive divergences

The cases where $E_e = 0$ and P_e is odd have been omitted, since they vanish by Furry's theorem. The last column refers to the sections in which the lowest order examples are evaluated.

E_e	P_e	κ	Type of Diagram	Lowest Order Discussed in Section
0	2	2	Photon self-energy	9–5
0	4	0	Scattering of light by light	13–1
2	0	1	Electron self-energy	9–4
2	1	0	Vertex part	9–6

logarithmic part of that case belongs to class (d). The scattering of light by light (second line, Table 10–1) belongs also to class (c). The remaining divergences are in class (d). We note incidentally that all the "serious" divergences, [class (d)], are actually only logarithmic: the linear divergence of the electron self-energy (third line of Table 10–1) reduces to a logarithmic divergence upon symmetrical integration (*cf.* Section 9–4).

In the following sections we shall assume that we have disposed of the vacuum diagrams of class (a) and the diagrams of class (b) and (c) in the manner discussed previously (Sections 9–3 and 9–2). We shall therefore be concerned only with the "serious" divergences. There are three "serious" primitive divergences: the logarithmic part of the photon self-energy, the electron self-energy (which consists of two parts corresponding to δm and B in Section 9–4), and the vertex part. Since we shall need to refer to these quite often, we introduce the abbreviations *SE-parts* and *V-parts*.

10–2 Irreducible and proper diagrams. The problem before us is to prove that the three divergences of class (d) can all be absorbed into the charge and mass renormalization. We have proved in Sections 9–4, 9–5, and 9–6 that this is possible for the SE- and V-parts of second order in e. It is now necessary to extend this proof to all higher orders.

The main difficulty in the higher order case is that the SE-parts and V-parts may contain in their internal lines and corners insertions of further SE- and V-parts. In order to survey the possible types of diagrams obtained by repeated insertions of this sort, we define the concept of *reducible* and *irreducible diagrams*.

Consider any diagram M. A part of M which is connected to the rest of M by exactly two lines is called an *inserted SE-part*. Similarly, a part of M

(a) (b) (c)

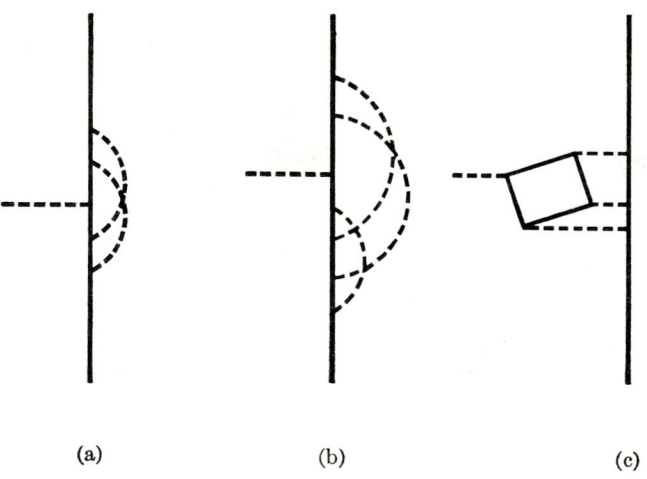

FIG. 10–2. Examples of irreducible V-diagrams.

which is connected to the rest of M by exactly two electron lines and one photon line is called an *inserted V-part*. To every M there corresponds a uniquely defined diagram M_s, called the *skeleton* of M, which is obtained from M by replacing every inserted SE-part by a line and every inserted V-part by a corner. A diagram or a diagram part is called *irreducible* if it is equal to its own skeleton. It is called *reducible* if this is not the case.

The only irreducible SE-diagrams are of the second order (Fig. 9–3 and 9–6). There are infinitely many irreducible V-diagrams. Three examples are shown in Fig. 10–2.

A reducible SE- or V-part can be either *proper* or *improper*. It is called improper if it can be separated into two disconnected parts by the omission of one single line. Diagrams for which this is not the case are called proper. An irreducible SE- or V-part is always proper, but a proper SE- or V-part is not necessarily irreducible.

An inserted electron SE-part W replaces the propagation function $S(p) = -(i\not{p} + m)^{-1}$ for a single electron line by a function $S(p)\Sigma(W,p)S(p)$. The divergent operator $\Sigma(W,p)$ represents the SE-part of W without the external lines. For example, the second order Σ is shown in Fig. 10–3(a), and its analytical expression for this order was calculated in Eqs. (9–16), (9–25), and (9–26).

Similarly, a photon SE-part W' when inserted into a photon line replaces the propagation function $D(k) = 1/k^2$ by $D(k)\Pi(W',k)D(k)$. The divergent function $\Pi(k)$ represents the analytical expression corresponding to a photon SE-part without its external photon lines. The diagram for the second-order case is shown in Fig. 10–3(b) and its analytical expression for this case was given in Eq. (9–63) with $\Pi(k) = -k^2 C(k^2)$.

Finally, a V-part inserted into a corner replaces the γ_μ of that corner by an operator $\Lambda_\mu(V,p',p)$, which, in general, depends on the momentum

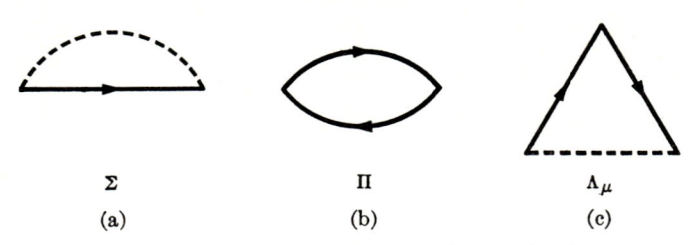

FIG. 10–3. Σ, Π, and Λ-parts of second order.

variables p', p associated with the external electron lines of the vertex-part. The diagram for the second order Λ_μ was given in Fig. 10–3(c). Its analytical expression was calculated in Eqs. (9–87), (9–90), and (9–91), for that case.

The *"true"* propagation function $S'(p)$ for an electron line is obtained by adding the propagation functions for all inserted electron SE-parts to the propagation function for a single electron line. We obtain thus the result

$$S'(p) = S(p) + S(p)\Sigma(p)S(p) \qquad (10\text{–}4)$$

where $\Sigma(p)$ is the sum of all $\Sigma(W,p)$:

$$\Sigma(p) = \sum_{\text{all } W} \Sigma(W,p). \qquad (10\text{–}5)$$

Similarly, the *"true"* propagation function $D'(k)$ for a photon line is obtained by adding the propagation function for all inserted photon SE-parts to the propagation function of a single photon line

$$D'(k) = D(k) + D(k)\Pi(k)D(k) \qquad (10\text{–}6)$$

with

$$\Pi(k) = \sum_{\text{all } W'} \Pi(W',k). \qquad (10\text{–}7)$$

Finally, the *"true"* vertex part $\Gamma_\mu(p',p)$ is obtained by adding all the proper vertex parts* to the operator γ_μ for a single corner.

$$\Gamma_\mu(p',p) = \gamma_\mu + \Lambda_\mu(p',p), \qquad (10\text{–}8)$$

$$\Lambda_\mu(p',p) = \sum_{\text{all proper } V} \Lambda_\mu(V,p',p). \qquad (10\text{–}9)$$

The functions S', D', Γ_μ are, of course, all highly divergent. Not only are their irreducible parts themselves divergent, but they contain all possible SE- and V-insertions, which introduce additional divergences.

The two Eqs. (10–4) and (10–6) involve the SE-parts Σ and Π. These include the contribution from all SE-parts, proper and improper. In the

---

* We shall never use improper vertex parts, so that we shall always mean *proper* vertex parts, even when not specifically so indicated.

following it will be more convenient to have relations which involve only the SE-parts Σ^* and Π^*, defined as the sum over all proper SE-parts.

$$\Sigma^*(p) = \sum_{\text{all proper } W} \Sigma(W,p), \qquad (10\text{--}5^*)$$

$$\Pi^*(k) = \sum_{\text{all proper } W'} \Pi(W',k). \qquad (10\text{--}7^*)$$

A general SE-part is either proper or it is a proper SE-part joined by a single line to another SE-part which itself may be either proper or improper. This property can be expressed in the two equations

$$\begin{aligned} S'(p) &= S(p) + S(p)\Sigma^*(p)S'(p), \\ S'(p) &= S(p) + S'(p)\Sigma^*(p)S(p), \end{aligned} \qquad (10\text{--}4^*)$$

for electron SE-parts, and

$$\begin{aligned} D'(k) &= D(k) + D(k)\Pi^*(k)D'(k), \\ D'(k) &= D(k) + D'(k)\Pi^*(k)D(k), \end{aligned} \qquad (10\text{--}6^*)$$

for photon SE-parts.

Two equivalent forms of these relations are obtained by solving them for Σ^* and Π^*:

$$\Sigma^* = S^{-1} - S'^{-1}, \qquad (10\text{--}10)$$

$$\Pi^* = D^{-1} - D'^{-1}. \qquad (10\text{--}11)$$

The "true" propagation functions S' and D', and the vertex-operators Γ_μ, enable us to calculate the radiative corrections to any order for every irreducible diagram. But in the present form they are actually useless for this purpose, since they contain all the divergent (or cut-off dependent) parts, which have no physical significance.†

It must be shown that certain finite terms can be separated from the functions S', D', and Γ_μ in such a way that the remaining infinite parts can be absorbed into mass and charge renormalization, while the finite parts, which cannot be so absorbed, describe observable effects. We explain this separation first for the irreducible and then for reducible diagrams.

---

† It may not be superfluous to remind the reader with a mathematical bent of mind that the manipulations of infinite quantities which we are doing here can always be avoided by introducing a suitable cut-off in momentum space. This can be done in a Lorentz- and gauge-invariant manner, e.g., by the method of regulators (cf. Section 10–9). Therefore, all the mathematical operations which are needed for carrying out the renormalization procedure to any order can be rigorously defined. It is only necessary to replace the word "infinite" by "cut-off dependent" and "finite" by "cut-off independent." Actually, there is no need to do so, since the infinite (or cut-off dependent) values of these integrals are irrelevant for the theory. We merely need the algebraic consequences of a uniformly applicable set of rules for handling these expressions.

10-3 Separation of divergences in irreducible parts.

We begin by separating the divergent parts in the *irreducible* SE- and V-parts. For the second-order SE- and V-parts the method of separation was discussed in great detail in Sections 9-4, 9-5, and 9-6, and the finite parts were evaluated in closed form [cf. Eqs. (9-26) and (9-66)]. The result for the second-order irreducible diagrams could be written in the form

$$\Sigma = A - S^{-1}B + S^{-2}\Sigma_f, \qquad (10\text{--}12)$$

$$\Pi = A' - D^{-1}C + D^{-2}\Pi_f, \qquad (10\text{--}13)$$

$$\Lambda_\mu = L\gamma_\mu + \Lambda_{\mu f}, \qquad (10\text{--}14)$$

where Σ_f, Π_f, and $\Lambda_{\mu f}$ were shown to be expressible as absolutely convergent integrals. Since all irreducible SE-diagrams are of second order, the task of separating the finite parts in the irreducible SE-diagrams is thereby accomplished.

Regarding the V-diagrams, on the other hand, the second order V-part is only one of an infinite class of irreducible V-parts. However, we know from the general theory of Section 10-1 that no matter what the order of the irreducible V-part may be, the degree of divergence is always logarithmic* [cf. Eq. (10-3)]. From this follows that for every irreducible V-part the divergent part can always be separated in the form of Eq. (10-14). The constant L as well as the finite part $\Lambda_{\mu f}$ will, of course, depend on the type of V-part.

Just as for the second order case, the separation of Λ_μ into $L\gamma_\mu$ and $\Lambda_{\mu f}$ is made unambiguous by the condition that $\Lambda_{\mu f}(p',p)$ vanishes when it operates on free-particle wave functions and p' is set equal to p. We can obtain it from $\Lambda_\mu(p',p)$ by carrying out the following operations. Bring all terms in p' to the left and all terms in p to the right of Λ_μ and then substitute $-m$ for ip and ip'. The resulting expression is denoted by $\Lambda_\mu(p_0,p_0)$.

This operation of substituting "free-particle values" for p is quite an important step in the following considerations. The notation which we have employed for this operation has been widely used in the literature, although it is apt to be misleading. There exists, in fact, no vector p_0 for which the equation $ip_0 + m = 0$ holds as a matrix relation. Furthermore, the result of this operation is actually independent of the momentum vectors, since this substitution also implies that $p^2 \equiv p^2$ is to be replaced by $-m^2$. Thus, after this operation is performed, we obtain an expression which is a function of m alone. We shall continue to use this notation, since it permits a simple means of indicating the above-mentioned process of substitution.

---

* The reader can easily verify that all irreducible V-parts are primitively divergent.

The function $\Lambda_\mu(p_0,p_0)$ is independent of p and p', and it has the transformation properties of a vector. It can therefore only be of the form

$$\Lambda_\mu(p_0,p_0) = L\gamma_\mu. \tag{10-15}$$

The same substitution and notation can be used for the separation of the infinite constants A and A' from Σ and Π,

$$A = \Sigma(p_0), \tag{10-16}$$

$$A' = \Pi(k_0). \tag{10-17}$$

Thus, Eq. (10-16) means: in the function $\Sigma(p)$ bring all terms involving p either to the left or to the right and then substitute $i p \rightarrow -m$. The result is the constant A, independent of p. This is the logarithmically divergent self-energy of the electron to second order in e. Similarly, Eq. (10-17) is obtained by substituting $k^2 = 0$. The result is the quadratically divergent self-energy of the photon to second order in e.

The constant A in Eq. (10-16) will be removed by mass renormalization. The constant A' is zero, since it is not gauge-invariant.* (In the neutral vector meson method it is removed by photon mass renormalization, *cf.* Section 9-5.)

10-4 Separation of divergences in reducible parts. Next in the proof is the consistent definition of the finite terms in *reducible* V- and SE-parts. We begin with the reducible V-parts. Let V be any V-part and V_s its skeleton. V is obtained from V_s by inserting various V- and SE-parts into the corners and lines of V_s. All these insertions are divergent. The separation of the finite parts in V now proceeds by induction. Let us assume the degree of V (i.e., the number of corners excluding the vertex itself) to be $2n$ and let us further assume that we have succeeded in separating the finite terms in all V- and SE-diagrams of degree less than $2n$. The finite term of V is then obtained in two steps.

In the first step, we separate the finite terms of all insertions. This means that for each corner in V_s we replace γ_μ by γ_μ plus the *finite part* of the inserted V-part at that corner. For each internal line of V_s we replace the propagation function D or S by D or S plus the *finite part* of the SE-part inserted at that line.

In the second step, we separate the finite terms from the divergent integral of the V_s-part with these finite insertions. The degree of divergence can be estimated with the method of Section 10-1. Since all V_s are primitive divergences, the integral is always logarithmically divergent. We call this last integral, after all the insertions are replaced by finite terms, the

---

* This means that any gauge-invariant cutoff procedure would give $A' = 0$.

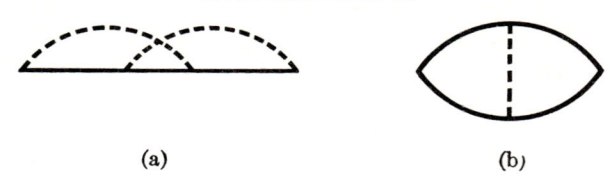

Fig. 10-4. Examples of SE-parts with overlapping divergence.

skeleton integral, and the remaining divergence the *skeleton divergence*. The idea of this nomenclature is to express the fact that the degree of the skeleton divergence is determined entirely by the topological structure of the skeleton and not by the insertions. These quantities are indicated by a subscript s.

The finite terms of the skeleton are separated by the same procedure that was used for deriving Eq. (10-14):

$$\Lambda_{\mu s}(p',p) = L_s \gamma_\mu + \Lambda_{\mu f}(p',p), \quad L_s = \tfrac{1}{4}\gamma^\mu \Lambda_{\mu s}(p_0,p_0). \quad (10\text{-}18)$$

The function $\Lambda_{\mu f}(p',p)$ is the finite part of the reducible vertex part V. This part, as well as the divergent part L_s, depends on the irreducible V-part in which the insertions were made as well as the finite part of these insertions.

Next we proceed to separate the finite terms in reducible proper SE-parts. Let W be a reducible proper electron SE-part of degree $2n$. Its skeleton W_s is the second order SE-part discussed in Section 9-4 (there is no other irreducible SE-part). W is obtained from W_s by insertion of V- and SE-parts into its corners and lines. All these insertions are of degree less than $2n$.

The separation of the finite term in this case is more complicated than for the V-parts, because an insertion of a V-part in one corner of W_s may also be looked upon as an insertion of a V-part in the other corner of W_s. This situation causes separate divergences for each way the V-part is inserted into the SE-part. We speak in this case of an *overlapping divergence*. They occur only in SE-parts, but never in V-parts. The simplest examples of overlapping divergences are shown in Fig. 10-4. The analytical expression for the case of Fig. 10-4(a), for instance, has the form

$$\Sigma(W,p) = -\frac{e^4}{(2\pi)^4} \iint d^4k_1 d^4k_2 \gamma^\mu S(p - k_1)$$

$$\times D(k_1) \gamma^\nu S(p - k_1 - k_2) D(k_2) \gamma_\mu S(p - k_2) \gamma_\nu. \quad (10\text{-}19)$$

Inspection of this expression shows that it is divergent if either of the two integration variables k_1 or k_2 is kept fixed; i.e., the integral is not primitively divergent. Thus, even if we separate a finite part from one vertex there still remains an infinite part from the other one.

For higher order insertions the situation becomes even more complicated, because such insertions can be distributed over the two corners in many different ways. There occurs then a divergence for each of the different ways the insertion is distributed over the corners. This situation is one of the major difficulties in the proof.†

It is possible to avoid this complication altogether if we define an operator $\Lambda_\mu(p)$ by the relation

$$\frac{\partial \Sigma^*(p)}{\partial p^\mu} \equiv i\Lambda_\mu(p), \qquad (10\text{--}20)$$

where

$$\Sigma^*(p) = \sum_{\text{all proper } W} \Sigma(W,p).$$

This operator can be expressed in terms of the $\Lambda_\mu(p',p)$ associated with the sum over all the vertex parts of the given order $2n$.

$$\Lambda_\mu(p) = \Lambda_\mu(p,p). \qquad (10\text{--}21)$$

This is Ward's identity.‡

The proof of this identity is based on the relation

$$\frac{\partial}{\partial p^\mu} S(p) = iS(p)\gamma_\mu S(p), \qquad (10\text{--}22)$$

which is a direct consequence of the definition $S(p) = -(i\boldsymbol{p} + m)^{-1}$. In terms of the Feynman-Dyson diagrams, Eq. (10–22) can be interpreted to mean that the differentiation of an electron line is equivalent to attaching to it a photon line which has zero momentum.

Consider first any proper electron SE-part W. With every such W we can associate a class of V-parts. The members V_i of this class are obtained from W by attaching a photon line at any of the electron lines in V_i.

Let us now consider the analytical expression $\Sigma(W,P)$ associated with the SE-part. W may consist of an open electron path together with a number of closed loops with an even number of corners (Furry's theorem!) attached to each other by internal photon lines. We can always choose the integration variables in such a way that only the open path carries the external variable p. It then contributes a number of factors to the integrand of $\Sigma(W,p)$ of the form

$$\Sigma(W,p) = \int \cdots \gamma^\mu S(p-k) \gamma^\nu S(p-l) \cdots d^4k\, d^4l \cdots.$$

---

† Dyson has given a special method for separating finite parts in this case [see F. J. Dyson, *Phys. Rev.* **75**, 1736 (1949), especially Section VII]. However, this method leads to difficulties in the later stages of the proof which were not completely solved in that paper. The most satisfactory treatment of the overlapping divergences is due to Ward (*loc. cit.*).

‡ J. C. Ward, *Phys. Rev.* **78**, 182 (1950).

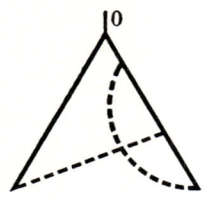

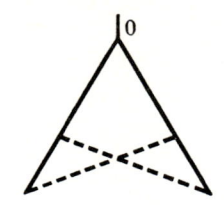

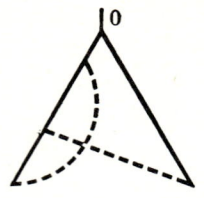

FIG. 10–5. The three V-parts obtained by differentiating the SE-part of Fig. 10–4(a).

Upon differentiation of this expression with respect to p^μ, each of these factors contributes a term of the form of Eq. (10–22). Thus, $\partial\Sigma(W,p)/\partial p^\mu$ is a sum of terms, each being proportional to one of the V-parts V_i for the special values $p' = p$ of the external momentum vectors. For instance, the Σ of Fig. 10–4(a) gives rise to the three vertices of Fig. 10–5. The sum contains all the V_i for which the external photon line is attached to the open path. We can complete the sum by including also the V_i for which the external photon line is attached to a closed loop, since they all vanish by Furry's theorem. In this way, we obtain

$$\frac{\partial\Sigma(W,p)}{\partial p^\mu} = i \sum_{V_i} \Lambda_\mu(V_i,p,p). \tag{10–23}$$

The summation on the right side is extended over the entire class of V-parts V_i associated with the given SE-part W. If we sum Eq. (10–23) over all proper W-parts, we find Ward's identity, Eq. (10–21),

$$\frac{\partial\Sigma^*(p)}{\partial p^\mu} \equiv i\Lambda_\mu(p) = i\Lambda_\mu(p,p).$$

This identity is of considerable importance, for it removes the ambiguity associated with overlap divergences, and *at the same time*—as we shall see—permits us to reduce the problem of separating divergencies of SE-parts to the simpler problem of separating them in V-parts.

Instead of separating the finite part in Σ^* with all the insertions, we separate them in $\partial\Sigma^*/\partial p^\mu \equiv i\Lambda_\mu(p)$. But this has been done already (Eq. 10–18). We merely need to put $p = p'$ and sum over all the irreducible V-parts up to the given order $2n$. In this way, we find for $\Lambda_\mu(p)$ an expression of the form:

$$\Lambda_\mu(p) \equiv -i\frac{\partial\Sigma^*(p)}{\partial p^\mu} = L\gamma_\mu + \Lambda_{\mu f}(p).$$

We can now also obtain the separation of the finite part in $\Sigma^*(p)$ by integrating this equation from p' to p, giving

$$\Sigma^*(p) - \Sigma^*(p') = i(\not{p} - \not{p}')L + F(p) - F(p'), \tag{10–24}$$

where
$$\frac{\partial F(p)}{\partial p^\mu} = i\Lambda_{\mu f}(p).$$

When we carry out the substitution $p' \to p_0$ in Eq. (10-24) (*cf.* fineprint remark in Section 10-3), we obtain

$$\Sigma^*(p) = \Sigma^*(p_0) + (i p\!\!\!/ + m)L + F(p) - F(p_0). \tag{10-25}$$

Defining the self-energy
$$A = \Sigma^*(p_0)$$
and the finite function
$$\Sigma_f(p) = S^2(p)[F(p) - F(p_0)], \tag{10-26}$$

Eq. (10-25) becomes
$$\Sigma^*(p) = A - S^{-1}(p)B + S^{-2}(p)\Sigma_f(p) \tag{10-27}$$
with
$$B = L \tag{10-28}$$

The result is worth noting. Equation (10-27) shows that the separation of infinite and finite parts in proper SE-parts of any order has exactly the same form as in the irreducible second order SE-part. The constants A and B, in the general case, are power series in α with logarithmically divergent coefficients. Furthermore, we have demonstrated that *the coefficient B of the second term is equal to L up to any order in α*. This result was verified in Section 9-6 (Eq. 9-92) by direct evaluation to second order in α.

Having accomplished the separation of finite terms in proper reducible electron SE-parts, we turn to the same problem for the photon SE-parts. Here, too, the overlap divergences can occur. The simplest example is given in Fig. 10-4(b). We avoid this difficulty in an analogous manner. We define the function $\Delta_\mu(k)$ by

$$\frac{\partial \Pi^*(k)}{\partial k^\mu} \equiv i\Delta_\mu(k). \tag{10-29}$$

This function is associated with diagrams containing only three external photon variables, one of which has momentum zero. These *three-photon V-parts* are linearly divergent expressions which reduce to logarithmically divergent ones after symmetrical integrations.

In our enumeration of primitive divergences we did not include such diagrams because of Furry's theorem. We showed in Section 8-4 that to every diagram with an odd number of external photon lines attached to a closed loop, there corresponds another one which is obtained by reversing all the arrows in the electron path which forms the closed loop, and that the sum of the two vanishes. In the expression $\Delta_\mu(k)$ this summation is not carried out, hence Furry's theorem does not apply.

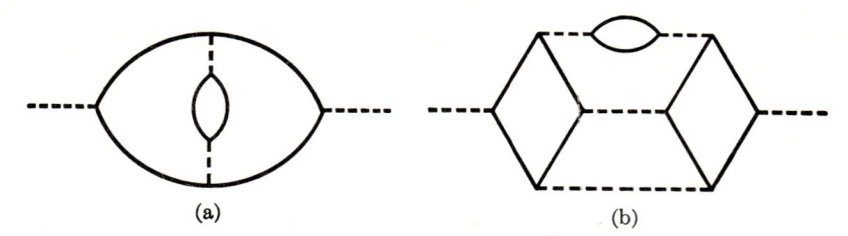

Fig. 10–6. Examples of two classes of proper photon SE-diagrams: (a) both external photon lines are attached to the same loop; (b) each external photon line is attached to a different loop.

Consider any proper photon SE-part W'. It contains a number of closed loops. We can distinguish two classes of photon SE-diagrams. In the first class there is one closed loop attached to both external photon lines. In the second class there are two different closed loops each attached to only one of the external photon lines. There may be additional closed loops in either case not attached to the external lines. We give one example of each kind in Fig. 10–6. Let us consider the SE-diagrams in the first class. We obtain a complete set of three-photon V-parts V_i' associated with all SE-parts W' in the first class by attaching in all possible ways one photon line of momentum zero to all the internal lines carrying the external momentum k. It is important to note that the integration variables can always be chosen such that only internal *electron* lines carry the momentum k. As examples, we show in Fig. 10–7 the three-photon V-part resulting from the second order photon SE-part, and in Fig. 10–8 the two fourth-order three-photon V-parts resulting from the overlap divergence of the fourth-order photon SE-part in Fig. 10–4(b).

The differentiation of the integrals associated with the photon SE-diagrams of the *second* class corresponds to insertions of zero-momentum photon lines in internal electron lines *as well as in internal photon lines* which carry the momentum k. No choice of variables can avoid this situation. The resulting corners where three photon lines join are of a topological structure not encountered previously in our diagrams. This corner contains a factor $-2k_\mu$ instead of the usual factor $i\gamma_\mu$ appearing in ordinary corners and upon differentiation of internal electron lines.

It is now easily seen that in the case of overlap divergences the addition of a zero-momentum photon line removes the divergence in the part to which it is added. This is valid for both classes of diagrams, for there is always an overlap of irreducible (and therefore primitively divergent) V-parts. Such a V-part is logarithmically divergent. Upon differentiation (addition of a photon line) the convergence improves by exactly one power of the internal momentum and that is just sufficient to make a V-part convergent. There remains therefore only one of the two overlapping diver-

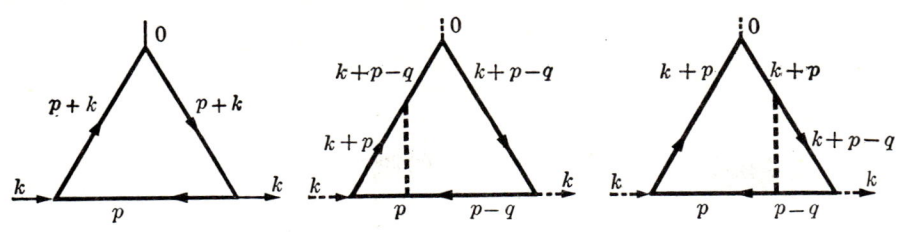

FIG. 10–7. Three-photon V-part of second order.

FIG. 10–8. Three-photon V-parts of fourth order.

gences which is a V-part and is easily separated. It follows that every three-photon V-part can be separated in the form

$$\Delta_\mu(k) = 2ik_\mu C + \Delta_{\mu f}(k). \tag{10-30}$$

Following the same arguments which led to Eq. (10–27) for the electron SE-part, we find

$$\Pi^*(k) = -D^{-1}(k)C + D^{-2}(k)\Pi_f(k), \tag{10-31}$$

$$\Pi_f(k) = D^2(k)[G(k) - G(k_0)], \tag{10-32}$$

$$i\Delta_{\mu f}(k) = \frac{\partial G(k)}{\partial k^\mu}. \tag{10-33}$$

The constant C is logarithmically divergent. Its second-order expression was given in Eq. (9–64). In the general case, it will be a power series in α with logarithmically divergent coefficients. The finite function $\Pi_f(k)$ describes the observable effects of vacuum polarization.

We have thus demonstrated how one can separate the finite and physically significant parts in the reducible SE- and V-parts of order n in α once the separation of finite terms in such parts of lower order has been accomplished.* Since for $n = 1$ this problem was explicitly solved in the preceding sections, we have now a recursive scheme by which it is possible to carry out this separation for any arbitrary order n of the diagrams.

These expressions (correct to order α^n) can be used for further calculation of the finite functions to the next higher order, α^{n+1}, by repeating this procedure for reducible V-parts and proper SE-parts of order α^{n+1}.

---

* In this connection the role played by Ward's identity must be put into proper perspective. The solution of the overlap divergence problem lies in the differentiation process. The fact that in some cases one is thereby led to other physically meaningful expressions (V-parts) is convenient, but not essential. Consequently, the equality $B = L$ is also not essential for the success of the renormalization process. Indeed, no use was made of it in Dyson's original paper.

The finite functions Σ_f, Π_f, and $\Lambda_{\mu f}$ are thus obtained as power series in α with finite coefficients. The question arises whether these power series are, in fact, convergent. The answer to this question is not known.* For the proof of the renormalization procedure the convergence of these expressions is not essential. It can be carried out for expressions calculated up to an arbitrary large but fixed power N of α. We shall therefore ignore this point from now on and use the language appropriate for exact expressions.

With the aid of the finite functions Σ_f, Π_f, and $\Lambda_{\mu f}$ we can define the finite propagation functions S_0 and D_0 and the vertex parts $\Gamma_{\mu 0}$. For instance, for S_0 we obtain from Eqs. (10–10), (10–26), and (10–27)

$$S_0^{-1}(p) = S^{-1}(p) - [F(p) - F(p_0)], \tag{10-34}$$

where

$$\frac{\partial F(p)}{\partial p^\mu} = i\Lambda_{\mu f}(p).$$

Similarly,

$$D_0^{-1}(k) = D^{-1}(k) - [G(k) - G(k_0)], \tag{10-35}$$

where

$$\frac{\partial G(k)}{\partial k^\mu} = i\Delta_{\mu f}(k).$$

Finally, the finite vertex-part is defined by

$$\Gamma_{\mu 0}(p',p) = \gamma_\mu + \Lambda_{\mu f}(p',p). \tag{10-36}$$

For later use we bring Eqs. (10–34) and (10–35) into a different form by differentiating them with respect to p^μ and k^μ, respectively. The result can be written

$$i\frac{\partial}{\partial p^\mu} S_0^{-1}(p) = \Gamma_{\mu 0}(p), \tag{10-34a}$$

$$i\frac{\partial}{\partial k^\mu} D_0^{-1}(k) = W_{\mu 0}(k), \tag{10-35a}$$

$$\Gamma_{\mu 0}(p) \equiv \gamma_\mu + \Lambda_{\mu f}(p), \tag{10-37}$$

$$W_{\mu 0}(k) \equiv 2ik_\mu + \Delta_{\mu f}(k). \tag{10-38}$$

---

* The prospect that the series converges is not very good. An example of a simplified theory which diverges after renormalization was discussed by W. Thirring, *Helv. Phys. Acta* **26**, 33 (1953); C. A. Hurst, *Proc. Camb. Phil. Soc.* **48**, 625 (1952); and A. Petermann, *Archives des Science de Genève* **6**, 5 (1953), *Phys. Rev.* **89**, 1160 (1953). An argument based on the analyticity of the S-matrix near $e = 0$ was presented by F. J. Dyson, *Phys. Rev.* **85**, 631 (1952). Although none of these or similar arguments presented to date *prove* the divergence of the renormalized series, they make it at least quite plausible.

10-5 Mass renormalization.
We first dispose of the term A in Eq. (10–27). The constant A is a power series in α with logarithmically divergent coefficients. We define the quantity

$$\delta m = A, \tag{10-39}$$

and the observable finite mass m_0,

$$m_0 = m + \delta m. \tag{10-40}$$

In the operator which expresses the total energy of the uncoupled field, we replace m by $m_0 - \delta m$, retain the term with m_0 in this operator, and include the term with δm in the interaction operator.

The total contribution to the proper electron SE-part consists now of two terms, one calculated from δm and the other calculated in Eq. (10–27). For this expression we therefore obtain

$$\Sigma^*(p) = (A - \delta m) - S^{-1}(p)B + S^{-2}(p)\Sigma_f(p). \tag{10-41}$$

The first term vanishes because of Eq. (10–39). Thus,

$$\Sigma^*(p_0) \equiv A - \delta m = 0. \tag{10-42}$$

All the quantities in Eq. (10–41) are now expressed in terms of the observable mass m_0 rather than m.

10-6 Charge renormalization.
Having carried out the mass renormalization, the remaining task is to show that all the infinities which are left (*viz.*, $B = L$ and C) can be absorbed into the charge renormalization.

We recall the definition of the infinite constants L and C. The constant L is obtained from $\Lambda_\mu(V_s,p',p)$ for any irreducible V-part V_s as follows. Substitute the finite functions D_0 and S_0 for all internal lines of V_s; substitute the finite vertex function $\Gamma_{\mu 0}$ for all corners; finally sum over all V_s. The operator which is so obtained is denoted by $\Lambda_{\mu s}(p',p)$. From the general discussion which led to Eq. (10–18), we know that

$$\Lambda_{\mu s}(p',p) = L\gamma_\mu + \Lambda_{\mu f}(p',p), \tag{10-43}$$

where

$$L\gamma_\mu = \Lambda_{\mu s}(p_0,p_0) \equiv \sum_{\text{all } V_s} \Lambda_{\mu s}(V_s,p_0,p_0). \tag{10-44}$$

The constant L is a power series in α with logarithmically divergent coefficients. Specializing to $p' = p$, we also have

$$\Lambda_{\mu s}(p) = L\gamma_\mu + \Lambda_{\mu f}(p). \tag{10-45}$$

In a similar way, we find

$$\Delta_{\mu s}(k) = 2ik_\mu C + \Delta_{\mu f}(k) \tag{10-46}$$

with

$$2ik_\mu C = \Delta_{\mu s}(k_0) \equiv \sum_{\text{all } V_s'} \Delta_{\mu s}(V_s',k_0). \tag{10-47}$$

The sum in Eq. (10–47) is extended over all irreducible photon vertices defined in Section 10–4.

We are now prepared to prove the fundamental relations

$$\Gamma_\mu(e) = \frac{1}{Z'} \Gamma_{\mu 0}(e_0), \qquad (10\text{–}48)$$

$$S'(e) = Z' S_0(e_0), \qquad (10\text{–}49)$$

$$D'(e) = Z D_0(e_0), \qquad (10\text{–}50)$$

$$W_\mu(e) = \frac{1}{Z} W_{\mu 0}(e_0), \qquad (10\text{–}51)$$

with

$$e_0 = \sqrt{Z}\, e. \qquad (10\text{–}52)$$

The meaning of these equations is the following. We assert that it is possible to choose Z and Z' in such a manner that these relations are consistent with the defining properties Eqs. (10–34), (10–35), and (10–36).

Let us first examine the expression for $\Lambda_\mu(e)$. Each irreducible V-part of degree $2n$ has n photon lines, $2n$ electron lines, and $2n+1$ corners. Counting the number of Z-factors introduced by the substitution of the finite functions from Eqs. (10–48), (10–49), and (10–50), and recalling the definition of $\Lambda_{\mu s}$ [cf. Eq. (10–18)], we find

$$\Lambda_\mu(e) = \frac{1}{Z'} \Lambda_{\mu s}(e_0). \qquad (10\text{–}53)$$

Similarly,

$$\Delta_\mu(e) = \frac{1}{Z} \Delta_{\mu s}(e_0). \qquad (10\text{–}54)$$

Adding γ_μ to Eq. (10–53), we find with Eqs. (10–8) and (10–43),

$$\Gamma_\mu(e) = \frac{1}{Z'} (L(e_0)\gamma_\mu + \Lambda_{\mu f}(e_0)) + \gamma_\mu. \qquad (10\text{–}55)$$

In an analogous way, we derive the relation

$$W_\mu(e) = \frac{1}{Z} (C(e_0) 2ik_\mu + \Delta_{\mu f}(e_0)) + 2ik_\mu. \qquad (10\text{–}56)$$

The last two equations are identical with Eqs. (10–48) and (10–51), respectively, provided that

$$Z' = 1 - L(e_0), \qquad (10\text{–}57)$$

$$Z = 1 - C(e_0). \qquad (10\text{–}58)$$

It is important to realize the meaning of Eqs. (10–55) and (10–56) when Eqs. (10–48) and (10–51) are inserted on the left side. They express the equality of the finite parts obtained in two different ways: once by separating and omitting the infinite parts and replacing the argument e of the remainder by e_0, and once by renormalization of the whole expressions according to Eqs. (10–48) to (10–52). The equations (10–57) and (10–58) are the necessary and sufficient conditions that these equalities hold.

As a consistency check, we use Ward's identity. We substitute e_0 for e in Eqs. (10–34a) and (10–35a) (omitting the argument p for simplicity) and obtain, with Eqs. (10–53) and (10–54),

$$i\frac{\partial}{\partial p^\mu} S_0^{-1}(e_0) = \Gamma_{\mu 0}(e_0) = Z'\Gamma_\mu(e) = Z'i\frac{\partial}{\partial p^\mu} S'^{-1}(e), \qquad (10\text{–}59)$$

$$i\frac{\partial}{\partial k^\mu} D_0^{-1}(e_0) = W_{\mu 0}(e_0) = ZW_\mu(e) = Zi\frac{\partial}{\partial k^\mu} D'^{-1}(e). \qquad (10\text{–}60)$$

We write this result in the form

$$\frac{\partial}{\partial p^\mu}(S_0^{-1}(e_0) - Z'S'^{-1}(e)) = 0, \qquad (10\text{–}61)$$

$$\frac{\partial}{\partial k^\mu}(D_0^{-1}(e_0) - ZD'^{-1}(e)) = 0. \qquad (10\text{–}62)$$

These equations can be integrated. The integration constant is zero in both cases because of $\Sigma^*(p_0) = 0$ and $\Pi(k_0) = 0$ (mass renormalization!). Therefore,

$$S'(e) = Z'S_0(e_0),$$

$$D'(e) = ZD_0(e_0),$$

which are the desired relations (10–49) and (10–50).

We finally note the expression for δm in terms of the renormalized e_0. From the structure of the self-energy diagrams,

$$\delta m \equiv A(e) = \frac{A(e_0)}{Z'}. \qquad (10\text{–}63)$$

10–7 Wave function renormalization. Up to this point we have paid attention only to the effect of renormalization on the SE- and V-parts. When we calculate general matrix elements for the transition between free-particle states we must also consider the effect of renormalization on the free-particle *wave functions*, u, v, e_μ. Their renormalization is equivalent to the renormalization of the corresponding free-particle operators $\varphi(p)$ and $a_\mu(\mathbf{k})$. In the following it will be more convenient to work with the latter.

The insertion of the radiative corrections in an external electron line results in a change of the free-particle operator $\varphi(p)$ into

$$\varphi'(p) = \varphi(p) - BSS^{-1}\varphi(p). \tag{10-64}$$

The finite part of $\Sigma(p)$ contributes nothing for free-particle operators. The expression obtained in this manner is ambiguous and cannot be used for the determination of the renormalization factor of the field operators. This factor can, however, be determined by the same procedure that is used in the second-order case (Section 9–4). There we have pointed out that consistency of the definition of the renormalized propagation functions requires that the contraction symbol formed with the renormalized wave function be equal to the renormalized propagation function. This means we must define the renormalized operator $\varphi_0(p)$ by setting

$$\varphi'(p) = \sqrt{Z'}\,\varphi_0(p). \tag{10-65}$$

The new operators $\varphi_0(p)$ are the renormalized operators including all the radiative corrections. However, since

$$(i\not{p} + m)\varphi_0(p) = 0, \tag{10-66}$$

the new operators $\varphi_0(p)$ can now be identified with the free-particle operators and assumed to be normalized in the usual manner.

The same kind of reasoning which we have just employed may be used to show that the renormalized photon operators must be defined by

$$a_\mu'(\mathbf{k}) = \sqrt{Z}\,a_{\mu 0}(\mathbf{k}). \tag{10-67}$$

10–8 Sufficiency proof. The final step in the proof consists in showing that the renormalization procedure which we have obtained from the divergent diagrams is just sufficient to eliminate the divergent factors from any arbitrary matrix element.

Thus, consider a matrix element $M_n(e)$ of order n, derived from an irreducible diagram with exactly n corners. The "exact" matrix element $M_n'(e)$ for this process would be obtained from $M(e)$ by replacing each S in M_n by S', each D by D', and each γ_μ by Γ_μ. Let $M_{n0}(e_0)$ be the matrix element obtained from M_n by substituting the finite parts S_0, D_0, and $\Gamma_{\mu 0}$ for the propagation functions and V-parts, expressed in terms of the renormalized charge e_0. We wish to show that

$$M_n'(e) = M_{n0}(e_0) \tag{10-68}$$

if

$$e_0 = \sqrt{Z}\,e.$$

This means that the correct, and physically observable, matrix element is obtained from the divergent one by dropping the divergent parts and renormalizing the charge. In other words, the entire effect of the divergent parts on M_n can be expressed as a change in the coupling constant e.

To prove Eq. (10–68), we assume that the skeleton of M_n is a diagram with P_e and P_i external and internal photon lines, E_e and E_i external and internal electron lines, and n corners. From the topological structure of the diagrams, we obtain, as in Section 10–1,

$$n = 2P_i + P_e = E_i + \tfrac{1}{2}E_e. \tag{10–69}$$

In an obvious symbolic notation which indicates only the powers of the various functions, we write for $M_n'(e)$:

$$M_n'(e) \sim e^n (\Gamma)^n (S')^{E_i} (D')^{P_i} (\bar{\varphi}'\varphi')^{\frac{1}{2}E_e} a'^{P_e}.$$

When we extract the Z-factors in accordance with Eqs. (10–48), (10–49), (10–50), and (10–51), we obtain

$$M_n'(e) \sim (Z')^{(-n+E_i+\frac{1}{2}E_e)} (Z)^{(-\frac{n}{2}+P_i+\frac{1}{2}P_e)} M_{n0}(e_0).$$

Because of Eqs. (10–69) the two exponents of the Z-factors are zero and Eq. (10–68) is established.*

With this result, it is finally proved that the program of renormalization can be carried through in quantum electrodynamics up to any desired order of e. It is therefore justified to omit all properly separated infinite terms in the applications of the theory, and to identify e with the observable charge. The results of this procedure are identical with those obtained by renormalization.

10–9 Regulators. The idea of eliminating the divergences from the theory of the electron by the assumptions of a finite elementary charge distribution is as old as the theory itself. It is well known that a classical Lorentz electron of finite radius will yield a finite self-energy. However, apart from the arbitrariness of such an assumption, other difficulties are thereby introduced as, for example, the nonvanishing self-stress. The associated problem of the stability of the electron finds its most accurate formulation in the transformation law of the energy-momentum tensor, which in this case could contradict relativistic invariance (cf. Section 16–4).

---

* We note that all renormalized S-matrix elements can be thought of as originating in a renormalized interaction operator obtained by the following steps:

$$ie\bar{\psi}\gamma_\mu\psi a^\mu \rightarrow ie\bar{\psi}'\Gamma_\mu'\psi' a'^\mu = ie_0\bar{\psi}_0\Gamma_{\mu 0}(e_0)\psi_0 a_0^\mu.$$

In the last expression all Z-factors cancel.

It was many years before it was recognized that a finite electronic charge distribution can also be introduced in a relativistic manner.* Such a theory is essentially equivalent to one in which the fields produced by the electronic charge are averaged over space time. The corresponding "shape factor" or "smearing-out function" still remained to be chosen from a rather large class of functions. It was therefore of considerable interest that this function can be chosen such that it can be interpreted as the interaction of the electron with other fields of various rest masses, coupling constants, and spins.† The applications of these ideas to quantum electrodynamics received considerable attention through Pais' f-field,‡ which, however, was soon found to be insufficient to remove all divergences from the theory. Similar ideas were used with some success by Feynman and others.§ However, the first consistent method of "auxiliary" fields or *regulators* was given by Pauli and Villars,** who showed that a Lorentz and gauge-invariant introduction of a sufficient number of these fields can remove the divergences of both the electron and the photon self-energy from the S-matrix, at least in lowest order. Soon it became apparent that the method works to all orders in the expansion of the coupling constant and can easily be extended to meson theories. This method introduces the auxiliary fields in the S-matrix; i.e., it regularizes the S-matrix, which is sufficient for the evaluation of scattering cross sections and related observables. However, in order to show that the old problem of the self-stress can be eliminated too, it is necessary to regularize the Lagrangian †† rather than the S-matrix. The electron field, the electromagnetic field, and all the regulator fields can thus be introduced into the theory from the very beginning, thereby giving a complete and consistent picture.¶

The method of regulators considers the interaction of $(n_e + 1)$ fields of spin $\frac{1}{2}$, masses m_i, and coupling constants ec_i ($i = 0, 1, 2, \cdots, n_e$), with $(n_p + 1)$ neutral vector meson fields of masses M_i ($i = 0, 1, \cdots, n_p$). One chooses $c_0 = 1$, $M_0 = 0$, and m_0 equal to the electron mass, so that the fields with subscript zero refer to the electron and photon fields. The $2n_e + n_p$

---

* H. McManus, *Proc. Roy. Soc.* **A195**, 323 (1948).

† F. Bopp, *Ann. Physik* **42**, 573 (1943); E. C. G. Stückelberg, *Helv. Phys. Acta* **14**, 51 (1941), where a scalar compensating field is introduced; see also A. Landé and L. H. Thomas, *Phys. Rev.* **60**, 121, 514 (1941) and **65**, 175 (1944), and B. Podolsky, *Phys. Rev.* **62**, 68 (1942).

‡ A. Pais, *Kon. Ned. Akad. v. Wet. Verb.* D1, **19**, (1947); independently, S. Sakata, *Prog. Theor. Phys.* **2**, 145 (1947), worked along similar lines.

§ R. P. Feynman, *Phys. Rev.* **74**, 939 and 1430 (1948), where classical and quantum electrodynamics are investigated, respectively. E. C. G. Stückelberg and D. Rivier, *Phys. Rev.* **74**, 218 and 986 (1948).

** W. Pauli and F. Villars, *Rev. Mod. Phys.* **21**, 434 (1949). This work grew out of earlier investigations by J. Rayski, *Phys. Rev.* **75**, 1961 (1949).

†† F. Rohrlich, *Phys. Rev.* **77**, 357 (1950). See also Section 16–4.

¶ See, e.g., S. N. Gupta, *Proc. Phys. Soc.* (London) **A66**, 129 (1953).

other constants must have such values that all S-matrix elements are convergent.

At this point, one might ask whether these masses and coupling constants have anything to do with the observed mesons. Unfortunately, there seems to be no possible relation, because, as a necessary convergence condition, some of the coupling constants of the spin one-half meson fields must be imaginary, so that these fields can have no realistic physically meaningful interpretation. It can be shown that choices other than the above types of fields do not change this situation.† We are therefore led to the conclusion that the regulators must be considered as a *purely formal means* of an invariant cutoff.‡

This conclusion implies that the regulator method is physically no improvement over the earlier cut-off functions and suffers from the same arbitrariness. Nevertheless, the method is often used in applications because of its simplicity. In that case it is convenient to choose the c_i, m_i, and M_i as follows. For a given matrix element the convergence conditions establish certain relations between these parameters, the physically meaningless divergent integrals are thereby either made zero or are made finite and are then removed by renormalization. In the remaining finite terms the masses m_i and M_i for $i = 1, \cdots, n_e$ and $i = 1, \cdots, n_p$, respectively, are made to approach infinity. One argues that all observable effects vanish in the limit as the intermediate states involve infinitely heavy particles. The result of this procedure is, therefore, that only those finite, observable effects survive which take place via the intermediate states of m_0 and M_0 = 0. These are identical with the finite matrix elements obtained without regularization in the previous sections.

The following two examples of the electron and photon self-energies will illustrate these remarks and show how the method of regulators works. Following Eq. (9–9) for the electron self-energy, we find, on introducing regulators,

$$\sum_{i=0}^{n_p} \frac{ie^2}{(2\pi)^4} c_i^2 \int \gamma_\mu \frac{i(\boldsymbol{p} - \boldsymbol{k}) - m}{(p - k)^2 + m^2} \gamma^\mu \frac{1}{k^2 + M_i^2} d^4k. \qquad (10\text{–}70)$$

This expression involves only addition neutral vector meson fields, since

---

† For example, Gupta (*loc. cit.*) constructs regulators with real coupling constants at the cost of working with auxiliary spin one-half fields satisfying Bose-Einstein statistics.

‡ More specifically, the main difficulty seems to be the elimination of the divergence of the constant C, the "true" charge renormalization. No known realistic field yields a compensating sign of the coupling constant. [*Cf.* D. Feldman, *Phys. Rev.* **76**, 1369 (1949)]. The most general investigation into the possible reality of the auxiliary fields was carried out by A. Pais and G. E. Uhlenbeck, *Phys. Rev.* **79**, 145 (1950). The reader is referred to that paper for a deeper analysis of this problem.

the only original integration was to be carried out over the intermediate *photon* momentum.

Each term in the expression Eq. (10–70) can be transformed in the same manner as Eq. (9–9), leading to the result, analogous to Eq. (9–17),

$$A(M_i) = \frac{\alpha m}{2\pi}\left(\frac{1}{i\pi^2}\int d^4k \int_0^1 dx \, \frac{1+x}{[k^2 + m^2x^2 + M_i^2(1-x)]^2} - \frac{1}{4}\right). \quad (10\text{--}71)$$

We transform the integral with respect to x by partial integration analogous to Eq. (9–20):

$$\int_0^1 dx \, \frac{1+x}{[k^2 + m^2x^2 + M_i^2(1-x)]^2}$$

$$= \frac{3}{2}\frac{1}{(k^2+m^2)^2} + \int_0^1 \frac{(4m^2x - 2M_i^2)\left(x + \dfrac{x^2}{2}\right)}{[k^2 + m^2x^2 + M_i^2(1-x)]^3} dx.$$

The second term on the right can be integrated with respect to k with the help of Eq. (A5–12), giving

$$\frac{1}{i\pi^2}\int d^4k \int_0^1 \frac{(4m^2x - 2M_i^2)\left(x + \dfrac{x^2}{2}\right)}{[k^2 + m^2x^2 + M_i^2(1-x)]^3} dx = \int_0^1 \frac{(2x - \rho_i)\left(x + \dfrac{x^2}{2}\right)}{x^2 + \rho_i(1-x)} dx,$$

where $\rho_i = M_i^2/m^2$. The remaining integral can be done explicitly. Since we want to go to the limit $\rho_i \to \infty$, we need only the terms of highest order in ρ_i,

$$\int_0^1 \frac{(2x - \rho_i)\left(x + \dfrac{x^2}{2}\right)}{x^2 + \rho_i(1-x)} dx = -\frac{3}{2}\ln \rho_i + O(1). \quad (10\text{--}72)$$

For $M_0 \to 0$, i.e., $\rho_0 \to 0$, the integral gives $\frac{5}{2}$. Combining these results, we find, with $c_0 = 1$,

$$A = \sum_0^{n_p} c_i A(M_i)$$

$$= \frac{\alpha}{2\pi} m \left[\sum_{i=0}^{n_p} c_i^2 \left(\frac{3}{2}D - \frac{1}{4}\right) + \frac{5}{2} + \sum_{i=1}^{n_p} c_i^2 \left(-\frac{3}{2}\ln \rho_i + O(1)\right)\right]. \quad (10\text{--}73)$$

Therefore, A will be finite, provided

$$\sum_{i=0}^{n_p} c_i^2 = 0, \quad \sum_{i=1}^{n_p} c_i^2 \ln \frac{M_i}{m} < \infty. \quad (10\text{--}74)$$

These conditions can actually be met by using two masses, M_1 and M_2, only. Of course, it is always possible to choose the parameters such that $A = 0$, but there is no need to do so, since A can be removed by mass renormalization.

It is easily seen that the conditions (10–74) are just sufficient to assure the finiteness of the constant B also. Since $B = L$, the same procedure will give a finite value for the second order V-part. The finite part Σ_f is exactly reproduced in the limit $M_i \to \infty$ for $i \neq 0$.

In the case of the photon self-energy, the intermediate state consists of virtual electron states, so that only auxiliary spin one-half fields need be added. But since we are dealing with a quadratic divergence in this case, we must satisfy the conditions

$$\sum_{i=0}^{n_e} c_i^2 m_i^2 = 0, \quad \sum_{i=0}^{n_e} c_i^2 = 0, \quad \sum_{i=1}^{n_e} c_i^2 \ln \frac{m_i}{m} < \infty. \qquad (10\text{--}75)$$

As can be shown by an argument similar to the one just outlined for the electron self-energy, these conditions are necessary and sufficient to yield a gauge-invariant and finite photon self-energy part; i.e., it gives zero for $A' = \delta\kappa^2$ (Eq. 9–59), a finite value for C (Eq. 9–64), and reproduces exactly Π_f (Eq. 9–66).

In summary, the method of regulators is a convenient Lorentz and gauge-invariant cutoff procedure which has no realistic interpretation and has the same inherent arbitrariness as any other invariant cutoff procedure. That it permits a justification for the manipulation of infinite quantities, as occurred in the previous sections, is its only value. It gives us no further physical insight, nor is it necessary for the consistency of the theory.

CHAPTER 11

THE PHOTON-ELECTRON SYSTEM

In this and the following two chapters we consider applications of the theory developed thus far to certain special systems.

It is convenient to classify systems according to the number and type of particles in the *initial* state. (The number and type of particles in the final state may be different.) Thus, the photon-electron system is defined to be a system with exactly one photon and one electron in its initial state. If we restrict ourselves to only two particles in the initial state, as we shall do in the following, we can distinguish photon-electron, electron-electron, and photon-photon systems. These three systems will be discussed in Chapters 11, 12, and 13, respectively.

The application of the theory to special systems is considerably simplified by the symmetry properties of the S-matrix under charge conjugation. According to Section 7–6(f), the charge conjugation operator γ commutes with the S-operator, and a state vector and its charge conjugate are related by

$$\omega^c = \gamma\omega. \tag{11-1}$$

It follows that the S-matrix element for a process, S_{fi}, is related to the S-matrix element for the charge-conjugate process, $S_{fi}{}^c$, by

$$S_{fi}{}^c = (\gamma\omega_f, S\gamma\omega_i) = (\omega_f, \gamma^{-1}S\gamma\omega_i) = (\omega_f, S\omega_i) = S_{fi}. \tag{11-2}$$

The matrix elements of charge conjugate processes are equal. The photon-positon scattering cross section is therefore equal to the photon-negaton scattering cross section, and the Møller cross section for negaton-negaton collisions is also valid for positon-positon collisions. These examples show that the charge symmetry of the theory reduces considerably the number of processes which must be calculated.

The photon-electron system can have two kinds of final states, those in which there is only one electron present (and one or more photons), and those in which there are also one or more pairs present. Processes leading to the former kind of final states may be called *photon-electron scattering*, whereas processes in which pairs are produced can be referred to as *pair production in photon-electron collisions*, or *pair production by photons in the field of an electron*. The latter designation is less appropriate, since it implies that the electron is initially at rest, which is not necessarily the case, and gives the erroneous impression that only the static field of the electron contributes to the process.

The following three sections will be devoted to scattering, whereas the last section of this chapter deals with pair production.

11-1 Compton scattering.
The scattering of a photon by an electron is called Compton scattering,* after A. H. Compton, who discovered it experimentally. The relativistic cross section was first calculated by Klein and Nishina, and by Tamm† to lowest order in e.

The separation of the lowest order from the higher order contributions is actually an idealization which does not correspond exactly to the physical situation. In all actual physical measurements the energy of the final state is determined only within a certain latitude depending on the energy resolution of the experimental equipment. It is therefore impossible to determine with certainty whether the final state contains exactly one photon or whether it contains in addition a number of very soft photons.

We call the electron-photon scattering process with two photons in the final state the *double Compton scattering*. The matrix element for this process is higher by one order in e than the matrix element for single Compton scattering. The cross section for the double Compton scattering exhibits the characteristic logarithmic divergence previously mentioned in Section 9-2, class (b). A similar divergence appears in the cross section which includes the first radiative correction to Compton scattering. We shall show in Section 16-1 that the two divergences exactly cancel. This is a special case of a general theorem which applies to all processes with infrared divergences.

In this section we shall discuss single Compton scattering. An electron of momentum p, and a photon of momentum k and polarization e, scatter into an electron of momentum p', and a photon of momentum k' and polarization e'. The cross section is given by Eqs. (8-50) and (8-49),‡

$$\sigma = (2\pi)^2 \frac{\epsilon\omega}{|p \cdot k|} \mathsf{S}_f \mathsf{S}_i \delta(p' + k' - p - k)|(f|M|i)|^2, \qquad (11\text{-}3)$$

where $\epsilon = p^0$, $\omega = k^0$.

The matrix element $M_{fi}^{(2)}$ is obtained from $S_{fi}^{(2)}$ by means of the relations (8-28) and (8-29). The Feynman-Dyson diagrams for momentum space were discussed in Section 8-3 (see Fig. 8-5) and the S-matrix element was also obtained there (Eq. 8-25) for the case of photon-negaton scattering.

---

* A. H. Compton, *Phys. Rev.* **21**, 715 (1923).

† O. Klein and Y. Nishina, *Z. Physik* **52**, 853 (1929); I. Tamm, *ibid.* **62**, 545 (1930).

‡ Here and in the following we use the simplified notation $A \cdot B = A_\mu B^\mu$, $A^2 = A_\mu A^\mu$.

Therefore,

$$(p'k'e'|M^{(2)}|pke) = \frac{ie^2}{8\pi^2} \cdot \frac{m}{\sqrt{\epsilon\epsilon'\omega\omega'}} \bar{u}(\mathbf{p}')$$

$$\times \left(e' \frac{i(p'+k') - m}{(p'+k')^2 + m^2} e + e \frac{i(p - k') - m}{(p - k')^2 + m^2} e' \right) u(\mathbf{p}). \quad (11\text{-}4)$$

In order to evaluate the cross section, we note that the δ-function in Eq. (11-3) and the fact that we are dealing with free particles in the initial and final states,

$$p^2 + m^2 = 0, \quad p'^2 + m^2 = 0, \quad k^2 = 0, \quad k'^2 = 0,$$

permit us to express all products of the momenta by m and the two scalars

$$\kappa = -p \cdot k = -p' \cdot k', \quad \kappa' = -p \cdot k' = -p' \cdot k. \quad (11\text{-}5)$$

For instance, the denominators in Eq. (11-4) become

$$(p + k)^2 + m^2 = -2\kappa, \quad (p - k')^2 + m^2 = 2\kappa'. \quad (11\text{-}6)$$

Also,

$$\kappa' - \kappa = k \cdot k' = p \cdot p' + m^2. \quad (11\text{-}7)$$

This last equation is important, because it permits the determination of ω' in terms of the scattering angle θ and the given initial state,

$$\omega\omega' \cos\theta = \mathbf{k} \cdot \mathbf{k}',$$

where $\mathbf{k}$ and $\mathbf{k}'$ are the photon momentum three-vectors. Let α and α' be the angles between the incident electron and the incident and outgoing photon, respectively, and let

$$|\mathbf{p}| = \beta\epsilon, \quad \epsilon = \gamma m, \quad \gamma = \frac{1}{\sqrt{1-\beta^2}}, \quad \beta = v.$$

With this notation, Eq. (11-7) yields

$$\frac{\omega'}{\omega} = \frac{1 - \beta\cos\alpha}{1 + (\omega/\epsilon)(1 - \cos\theta) - \beta\cos\alpha}. \quad (11\text{-}8)$$

The angle φ is the angle between the planes formed by $\mathbf{k},\mathbf{p}$ and $\mathbf{k},\mathbf{k}'$. The angle α' can be expressed in terms of α, θ, and φ:

$$\cos\alpha' = \cos\alpha \cos\theta + \sin\alpha \sin\theta \cos\varphi.$$

We consider a scattering experiment in which the initial and final spin states are not observed, as is usually the case. In Eq. (11-3) $\bar{S}_i$ will then be the average over the initial spin states, and S_f will consist of the sum over the final spin states and an integration over the final energy-momentum space, τ_f. These spin sums are most conveniently carried out by use

COMPTON SCATTERING

of the projection operators $\Lambda_\pm$ (Eq. 3–77) and the completeness relation (3–93), as is explained in the Appendix, Section A2–5. The result is a trace. We find from Eqs. (11–3) and (11–4):

$$\sigma = \left(\frac{e^2}{4\pi}\right)^2 \frac{1}{2\kappa\epsilon'\omega'} \int d\tau_f \delta(p' + k' - p - k) X, \qquad (11\text{–}9)$$

where

$$X = m^2 \sum_{\text{spins}} |\bar{u}(\mathbf{p}')(\cdots)u(\mathbf{p})|^2,$$

$$X = \frac{m^2}{4} \operatorname{Tr}\left\{\left[\frac{1}{\kappa}e'(ip + ik - m)e - \frac{1}{\kappa'}e(ip - ik' - m)e'\right]\Lambda_-(p)\right.$$

$$\left. \times \left[\frac{1}{\kappa}e(ip + ik - m)e' - \frac{1}{\kappa'}e'(ip - ik' - m)e\right]\Lambda_-(p')\right\}. \qquad (11\text{–}10)$$

This trace can be written

$$X = \frac{m^2}{\kappa^2} A + \frac{m^2}{\kappa'^2} B - \frac{m^2}{\kappa\kappa'}(C + D), \qquad (11\text{–}11)$$

where

$$\begin{aligned}
A &= (e'(ip - m + ik)e\Lambda_-(p)e(ip - m + ik)e'\Lambda_-(p')), \\
B &= (e(ip - m - ik')e'\Lambda_-(p)e'(ip - m - ik')e\Lambda_-(p')), \\
C &= (e'(ip - m + ik)e\Lambda_-(p)e'(ip - m - ik')e\Lambda_-(p')), \\
D &= (e(ip - m - ik')e'\Lambda_-(p)e(ip - m + ik)e'\Lambda_-(p')).
\end{aligned} \qquad (11\text{–}12)$$

In this expression we have employed the notation of Section A2–4 for the trace

$$\tfrac{1}{4}\operatorname{Tr}(a_1 a_2 \cdots a_n) \equiv (a_1 a_2 \cdots a_n).$$

We observe that the substitution

$$e \leftrightarrow e', \quad k \leftrightarrow -k'$$

will leave the trace (11–11) unaltered, for it transforms A into B, C into D, and, conversely, B into A, D into C. We also notice that $C = D$. It is therefore sufficient to evaluate A and C only. Repeated application of Eq. (A2–80) yields*

$$X = \frac{1}{2}\left(\frac{\kappa}{\kappa'} + \frac{\kappa'}{\kappa}\right) - 1$$

$$+ 2\left(e\cdot e' + \frac{e\cdot p\; e'\cdot p'}{\kappa} - \frac{e\cdot p'\, e'\cdot p}{\kappa'}\right)^2 \qquad (11\text{–}13)$$

---

* We are indebted to Dr. K. Tanaka for communicating this result to us.

When the polarizations of the photons are not observed, this expression must be summed over the final and averaged over the initial polarization directions. Actually, it is just as easy to repeat the trace calculation (11–12) using the method of Section A2–6 (Eq. A2–91). We find

$$\bar{X} = \frac{1}{2}\sum_{\text{pol}} X = \frac{\kappa}{\kappa'} + \frac{\kappa'}{\kappa} + 2\left(\frac{m^2}{\kappa} - \frac{m^2}{\kappa'}\right) + \left(\frac{m^2}{\kappa} - \frac{m^2}{\kappa'}\right)^2. \quad (11\text{–}14)$$

The particular experiment of interest will determine over which variables the integration $\int d\tau_f$ in Eq. (11–9) must be carried out and what its limits are. For example, we may ask for the cross section per unit solid angle for scattering into a differential angular interval between θ and $\theta + d\theta$, and φ and $\varphi + d\varphi$. In this case, we write

$$d\tau_f = d^3p'd^3k' = d^3p'\omega'^2 d\omega' d\Omega,$$

and we do *not* integrate over the solid angle $d\Omega$ of the emerging photon. The integration over d^3p' simply replaces $\mathbf{p}'$ by $\mathbf{p} + \mathbf{k} - \mathbf{k}'$ everywhere in the integrand, since σ in (11–4) contains a δ-function over the three-momenta. Similarly, the integration over ω' is very simple:

$$\int \omega'^2 d\omega' \delta(\epsilon' + \omega' - \epsilon - \omega) X = \omega'^2 \frac{d\omega'}{d(\omega' + \epsilon')} X. \quad (11\text{–}15)$$

In this result ω' can be regarded as a function of θ, φ, and the initial state, and the derivative is to be taken for constant θ, φ. Since

$$\epsilon'^2 = m^2 + (\mathbf{p} + \mathbf{k} - \mathbf{k}')^2$$

$$= \epsilon^2 + \omega^2 + \omega'^2 + 2\beta\epsilon\omega\cos\alpha - 2\beta\epsilon\omega'\cos\alpha' - 2\omega\omega'\cos\theta,$$

we find, upon differentiation and use of Eq. (11–8),

$$\frac{d(\omega' + \epsilon')}{d\omega'} = 1 + \frac{\omega' - \beta\epsilon\cos\alpha' - \omega\cos\theta}{\epsilon'} = \frac{\epsilon\omega}{\epsilon'\omega'}(1 - \beta\cos\alpha). \quad (11\text{–}16)$$

When this factor is inserted in Eq. (11–15) and the result combined with Eq. (11–9), one finds the differential cross section per unit solid angle for Compton scattering:

$$\frac{d\sigma}{d\Omega} = \frac{1}{2\epsilon^2}\left(\frac{e^2}{4\pi}\right)^2\left(\frac{\omega'}{\omega}\right)^2\frac{X}{(1-\beta\cos\alpha)^2} = \frac{r_0^2 X}{2\gamma^2(1-\beta\cos\alpha)^2}\left(\frac{\omega'}{\omega}\right)^2. \quad (11\text{–}17)$$

In this formula X is given by Eq. (11–13) and ω' by Eq. (11–8). When one is not interested in the polarizations, X is to be replaced by $\bar{X}$ of Eq.

(11–14). The order of magnitude of this cross section is determined by the classical electron radius

$$r_0 = \frac{e^2}{4\pi m} \sim 2.8 \times 10^{-13} \text{ cm.} \tag{11-18}$$

The cross section (11–17) simplifies considerably if we calculate it in the rest system of the initial or final electron. In most experiments the initial electron is practically at rest in the laboratory system. In this case, $p = (m,0,0,0)$ and $e \cdot p = e' \cdot p = 0$. The last relation requires a special choice of gauge such that $e_0 = e_0' = 0$. Since the theory is gauge-invariant, this choice will simplify the calculation without loss of generality in the final result.

It follows from these remarks and Eq. (A2–82) that in the rest frame of the incident electron Eq. (11–12) gives

$$m^2 A = \tfrac{1}{2}\kappa\kappa' + \kappa e' \cdot k e' \cdot p',$$

$$m^2 B = \tfrac{1}{2}\kappa\kappa' + \kappa' e \cdot k' e \cdot p', \tag{11-19}$$

$$m^2 C = m^2 D = \tfrac{1}{2}\kappa\kappa' + \tfrac{1}{2}\kappa' e' \cdot k e' \cdot p' + \tfrac{1}{2}\kappa e \cdot k' e \cdot p' - \kappa\kappa'(e \cdot e')^2.$$

Consequently,

$$X = \frac{1}{2}\left(\frac{\kappa}{\kappa'} + \frac{\kappa'}{\kappa}\right) - 1 + 2(e \cdot e')^2. \tag{11-20}$$

This equation is easily seen to be a special case of Eq. (11–13). We need only to put $e \cdot p = e' \cdot p = 0$. It is to be noted that Eq. (11–20) is therefore not gauge-invariant; but the general expression (11–13) *is* gauge-invariant. These assertions can be verified as follows.

The trace X is a homogeneous polynomial of second degree in both e and e' [*cf.* Eq. (11–12)]. This homogeneity was lost in the evaluation of the trace, because use was made of the equations $e^2 = e'^2 = 1$; but it can easily be restored by multiplying the first two terms of (11–13) and (11–20) by $e^2 e'^2$. In this form X can be checked for gauge invariance. We simply replace e and e' by k/m and k'/m, and make use of $k^2 = k'^2 = 0$. If the expression is gauge-invariant, it must vanish under this substitution. The trace (11–13) indeed vanishes, but (11–20) gives $2(k \cdot k')^2/m^2 \neq 0$.

The differential cross section (11–17) in the laboratory system, i.e., the rest system of the incident electron ($\beta = 0$, $\gamma = 1$, $\kappa = m\omega$, $\kappa' = m\omega'$), simplifies to

$$\frac{d\sigma}{d\Omega} = \frac{r_0^2}{4}\left(\frac{\omega'}{\omega}\right)^2\left(\frac{\omega}{\omega'} + \frac{\omega'}{\omega} - 2 + 4(e \cdot e')^2\right), \quad \text{(L-system)}. \tag{11-21}$$

In this form the cross section is known as the *Klein-Nishina formula*. The energy ω' of the outgoing photon is determined by Eq. (11-8) which, for this choice of the reference frame, reduces to

$$\frac{\omega'}{\omega} = \frac{1}{1 + \dfrac{\omega}{m}(1 - \cos\theta)}. \tag{11-22}$$

The dependence on the polarization vectors of the photon is contained in the last term. For simplicity, we have derived the formula for real polarization vectors corresponding to states of linear polarization. In the case of general polarization, the polarization vectors e, e' for initial and final state photons are complex, and we obtain instead of $(e \cdot e')^2$ the quantity $|e \cdot e'^*|^2$. If the polarization of the incident and final photons is not observed, the appropriate cross section is obtained by averaging the initial polarization states and summing over the final polarization states, as is done in Eq. (11-14). The result of this operation is to replace $(e \cdot e')^2$ by $\frac{1}{2}(1 + \cos^2\theta)$, while the terms independent of e and e' are multiplied by 2. The cross section becomes

$$d\sigma = \frac{r_0^2}{2}\left(\frac{\omega'}{\omega}\right)^2\left(\frac{\omega}{\omega'} + \frac{\omega'}{\omega} - \sin^2\theta\right)d\Omega, \quad \text{(L-system)}, \tag{11-23}$$

where ω' is given by Eq. (11-22). The same result can be obtained from Eq. (11-14) using (11-22) and putting $\kappa = \omega m$, $\kappa' = \omega' m$.

The total cross section σ, which is independent of the choice of the coordinate system, can be calculated more easily from Eq. (11-23) than from Eq. (11-17). On integration over the angles, we find

$$\sigma(\omega) = 2\pi r_0^2 \left\{\frac{1+\omega}{\omega^3}\left[\frac{2\omega(1+\omega)}{1+2\omega} - \ln(1+2\omega)\right]\right.$$

$$\left. + \frac{\ln(1+2\omega)}{2\omega} - \frac{1+3\omega}{(1+2\omega)^2}\right\}. \tag{11-24}$$

In this expression ω is measured in units of m.

An important special case of these results is the nonrelativistic limit of photon-electron scattering in which the energy of the incident photon is considered small, $\omega \ll m$. In this limit, $\omega' = \omega$, and the scattering is elastic. From Eq. (11-21), we find

$$\frac{d\sigma}{d\Omega} = r_0^2 |e \cdot e'|^2 = r_0^2 \cos^2\Theta, \tag{11-25}$$

where Θ is the angle between the incident and outgoing directions of polarization. In the unpolarized case Eq. (11–23) gives

$$\frac{d\sigma}{d\Omega} = r_0^2 \frac{1 + \cos^2 \theta}{2}, \quad (\text{NR}). \tag{11–26}$$

The total cross section follows from Eq. (11–24) or, more simply, by integration of (11–26),

$$\sigma = \frac{8\pi}{3} r_0^2, \quad (\text{NR}). \tag{11–27}$$

These formulas were first obtained by J. J. Thomson,* and the nonrelativistic limit of Compton scattering is therefore usually called *Thomson scattering*.

If the energy of the incident photon is much larger than the electron rest energy (extreme relativistic limit), the differential cross section (11–23) becomes approximately

$$\frac{d\sigma}{d\Omega} = \frac{r_0^2}{4} \frac{m}{\omega} \frac{1}{\sin^2 (\theta/2)}, \quad (\text{ER}). \tag{11–28}$$

This formula is valid only for $\theta^2 \gg 2m/\omega$. The total cross section follows from (11–24):

$$\sigma = \pi r_0^2 \frac{m}{\omega} \left(\ln \frac{2\omega}{m} + \frac{1}{2} \right), \quad (\text{ER}). \tag{11–29}$$

11–2 Double Compton scattering. As explained in the introduction to Section 11–1, double Compton scattering is the lowest order approximation to photon-electron scattering in which the final state consists of an electron and two photons. As in the case of single Compton scattering, the cross section for this process is, in general, a meaningless (divergent) quantity, because the possibility of the presence in the final state of an undetermined number of very soft photons can never be excluded; but in lowest (third) order of the iteration solution of the S-operator in which this process can take place not more than two photons can be created in the final state, such that the result is well defined. We shall not consider radiative corrections to double Compton scattering, but shall restrict ourselves to the lowest order.

---

* J. J. Thomson, *Conduction of Electricity through Gases*, 3rd Edition, Cambridge University Press, Cambridge, England, 1933, Vol. II, p. 256.

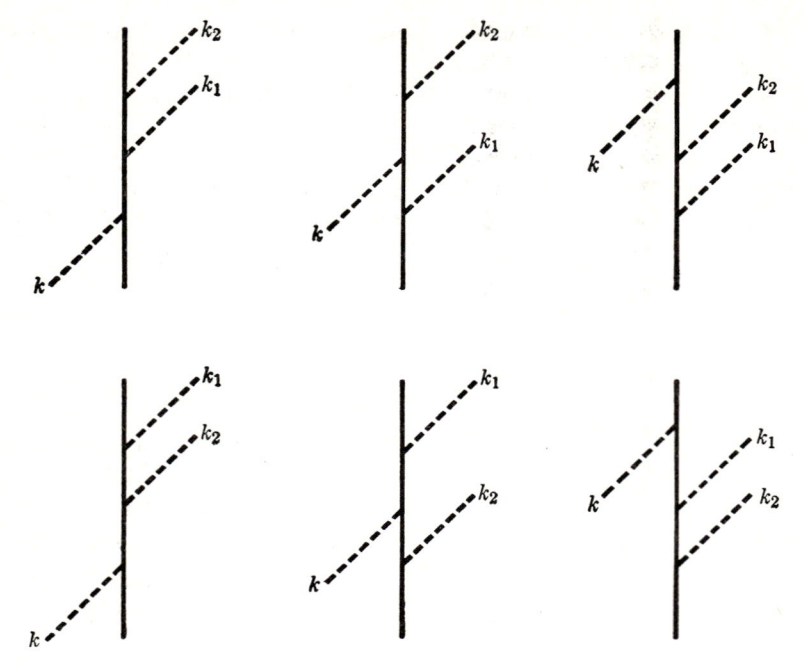

FIG. 11-1. The double Compton effect diagrams.

The diagrams for double Compton scattering are shown in Fig. 11-1. The three diagrams in the second row of this figure differ from those in the first row only by the interchange of the two outgoing photons k_1 and k_2. Let the ingoing photon have momentum k; let the electron be a negaton of momentum p and p' in its initial and final states, respectively. The matrix element becomes

$$(p'k_1k_2|M^{(3)}|pk) = \frac{e^3}{(2\pi)^{7/2}} \cdot \frac{m}{\sqrt{8\epsilon\epsilon'\omega\omega_1\omega_2}} \bar{u}(p')Nu(p),$$

$$N = e_2 \frac{i(\boldsymbol{p}' + \boldsymbol{k}_2) - m}{(p' + k_2)^2 + m^2} e_1 \frac{i(\boldsymbol{p} + \boldsymbol{k}) - m}{(p + k)^2 + m^2} e$$

$$+ e_2 \frac{i(\boldsymbol{p}' + \boldsymbol{k}_2) - m}{(p' + k_2)^2 + m^2} e \frac{i(\boldsymbol{p} - \boldsymbol{k}_1) - m}{(p - k_1)^2 + m^2} e_1$$

$$+ e \frac{i(\boldsymbol{p}' - \boldsymbol{k}) - m}{(p' - k)^2 + m^2} e_2 \frac{i(\boldsymbol{p} - \boldsymbol{k}_1) - m}{(p - k_1)^2 + m^2} e_1$$

$$+ (1 \leftrightarrow 2). \qquad (11\text{-}30)$$

The symbol $(1 \leftrightarrow 2)$ indicates that the first three lines should be repeated with the indices 1 and 2 interchanged. The general expression for the cross section is, according to Section 8–5, formally the same as for the Compton effect, Eq. (11–3).

The summation over the spin directions of the initial and final states can be performed in exactly the same way as in the Compton effect, but involves rather laborious algebra. The final result is considerably simplified if we assume that the incident and outgoing photons are unpolarized, so that we shall also sum over the polarizations. Then we can write Eq. (11–3) in the form

$$\sigma = \frac{\alpha r_0^2}{(4\pi)^2} \int \frac{X}{|p \cdot k| \epsilon' \omega_1 \omega_2} \delta(p' + k_1 + k_2 - p - k) d^3k_1 d^3k_2 d^3p', \quad (11\text{–}31)$$

where $\alpha = e^2/4\pi \simeq 1/137$ is the fine-structure constant, and

$$X = m^4 \sum_{\text{pol}} \text{Tr}\{N\Lambda_-(p)\bar{N}\Lambda_-(p')\}, \quad (11\text{–}32)$$

where N is defined as in Eq. (11–30). We note that Eq. (11–31) is completely symmetrical in k_1 and k_2.

X can be expressed in terms of the quantities

$$m^2\kappa_1 = -p \cdot k_1, \quad m^2\kappa_2 = -p \cdot k_2, \quad m^2\kappa_3 = +p \cdot k;$$
$$m^2\kappa_1' = +p' \cdot k_1, \quad m^2\kappa_2' = +p' \cdot k_2, \quad m^2\kappa_3' = -p' \cdot k. \quad (11\text{–}33)$$

Using the abbreviations

$$a = \sum_1^3 \frac{1}{\kappa_i}, \quad b = \sum_1^3 \frac{1}{\kappa_i'}, \quad c = \sum_1^3 \frac{1}{\kappa_i \kappa_i'},$$
$$x = \sum_1^3 \kappa_i, \quad y = \sum_1^3 \kappa_i', \quad z = \sum_1^3 \kappa_i \kappa_i', \quad (11\text{–}34)$$
$$A = \kappa_1 \kappa_2 \kappa_3, \quad B = \kappa_1' \kappa_2' \kappa_3',$$
$$\rho = \sum_1^3 \left(\frac{\kappa_i}{\kappa_i'} + \frac{\kappa_i'}{\kappa_i}\right),$$

we find,* since the conservation law yields $x = y$,

$$X = 2(ab - c)[(a + b)(x + 2) - (ab - c) - 8]$$
$$- 2x(a^2 + b^2) - 8c + \frac{4x}{AB}\Bigg[(A + B)(x + 1) - (aA + bB)$$
$$\times \left(2 + z\frac{1-x}{x}\right) + x^2(1-z) + 2z\Bigg] - 2\rho[ab + c(1-x)]. \quad (11\text{–}35)$$

---
* F. Mandl and T. H. R. Skyrme, *Proc. Roy. Soc.* **A215**, 497 (1952).

We are interested in the differential cross section for the emission of photons 1 and 2 into the solid angles $d\Omega_1$ and $d\Omega_2$ in the directions θ_1, φ_1 and θ_2, φ_2. The angles refer to a polar coordinate system whose axis is the incident photon direction. The azimuths φ are conveniently measured from the plane of incidence which is formed by the vectors $\mathbf{p}$ and $\mathbf{k}$. Let the angle between $\mathbf{p}$ and $\mathbf{k}$ be α. The expression (11–31) can be integrated over $d^3p' d\omega_2$, which eliminates the δ-function and produces a factor $d\omega_2/d(\epsilon' + \omega_1 + \omega_2)$. This differential quotient is to be evaluated for constant $\mathbf{p}$, $\mathbf{k}$, $\mathbf{k}_1$ and θ_2, φ_2. The conservation law of momenta gives

$$\epsilon'^2 = \epsilon^2 + \omega^2 + \omega_1^2 + \omega_2^2 + 2(|\mathbf{p}|\omega \cos \alpha - |\mathbf{p}|\omega_1 \cos \theta_1' - |\mathbf{p}|\omega_2 \cos \theta_2'$$
$$+ \omega_1\omega_2 \cos \theta_{12} - \omega\omega_1 \cos \theta_1 - \omega\omega_2 \cos \theta_2),$$

where θ_{12} is the angle between $\mathbf{k}_1$ and $\mathbf{k}_2$, such that

$$\frac{d(\epsilon' + \omega_1 + \omega_2)}{d\omega_2} = 1 + \frac{d\epsilon'}{d\omega_2}$$

$$= \frac{1}{\epsilon'}[\epsilon'(1 - \beta \cos \theta_2') + \omega(1 - \cos \theta_2) - \omega_1(1 - \cos \theta_{12})].$$
(11–36)

The angles θ_1', θ_2', and θ_{12} can be expressed in terms of α, θ_1, θ_2, φ_1, and φ_2,

$$\cos \theta_1' = \cos \alpha \cos \theta_1 + \sin \alpha \sin \theta_1 \cos \varphi_1,$$
$$\cos \theta_2' = \cos \alpha \cos \theta_2 + \sin \alpha \sin \theta_2 \cos \varphi_2, \quad (11\text{–}37)$$
$$\cos \theta_{12} = \cos \theta_1 \cos \theta_2 + \sin \theta_1 \sin \theta_2 \cos (\varphi_1 - \varphi_2).$$

Combining these results, we find from Eq. (11–31) for the differential cross section of double Compton scattering,

$$d\sigma_D = \alpha r_0^2 X \cdot \frac{d\Omega_1}{4\pi} \cdot \frac{d\Omega_2}{4\pi}$$
$$\times \frac{\omega_1 \omega_2 d\omega_1}{\omega\epsilon(1 - \beta \cos \alpha)[\epsilon(1 - \beta \cos \theta_2') + \omega(1 - \cos \theta_2) - \omega_1(1 - \cos \theta_{12})]}.$$
(11–38)

In this formula X is given by Eq. (11–35), and all quantities are considered to be functions of $|\mathbf{p}|$, ω, α; ω_1, θ_1, φ_1; θ_2, and φ_2. The angles θ_2' and θ_{12} are thus to be expressed by Eqs. (11–37), and ω_2 follows from the conservation laws:

$$\omega_2 = \frac{\omega\epsilon(1 - \beta \cos \alpha) - \omega_1\epsilon(1 - \beta \cos \theta_1') - \omega_1\omega(1 - \cos \theta_1)}{\epsilon(1 - \beta \cos \theta_2') + \omega(1 - \cos \theta_2) - \omega_1(1 - \cos \theta_{12})}. \quad (11\text{–}39)$$

The physical implications of the result (11–38) can best be seen from several special cases.* For this purpose we assume the electron to be initially at rest ($\mathbf{p} = 0$, $\epsilon = m$), as can be assumed to be the case in most experiments. Equations (11–35) and (11–38) thereby simplify considerably. The cross section $d\sigma_D$ becomes a function of ω, ω_1, θ_1, θ_2, and φ_2 only, since we can put $\varphi_1 = 0$.

The first case which we want to consider is the emission of both photons in the nearly forward direction. With $\theta_1 \ll 1$, $\theta_2 \ll 1$, and $\omega \gg \epsilon' - m$, we find

$$d\sigma_D(\omega,\omega_1,\theta_1,\theta_2,\varphi_2) = \frac{\alpha r_0^2}{2\pi^2} \frac{d\omega_1 d\Omega_1 d\Omega_2}{m^2} \left[\omega_1^2 + \omega_2^2 + \left(\frac{\omega_1 \omega_2}{\omega}\right)^2 \right]$$

$$\times \left[\frac{1}{\omega_1}(1 - \cos\theta_2) + \frac{1}{\omega_2}(1 - \cos\theta_1) - \frac{1}{\omega}(1 - \cos\theta_{12}) \right]. \quad (11\text{--}40)$$

This expression exhibits the "infrared divergence" for either $\omega_1 \to 0$ or $\omega_2 \to 0$. When one tries to integrate over the respective energies one encounters a logarithmic divergence (see Section 16–1). This apparent difficulty will be discussed in detail in the following section in connection with the radiative corrections to Compton scattering. Equation (11–40) also shows that the double Compton effect vanishes when both photons are to be emitted exactly forward ($\theta_1 = \theta_2 = 0$).

As a second special case, we assume that one of the emitted photons is very soft ($\omega_2 \ll \omega_1$, $\omega_2 \ll m$). Equation (11–38) reduces to

$$d\sigma_D = \frac{\alpha}{\pi} d\sigma_C(\omega,\omega_1) \cdot \frac{d\omega_2}{\omega_2} \cdot \frac{d\Omega_2}{4\pi} \cdot \left\{ \frac{2(m + \omega - \omega_1)}{m + \omega(1 - \cos\theta_2) - \omega_1(1 - \cos\theta_{12})} \right.$$

$$\left. - 1 - \frac{m^2}{[m + \omega(1 - \cos\theta_2) - \omega_1(1 - \cos\theta_{12})]^2} \right\}, \quad (11\text{--}41)$$

where $d\sigma_C(\omega,\omega_1)$ is the differential cross section for (single) Compton scattering, Eq. (11–23), which can be written, using Eq. (11–22),

$$d\sigma_C(\omega,\omega_1) = \pi r_0^2 \frac{m d\omega_1}{\omega^2} \left[\frac{\omega}{\omega_1} + \frac{\omega_1}{\omega} + \left(\frac{m}{\omega_1} - \frac{m}{\omega}\right)^2 - 2\left(\frac{m}{\omega_1} - \frac{m}{\omega}\right) \right]. \quad (11\text{--}42)$$

The use of Eq. (11–22) is indeed justified, because we find, in analogy to Eq. (11–39),

$$\omega_1 = \frac{\omega\epsilon(1 - \beta\cos\alpha) - \omega_2\epsilon(1 - \beta\cos\theta_2') - \omega\omega_2(1 - \cos\theta_2)}{\epsilon(1 - \beta\cos\theta_1') + \omega(1 - \cos\theta_1) - \omega_2(1 - \cos\theta_{12})}.$$

---

* The following discussion was first given by Mandl and Skyrme (*loc. cit.*). The reader is referred to their paper for further details.

In the limit $\mathbf{p} = 0$, $\epsilon = m$, $\omega_2 \ll \omega_1$, $\omega_2 \ll m$, this reduces to

$$\omega_1 = \frac{\omega m}{m + \omega(1 - \cos\theta_1)}.$$

The single and the double Compton effects with one very soft photon can be compared when we integrate over all possible directions of the soft photon. Equation (11–41) becomes

$$d\sigma_D(\omega,\omega_1) = \frac{2\alpha}{\pi} d\sigma_C(\omega,\omega_1) \frac{d\omega_2}{\omega_2} \cdot \left\{ \frac{m + \omega - \omega_1}{\sqrt{(m + \omega - \omega_1)^2 - m^2}} \right.$$

$$\left. \times \ln\left[1 + \frac{\omega}{m} - \frac{\omega_1}{m} + \sqrt{\left(1 + \frac{\omega}{m} - \frac{\omega_1}{m}\right)^2 - 1}\right] - 1 \right\}. \quad (11\text{–}43)$$

Since ω_2 is assumed to be much smaller than ω_1 and m, the energy of the outgoing electron is approximately

$$\epsilon' = m + \omega - \omega_1,$$

so that this result can be written with $\beta' = |\mathbf{p}'|/\epsilon'$,

$$d\sigma_D = \frac{\alpha}{\pi}\left(\frac{1}{\beta'} \ln\frac{1 + \beta'}{1 - \beta'} - 2\right)\frac{d\omega_2}{\omega_2} d\sigma_C. \quad (11\text{–}44)$$

In the nonrelativistic limit $\omega \ll m$, this can be written as

$$d\sigma_D = \frac{2\alpha}{3\pi} \beta'^2 \frac{d\omega_2}{\omega_2} d\sigma_C$$

$$= \frac{4\alpha}{3\pi}\left(\frac{\omega}{m}\right)^2 (1 - \cos\theta_1) \frac{d\omega_2}{\omega_2} d\sigma_C, \quad (\text{NR}). \quad (11\text{–}45)$$

When one of the emitted photons (ω_2) is very soft, the double Compton scattering cross section differs from the Compton cross section only by a multiplicative function. This situation is an example of a general theorem which will be discussed in Section 16–1 in connection with the problem of the infrared divergence.

We can regard ω_1 in Eq. (11–43) as a function of ω and θ_1:

$$d\sigma_C(\omega,\theta_1) = \frac{r_0^2}{2} f_C(\omega,\theta_1) d\Omega_1,$$

$$d\sigma_D(\omega,\theta_1) = \frac{2\alpha}{\pi} \cdot \frac{d\omega_2}{\omega_2} \cdot \frac{r_0^2}{2} f_D(\omega,\theta_1) d\Omega_1.$$

(11–46)

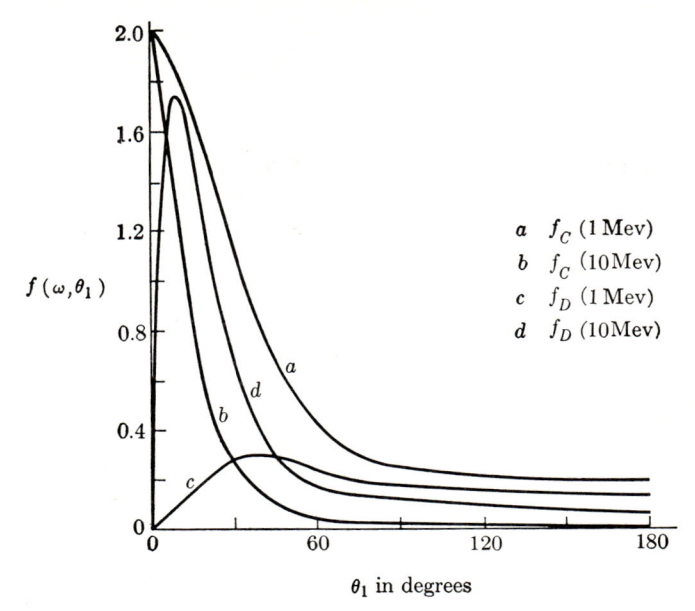

FIG. 11-2. Angular distribution of the hard photon in single and double Compton scattering. [After F. Mandl and T. H. R. Skyrme, *Proc. Roy. Soc.* **215**, 497 (1952).]

The two functions f_C and f_D were plotted vs. θ_1 by Mandl and Skyrme for 1 and 10 Mev, and are shown in Fig. 11-2. They are of comparable orders of magnitude. Equations (11-46) indicate the essential features of this case. With increasing energy double Compton scattering becomes increasingly more important compared with single Compton scattering. This is valid for all angles except at very small angles, since double scattering always vanishes for $\theta_1 = 0$. The peak near small θ_1 for double scattering also increases with energy.

Further inspection of Eq. (11-41) also shows that there is a strong preference for emitting the two photons into small forward angles. In particular, the emission of one photon forward and the other to the side is much more probable than the emission of both photons to the side. These properties become more pronounced with increasing energy.*

11-3 Radiative corrections to Compton scattering. The first radiative corrections to (single) Compton scattering (Section 11-1) separate into the corrections to each of the two diagrams Fig. 8-5. The corrections to the

---

* The double Compton scattering was first observed experimentally by I. F. Boekelheide, Ph.D. Thesis, June 1952, State University of Iowa. The agreement with the theory is satisfactory. See also P. E. Cavanaugh, *Phys. Rev.* **87**, 1131 (1952).

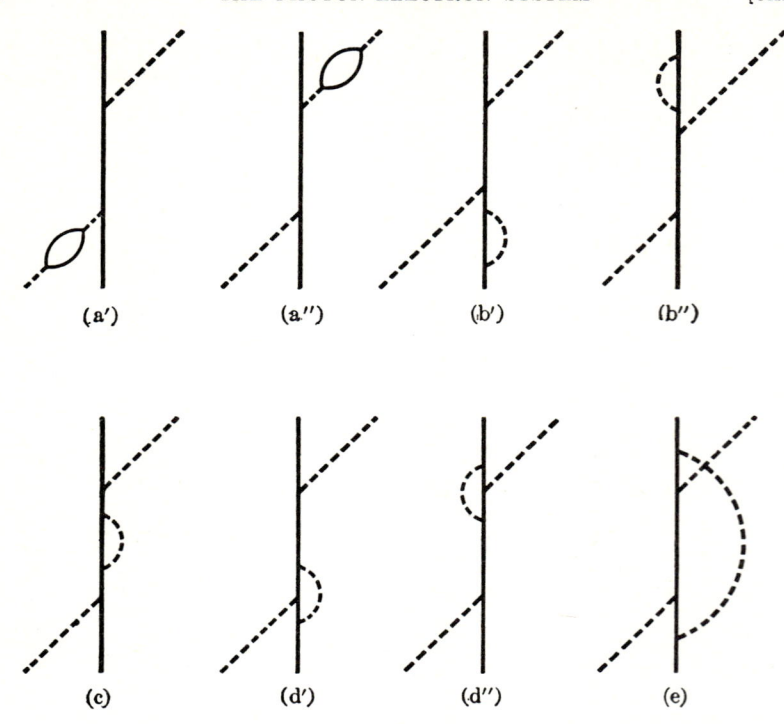

FIG. 11-3. Radiative corrections to Compton scattering.

first diagram are shown in Fig. 11–3. The corrections to the second diagram differ from these only in the interchange of the emission and absorption points of k and k', as one follows the electron path along its arrow. Therefore, the corrections to the second diagram of Fig. 8–5 can be obtained from those of the first diagram by the interchange of e and e', and of k and $-k'$.

The matrix elements which correspond to the diagrams of Fig. 11–3 can be written down with the help of Sections 8–3 and 9–4 to 9–6. They will all be of the form

$$(p'k'e'|M^{(4)}|pke) = \frac{ie^2}{8\pi^2} \frac{m}{\sqrt{\epsilon\epsilon'\omega\omega'}} \bar{u}(\mathbf{p}')Qu(\mathbf{p}), \qquad (11\text{–}47)$$

provided we assume the electron to be a negaton.

The diagrams (a) of Fig. 11–3 contain photon self-energy parts in external photon lines; their contribution therefore gives only a constant multiplicative factor, $-\frac{1}{2}C$, for each external photon line [cf. Eq. (9–72)]. These factors will be absorbed in the charge renormalization. The observable contributions from the diagrams (a) are therefore zero:

$$Q_a = 0. \qquad (11\text{–}48)$$

The diagrams (b) contain electron self-energy parts, $\Sigma(p)$, in second order.

11-3] RADIATIVE CORRECTIONS TO COMPTON SCATTERING

These are inserted in external electron lines. According to Section 9–4, they will contribute only to the wave function renormalization once the mass renormalization has been carried out. Indeed, we find

$$Q_b = \Sigma(p')S_c(p')e'S_c(p+k)e + e'S_c(p'+k')eS_c(p)\Sigma(p),$$

where

$$S_c(p) = -(i p + m)^{-1}.$$

Using Eq. (9–16) and $p' + k' = p + k$, we obtain, after mass renormalization,

$$Q_b = -B((i p' + m)(i p' + m)^{-1}e'S_c(p+k)e$$
$$+ e'S_c(p+k)e(i p + m)^{-1}(i p + m))$$
$$- (i p' + m)\Sigma_f(p')e'S_c(p+k)e - e'S_c(p+k)e\Sigma_f(p)(i p + m).$$

Since Q_b is to be multiplied by $\bar{u}(p')$ on the left and by $u(p)$ on the right, the expression multiplying B is undetermined, but the remaining terms vanish. The undetermined expression is an example of the general Eq. (9–36). There it was found with a consistency argument that each external electron line contributes $-\frac{1}{2}B$ [cf. Eq. (9–38)]. We obtain, therefore,

$$Q_b = -Be'S_c(p+k)e. \tag{11-49}$$

Similarly, diagram (c), after mass renormalization, gives

$$Q_c = -Be'S_c(p+k)e + e'\Sigma_f(p+k)e. \tag{11-50}$$

The diagrams (d) can easily be expressed by the vertex part of Section 9–6. Equation (9–87) yields

$$Q_d = [Le' + e_\mu'\Lambda_f^\mu(p', p+k)]S_c(p+k)e$$
$$+ e'S_c(p+k)[Le + e_\mu\Lambda_f^\mu(p+k, p)]. \tag{11-51}$$

Finally, diagram (e) gives

$$Q_e = \frac{ie^2}{(2\pi)^4}\int \gamma_\mu S_c(p-k'')e'S_c(p+k-k'')eS_c(p'-k'')\gamma^\mu D_c(k'')\,d^4k''. \tag{11-52}$$

The sum

$$\bar{u}(\mathbf{p}')Q_1 u(\mathbf{p}) = \bar{u}(\mathbf{p}')(Q_a + Q_b + Q_c + Q_d + Q_e)u(\mathbf{p})$$

constitutes the second-order radiative correction of the factor

$$\bar{u}(\mathbf{p}')e'S_c(p+k)eu(\mathbf{p}).$$

We see that the terms involving the infinite constants B in Eqs. (11–49) and (11–50) exactly cancel the terms involving the constants L in Eq.

(11–51), since by Eq. (9–92)
$$B = L.$$

Therefore, we see here explicitly that only convergent terms remain after mass and charge renormalization. We find for the second-order radiative corrections the observable terms

$$Q_{1f} = e'\Sigma_f(p+k)e + e_\mu'\Lambda_f{}^\mu(p',p+k)S_c(p+k)e$$
$$+ e'S_c(p+k)\Lambda_f{}^\mu(p-k,p)e_\mu + Q_e. \quad (11\text{–}53)$$

The infinite terms contribute only to the renormalization of the Compton effect. The renormalization terms of the corrections will appear only in higher order.

The radiative corrections Q_{2f} to the second Compton diagram of Fig. 8–5 can be obtained in a completely analogous way, but they can be written down immediately by following the remarks at the beginning of this section and carrying out the interchanges

$$e' \leftrightarrow e, \quad k \leftrightarrow -k'$$

in the terms associated with the first diagram. Combining the result with Eq. (11–47), we find

$$(p'k'e'|M^{(2)} + M^{(4)}|pke) = \frac{i\alpha}{2\pi}\frac{m}{\sqrt{\epsilon\epsilon'\omega\omega'}}[\bar{u}(\mathbf{p}')Cu(\mathbf{p}) + \bar{u}(\mathbf{p}')Qu(\mathbf{p})], \quad (11\text{–}54)$$

where

$$C = e'S_c(p+k)e + eS_c(p-k')e'$$

describes the Compton effect, and

$$Q = Q_{1f} + Q_{2f} \quad (11\text{–}55)$$

describes its second-order corrections. Since Σ_f, $\Lambda_f{}^\mu$, and Q_e are all of order e^2, it follows that Q is of order e^2 compared with C. The cross section, which can be calculated in exactly the same way as the (second-order) Compton effect [cf. Eqs. (11–3) and (11–9)], will contain the Compton cross section, Eq. (11–17), as well as corrections to it of orders α and α^2. The corrections of order α arise from the cross term of C with Q, and the corrections of order α^2 arise from the square of $\bar{u}Qu$. However, there are also corrections of order α^2 which arise from the cross terms between C and the 6th-order diagrams (i.e., from the 4th-order corrections to Compton scattering). Since these were not included in Eq. (11–54), the above α^2 terms are incomplete. It follows that only the α terms of (11–54) are significant. The result will be of the form

$$d\sigma_S = d\sigma_C(1 + \alpha\delta_1). \quad (11\text{–}56)$$

The cross section for photon-electron scattering with a final state of one electron and one photon, $d\sigma_S$, is here represented as the Compton scattering cross section, $d\sigma_C$, and its second order correction, $\alpha\delta_1 \cdot d\sigma_C$. Higher order corrections are neglected.

Thus far we have ignored the infrared divergences contained in Σ_f and $\Lambda_f{}^\mu$, and therefore also in δ_1. However, this is not a difficulty, because the cross section (11–56) is not a physically meaningful quantity: as was mentioned previously (*cf.* the introductory remarks of Section 11–1), it is impossible in principle to assure experimentally that photon-electron scattering occurs with only a single photon in the final state. Instead, the physical situation calls for the possibility of simultaneous emission of other photons—undetermined in number—whose total energy does not exceed a certain upper limit, $\omega_{\max}$. This limit depends on the finite limits of errors in the particular measurements.

Within our approximation only one additional photon can be emitted, since double Compton scattering is of the same order as the second-order corrections to single Compton scattering. We therefore recall the double Compton scattering cross section with one very soft photon of energy ω_2, Eq. (11–45), and integrate it over ω_2 from zero to $\omega_{\max}$. The result will give the required probability of the presence of an additional soft photon of energy less than $\omega_{\max}$, which must be added to the right side of Eq. (11–56) to give a physically meaningful result,

$$d\sigma_S = d\sigma_C(1 + \alpha\delta_1) + d\sigma_D(\omega_{\max}).$$

Further inspection of this procedure shows that the integration of Eq. (11–45) over ω_2 involves an infrared divergence which *exactly cancels* the infrared divergences in δ_1. Indeed, $d\sigma_D(\omega_{\max})$ is of the form $d\sigma_C \cdot \alpha\delta_2$, such that

$$d\sigma_S = d\sigma_C(1 + \alpha\delta_1 + \alpha\delta_2) = d\sigma_C(1 + \alpha\delta). \qquad (11\text{–}57)$$

The resulting cross section $d\sigma_S$ depends logarithmically on $\omega_{\max}$, which is essentially the energy resolution of the experiment. It is clear that $d\sigma_S$ *decreases* when $\omega_{\max}$ is made smaller, since more double scattering is thereby excluded.

For a detailed discussion of this problem the reader is referred to a paper by Brown and Feynman.* These authors evaluate the integrals and traces, and they show explicitly how the two infrared divergences in δ_1 and δ_2 cancel. A discussion of the extreme relativistic and nonrelatitivistic limits of the cross section is also given in their work.

One may ask whether any general statements can be made about photon-electron scattering, valid for all radiative corrections. The following theorems are examples of such statements.

---

* L. M. Brown and R. P. Feynman, *Phys. Rev.* **85**, 231 (1952). See also M. R. Schafroth, *Helv. Phys. Acta* **22**, 50 (1949) and **23**, 542 (1950). The latter paper contains errors.

THEOREM 1. *As a consequence of Lorentz and gauge invariance, the sum of Compton scattering and all its radiative corrections differs from Thomson scattering at most by a multiplicative constant in the nonrelativistic limit.*

Proof. Because of gauge invariance, we can choose $e_0 = e_0' = 0$, so that $e \cdot p = e' \cdot p = 0$ in the nonrelativistic limit; therefore, using $p + k = p' + k'$, $e \cdot p' = e \cdot p + O(\omega) = O(\omega)$, $e' \cdot p' = O(\omega)$. The matrix element for Compton scattering and all its radiative corrections is homogeneous and linear in both e and e'. In the nonrelativistic limit it is therefore of the form

$$\text{const. } \bar{u}(\mathbf{p}')[A e \cdot e' + B e e' + C e' e + O(\omega)] u(\mathbf{p}),$$

provided we use $(i\not{p} + m)u(\mathbf{p}) = 0$ and $\bar{u}(\mathbf{p}')(i\not{p}' + m) = 0$. A, B, and C are constants. Since to each radiative correction of the first diagram, Fig. 8–5, there is a corresponding one of the second diagram, the above matrix element must be invariant under the simultaneous interchanges

$$e \leftrightarrow e', \quad k \leftrightarrow -k'.$$

Therefore, $B = C$, and the matrix element can be written

$$\text{const. } \bar{u}(\mathbf{p}') u(\mathbf{p}) e \cdot e' + O(\omega).$$

This result leads to a cross section

$$\frac{d\sigma}{d\Omega} = \text{const. } r_0^2 |e \cdot e'|^2,$$

which differs from Thomson scattering, Eq. (11–25), only by a multiplicative constant.

THEOREM 2. *The multiplicative constant of Theorem 1 is exactly the renormalization factor which is necessary to renormalize the mass and charge in $r_0^2 = e^2/(4\pi m)$ of Thomson scattering; i.e., the observable effects of the radiative corrections to Compton scattering vanish in the nonrelativistic limit.*

Proof.[*] The nth order correction to Compton scattering can be obtained by attaching two external photon lines in all possible ways to the nth order electron self-energy diagrams. These diagrams consist of one open electron path and possibly one or more closed electron paths, i.e., closed loops. When the external photon lines are attached to closed loops, the resulting diagrams will either contain loops with an odd number of corners, or they will contain one loop in which the number of corners was increased by two. In the former case the diagrams give no contribution because of Furry's theorem; in the latter case their contribution will vanish as k, the momentum of one of the external photon lines, approaches zero (*cf.* Section 8–4).

---

[*] A similar proof was given by W. Thirring, *Phil. Mag.* **41**, 1193 (1950).

On the other hand, where the external photon lines are attached to the open electron path, the electron self-energy matrix elements will be modified as follows.

In the low-frequency limit of the external photon line, every electron propagation function $S_c(p)$ to which such a line is being attached is thereby replaced by

$$S_c(p) \rightarrow \text{const.} \; S_c(p) e S_c(p) = -i \, \text{const.} \; e_\mu \frac{\partial}{\partial p_\mu} S_c(p).$$

The sum of the matrix elements for Compton scattering and all its radiative corrections in this limit therefore contains the term

$$S_c^{-1}(p) e_\mu e_\nu' \frac{\partial^2 S_c'(p)}{\partial p_\mu \partial p_\nu} S_c^{-1}(p),$$

where $S_c'(p)$ is given by Eq. (10–10) and becomes, after renormalization,

$$S_0(p) = S_c(p) + \Sigma_f(p) S_0(p) S_c^{-1}(p),$$

by Eq. (10–41). The first term will give

$$\frac{\partial^2 S_c(p)}{\partial p_\mu \partial p_\nu} = S_c(p) \gamma^\mu S_c(p) \gamma^\nu S_c(p) + S_c(p) \gamma^\nu S_c(p) \gamma^\mu S_c(p),$$

which leads to the nonrelativistic limit of Compton scattering. The second term is finite and will vanish when multiplied on both sides by S_c^{-1} and by the wave function $u(\mathbf{p})$. This means that *the radiative corrections do not contribute observable terms to Thomson scattering* in the nonrelativistic limit. But they do contribute renormalization terms. We can now appeal to the general proof of the consistency of the renormalization procedure (Chapter 10), and we conclude that these renormalization terms must exactly renormalize the factor e^2/m into e_0^2/m_0, where e_0 and m_0 are the observable charge and mass of the electron.

11–4 Pair production in photon-electron collisions. The lowest-order diagrams for this process are shown in Fig. 11–4. To these diagrams one

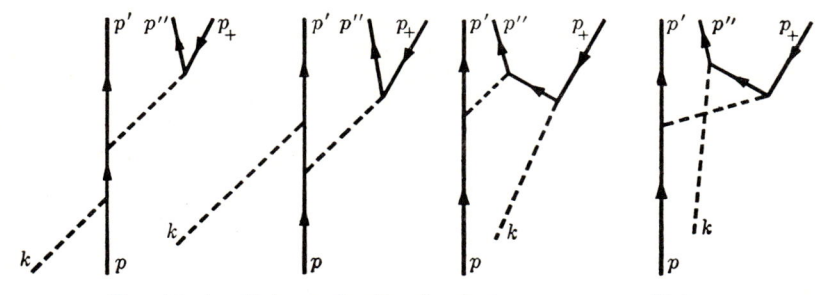

Fig. 11–4. Pair production in photon-negaton collision.

must add the corresponding diagrams in which the two final electrons of the same charge (negatons in the figure) are interchanged. The matrix elements of the latter will also differ in sign from the matrix elements of the diagrams shown.

With the aid of Section 8–3, we find

$$(p'p''p_+|M^{(3)}|pke) = \frac{e^3}{(2\pi)^{7/2}} \cdot \frac{m^2}{\sqrt{2\omega\epsilon\epsilon'\epsilon''\epsilon_+}}$$

$$\times \left[\bar{u}(\mathbf{p}'')\gamma_\mu v(\mathbf{p}_+) \frac{1}{(p''+p_+)^2} \bar{u}(\mathbf{p}')C^\mu(p',p)u(\mathbf{p}) \right.$$

$$- \bar{u}(\mathbf{p}')\gamma_\mu v(\mathbf{p}_+) \frac{1}{(p'+p_+)^2} \bar{u}(\mathbf{p}'')C^\mu(p'',p)u(\mathbf{p})$$

$$- \bar{u}(\mathbf{p}'')\gamma_\mu u(\mathbf{p}) \frac{1}{(p''-p)^2} \bar{u}(\mathbf{p}')C^\mu(p',-p_+)v(\mathbf{p}_+)$$

$$\left. + \bar{u}(\mathbf{p}')\gamma_\mu u(\mathbf{p}) \frac{1}{(p'-p)^2} \bar{u}(\mathbf{p}'')C^\mu(p'',-p_+)v(\mathbf{p}_+) \right], \quad (11\text{–}58)$$

where

$$C^\mu(p',p) = \gamma^\mu \frac{i(\not{p}+\not{k})-m}{(p+k)^2+m^2} e + e \frac{i(\not{p}'-\not{k})-m}{(p'-k)^2+m^2} \gamma^\mu \quad (11\text{–}59)$$

is the characteristic factor which enters the Compton scattering matrix element. The cross section follows from Eq. (8–49),

$$d\sigma = (2\pi)^2 \frac{\omega\epsilon}{|p\cdot k|} S_f \bar{S}_i \delta(p'+p''+p_+-p-k)|(p'p''p_+|M^{(3)}|pke)|^2$$

$$= \frac{\alpha}{(2\pi)^2} r_0^2 \frac{\beta'\beta''\epsilon'\epsilon''d\epsilon'}{|p\cdot k|\epsilon_+} \cdot \frac{d\epsilon''}{d(\epsilon'+\epsilon''+\epsilon_+)} X d\Omega' d\Omega''. \quad (11\text{–}60)$$

This expression is obtained with the notation $|\mathbf{p}'| = \beta'\epsilon'$, $|\mathbf{p}''| = \beta''\epsilon''$, $r_0 = \alpha/m$, and resulted from a summation over final spin and polarization states and an average over initial spin and polarization states, following Sections A2–5 and A2–6. The integration over the δ-function of momentum and energy eliminated p_+ and replaced $d^3p'' = \beta''\epsilon''^2 d\epsilon'' d\Omega''$ by

$$\beta''\epsilon''^2 d\Omega'' \frac{d\epsilon''}{d(\epsilon'+\epsilon''+\epsilon_+)}.$$

The dimensionless trace X can be written in the form

$$X = \left\{ \frac{x(p'p,p'p)}{[(p'-p)^2]^2} + \frac{x(p''-p_+,p'p)}{(p''+p_+)^2(p'-p)^2} - \frac{x(p'-p_+,p'p)}{(p'+p_+)^2(p'-p)^2} \right.$$

$$\left. - \frac{x(p''p,p'p)}{(p''-p)^2(p'-p)^2} + (p \leftrightarrow -p_+) \right\} + (p' \leftrightarrow p''). \quad (11\text{-}61)$$

The symbol $(p \leftrightarrow q)$ means "the previous expression with p and q interchanged." The traces $x(p'p,p'p)$, etc., are as follows:

$$x(p'p,p'p) = \text{Tr}\,[\gamma_\mu \Lambda_-(p)\gamma_\nu \Lambda_-(p')]$$

$$\times \text{Tr}\,[C^\mu(p'',-p_+)\Lambda_+(p_+)\overline{C^\nu(p'',-p_+)}\Lambda_-(p'')],$$

$$x(p''-p_+,p'p) = \text{Tr}\,[\gamma_\mu \Lambda_+(p_+)\overline{C^\nu(p'',-p_+)}\Lambda_-(p'')]$$

$$\times \text{Tr}\,[C^\mu(p',p)\Lambda_-(p)\gamma_\nu \Lambda_-(p')], \quad (11\text{-}62)$$

$$x(p'-p_+,p'p) = \text{Tr}\,[\gamma_\mu \Lambda_+(p_+)\overline{C^\nu(p'',-p_+)}\Lambda_-(p'')$$

$$\times C^\mu(p'',p)\Lambda_-(p)\gamma_\nu \Lambda_-(p')],$$

$$x(p''p,p'p) = (p' \leftrightarrow p)(p'' \leftrightarrow -p_+) \text{ acting on } x(p'-p_+,p'p).$$

As in Section A2–5, the bar indicates that the γ-matrices in that expression are to be written in reverse order.

The symmetry properties of these traces are important. They are connected with the fact that a more symmetrical process, *viz.*, bremsstrahlung in electron-electron collisions, is described by exactly the same formulas with only a different interpretation of the momenta involved (*cf.* Section 12–3). These symmetry properties enable one to reduce the evaluation of X from a sum of sixteen expressions to a sum of four expressions, only three of which are essentially different (Eqs. 11–62). The actual evaluation of these expressions is not difficult, but is very laborious. They were first evaluated correctly by V. Votruba.* The reader will be spared the sight of his results. The final expressions are much too long and complicated to be of practical use. A general analytic integration seems therefore rather remote. However, Votruba was able to estimate the nonrelativistic and the extreme relativistic limits of the total cross section.

In the center-of-momentum system the initial electron will have an energy $\epsilon_0 = \sqrt{\omega_0^2 + m^2}$, where ω_0 is the energy of the incident photon. At

* V. Votruba, *Bull. Int. Acad. Tcheque des Sciences* **49**, 19 (1948) and *Phys. Rev.* **73**, 1468 (1948).

threshold, the final electrons will have zero momentum, so that

$$\omega_0 + \sqrt{\omega_0^2 + m^2} = 3m$$

or

$$\omega_0 = \tfrac{4}{3}m. \tag{11-63}$$

This is the threshold energy in the center-of-momentum system. A transformation to the rest system of the initial electron shows easily the general relation

$$\omega_0 = \frac{\omega}{\sqrt{1 + 2\omega/m}}, \tag{11-64}$$

where ω is the photon energy in the rest system. In that system the threshold energy is therefore

$$\omega = 4m, \tag{11-65}$$

corresponding to a final state in which all three electrons have equal momenta $|\mathbf{p}| = \tfrac{4}{3}m$ and emerge in the forward direction. The following remarks apply to this system.

By nonrelativistic limit is meant the case of

$$\omega - 4m \ll m.$$

In this limit the total cross section can be evaluated and yields

$$\sigma(\omega) = \frac{\pi\sqrt{3}}{4 \cdot 3^5} \alpha r_0^2 \left(\frac{\omega}{m} - 4\right)^2, \quad \text{(NR)}. \tag{11-66}$$

The numerical factor is not precise, but represents the estimate of an integral.

In the extreme relativistic limit ($\omega \gg 4m$) the total cross section is*

$$\sigma(\omega) = \alpha r_0^2 \left(\frac{28}{9} \ln \frac{2\omega}{m} - \frac{100}{9}\right), \quad \text{(ER)}. \tag{11-67}$$

As is to be expected, the contributions to the logarithmic term come exclusively from collisions with small momentum transfer. In a photon-negaton collision of this type one of the final negatons will have very large, the other very small, momentum. The constant term contains no contributions from final states in which all three electrons have large momenta, because these final states are relatively less likely to occur. In this limit, final states in which the positon has small momentum whereas the two negatons have large momenta contribute only terms proportional to $\ln(\omega/m)/(\omega/m)$ and are not included in Eq. (11-67).

---

* This estimate is given in the review paper by J. Joseph and F. Rohrlich, *Rev. Mod. Phys.* **30**, 354 (1958).

The estimate (11–67) contributes a lower bound. It neglects contributions from momentum transfers which are of the order of the electron mass. These are difficult to evaluate, but are known to be positive.

This estimate was improved by Suh and Bethe* who proved that in the high energy limit the momentum distribution of the recoiling electron is independent of the mass of the recoiling particle, when that recoil momentum is of the order of magnitude of the recoil particle rest mass or near its minimum value. This conclusion permits one to use the results of Borsellino† for the high energy limit. He computed pair production in the field of a particle of unspecified mass M. This calculation with $M = m$ differs from Votruba's work in the neglect of exchange as well as in the neglect of the first two diagrams of Fig. 11–4. These two effects are expected to compensate each other in part and, in any case, are expected to be small for large ω/m. The work of Suh and Bethe proved this expectation to be correct to order m/ω.

Thus, in the extreme relativistic limit the pair production cross section in the field of an electron exceeds the lower bound (11–67) and eventually approaches the same value as for pair production in the field of a heavy particle (see Eq. (15–112) of Chapter 15).

More recently a complete exact evaluation of the traces in (11–61) has been carried out, followed by a numerical integration to yield the total cross section.‡ Mork thus proved that the previous approximations were correct. The experimental data are now accurate enough to confirm these results.

Finally, E. Haug§ succeeded in carrying out the integrations exactly analytically to this order of perturbation expansion. By means of the substitution law (Section 8–5) he thereby obtained the differential cross sections for both pair production in photon–electron collisions ("trident production") and bremsstrahlung in electron–electron collisions (Section 12–3). His results confirm the earlier approximations and Mork's numerical work. In particular, the exact total pair production cross section reaches the Borsellino cross section at $\omega/m \sim 18$ and even rises slightly above (by about 1%) in the energy region $35 \lesssim \omega/m \lesssim 80$.

A further conclusion of the work of Suh and Bethe and of Mork is that the Weizsäcker-Williams approximation used by Wheeler and Lamb is justified in computing pair production in the field of bound electrons.**

---

* K. S. Suh and H. A. Bethe, *Phys. Rev.* **115**, 672 (1959).
† A. Borsellino, *Helv. Phys. Acta* **20**, 136 (1947) and *Nuovo Cim.* **4**, 12 (1947).
‡ K. J. Mork, *Phys. Rev.* **160**, 1065 (1967).
§ E. Haug, *Z. F. Naturf.* **30**a, 1099 (1975). Other recent work is referred to in this paper.
** J. A. Wheeler and W. E. Lamb, *Phys. Rev.* **55**, 858 (1939) and **101**, 1836 (1956).

CHAPTER 12

THE ELECTRON-ELECTRON SYSTEM

According to our definition (*cf.* the introduction to Chapter 11) the electron-electron systems are specified by their initial states, in which there are either two negatons, two positons, or one negaton and one positon. The last case is of special interest, since it permits bound states; these states, however, are not stable. The bound structure which consists of one negaton and one positon is usually called *positronium*. We shall discuss it and its modes of decay in Sections 12-5 and 12-6 of this chapter. Sections 12-1 to 12-4 are devoted to scattering problems.

As in the photon-electron system, the scattering of two electrons into a final state of exactly two electrons and no photon is only an approximate description of this process. This is so because the scattering of a charged particle through a finite angle will always be accompanied by an undetermined number of very low energy photons. We obtain correct results only in the approximation in which the possible emission of photons in the final state can be consistently neglected. This occurs only in lowest order and when the scattering angle is not too small. This lowest order approximation is usually called *Møller scattering* when referred to two negatons or two positons (*cf.* Section 12-1), and *Bhabha scattering* when referred to one negaton and one positon (*cf.* Section 12-2). The radiative corrections to this approximation require the knowledge of the energy resolution of the experiment which determines the maximum energy of possible photons in the final state. The effect of *bremsstrahlung in electron-electron collisions* is therefore strictly not separable from that of scattering. However, since we shall not calculate radiative corrections to Møller or Bhabha scattering, we shall find it convenient to discuss bremsstrahlung independently (*cf.* Section 12-3).

Of considerable interest are electron-electron systems in which the final state does not contain electrons. Because of charge conservation only the negaton-positon systems can be of this type. Conservation of energy and momentum then requires that the final state consist of at least two photons. This problem is discussed in Section 12-4 for free electrons and in Section 12-6 for bound electrons (positronium). Section 12-5 is devoted to a discussion of the symmetry properties of positronium and the selection rules which govern its decay modes.

12-1 Møller scattering. Consider two electrons of four-momenta p_1 and p_2 in the initial state. Let the final states be p_1' and p_2'. The diagrams for this process are shown in Fig. 8–4 and the corresponding matrix element is given in Eq. (8–24). Following Eqs. (8–28) and (8–29), we write

$$(p_1'p_2'|M^{(2)}|p_1p_2) = \frac{e^2}{4\pi^2 i} \frac{m^2}{\sqrt{\epsilon_1'\epsilon_2'\epsilon_1\epsilon_2}} \left[\frac{\bar{u}(\mathbf{p}_1')\gamma^\mu u(\mathbf{p}_1)\bar{u}(\mathbf{p}_2')\gamma_\mu u(\mathbf{p}_2)}{(p_1'-p_1)^2} \right.$$

$$\left. - \frac{\bar{u}(\mathbf{p}_2')\gamma^\mu u(\mathbf{p}_1)\bar{u}(\mathbf{p}_1')\gamma_\mu u(\mathbf{p}_2)}{(p_2'-p_1)^2} \right]. \quad (12\text{–}1)$$

The cross section follows from Eq. (8–49):

$$\sigma = (2\pi)^2 \frac{\epsilon_1 \epsilon_2}{\sqrt{(p_1 \cdot p_2)^2 - m^4}} S_f \bar{S}_i \delta(p_1' + p_2' - p_1 - p_2)|(f|M^{(2)}|i)|^2. \quad (12\text{–}2)$$

If we assume that the spins are not observed, which is the situation most frequently encountered in the experiments, the expression (12–2) becomes, with (12–1),

$$\sigma = 4\alpha^2 \int \frac{\delta(p_1' + p_2' - p_1 - p_2) X d^3p_1' d^3p_2'}{\epsilon_1'\epsilon_2'\sqrt{(p_1 \cdot p_2)^2 - m^4}}, \quad (12\text{–}3)$$

where the trace X is obtained by summing over the final spin states and averaging over the initial spin states in the square of the absolute value of the matrix element (12–1). Using the technique of projection operators as explained in Section A2–5, we find

$$X = \frac{m^4}{(p_1'-p_1)^4} A + \frac{m^4}{(p_2'-p_1)^4} B - \frac{m^4}{(p_1'-p_1)^2(p_2'-p_1)^2} (C+D),$$

$$(12\text{–}4)$$

where

$$\begin{aligned}
A &= 4 \cdot \tfrac{1}{4}\text{Tr}\,[\gamma^\mu\Lambda_-(p_1)\gamma^\nu\Lambda_-(p_1')] \cdot \tfrac{1}{4}\text{Tr}\,[\gamma_\mu\Lambda_-(p_2)\gamma_\nu\Lambda_-(p_2')], \\
B &= 4 \cdot \tfrac{1}{4}\text{Tr}\,[\gamma^\mu\Lambda_-(p_1)\gamma^\nu\Lambda_-(p_2')] \cdot \tfrac{1}{4}\text{Tr}\,[\gamma_\mu\Lambda_-(p_2)\gamma_\nu\Lambda_-(p_1')], \\
C &= \tfrac{1}{4}\text{Tr}\,[\gamma^\mu\Lambda_-(p_1)\gamma^\nu\Lambda_-(p_2')\gamma_\mu\Lambda_-(p_2)\gamma_\nu\Lambda_-(p_1')], \\
D &= \tfrac{1}{4}\text{Tr}\,[\gamma^\nu\Lambda_-(p_1)\gamma^\mu\Lambda_-(p_1')\gamma_\nu\Lambda_-(p_2)\gamma_\mu\Lambda_-(p_2')].
\end{aligned} \quad (12\text{–}5)$$

These expressions show that the substitutions $p_1' \leftrightarrow p_2'$ transform A into B, and B into A. Furthermore, $C = D$. It is thus sufficient to evaluate A and C. The calculation is of the same type as that which led to the formulas for the Compton effect in Section 11–1. In a similar way, we introduce the inner products

$$\begin{aligned}
-m^2\kappa &= p_1 \cdot p_2 = p_1' \cdot p_2', \\
-m^2\lambda &= p_1 \cdot p_1' = p_2 \cdot p_2', \\
-m^2\mu &= p_1 \cdot p_2' = p_2 \cdot p_1'.
\end{aligned} \quad (12\text{–}6)$$

The result of the evaluation of the traces (12–5) in the way indicated in Eq. (A2–82) can be expressed in terms of these three quantities. We find

$$A = \tfrac{1}{2}(\kappa^2 + \mu^2 - 2\lambda + 2),$$
$$B = \tfrac{1}{2}(\kappa^2 + \lambda^2 - 2\mu + 2), \qquad (12\text{–}7)$$
$$C = D = -\tfrac{1}{2}(\kappa^2 - \kappa - \lambda - \mu + 1).$$

The integration over the δ-function in Eq. (12–3) is done in the usual way and yields

$$d\sigma = 4\alpha^2 \frac{\beta_1' X d\Omega_1'}{\sqrt{(p_1 \cdot p_2)^2 - m^4}} \cdot \frac{\epsilon_1'}{\epsilon_2'} \cdot \frac{d\epsilon_1'}{d(\epsilon_1' + \epsilon_2')}. \qquad (12\text{–}8)$$

The last factor is evaluated by use of the conservation laws:

$$\frac{d(\epsilon_1' + \epsilon_2')}{d\epsilon_1'} = \frac{1}{\beta_1' \epsilon_2'}(\beta_1' E - |\mathbf{P}| \cos \theta_1'). \qquad (12\text{–}9)$$

The four-vector $P = (E, \mathbf{P})$ describes the motion of the center of momentum of the system, and the direction of $\mathbf{P}$ is assumed to be the polar axis of our coordinate system. The differential cross section for Møller scattering follows from these results and Eqs. (12–4) and (12–7):

$$d\sigma = 4r_0^2 X \frac{m^2 \beta_1'^2 d\Omega_1'}{\sqrt{(p_1 \cdot p_2)^2 - m^4}} \cdot \frac{\epsilon_1'}{\beta_1' E - |\mathbf{P}| \cos \theta_1'}. \qquad (12\text{–}10)$$

This formula is completely general and will be specialized in the following to the center-of-momentum system, and to the rest system of one of the initial electrons.

Consider first the center-of-momentum system (CM-system). It is characterized by the relations

$$\mathbf{p}_1 = -\mathbf{p}_2 = \mathbf{p},$$
$$\mathbf{p}_1' = -\mathbf{p}_2' = -\mathbf{p}',$$
$$\epsilon_1' = \epsilon_2' = \epsilon_1 = \epsilon_2 = \epsilon, \qquad (12\text{–}11)$$
$$|\mathbf{p}| = \beta \epsilon, \quad \epsilon = \gamma m,$$
$$-\mathbf{p} \cdot \mathbf{p}' = (\gamma^2 - 1)m^2 \cos \theta.$$

The differential cross section (12–10) in this coordinate system becomes

$$d\sigma = r_0^2 X \frac{d\Omega}{\gamma^2}, \quad \text{(CM-system)}. \qquad (12\text{–}12)$$

12-1] MØLLER SCATTERING

The trace X is obtained by substituting Eqs. (12–11) into Eqs. (12–4) and (12–7):

$$\kappa = 2\gamma^2 - 1,$$
$$\lambda = \gamma^2 - (\gamma^2 - 1)\cos\theta = \gamma^2(1 - \beta^2\cos\theta),$$
$$\mu = \gamma^2 + (\gamma^2 - 1)\cos\theta = \gamma^2(1 + \beta^2\cos\theta).$$

We find

$$X = \frac{1}{(2\beta\gamma)^4}\left[\frac{A}{\sin^4(\theta/2)} + \frac{B}{\cos^4(\theta/2)} - \frac{2C}{\sin^2(\theta/2)\cos^2(\theta/2)}\right], \quad (12\text{–}13\text{a})$$

$$X = \frac{1}{(\gamma^2 - 1)^2}\left[\frac{(2\gamma^2 - 1)^2}{\sin^4\theta} - \frac{2\gamma^4 - \gamma^2 - \frac{1}{4}}{\sin^2\theta} + \frac{(\gamma^2 - 1)^2}{4}\right],$$

$$\text{(CM-system).} \quad (12\text{–}13\text{b})$$

For applications it is usually more convenient to know the cross sections in the rest system of one of the incident electrons. This system is often, to good approximation, the same as the laboratory system.

Writing, for a moment, primes on the variables which refer to the laboratory systems, we have the following relations between the two sets of variables:

$$\cos\theta = \frac{2 - (\gamma' + 3)\sin^2\theta'}{2 + (\gamma' - 1)\sin^2\theta'} \equiv x,$$

$$\gamma' = 2\gamma^2 - 1, \quad d\Omega = \frac{8\cos\theta'(\gamma' + 1)}{[2 + (\gamma' - 1)\sin^2\theta']^2}\,d\Omega'. \quad (12\text{–}14)$$

If we carry through this substitution and, for simplicity, omit the primes in the final result, we obtain the differential cross section for the laboratory system (L-system).

$$d\sigma = r_0^2\left(4\frac{\gamma + 1}{\beta^2\gamma}\right)^2\frac{\cos\theta\,d\Omega}{[2 + (\gamma - 1)\sin^2\theta]^2}$$

$$\times\left[\frac{4}{(1 - x^2)^2} - \frac{3}{1 - x^2} + \left(\frac{\gamma - 1}{2\gamma}\right)^2\left(1 + \frac{4}{1 - x^2}\right)\right],$$

$$\text{(L-system).} \quad (12\text{–}15)$$

This is the formula of Møller.* The scattering of two identical particles without spin satisfying the exclusion principle (a physical impossibility), gives just the first two terms. The last term in Eq. (12–15) may thus be regarded as the effect of the spin.

---

* C. Møller, *Ann. d. Physik* **14**, 568 (1932).

In applications to energy-loss problems it is advantageous to express the differential cross section (12–15) in terms of the relative energy transfer $w = W/T$, where W is the energy lost in the collision by the incident negaton, and $T = m(\gamma - 1)$ is the incident kinetic energy. The simple relation between $x = \cos\theta$ (Eq. 12–14) and w,

$$x = 1 - 2w,$$

enables us to rewrite Eq. (12–15) in the form

$$d\sigma = \frac{2\pi r_0^2}{\beta^2} \cdot \frac{m}{T}\left[\frac{1}{w^2} + \frac{1}{(1-w)^2} + \left(\frac{\gamma-1}{\gamma}\right)^2 - \frac{2\gamma-1}{\gamma^2}\frac{1}{w(1-w)}\right]dw,$$

(L-system). (12–15w)

This result is invariant under the transformation $w \to 1 - w$, because of the symmetry of the final state.

The total cross section is obtained by integrating the differential cross section over all angles and dividing the result by two. This factor $\frac{1}{2}$ is necessary because the two particles are identical. In the angular distribution each state is counted exactly twice if the direction of scattering is varied over the entire sphere. If we integrate the formula (12–15w), the integration is to be carried out from $w = 0$ to $w = \frac{1}{2}$; one can interpret this by identifying the faster outgoing electron as the scattered electron, the slower one as the recoil electron.

Actually, the expressions (12–13) or (12–15) cannot be integrated over all angles, because the integral diverges at $\theta = 0$ and π. As mentioned in the introduction, this divergence is connected with the physically unrealizable requirement implied in the derivation of these expressions, that the two electrons scatter without emission of photons. Very low energy photon emission cannot be neglected when the momentum transfer becomes very small ($\theta \to 0$). However, because the two electrons are indistinguishable, the case $\theta \to \pi$ will lead to the same difficulty. Details of the problem of low energy photon emission will be found in Section 16–1.

In conclusion of this discussion, the NR limit is worth noting. In the NR limit $x = \cos 2\theta'$; i.e., the scattering angle in the CM-system, θ, is twice the angle in the laboratory system, θ'. For the former we find from Eqs. (12–12) and (12–13a),

$$\frac{d\sigma}{d\Omega} = \frac{r_0^2}{(2\beta)^4}\left[\frac{1}{\sin^4(\theta/2)} + \frac{1}{\cos^4(\theta/2)} - \frac{1}{\sin^2(\theta/2)\cos^2(\theta/2)}\right];$$

(NR, CM-system),

for the latter we obtain, omitting primes,

$$\frac{d\sigma}{d\Omega} = \frac{r_0^2}{\beta^4} 4 \cos\theta \left[\frac{1}{\sin^4\theta} + \frac{1}{\cos^4\theta} - \frac{1}{\sin^2\theta \cos^2\theta} \right], \quad \text{(NR, L-system)}.$$

The variables β and θ refer to the respective systems.

These formulas are simply the Rutherford cross section with exchange.*

12–2 Bhabha scattering.† This process is by its nature very similar to Møller scattering. Let the momentum variables in the initial state be p and q for the negaton and positon, respectively. Let the final states be characterized by p' and q'. The Feynman diagrams for the process are shown in Fig. 12–1.

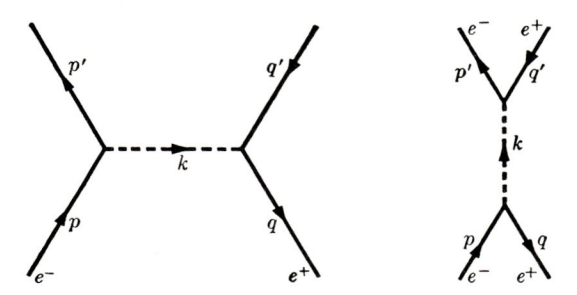

FIG. 12–1. Negaton-positon or Bhabha scattering diagrams in lowest order.

A comparison between Fig. 12–1 and Fig. 8–4 shows the essential difference between positon-negaton and negaton-negaton scattering. Whereas the first diagrams correspond to each other, the second diagram in Fig. 8–4 arises from the exclusion principle, which does not affect positon-negaton scattering. On the other hand, the latter process can take place via the virtual annihilation and re-creation of the negaton-positon pair. This effect has no analogue in the negaton-negaton system. The "exchange effect" in the latter system is replaced by the "annihilation effect" in the positon-negaton system.

The kinematic equivalence of the two systems will give rise to the same general expression for the cross section. But the topological relationship of the associated diagrams goes much deeper than that. Indeed, it bridges the apparent difference which was just mentioned, so that the exchange and annihilation effects emerge as two different physical consequences of the same mathematical structure. For, according to the *substitution* law of Section 8–5, the diagrams of Figs. 8–4 and 12–1 imply that the matrix

---

* N. F. Mott, *Proc. Roy. Soc.* **126**, 259 (1930).
† H. J. Bhabha, *Proc. Roy. Soc.* **A154**, 195 (1935).

element for Bhabha scattering can be obtained from that of Møller scattering by the substitutions

$$p_1 \to p, \quad p_1' \to p', \quad p_2 \to -q', \quad p_2' \to -q. \qquad (12\text{-}16)$$

This substitution is valid for *both* diagrams. It is almost trivial for the first one, but it may appear surprising for the second diagram, where it establishes a link between the symmetry properties required by the exclusion principle and the creation and annihilation character of the negaton with its antiparticle, the positon.

The evaluation of the matrix elements, therefore, does not require new trace calculations, but only the substitution of Eq. (12-16) into the results for Møller scattering. In particular, the definitions (12-6) now read

$$\begin{aligned} -m^2\kappa &= -p \cdot q' = -p' \cdot q, \\ -m^2\lambda &= p \cdot p' = q \cdot q', \\ -m^2\mu &= -p \cdot q = -p' \cdot q', \end{aligned} \qquad (12\text{-}17)$$

and Eqs. (12-7) remain formally unchanged. The differential cross section for Bhabha scattering is in analogy to Eqs. (12-3), (12-8), and (12-10), using (8-49):

$$\begin{aligned} d\sigma &= 4\alpha^2 \int \frac{\delta(p' + q' - p - q) X d^3 p' d^3 q'}{\epsilon_+' \epsilon_-' \sqrt{(p \cdot q)^2 - m^4}} \\ &= 4\alpha^2 \frac{\beta_-' X d\Omega_-'}{\sqrt{(p \cdot q)^2 - m^4}} \cdot \frac{\epsilon_-'}{\epsilon_+'} \cdot \frac{d\epsilon_-'}{d(\epsilon_+' + \epsilon_-')} \\ &= 4r_0^2 X \frac{m^2 \beta_-'^2 d\Omega_-'}{\sqrt{(p \cdot q)^2 - m^4}} \cdot \frac{\epsilon_-'}{\beta_-' E - |\mathbf{P}| \cos\theta_-'}. \end{aligned} \qquad (12\text{-}18)$$

The subscripts $\pm$ refer to positons and negatons; the primed and unprimed quantities indicate the final and initial states, respectively; $P = (E,\mathbf{P})$ is again the energy-momentum vector of the center of momentum. The trace X follows from (12-4):

$$X = \frac{m^4}{(p'-p)^4} A + \frac{m^4}{(p+q)^4} B - \frac{m^4}{(p'-p)^2(p+q)^2}(C+D), \qquad (12\text{-}19)$$

where A, B, C, and D are given by Eqs. (12-7) and (12-17).

We specialize to the center-of-momentum system characterized by

$$\begin{aligned} \mathbf{p} &= -\mathbf{q}, \quad \mathbf{p}' = -\mathbf{q}', \\ \epsilon_+ &= \epsilon_- = \epsilon_+' = \epsilon_-' = \epsilon, \\ |\mathbf{p}| &= \beta\epsilon, \quad \epsilon = \gamma m, \\ -p \cdot q' &= (\gamma^2 - 1)m^2 \cos\theta. \end{aligned} \qquad (12\text{-}20)$$

The differential cross section becomes

$$d\sigma = r_0^2 X \frac{d\Omega}{\gamma^2}, \quad \text{(CM-system)}. \tag{12-21}$$

The products (12–17) simplify to

$$-\kappa = \gamma^2 + (\gamma^2 - 1)\cos\theta = \gamma^2(1 + \beta^2 \cos\theta),$$

$$-\lambda = \gamma^2 + (\gamma^2 - 1)\cos\theta = -\gamma^2(1 - \beta^2 \cos\theta),$$

$$-\mu = 2\gamma^2 - 1.$$

Insertion of these expressions into (12–7) enables the evaluation of X, Eq. (12–19). The result can be combined with (12–21) and yields the Bhabha cross section in the center-of-momentum system:

$$\frac{d\sigma}{d\Omega} = \frac{r_0^2}{\gamma^2} \cdot \frac{1}{(2\gamma)^4} \left\{ \frac{A}{[\beta \sin(\theta/2)]^4} + B + \frac{2C}{[\beta \sin(\theta/2)]^2} \right\}, \tag{12-22a}$$

$$\frac{d\sigma}{d\Omega} = \frac{r_0^2}{16\gamma^2} \cdot \left\{ \frac{1}{[\beta\gamma \sin(\theta/2)]^4} \left[1 + \left(2\beta\gamma \cos\frac{\theta}{2}\right)^2 + 2(\beta\gamma)^4 \left(1 + \cos^4\frac{\theta}{2}\right) \right] \right.$$

$$- \frac{1}{\gamma^2[\beta\gamma \sin(\theta/2)]^2} \left[3 + 2\left(2\beta\gamma \cos\frac{\theta}{2}\right)^2 + \frac{1}{4}\left(2\beta\gamma \cos\frac{\theta}{2}\right)^4 \right]$$

$$\left. + \frac{1}{\gamma^4}[3 + 4(\beta\gamma)^2 + (\beta\gamma)^4(1 + \cos^2\theta)] \right\}, \quad \text{(CM-system)}. \tag{12-22b}$$

A comparison of Eq. (12–22a) with Eqs. (12–13a) and (12–12) shows the similarity between Møller scattering and Bhabha scattering. One can easily see that the trace A is identical in the two cases. Therefore, the respective first terms in the cross sections are identical. The B term is the exchange term in Møller scattering and it is the annihilation term in Bhabha scattering. The C terms are the respective interference terms, since both exchange and annihilation effects are added in the scattering amplitude, rather than in the scattering probability. The B and C terms in Eq. (12–22b) and the following equations are interchanged.

The relation between the center-of-momentum system and the laboratory system in which the negaton is initially at rest is the same as in Møller scattering, Eq. (12–14). The differential cross section (12–22) can therefore be transformed by this substitution:

$$\frac{d\sigma}{d\Omega} = r_0^2 \left(2\frac{\gamma+1}{\beta^2\gamma}\right)^2 \frac{\cos\theta}{[2+(\gamma-1)\sin^2\theta]^2}$$

$$\times \left\{\frac{4}{(1-x)^2}\left[1 - \frac{\gamma^2-1}{2\gamma^2}(1-x) + \frac{1}{2}\left(\frac{\gamma-1}{2\gamma}\right)^2(1-x)^2\right]\right.$$

$$-\frac{2}{1-x}\left(\frac{\gamma-1}{\gamma+1}\right)\left[\frac{2\gamma+1}{\gamma^2} + \frac{\gamma^2-1}{\gamma^2}x + \left(\frac{\gamma-1}{2\gamma}\right)^2(1-x)^2\right]$$

$$\left. + \left(\frac{\gamma-1}{\gamma+1}\right)^2\left[\frac{1}{2} + \frac{1}{\gamma} + \frac{3}{2\gamma^2} - \left(\frac{\gamma-1}{2\gamma}\right)^2(1-x^2)\right]\right\},$$

(L-system). (12–23)

All quantities in this formula refer to the laboratory system and

$$x = \frac{2-(\gamma+3)\sin^2\theta}{2+(\gamma-1)\sin^2\theta}. \qquad (12\text{–}24)$$

As in Møller scattering, we can introduce the relative energy transfer $w = W/T = \frac{1}{2}(1-x)$, and find

$$d\sigma = \frac{2\pi r_0^2}{\beta^2} \cdot \frac{m}{T} \left\{\frac{1}{w^2} - \frac{\gamma^2-1}{\gamma^2}\frac{1}{w} + \frac{1}{2}\left(\frac{\gamma-1}{\gamma}\right)^2\right.$$

$$-\left(\frac{\gamma-1}{\gamma+1}\right)\left[\frac{\gamma+2}{\gamma}\frac{1}{w} - 2\frac{\gamma^2-1}{\gamma^2} + w\left(\frac{\gamma-1}{\gamma}\right)^2\right]$$

$$\left.+\left(\frac{\gamma-1}{\gamma+1}\right)^2\left[\frac{1}{2} + \frac{1}{\gamma} + \frac{3}{2\gamma^2} - \left(\frac{\gamma-1}{\gamma}\right)^2 w(1-w)\right]\right\} dw, \quad (12\text{–}23w)$$

where β, γ, and T in this equation refer to the L-system. The total cross section is obtained by integration of Eqs. (12–22) or (12–23) over all angles. There is no factor of two as in Møller scattering, because the particles are distinguishable (by their charge). Accordingly, the integration over the relative energy transfer in Eq. (12–23w) is to be carried out from 0 to 1. However, just as in Møller scattering, the differential cross section diverges as the scattering angle approaches zero, so that it cannot be integrated completely. There is, of course, no difficulty at $\theta = \pi$, in contradistinction to Møller scattering. The physical reason for this divergence is the same as before (cf. Sections 12–1 and 16–1).

Finally, we note the NR limit of the angular distribution. The scattering angle in the CM-system is again twice that in the L-system, and Eq. (12–22) gives

$$d\sigma = \frac{r_0^2}{16\beta^4 \sin^4(\theta/2)} d\Omega, \quad \text{(NR, CM-system)}.$$

This is exactly the first term of the corresponding Møller scattering cross section.

The theoretical results of this and the previous section are found to be in excellent agreement with observations. At suitable energies and angles the spin and exchange terms in Møller scattering and the annihilation terms in Bhabha scattering can all be measured and they are confirmed quantitatively.*

12–3 Bremsstrahlung in electron-electron collisions. The process of bremsstrahlung in electron-electron collisions is very closely related to that of pair production in photon-electron collisions. For consider the diagrams of this process. They are shown for negaton-negaton collisions in Fig. 12–2.

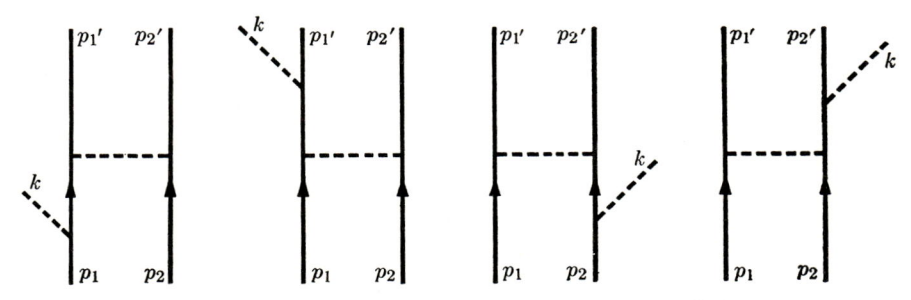

FIG. 12–2. Bremsstrahlung in negaton-negaton collisions.

The matrix element is the sum of these diagrams and the diagrams in which the momenta of the two final negatons are interchanged. A comparison of Figs. 11–4 and 12–2 shows that there is a one-to-one correspondence between the diagrams of pair production and bremsstrahlung. In fact, according to the substitution law of Section 8–5, the associated matrix elements are related as follows:

Pair production in photon-negaton collisions	p	p'	$-p_+$	p''	k
Bremsstrahlung in negaton-negaton collision	p_1	p_1'	p_2	p_2'	$-k$

(12–25)

---

* See, e.g., the beautiful experiments by A. Ashkin, L. A. Page, and W. M. Woodward, *Phys. Rev.* **94**, 357 (1954).

Of course, one can equally well obtain the matrix element for bremsstrahlung in *positon*-negaton collisions. It is only necessary to replace the substitutions $-p_+ \to p_2$ and $p'' \to p_2'$ of (12-25) by $p_+ \to q'$ and $p'' \to -q$. That the virtual annihilation in positon-negaton collisions is thereby correctly described can be seen from the very similar correspondence of Møller and Bhabha scattering in the last two sections. An analogous method yields bremsstrahlung in positon-positon collisions.

The cross section now follows in the usual way:

$$d\sigma = (2\pi)^2 \frac{\epsilon_1 \epsilon_2}{\sqrt{(p_1 \cdot p_2)^2 - m^4}} S_f \bar{S}_i \delta(p_1' + p_2' + k - p_1 - p_2) |(f|M|i)|^2; \quad (12\text{-}26)$$

Summing over the final spins and polarizations and averaging over the initial spin directions, we find

$$d\sigma = \frac{\alpha r_0^2}{(2\pi)^2} \frac{\beta_1' \epsilon_1' \omega d\omega}{\epsilon_2' \sqrt{(p_1 \cdot p_2)^2 - m^4}} \frac{d\epsilon_1'}{d(\epsilon_1' + \epsilon_2' + \omega)} \cdot d\Omega_1' d\Omega_k X, \quad (12\text{-}27)$$

where the trace X is obtained from the corresponding trace for pair production (11-61) by the substitution (12-25):

$$X = \left\{ \frac{x(p_1'p_1, p_1'p_1)}{[(p_1' - p_1)^2]^2} + \frac{x(p_2'p_2, p_1'p_1)}{(p_2' - p_2)^2 (p_1' - p_1)^2} \right.$$

$$\left. - \frac{x(p_1'p_2, p_1'p_1)}{(p_1' - p_2)^2 (p_1' - p_1)^2} - \frac{x(p_2'p_1, p_1'p_1)}{(p_2' - p_1)^2 (p_1' - p_1)^2} + (p_1 \leftrightarrow p_2) \right\}$$

$$+ (p_1' \leftrightarrow p_2'). \quad (12\text{-}28)$$

The traces x are given in Eq. (11-62).

The trace X was evaluated by Hodes.* The substitution (12-25) which determines (12-28) from (11-61) is here used as a very valuable cross check which verifies the correctness of the independent work of Votruba and Hodes.

An integration of the result (12-27) is very difficult and has been carried out only in the special case of photon emission in the rest system of one of the initial negatons and in the forward direction (*cf.* I. Hodes, *loc. cit.*).** As is to be expected, the results show that this case can be well approximated in the ER limit by assuming small momentum transfers of the incident negaton and by neglecting the exchange effect. In this limit the total cross

---

* I. Hodes, unpublished Ph.D. thesis, University of Chicago, December 1953.
** See however the recent work by E. Haug, *loc. cit.* p. 251.

section is therefore of the form

$$\sigma = 4\alpha r_0^2 \left(\ln \frac{2\epsilon_1}{m} - \text{const.} \right), \quad (\text{ER}), \qquad (12\text{--}29)$$

where ϵ_1 is the energy of the incident negaton in the rest system of negaton 2. The constant is of order one (*cf.* the case of electron bremsstrahlung in the nuclear field, Section 15–6). This result was first obtained by Wheeler and Lamb* and is completely analogous to the case of pair production.

12–4 Annihilation of free negaton-positon pairs. There are three essentially different kinds of pair annihilations. Their immediate end products are always photons. The first kind is the annihilation of a free positon with a free negaton in relative motion towards each other; we shall call this the *annihilation of free negaton-positon pairs* and shall discuss it in this section. Since the probability for the occurrence of this process increases with decreasing relative velocity, it will compete with the process of capture into the bound positronium state, which, in turn, can decay into photons. This *positronium annihilation* most likely occurs from one of the lowest energy levels (S-states) and will always be governed by *selection rules*. We shall devote Section 12–5 to these selection rules and to positronium annihilation. Finally, the third kind of annihilation takes place in the presence of an external field, e.g., the Coulomb field of a nucleus. This *pair annihilation in an external field* is of importance when positons annihilate with tightly bound negatons whose binding to the nucleus cannot be neglected. We shall calculate this process in Section 15–7. Let us remark here, though, that this latter process is the only one in which energy-momentum conservation does not exclude *one*-photon annihilation, i.e., pair annihilation with the emission of only one photon.

Returning to free pairs, we find that the probability for annihilation into $(n + 1)$ photons rather than n photons will be of order α times smaller. It follows that the *lifetime* of a free positon in a medium of free negatons is to a good approximation the two-photon annihilation time, τ_2, evaluated in lowest order of the iteration solution. The radiative corrections to τ_2 are of the same order of relative magnitude (*viz.*, fractions of one percent) as the three-photon annihilation lifetime, τ_3. Both are physically indistinguishable corrections to the lowest order of τ_2 and, at the present time, are too small to be observable.

The *differential cross sections* for two- and three-photon annihilation are, of course, separately observable, except that for any finite energy resolution ΔE of the measuring apparatus, the observation of two-photon annihilation also includes, in an indistinguishable way, three-photon annihilation when

---

* Cf. the references at the end of Chapter 11.

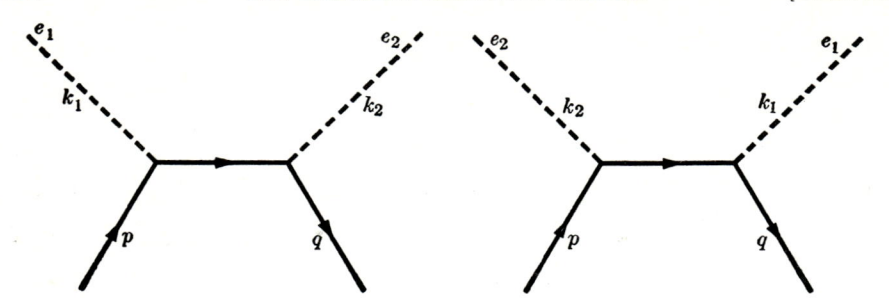

FIG. 12–3. Two-photon annihilation of a free pair.

one of the three photons has energy less than ΔE. Similar arguments apply to higher multiple photon annihilations. These facts are in satisfactory agreement with the occurrence of infrared divergences in the radiative corrections to τ_2 and τ_3, which cancel each other exactly. The situation is completely analogous to the case of the radiative corrections to Compton scattering and double Compton scattering. In fact, the corresponding diagrams are topologically related in the sense of Section 8–5. The case is different when suitable spin and polarization measurements can be performed, such that the existence of a third photon can be inferred from conservation laws. In particular, in positronium annihilation the knowledge of the initial positronium state always determines whether the number of decay photons is even or odd (*cf.* Section 12–6).

In this section we shall calculate the differential cross section for free-pair annihilation into two and three photons. As mentioned above, the corresponding diagrams are topologically related to other diagrams which we have already discussed, so that full use can be made of the substitution law of Section 8–5. The lowest order two-photon annihilation is related to the Compton effect, whereas three-photon annihilation is related to the double Compton effect. However, since the two-photon annihilation is simple enough, we shall take this opportunity to show explicitly how one is led to the substitution law.

Let $p = (\epsilon_-, \mathbf{p})$ and $q = (\epsilon_+, \mathbf{q})$ be the momentum four-vectors of the incident negaton and positon, and let $k_1 = (\omega_1, \mathbf{k}_1)$, e_1 and $k_2 = (\omega_2, \mathbf{k}_2)$, e_2 be the momentum and polarization four-vectors of the two emerging photons. The lowest-order diagrams are shown in Fig. 12–3. In the same way as in the previous sections, we find for the matrix element

$$(k_1 k_2)|M^{(2)}|pq) = -\frac{ie^2}{(2\pi)^2} \frac{m}{\sqrt{\epsilon_+ \epsilon_- 2\omega_1 2\omega_2}}$$

$$\times \bar{v}(\mathbf{q}) \left[\frac{e_2(i\not{p} - i\not{k}_1 - m)e_1}{2p \cdot k_1} + \frac{e_1(i\not{p} - i\not{k}_2 - m)e_2}{2p \cdot k_2} \right] u(\mathbf{p}). \quad (12\text{-}30)$$

The cross section follows from Section 8–6,

$$\sigma = \frac{(2\pi)^2 \epsilon_+ \epsilon_-}{\sqrt{(p \cdot q)^2 - m^4}} \, \mathsf{S}_f \bar{\mathsf{S}}_i \delta(k_1 + k_2 - p - q) |(k_1 k_2 | M^{(2)} | pq)|^2. \quad (12\text{–}31)$$

If we average over the spins of the pair (one is usually dealing with unpolarized negatons and positons), we find for the annihilation cross section

$$d\sigma = \frac{\alpha^2}{4\omega_1 \omega_2 \sqrt{(p \cdot q)^2 - m^4}} \int d\tau_f X' \delta(k_1 + k_2 - p - q), \quad (12\text{–}32)$$

$$-X' = \frac{m^2}{\kappa_1^2} A' + \frac{m^2}{\kappa_2^2} B' + \frac{m^2}{\kappa_1 \kappa_2} (C' + D'), \quad (12\text{–}33)$$

where

$$-\kappa_1 = p \cdot k_1 = q \cdot k_2, \quad -\kappa_2 = p \cdot k_2 = q \cdot k_1, \quad (12\text{–}34)$$

and

$$\begin{aligned} A' &= (e_2(ip - m - ik_1)e_1\Lambda_-(p)e_1(ip - m - ik_1)e_2\Lambda_+(q)), \\ B' &= (e_1(ip - m - ik_2)e_2\Lambda_-(p)e_2(ip - m - ik_2)e_1\Lambda_+(q)), \\ C' &= -(e_2(ip - m - ik_1)e_1\Lambda_-(p)e_2(ip - m - ik_2)e_1\Lambda_+(q)), \\ D' &= -(e_1(ip - m - ik_2)e_2\Lambda_-(p)e_1(ip - m - ik_1)e_2\Lambda_+(q)), \end{aligned} \quad (12\text{–}35)$$

in the notation of Section A2–4.

A comparison of these expressions with the trace X obtained in the calculation for Compton scattering shows the desired similarity. Indeed, the expressions A, B, C, D of Eq. (11–12) become the A', B', C', D' of Eq. (12–35) under the substitutions

$$p \to p, \quad p' \to -q, \quad e \to e_1, \quad e' \to e_2, \quad k \to -k_1, \quad k' \to k_2. \quad (12\text{–}36)$$

We can immediately take over the result (11–13) and (11–14) with these substitutions. This conclusion is therefore a special case of the general arguments which lead to the substitution law of Section 8–5.

The integration over the final states, $\int d\tau_f$, in (12–32) also proceeds similarly to the case of Compton scattering. Since

$$d\tau_f = d^3 k_2 d^3 k_1 = d^3 k_2 \omega_1^2 d\omega_1 d\Omega_1,$$

and we are interested in the differential cross section per unit solid angle in a given direction of one of the emerging photons, we do not want to integrate over $d\Omega$. The integration over $d^3 k_2$ replaces k_2 in the integrand by $p + q - k_1$; the integration over $d\omega_1$ results in a factor

$$\frac{d\omega_1}{d \cdot (\omega_1 + \omega_2)}.$$

266   THE ELECTRON-ELECTRON SYSTEM   [CHAP. 12

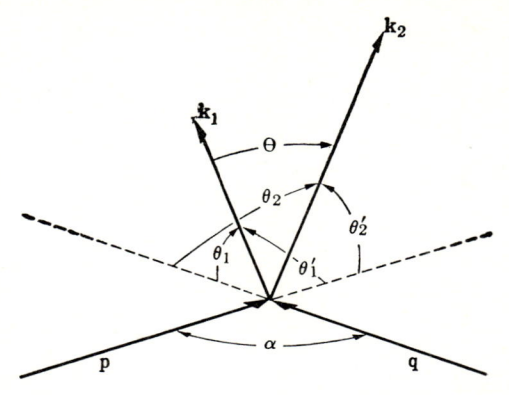

FIG. 12-4. The angles between the momentum vectors in two-photon annihilation.

In this expression ω_2 is a function of ω_1. We find from the conservation laws,

$$\omega_2{}^2 = \mathbf{p}^2 + \mathbf{q}^2 + \omega_1{}^2 + 2[|\mathbf{p}|\,|\mathbf{q}|\cos\alpha - \omega_1(|\mathbf{p}|\cos\theta_1' + |\mathbf{q}|\cos\theta_1)], \quad (12\text{-}37)$$

where θ_1 and θ_1' are the angles between k_1 and q_1, and between k_1 and p; α is the angle between p and q. The projection of the momenta parallel and perpendicular to k_1 gives

$$\omega_1 + \omega_2 \cos\Theta = |\mathbf{q}|\cos\theta_1 + |\mathbf{p}|\cos\theta_1',$$
$$\omega_2 \sin\Theta = |\mathbf{q}|\sin\theta_1 + |\mathbf{p}|\sin\theta_1', \quad (12\text{-}38)$$

where Θ is the angle between $\mathbf{k}_1$ and $\mathbf{k}_2$ (cf. Fig. 12-4). Differentiating Eq. (12-37) with respect to ω_1 and keeping θ_1, θ_1', and the initial state $(\mathbf{p},\mathbf{q},\alpha)$ constant, we find, with Eq. (12-38),

$$\frac{d(\omega_1 + \omega_2)}{d\omega_1} = 1 - \cos\Theta. \quad (12\text{-}39)$$

We can now combine this result with the trace (11-13) and the substitution (12-36). The differential cross section for pair annihilation in flight follows then from Eq. (12-32):

$$d\sigma = \frac{1}{4}r_0{}^2\,\frac{\omega_1}{\omega_2}\,\frac{m^2 X\,d\Omega_1}{\sqrt{(p\cdot q)^2 - m^4}\,(1 - \cos\Theta)}, \quad (12\text{-}40)$$

where

$$X = \frac{1}{2}\left(\frac{\kappa_1}{\kappa_2} + \frac{\kappa_2}{\kappa_1}\right) + 1 - 2(e_1 \cdot e_2)^2 - 2\left[\frac{(e_1 \cdot p)^2 (e_2 \cdot q)^2}{\kappa_1^2} + \frac{(e_2 \cdot p)^2 (e_1 \cdot q)^2}{\kappa_2^2}\right]$$
$$- 4\left[\left(\frac{e_1 \cdot p \, e_2 \cdot q}{\kappa_1} + \frac{e_1 \cdot q \, e_2 \cdot p}{\kappa_2}\right) e_1 \cdot e_2 + \frac{e_1 \cdot p \, e_2 \cdot p \, e_1 \cdot q \, e_2 \cdot q}{\kappa_1 \kappa_2}\right]. \quad (12\text{–}41)$$

The cross section (12–40) can be summed over the polarization directions of the emerging photons. The result will be a replacement of X by $\bar{X}$, which follows from Eq. (11–14) with the substitution (12–36):

$$\bar{X} = \sum_{\text{pol}} X = 2\left(\frac{\kappa_1}{\kappa_2} + \frac{\kappa_2}{\kappa_1}\right) + 4\left(\frac{m^2}{\kappa_1} + \frac{m^2}{\kappa_2}\right) - 2\left(\frac{m^2}{\kappa_1} + \frac{m^2}{\kappa_2}\right)^2. \quad (12\text{–}42)$$

As was shown in Section 8–5, a consistent application of the substitution law will introduce an over-all minus sign for each negaton that becomes a positon (or vice versa). The last two equations therefore differ in over-all sign from the corresponding expressions for Compton scattering.

These expressions are valid for any Lorentz frame. When the initial state $(\mathbf{p}, \mathbf{q}, \alpha)$ is known, Eq. (12–40) gives the angular distribution of photon 1 for any arbitrary direction of photon 2 and for arbitrary choice of e_1 and e_2. For, if we choose the direction of $\mathbf{q}$ as polar axis and measure azimuths relative to the azimuth of $\mathbf{p}$, the direction of $\mathbf{k}_1$ will be determined by θ_1 and φ_1, the direction of $\mathbf{k}_2$ by θ_2 and φ_2. When θ_2 and φ_2 are chosen, Θ can be expressed by $\theta_1, \theta_2, \varphi_1, \varphi_2$:

$$\cos \Theta = \cos \theta_1 \cos \theta_2 - \sin \theta_1 \sin \theta_2 \cos (\varphi_1 - \varphi_2).$$

Also, θ_1' and θ_2' are known:

$$\cos \theta_1' = \cos \alpha \cos \theta_1 - \sin \alpha \sin \theta_1 \cos \varphi_1,$$
$$\cos \theta_2' = \cos \alpha \cos \theta_2 - \sin \alpha \sin \theta_2 \cos \varphi_2,$$

and the energies ω_1 and ω_2 can be determined from the total energy of the system and from Eqs. (12–34):

$$\omega_1 + \omega_2 = m(\gamma_+ + \gamma_-),$$
$$\frac{\omega_1}{\omega_2} = \frac{\gamma_-(1 - \beta_- \cos \theta_2')}{\gamma_+(1 - \beta_+ \cos \theta_1)}, \quad (12\text{–}43)$$
$$|\mathbf{p}| = \beta_- \epsilon_-, \quad |\mathbf{q}| = \beta_+ \epsilon_+, \quad \epsilon_\pm = \gamma_\pm m.$$

Thus, $d\sigma$ in (12–40) can be expressed in terms of $\theta_1, \varphi_1, \theta_2, \varphi_2$, the initial state, and the polarization directions.

The most important special case is the annihilation of positons by negatons at rest ($\mathbf{p} = 0$) when the polarizations are not specified. The conservation law in this case gives

$$\frac{1}{\omega_1} + \frac{1}{\omega_2} = \frac{1}{m}(1 - \cos\Theta). \tag{12-44}$$

The sum over the final polarization states in the rest system gives

$$\sum_{\text{pol}} (e_1 \cdot e_2)^2 = 1 + \cos^2\Theta$$

and a factor of 4 for those terms which are independent of e_1 and e_2. The square brackets of Eq. (12–41) vanish. In this way [or directly from Eq. (12–42)] we find, with $\mathbf{p} = 0$,

$$X = 2\left(\frac{\omega_1}{\omega_2} + \frac{\omega_2}{\omega_1}\right) + 4 - 2(1 + \cos^2\Theta)$$

$$= 2(\gamma_+ + 1)(1 - \cos\Theta) - 2(1 + \cos^2\Theta). \tag{12-45}$$

The differential cross section (12–40) simplifies therefore in the rest system of the negaton (L-system) for summed final polarization states, and becomes

$$d\sigma = \frac{1}{2}r_0^2 \frac{d\Omega_1}{\beta\gamma} \cdot \frac{\omega_1}{\omega_2}\left(1 + \gamma - \frac{1 + \cos^2\Theta}{1 - \cos\Theta}\right), \quad \text{(L-system)}, \tag{12-46}$$

$$\epsilon_+ = \gamma m, \quad |\mathbf{q}| = \beta\epsilon_+.$$

Equations (12–43) can now be written

$$\frac{\omega_1}{m} = \frac{1 + \gamma}{1 + \gamma(1 - \beta\cos\theta_1)}, \quad \frac{\omega_2}{m} = \frac{\gamma(1 + \gamma)(1 - \beta\cos\theta_1)}{1 + \gamma(1 - \beta\cos\theta_1)}. \tag{12-47}$$

They determine ω_1 and ω_2 as functions of β and θ_1. Equation (12–44) gives Θ in terms of these parameters, so that $d\Omega_1 = 2\pi \sin\theta_1 d\theta_1$, in Eq. (12–46), and $d\sigma$ can be expressed in terms of θ_1 only. The incident positon velocity, β, is given. We find the angular distribution of free-pair annihilation into two photons in the negaton rest system:

$$d\sigma = r_0^2\pi \frac{dx}{\beta\gamma^2(1 - \beta x)}\left[\gamma + 3 - \frac{[1 + \gamma(1 - \beta x)]^2}{\gamma(1 + \gamma)(1 - \beta x)} - \frac{2\gamma(1 + \gamma)(1 - \beta x)}{[1 + \gamma(1 - \beta x)]^2}\right],$$

$$x = \cos\theta_1, \quad \text{(L-system)}. \tag{12-48}$$

The frequency spectrum of the annihilation radiation can be obtained from Eq. (12–46) by use of Eq. (12–44). Since

$$d\cos\theta_1 = \frac{\gamma + 1}{\beta\gamma} \cdot \frac{m\,d\omega_1}{\omega_1^2},$$

we find

$$d\sigma = r_0^2\pi \frac{d\omega_1}{m} \frac{1}{(\beta\gamma)^2}\left[\frac{\omega_1}{\omega_2} + \frac{\omega_2}{\omega_1} + 2\frac{m^2}{\omega_1\omega_2}(1+\gamma) - \left(\frac{m^2(1+\gamma)}{\omega_1\omega_2}\right)^2\right],$$

$$\omega_1 + \omega_2 = m(1+\gamma), \quad \text{(L-system)}. \quad (12\text{--}49)$$

The total cross section, which is independent of the coordinate system, can be most easily obtained from the special cases (12–48) or (12–49). An elementary integration yields

$$\sigma = r_0^2\pi \frac{1}{\beta^2\gamma(\gamma+1)}\left[\left(\gamma + 4 + \frac{1}{\gamma}\right)\ln(\gamma + \sqrt{\gamma^2-1}) - \beta(\gamma+3)\right].$$

(12–50)

In this expression a factor $\frac{1}{2}$ was supplied, because in integrating over all directions of one of the photons we obtain each final state exactly twice. The result (12–50) was first obtained by Dirac.*

For small values of the momentum $|\mathbf{q}|$ of the incident positon Eq. (12–50) gives

$$\sigma = \frac{r_0^2\pi}{\beta}, \quad (\beta \ll 1), \quad \text{(NR)}. \quad (12\text{--}51)$$

This expression is seen to approach infinity as β approaches zero. But the number of annihilation processes per unit time remains finite, since the current of the incoming positons, $e\beta$, approaches zero in this limit. Thus, the number of annihilation processes per unit time approaches a constant,

$$P \equiv \frac{1}{\tau_2} = r_0^2\pi\rho, \quad (12\text{--}52)$$

where ρ is the density of negatons in the medium. In a solid the *lifetime* τ_2 is of the order of 10^{-10} sec. The slowing-down time for positons can be estimated† and is of the same order of magnitude as this decay time in flight. The resulting competition between free decay and positronium formation was pointed out previously.

An estimate of the annihilation probability of free positons involves considerations of energy loss and a corresponding integration of σ along the trajectory of the positon. The details of such a calculation can be found in a paper by Bethe.‡

---

* P. A. M. Dirac, *Proc. Camb. Phil. Soc.* **26**, 261 (1930).
† See, e.g., S. de Benedetti, C. E. Cowan, W. R. Konnecker, H. Primakoff, *Phys. Rev.* **77**, 205 (1950), Appendix.
‡ H. A. Bethe, *Proc. Roy. Soc.* **A150**, 129 (1935).

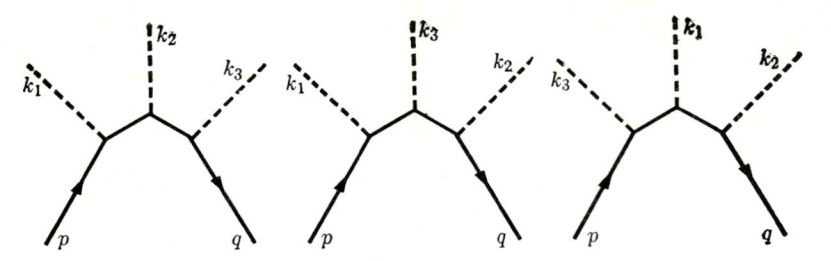

FIG. 12–5. Diagrams for three-quantum pair annihilation.

In the extreme relativistic limit Eq. (12–50) yields

$$\sigma = r_0^2 \pi \ln (2\gamma)/\gamma, \quad \text{(ER)}.$$

The cross section has a maximum and tends to zero for large energies.

The radiative corrections to two-quantum pair annihilations in flight can be obtained directly from the calculations of the radiative corrections to Compton scattering by the methods of Section 8–5. However, they are of little interest within the present accuracy of measurements and we shall not discusss them. A different situation prevails for the *three-quantum annihilation in flight*, although this process is of the same order as the radiative corrections. The reasons will become clear in Section 12–6 when these results are applied to positronium annihilation in which two-quantum annihilation may be forbidden by selection rules.

The diagrams for this process are shown in Fig. 12–5. To these must be added a similar set of diagrams in which k_1 and k_2 are interchanged, yielding a total of six diagrams.

The topological relationship of these diagrams to the double Compton effect is evident. Indeed, a comparison of Figs. 11–1 and 12–5 (with the additional three diagrams mentioned above) shows this relationship. Following Section 8–5, it is expressed by the substitution

$$k \to -k_3, \quad e \to e_3, \quad p' \to -q. \tag{12-53}$$

Therefore, the matrix element $(k_1 k_2 k_3 | M^{(3)} | pq)$ is identical with Eq. (11–30) provided we carry out this substitution and change accordingly $u(p')$ into $v(q)$. The expression for the cross section differs from the one for two-photon annihilation only in the replacement of $(k_1 k_2 | M^{(2)} | pq)$ by $(k_1 k_2 k_3 | M^{(3)} | pq)$. We find, therefore,

$$d\sigma = \frac{\alpha r_0^2}{16\pi^2} \int \frac{\delta(k_1 + k_2 + k_3 - p - q) X d^3 k_1 d^3 k_2 d^3 k_3}{\sqrt{(p \cdot q)^2 - m^4} \, \omega_1 \omega_2 \omega_3}, \tag{12-54}$$

where the trace X includes the sum over the polarizations. It is identical with the trace X for double Compton scattering, Eq. (11–35), except for the substitution (12–53) and a minus sign which arises from the replacement of $\Lambda_-(p')$ by $-\Lambda_+(q)$. Equations (11–33) now read

$$m^2\kappa_i = -p\cdot k_i, \quad m^2\kappa_i' = -q\cdot k_i, \quad (i = 1, 2, 3), \qquad (12\text{–}55)$$

whereas the definitions (11–34) and consequently also the form of (11–35) remain unchanged. It follows that X in Eq. (12–54) is given (except for an over-all minus sign) by Eqs. (11–35), (11–34), and (12–55).

The integration over the δ-function effectively replaces $d^3k_1 d^3k_2 d^3k_3$ by

$$d^3k_1 \omega_2^2 d\Omega_2 \frac{d\omega_2}{d(\omega_1 + \omega_2 + \omega_3)}.$$

The last factor follows easily from the conservation laws, keeping $\mathbf{p}$, $\mathbf{q}$, $\mathbf{k}_1$, and the direction of $\mathbf{k}_2$ fixed:

$$\frac{d(\omega_1 + \omega_2 + \omega_3)}{d\omega_2} = 1 + \frac{d\omega_3}{d\omega_2} = 1 + \frac{1}{2\omega_3}\frac{d}{d\omega_2}(\mathbf{p} + \mathbf{q} - \mathbf{k}_1 - \mathbf{k}_2)^2$$

$$= \frac{m}{\omega_3}\left[\gamma_+(1 - \beta_+\cos\theta_2) + \gamma_-(1 - \beta_-\cos\theta_2') - \frac{\omega_1}{m}(1 - \cos\theta_{12})\right],$$

where $\beta_\pm$, $\gamma_\pm$ have the same meaning as in Eq. (12–43), θ_i and θ_i' are the angles between $\mathbf{k}_i$ and the two incident directions of $\mathbf{q}$ and $\mathbf{p}$, and the θ_{ij} are the angles between the directions of $\mathbf{k}_i$ and $\mathbf{k}_j$.

With these results, the differential cross section for three-photon annihilation of a free pair, Eq. (12–54), becomes

$$d\sigma = \alpha r_0^2 \frac{X\omega_1\omega_2 d\omega_1}{\left\{m\sqrt{(p\cdot q)^2 - m^4}\,[\gamma_+(1 - \beta_+\cos\theta_2) + \gamma_-(1 - \beta_-\cos\theta_2') - (\omega_1/m)(1 - \cos\theta_{12})]\right\}} \cdot \frac{d\Omega_1}{4\pi} \cdot \frac{d\Omega_2}{4\pi}.$$

$$(12\text{–}56)$$

The special case in which the negaton is at rest is of particular interest:

$$\beta_- = 0, \quad \gamma_- = 1, \quad \beta_+ \equiv \beta, \quad \gamma_+ \equiv \gamma.$$

$$\kappa_i = \frac{\omega_i}{m}, \quad \sum_{i=1}^{3}\omega_i = m(1 + \gamma). \qquad (12\text{–}57)$$

The cross section becomes, in this case,

$$d\sigma = \alpha r_0^2 \frac{X\omega_1\omega_2 d\omega_1}{\beta\gamma m^3\left[1 + \gamma(1 - \beta\cos\theta_2) - \frac{\omega_1}{m}(1 - \cos\theta_{12})\right]} \cdot \frac{d\Omega_1}{4\pi} \cdot \frac{d\Omega_2}{4\pi}.$$

$$(12\text{–}58)$$

The expression X is somewhat simplified by the relations (12–57).

Considerable simplifications arise in the nonrelativistic limit where $\gamma = 1$. Equations (12–57) are augmented by

$$\kappa_i' = \kappa_i = \omega_i/m,$$

$$\sum_{i=1}^{3} \mathbf{k}_i = 0, \quad \sum_{i=1}^{3} \omega_i = 2m, \qquad (12\text{–}59)$$

and various relations which arise from these. The trace X simplifies to

$$X = \left(\frac{4m}{\omega_1\omega_2\omega_3}\right)^2 \cdot \frac{1}{2}\left[\left(\sum_{i=1}^{3} \omega_i^2\right)^2 - 2m^2 \sum_{i=1}^{3} \omega_i^2 + 4m\omega_1\omega_2\omega_3\right].$$

The square brackets in this expression can be written in a more symmetrical way. Evaluating $\left(\sum_{i=1}^{3} \omega_i^2\right)^2$ and $\left(\sum_{i=1}^{3} \omega_i\right)^2$, and using Eqs. (12–59), we find

$$\frac{1}{2}\left(\sum_{i=1}^{3} \omega_i^2\right)^2 = \sum_{i=1}^{3} \omega_i^4 - 4m^2 \sum_{i=1}^{3} \omega_i^2 + 8m(m^3 - \omega_1\omega_2\omega_3). \qquad (12\text{–}60)$$

We further notice that the three ω_i are the roots of the cubic equation

$$\omega^3 - 2m\omega^2 + \sum_{i<j} \omega_i\omega_j\omega - \omega_1\omega_2\omega_3 = 0.$$

When we insert ω_1, ω_2, and ω_3, respectively, for ω, we obtain three equations. The sum of these equations can be simplified by use of

$$4m^2 = \left(\sum_{i=1}^{3} \omega_i\right)^2 = \sum_{i=1}^{3} \omega_i^2 + 2\sum_{i<j} \omega_i\omega_j$$

and reduces to

$$\sum_{i=1}^{3} \omega_i^3 - 3\omega_1\omega_2\omega_3 - 3m\sum_{i=1}^{3} \omega_i^2 + 4m^3 = 0. \qquad (12\text{–}61)$$

With the help of Eqs. (12–60) and (12–61), X can be written in the symmetrical forms

$$X = \left(\frac{4m}{\omega_1\omega_2\omega_3}\right)^2 \left(m^2 \sum_{i=1}^{3} \omega_i^2 - 2m \sum_{i=1}^{3} \omega_i^3 + \sum_{i=1}^{3} \omega_i^4\right)$$

$$= \left(\frac{4m}{\omega_1\omega_2\omega_3}\right)^2 \sum_{i=1}^{3} \omega_i^2 (m - \omega_i)^2$$

$$= 4[(1 - \cos\theta_{12})^2 + (1 - \cos\theta_{23})^2 + (1 - \cos\theta_{31})^2], \qquad (12\text{–}62)$$

where we used

$$1 - \cos\theta_{ij} = 2\frac{\omega_k(1 - \omega_k)}{\omega_i\omega_j}, \quad (i, j, k = \text{cyclic perm. of } 1, 2, 3). \qquad (12\text{–}63)$$

The nonrelativistic limit of the three-photon annihilation cross section is therefore

$$d\sigma = 4\frac{\alpha r_0^2}{\beta} \cdot \frac{(1 - \cos\theta_{12})^2 + (1 - \cos\theta_{23})^2 + (1 - \cos\theta_{31})^2}{2m - \omega_1(1 - \cos\theta_{12})}$$

$$\times \frac{\omega_1 \omega_2 d\omega_1}{m^2} \cdot \frac{d\Omega_1}{4\pi} \cdot \frac{d\Omega_2}{4\pi}, \quad \text{(NR)}. \quad (12\text{--}64)$$

The frequency spectrum of the annihilation radiation in this limit is obtained from Eqs. (12–63) and the relation

$$\frac{d\Omega_2}{4\pi} = \frac{m(m - \omega_1)}{\omega_1 \omega_2} \cdot \frac{d\omega_2}{\omega_2},$$

and it is found to be

$$d\sigma = \frac{8\alpha r_0^2}{\beta} \cdot \left[\left(\frac{m - \omega_1}{\omega_2 \omega_3}\right)^2 + \left(\frac{m - \omega_2}{\omega_3 \omega_1}\right)^2 + \left(\frac{m - \omega_3}{\omega_1 \omega_2}\right)^2\right] d\omega_1 d\omega_2, \quad \text{(NR)}.$$

(12–65)

This relation, which arises here as a special case of Eq. (12–56), was first found by Ore and Powell.*

In the evaluation of the total cross section we must use Eqs. (12–59) to determine the limits of integration. We must also divide the result by 3!, since all six permutations of the three photons give the same final state:

$$\sigma = \frac{4}{3} \frac{\alpha r_0^2}{\beta} \int_0^m d\omega_1 \int_{m-\omega_1}^m d\omega_2 \left[\left(\frac{m - \omega_1}{\omega_2 \omega_3}\right)^2 + \left(\frac{m - \omega_2}{\omega_3 \omega_1}\right)^2 + \left(\frac{m - \omega_3}{\omega_1 \omega_2}\right)^2\right]$$

$$= \frac{4}{3} \cdot \frac{\alpha r_0^2}{\beta} \cdot (\pi^2 - 9), \quad \text{(NR)}. \quad (12\text{--}66)$$

When the density of negatons in the medium is ρ, the annihilation probability per unit time, which is the reciprocal *lifetime of a free positon under three-photon annihilation*, is

$$P = \frac{1}{\tau_3} = \beta\sigma\rho = \frac{4}{3}(\pi^2 - 9)\alpha r_0^2 \rho, \quad \text{(NR)}. \quad (12\text{--}67)$$

This lifetime is about 300 times larger than the nonrelativistic two-photon annihilation lifetime, Eq. (12–52), and is therefore too small to be observ-

---

* A. Ore and J. L. Powell, *Phys. Rev.* **75**, 1696 (1949). Previous work by E. M. Lifshitz, *Dokl. Akad. Nauk.* **60**, 211 (1948), and by D. Ivanenko and A. Sokolov, *Dokl. Akad. Nauk.* **61**, 51 (1948), is incorrect.

able; it could only be considered a correction to τ_2, of the same order as the radiative corrections to τ_2.

We remark that in the nonrelativistic limit, Eq. (12–65), the infrared divergence does not arise. This can be directly verified by taking the same limit in Eq. (11–38) and carrying out the substitution (12–53). This feature is actually to be expected because of the close relationship of τ_3 (Eq. 12–67) and the three-photon annihilation of positronium (cf. Section 12–6). The latter never occurs in conjunction with two-photon annihilation because of selection rules, such that there can be no cancellation of infrared divergences between the two processes.*

12–5 Positronium; selection rules. The bound system consisting of one negaton and one positon is called *positronium*.† Although a satisfactory theoretical understanding of this system cannot be obtained with the S-matrix in the iteration solution as it is discussed in the present chapter, it is convenient to study the positronium annihilation at this point, following the discussion of free-pair annihilation of the previous section. A complete and satisfactory discussion of the positronium atom cannot be given without a study of the relativistic two-body problem. As mentioned in the introduction, this problem could not be included here.

Here we need only the fact that the nonrelativistic limit of the theoretical description of positronium leads to a Schroedinger equation which differs from that for the hydrogen atom only in the reduced mass. The latter is half the electron mass, so that the Bohr radius of positronium is about twice that of hydrogen, whereas the binding energy in the ground state is

$$E \sim \frac{1}{2}\alpha^2 \frac{m}{2} \sim 6.8 \text{ ev}, \qquad (12\text{–}68)$$

which is about half that of hydrogen. Corresponding factors occur for the excited states.

It follows that, nonrelativistically, the orbital, spin, and total angular momentum operators $L^2, L_z, S^2, S_z, J^2, J_z$ are constants of the motion and commute with the energy-momentum operator of the system, P_μ. The usual spectroscopic notation is therefore applicable to these energy levels.

Relativistically, spin and orbital angular momenta will be coupled, so that L^2, L_z, and S_z are no longer constants of the motion, but S^2 still commutes with J^2 and J_z. To complete the set of commuting observables, it is convenient to choose the space inversion operator Π and the charge conjugation operator Γ. Their eigenvalues, Π' and Γ', are called the space parity and the charge parity.

---

* F. Rohrlich, *Phys. Rev.* **98**, 181 (1955).

† This name was suggested by A. E. Ruark, *Phys. Rev.* **68**, 278 (1945), who also pointed out some of the characteristic properties of this system.

These operators commute with P_μ (see Chapter 5), and, since positronium is a charge-symmetrical system, the positronium state vector ω can be chosen to be their simultaneous eigenstate. The operators P_μ, J^2, J_z, Π, and Γ constitute a complete set of observables.

Since our system consists of an even number of particles of spin $\frac{1}{2}$ (this includes also the possible occurence of virtual pairs) the operators Π and Γ commute:

$$\Sigma = \Pi\Gamma = \Gamma\Pi. \tag{12-69}$$

We shall show below that the operator Σ thus defined interchanges the spins of the two particles in positronium. In fact, as will also be shown, it is more convenient to work with Σ than with S^2. Since $\Gamma^2 = 1$ and $\Pi^2 = 1$, we also have $\Sigma^2 = 1$. The two eiegenvalues $+1$ and -1 correspond to the symmetrical (triplet) and the antisymmetrical (singlet) spin system.

The space-inversion operator also has the eigenvalues ± 1. We shall prove that they are related to the orbital angular momentum by

$$\Pi\omega = \Pi'\omega, \quad \Pi' = \epsilon(-1)^L, \tag{12-70}$$

where ϵ is the parity of the $L = 0$ state. For positronium $\epsilon = -1$; that is, this system transforms like a pseudoscalar in both the ^1S and ^3S states. The space parity is therefore $-(-1)^L$.

Since an interchange of spin and an inversion of space through the center of mass of the positronium system is evidently equivalent to an interchange of the charge of the two particles, it follows that

$$\Gamma' = \Sigma'\Pi' = \mp(-1)^L, \tag{12-71}$$

as can be concluded from Eq. (12-69). The upper and lower signs refer to the triplet and singlet systems, respectively. It follows in particular that the charge parity of the ground state (^1S) is $\Gamma' = +1$.

With these relations, we can classify the energy levels of positronium as is shown in Tables 12-1 and 12-2. We see from Table 12-2 that there are

TABLE 12-1

Quantum numbers for the singlet states of positronium

Term	J	Σ	Π	Γ
^1S	0	-1	-1	$+1$
^1P	1	-1	$+1$	-1
^1D	2	-1	-1	$+1$
^1F	3	-1	$+1$	-1

TABLE 12–2

Quantum numbers for the triplet states of positronium

Term	J	Σ	Π	Γ
${}^3P_0$	0	+1	+1	+1
${}^3S_1 + {}^3D_1\,(2)$	1	+1	−1	−1
${}^3P_1$	1	+1	+1	+1
${}^3P_2 + {}^3F_2\,(2)$	2	+1	+1	+1
${}^3D_2$	2	+1	−1	−1

two states for some sets of eigenvalues of J^2, Π, and Γ, when $\Sigma = +1$. Both are mixtures of states with orbital angular momentum $L = J - 1$ and $L = J + 1$, since these have the same space parity.

In order to prove the relations (12–70) and (12–71) and to obtain a more detailed insight into the symmetry properties of the positronium state vector, we proceed as follows. We expand ω in terms of the eigenstates ω_{2n} which designate the presence of exactly n negaton-positon pairs. The states with $n > 1$ correspond to the radiative corrections. Since all vectors ω_{2n} $(n = 1, 2, \cdots)$ are mutually orthogonal, it follows that all the symmetry properties which are valid for ω must also be valid for any ω_{2n}. In particular, we shall investigate ω_2. Thus, we write

$$\omega = \sum_{r,s} \iint d^3p\, d^3q\, \varphi(\mathbf{p},\mathbf{q};r,s) a_r^*(\mathbf{p}) b_s^*(\mathbf{q}) \omega_0 + \cdots \qquad (12\text{–}72)$$

and drop the higher order terms which are indicated by dots. The vacuum state vector ω_0 is defined by

$$a_r(\mathbf{p})\omega_0 = b_s(\mathbf{q})\omega_0 = 0, \quad \text{for all } r, s, \mathbf{p}, \mathbf{q}.$$

The function $\varphi(\mathbf{p},\mathbf{q};r,s)$ may be looked upon as the Schroedinger wave function in the momentum representation. Restricting ourselves to the center-of-momentum system ($\mathbf{P} = 0$), we can write

$$\omega = \sum_{r,s} \int d^3p\, \varphi(\mathbf{p},rs) a_r^*(\mathbf{p}) b_s^*(-\mathbf{p}) \omega_0, \qquad (12\text{–}73)$$

where we defined $\varphi(\mathbf{p},rs) \equiv \varphi(\mathbf{p},-\mathbf{p};r,s)$.

The effect of the charge conjugation operator on (12–73) follows from the defining property (5–58) and the relations (A2–76).

$$\begin{aligned}\Gamma a_r^*(\mathbf{p}) \Gamma^{-1} &= b_r^*(\mathbf{p}), \\ \Gamma b_s^*(\mathbf{q}) \Gamma^{-1} &= a_s^*(\mathbf{q}).\end{aligned} \qquad (12\text{–}74)$$

Since we can choose the vacuum state to satisfy†

$$\Gamma \omega_0 = \omega_0, \tag{12-75}$$

the charge conjugation Γ transforms the wave function φ into a new wave function $\Gamma\varphi$ which is defined by

$$\Gamma\omega = \sum_{r,s} \int d^3p\, \Gamma\varphi(\mathbf{p},rs) a_r^*(\mathbf{p}) b_s^*(-\mathbf{p}) \omega_0. \tag{12-76}$$

The new wave function $\Gamma\varphi$ can be expressed in terms of the original φ if we use the relations (12–74) and (12–75). We obtain

$$\Gamma\omega = \sum_{r,s} \int d^3p\, \varphi(\mathbf{p},rs) b_r^*(\mathbf{p}) a_s^*(-\mathbf{p}) \omega_0$$

$$= -\sum_{r,s} \int d^3p\, \varphi(-\mathbf{p},sr) a_r^*(\mathbf{p}) b_s^*(-\mathbf{p}) \omega_0. \tag{12-77}$$

From this result and Eq. (12–76) it follows that

$$\Gamma\varphi(\mathbf{p},rs) = -\varphi(-\mathbf{p},sr). \tag{12-78}$$

The operation of space inversion is evaluated in a similar manner. From the defining property of the space inversion transformation (Section 5–3), we obtain

$$\Pi a_r^*(\mathbf{p}) \Pi^{-1} = a_{-r}^*(-\mathbf{p}),$$
$$\Pi b_s^*(\mathbf{p}) \Pi^{-1} = -b_{-s}^*(-\mathbf{p}). \tag{12-79}$$

This shows that the space inversion reverses the direction of motion but leaves the spin direction unchanged, a result well known from nonrelativistic kinematics.

The opposite sign of the right sides of equations (12–79) is very important for the correct assignment of space parity to the positronium states. It stems from the property of the spinors u and v to transform with opposite parity under space inversion. This results in odd parity of S-states, even parity of P-states, and so on. The positronium, when compared with an elementary particle, can therefore be said to behave as if it were a pseudoscalar particle. Conversely, a pseudoscalar particle may be thought of as a composite system consisting of a particle and its antiparticle.

---

† Equation (12–75) states that the charge parity of the vacuum is even. This is consistent with the definition of the vacuum state, but it is an arbitrary convention. We could just as well assume a vacuum state of odd charge parity. Then the charge parity of all other states would be reversed. There is no observable difference between the two assumptions. The situation is exactly the same as in the case of the space parity, where this ambiguity is well known.

Since this property of the positronium states leads to observable effects and is therefore of great importance, and since it is contrary to the nonrelativistic notion of parity, a more detailed explanation of the origin of this property may be desirable. That the plane wave amplitudes u and v transform with opposite parity follows from general properties of the Lorentz group and its representation in the four-dimensional spinor space. If $L(\mathbf{v})$ is a proper Lorentz transformation corresponding to the velocity $\mathbf{v}$, and σ is the space reflection, we have

$$L(\mathbf{v})\sigma = \sigma L(-\mathbf{v}). \tag{12-80}$$

If, now, all the spinors u and v were of the same parity, then the representation $S(\sigma)$ of the space reflection in the four-dimensional spinor space would be the unit matrix I, and consequently would commute with all the other matrices. The relation (12–80) could therefore not hold for the representation of these transformations. The diagonal form of the 4×4 matrix $S(\sigma)$, which can contain only the elements $+1$ and -1 ($S^2 = 1$), cannot therefore be ± 1. Furthermore, the space inversions commute with the three-dimensional rotations associated with the two two-dimensional subspaces of the four-dimensional spinor space. Therefore, the diagonal form of $S(\sigma)$ must be of the form

$$S(\sigma) = \pm \begin{pmatrix} 1 & 0 & 0 & 0 \\ 0 & 1 & 0 & 0 \\ 0 & 0 & -1 & 0 \\ 0 & 0 & 0 & -1 \end{pmatrix},$$

of which we can choose the upper sign without loss of generality.

A more direct way to verify this assertion is to write down the parity transformation $S = -i\gamma_0$ in a special representation, for instance in the standard representation given in (A2-14). Writing similarly an explicit solution of the spinors u and v, we verify without difficulty that

$$Su_r(\mathbf{p}) = u_{-r}(-\mathbf{p}),$$
$$Sv_s(\mathbf{p}) = -v_{-s}(-\mathbf{p}), \tag{12-81}$$

which leads to (12–79).

In this case we have chosen the parity operator $-i\gamma_0$. It was pointed out before [*cf.* Eq. (5–25)] that a numerical factor ± 1 or $\pm i$ is arbitrary for the spinor matrix S which corresponds to the space inversion. The equations (12–81) correspond to a particular choice of this factor which makes the intrinsic parity even for negatons and odd for positons. As mentioned above, this ambiguity in the definition of S does not affect the conclusions concerning the parity of the positronium states.

If we combine (12–79) with the assumption that the space parity of the vacuum is even, we obtain, by the same procedure that led to (12–78), for the effect of the space-parity operator on the wave-function,

$$\Pi\varphi(\mathbf{p},rs) = -\varphi(-\mathbf{p},-r-s). \tag{12-82}$$

12-5] POSITRONIUM; SELECTION RULES

The combination of the two operations Γ and Π leads to a third operation $\Sigma = \Gamma\Pi = \Pi\Gamma$. It transforms the wave function according to

$$\Sigma\varphi(\mathbf{p},rs) = \varphi(\mathbf{p},-s-r). \tag{12-83}$$

This transformation has a simple interpretation. If we recall the definition of the spin orientation labeled by the indices r and s, we find that the transformation (12–83) of the wave function interchanges the spins of the negaton and the positon. The eigenvalues Σ' of the operator Σ are ± 1. For a given functional dependence on p there exist three linearly independent eigenvectors belonging to the eigenvalue $\Sigma' = +1$. These are therefore the triplet states. Similarly, the states with $\Sigma' = -1$ are the singlet states.

Equations (12–70) and (12–71) now follow easily from the results (12–78), (12–82), and (12–83).

In order to investigate the selection rules for transitions of positronium into a final state involving photons, we must recall that the charge conjugation operator changes the sign of the photon field [*cf.* Eq. (5–58)]:

$$\Gamma a_\mu(\mathbf{k})\Gamma^{-1} = -a_\mu(\mathbf{k}). \tag{12-84}$$

A state vector ω_n for a state of n photons will therefore satisfy

$$\Gamma\omega_n = (-1)^n \omega_n. \tag{12-85}$$

A radiative transition of positronium from a state with quantum numbers Σ', Π' to a state with Σ'', Π'', where the primed quantities indicate the spin and space parities, and which involves only a single photon, must satisfy

$$\Sigma'\Pi' = -\Sigma''\Pi'';$$

that is

$$\Sigma' = \Sigma'', \quad (L' - L'' \text{ odd})$$

or

$$\Sigma' = -\Sigma'', \quad (L' - L'' \text{ even}), \tag{12-86}$$

corresponding to transitions without and with change of spin. We note that these selection rules differ from those for the hydrogen atom, although the common electric dipole transitions are the same, i.e., $\Delta\Sigma = 0$, $\Delta L = \pm 1$.

For positronium annihilation into n photons, we find, from Eqs. (12–71) and (12–85),

$$\Gamma' = \Sigma'\Pi' = (-1)^n$$

or

$$\mp(-1)^L = (-1)^n \quad \text{for } \begin{pmatrix}\text{triplet states}\\ \text{singlet states}\end{pmatrix}. \tag{12-87}$$

Therefore, the positronium states of *even* charge parity, i.e., the singlet states of even L and the triplet states of odd L, decay only into an *even* number of photons, of which the two-photon decay is, of course, of primary importance. On the other hand, positronium annihilation into an *odd* number of photons (of which three is obviously the minimum) can take place only from an *odd* parity state, i.e., a singlet state of odd L or a triplet state of even L.*

Because of the fundamental character of the two-photon annihilation of positronium, we shall now consider the symmetry properties of this reaction in further detail. Consider the center of momentum system in which positronium is at rest and annihilation takes place into two photons of equal energy and opposite direction with momenta $\mathbf{k}$ and $-\mathbf{k}$.

The order of symmetry of the final states with one of the photons travelling in a specified direction $\mathbf{k}$ is less than the order of symmetry of the original positronium states at rest. Instead of the full-rotation group, there are only the rotations around the axis $\mathbf{k}$ and a rotation by the angle π around an axis normal to $\mathbf{k}$. We denote the latter operation by R. The Hermitian operator for the infinitesimal rotation around $\mathbf{k}$ is the angular momentum operator in the direction $\mathbf{k}$ and is denoted by J_3.

In order to describe the state vectors representing the final states of the system, we introduce the operators $a_+(\mathbf{k})$ and $a_-(\mathbf{k})$ for the circularly polarized photons in the direction $\mathbf{k}$. They are related to the operators for the transverse components $a_1(\mathbf{k})$ and $a_2(\mathbf{k})$ which were previously introduced in Section 2–8, especially Eq. (2–83):

$$a_\pm(\mathbf{k}) = \frac{1}{\sqrt{2}}(a_1(\mathbf{k}) \pm ia_2(\mathbf{k})),$$

and consequently,

$$a_1(\mathbf{k}) = \frac{1}{\sqrt{2}}(a_+(\mathbf{k}) + a_-(\mathbf{k})), \quad a_2(\mathbf{k}) = \frac{1}{i\sqrt{2}}(a_+(\mathbf{k}) - a_-(\mathbf{k})). \quad (12\text{–}88)$$

The expression (2–83) for the spin operator N_{12} in the direction of propagation becomes

$$N_{12}(\mathbf{k}) = a_+^*(\mathbf{k})a_+(\mathbf{k}) - a_-^*(\mathbf{k})a_-(\mathbf{k}), \quad (12\text{–}89)$$

which shows that $a_+^* a_+$ represents the operator describing the number of photons with their spins pointing in the direction $\mathbf{k}$, while $a_-^* a_-$ describes the number of photons with the spins pointing in the direction $-\mathbf{k}$.

---

* Special cases of these selection rules were noted by J. Pirenne, *Thèse*, Univ. of Paris (1944), esp. p. 124 ff., and J. A. Wheeler, *Ann. N. Y. Acad. Sci.* **48**, 219 (1946). The selection rules derived by L. Landau, *Dokl. Akad. Nauk.* **60**, 207 (1948), are incorrect. Space parity was discussed by C. N. Yang, *Phys. Rev.* **77**, 242 (1950), and charge parity by L. Wolfenstein and D. G. Ravenhall, *Phys. Rev.* **88**, 279 (1953). Our results agree with those reported by L. Michel, *Nuovo Cimento* **10**, 319 (1953).

The general state vector for two photons with specified and opposite propagation direction is given by

$$\omega = \sum_{r,s} c_{rs} a_r^*(\mathbf{k}) a_s^*(-\mathbf{k}) \omega_0, \tag{12-90}$$

where ω_0 is the state vector for the photon vacuum.

The four coefficients c_{rs} $(r,s = \pm)$ describes components of the state of the two-photon system under consideration. The four operators $M_{12} = J_3$, R, Π, and Γ induce certain linear transformations of the components c_{rs}, which we shall now determine.

We begin with the angular momentum J_3 in the direction $\mathbf{k}$. From (1–56) and (2–39), we find for the commutator of J_3 with a_1 and a_2,

$$i[J_3, a_\mu] = a_2 g_{\mu 1} - a_1 g_{\mu 2}, \quad (\mu = 1, 2),$$

which leads to†

$$[J_3, a_\pm^*] = \pm a_\pm^*.$$

Consequently,

$$J_3 \omega = \sum_{r,s} (r - s) c_{rs} a_r^*(\mathbf{k}) a_s^*(-\mathbf{k}) \omega_0. \tag{12-91}$$

The effect of the operator J_3 on c_{rs} is to replace c_{rs} by $J_3 c_{rs}$, defined by

$$J_3 c_{rs} = (r - s) c_{rs}, \quad (r,s = \pm). \tag{12-92}$$

The eigenvectors of J_3 are therefore the vectors of the form

$$\omega_{rs} = a_r^*(\mathbf{k}) a_s^*(-\mathbf{k}) \omega_0 \tag{12-93}$$

and the eigenvalues are $r - s$.

The operator R for the rotation around the axis 1 perpendicular to $\mathbf{k}$ by an angle π has the property‡

$$\begin{aligned} R a_1(\mathbf{k}) R^{-1} &= a_1(-\mathbf{k}), \\ R a_2(\mathbf{k}) R^{-1} &= a_2(-\mathbf{k}). \end{aligned} \tag{12-94}$$

Expressed in terms of the operators $a_\pm$, these relations become

$$R a_\pm(\mathbf{k}) R^{-1} = a_\pm(-\mathbf{k}).$$

Consequently,

$$R\omega = \sum_{r,s} c_{sr} a_r^*(\mathbf{k}) a_s^*(-\mathbf{k}) \omega_0. \tag{12-95}$$

---

† Note that these equations also hold when J_3 is replaced by N_{12}, since the component parallel to $\mathbf{k}$ of the orbital angular momentum commutes with a_μ.

‡ The sign of the component a_2 is not reversed in the second of equations (12–94) because the components do not refer to a fixed coordinate system in space, but to a coordinate system which depends on $\mathbf{k}$ in such a manner that the two directions labeled by 1 and 2, and the direction of $\mathbf{k}$, form a right-handed system.

The operation R, when expressed in terms of the components c_{rs}, consists therefore in the substitution

$$Rc_{rs} = c_{sr}. \tag{12-96}$$

It follows that ω_{++} and ω_{--} are eigenvectors of R with eigenvalues $+1$, while $\omega_{+-} \pm \omega_{-+}$ are eigenvectors with eigenvalues ± 1.

In a similar manner, we find for the parity and charge conjugation operations [cf. Eqs. (5–17), (5–18), (5–27), and (5–58), (5–62), respectively, for the definitions of these transformations]

$$\Pi c_{rs} = c_{-s-r}, \tag{12-97}$$

$$\Gamma c_{rs} = c_{rs}. \tag{12-98}$$

The last relation follows immediately from Eq. (12–84) and is equivalent to Eq. (12–85) for $n = 2$.

We summarize the result of this discussion in Table (12–3), where we list the eigenvalues of the operators J_3, R, Π, and Γ. The eigenvalues of R for the eigenvectors with $J_3 = \pm 2$ are not listed, because R is not diagonal in these states (R and J_3 do not commute).

TABLE 12–3

The eigenvalues of the symmetry operators J_3, R, Π, and Γ for the two-photon system

J_3	R	Π	Γ	State vector
0	+1	+1	+1	$(\omega_{++} + \omega_{--})$
0	+1	−1	+1	$(\omega_{++} - \omega_{--})$
+2	—	+1	+1	ω_{+-}
−2	—	+1	+1	ω_{-+}

When we compare this Table with Table 12–1, we conclude that the ground state of positronium, which is a $^1S_0$ state, decays into the state $(\omega_{++} - \omega_{--})$.† The polarization directions of the two emerging photons are therefore perpendicular to each other, as can be seen by returning to the linear polarizations using Eq. (12–88) [cf. also Eqs. (12–102) and (12–103)].

12–6 Positronium annihilation. In the approximation (12–72) positronium is regarded as a negaton and a positon with a suitable momentum distribution $\varphi(\mathbf{p},rs)$ which depends on the relative spin orientation. The

---

† C. N. Yang, *Phys. Rev.* **77**, 242 (1950).

function $\varphi(\mathbf{p})$ may therefore be considered as the wave function of the system in momentum space. The transition probability amplitude for positronium annihilation is then simply the free-pair annihilation amplitude integrated over this momentum distribution:

$$M = \sum_{r,s} \int M_{rs}(\mathbf{p},\mathbf{q})\delta(\mathbf{p}+\mathbf{q})\varphi(\mathbf{pq},rs)d^3p\,d^3q.$$

The matrix element $M_{rs}(\mathbf{p},\mathbf{q})$ is in lowest order given by the free-pair annihilation expression (12–30) for two-photon annihilation, and by Eq. (11–30) with the substitution (12–53) for three-photon annihilation.

Since the characteristic momenta in the bound system are of order αm, it is a good approximation to expand $M_{rs}(\mathbf{p},-\mathbf{p})$ in powers of $\mathbf{p}/m$ and to keep only the leading term. For S-states we can therefore write

$$M = \sum_{rs} M_{rs}(0,0) \int \varphi(\mathbf{p},rs)d^3p = \sum_{rs} M_{rs}(0,0)(2\pi)^{3/2} f_{rs}(0). \quad (12\text{–}99)$$

Here, $f_{rs}(0)$ is the value of $f_{rs}(\mathbf{r})$, the three-dimensional Fourier transform of $\varphi(\mathbf{p},rs)$, at the origin.

Let us consider the ^1S states first. Since these can annihilate into an even number of photons only, we must use Eq. (12–30) for two-photon annihilation. Each additional pair of photons in the final state will reduce the transition probability by one order of α^2. The corresponding corrections to the lifetime are much too small to be observable. Equations (12–30), after an easy calculation, yields*

$$M_{rs}(0,0) = -\frac{ie^2}{(4\pi)^2} \cdot \frac{2}{m^3} \bar{v}_s(-0)(\mathbf{e}_1 \cdot \mathbf{e}_2 m - \mathbf{e}_1 \times \mathbf{e}_2 \cdot \boldsymbol{\sigma}\mathbf{k} \cdot \boldsymbol{\gamma})u_r(0). \quad (12\text{–}100)$$

The positronium wave function for singlet states satisfies, according to Eq. (12–83), the relations

$$\varphi(\mathbf{p},++) = -\varphi(\mathbf{p},--) \equiv \varphi(\mathbf{p}), \quad \varphi(\mathbf{p},+-) = \varphi(\mathbf{p},-+) = 0, \quad (12\text{–}101)$$

which leads, of course, to the singlet wave function $[\varphi(++) - \varphi(--)]$. Therefore, Eq. (12–99) becomes

$$M = -\frac{ie^2}{4\sqrt{\pi}} \cdot \frac{f(0)}{m^3} \cdot \sum_r r\bar{v}_r(-0)(\mathbf{e}_1 \cdot \mathbf{e}_2 m - \mathbf{e}_1 \times \mathbf{e}_2 \cdot \boldsymbol{\sigma}\mathbf{k} \cdot \boldsymbol{\gamma})u_r(0).$$

---

* The notation $v(-0)$ indicates the plane wave spinor amplitude for the positon with momentum $\mathbf{q} = -\mathbf{p}$ in the limit $-\mathbf{p} \to 0$. The direction from which zero is approached must be indicated, since $v_s(-0) = v_{-s}(+0)$. This follows from Eqs. (A2–50) and (A2–55) and a suitable choice of phase factors.

In the Appendix [Eqs. (A2–64) and (A2–66)] it is shown that

$$rv_r(-0) = rv_{-r}(0) = \gamma_5 u_r(0),$$

so that the matrix element can be written

$$M = -\frac{ie^2}{4\sqrt{\pi}} \cdot \frac{f(0)}{m^3} \cdot \sum_r \bar{u}_r(0)\, \gamma_5(\mathbf{e}_1 \cdot \mathbf{e}_2 m - \mathbf{e}_1 \times \mathbf{e}_2 \cdot \boldsymbol{\sigma}\mathbf{k} \cdot \boldsymbol{\gamma}) u_r(0)$$

$$= -i\alpha\sqrt{\pi}\frac{f(0)}{m^3}\, \mathrm{Tr}[\gamma_5(\mathbf{e}_1 \cdot \mathbf{e}_2 m - \mathbf{e}_1 \times \mathbf{e}_2 \cdot \boldsymbol{\sigma}\mathbf{k} \cdot \boldsymbol{\gamma})\Lambda_-(0)].$$

The first term vanishes, since both γ_5 and $\gamma_5\gamma_0$ have vanishing traces. The second term involves the trace

$$\mathrm{Tr}[\gamma_5\sigma_k\gamma_l(1+i\gamma_0)] = i\,\mathrm{Tr}[\gamma_0\gamma_5\sigma_k\gamma_l] = 4g_{kl}.$$

The resultant value for M is therefore

$$M = \frac{2i\alpha\sqrt{\pi}}{m^3} \mathbf{e}_1 \times \mathbf{e}_2 \cdot \mathbf{k}\, f(0). \tag{12-102}$$

Since the two emerging photons have opposite directions, the vector product of the polarizations gives

$$\mathbf{e}_+(\mathbf{k}) \times \mathbf{e}_-(-\mathbf{k}) = \mathbf{e}_-(\mathbf{k}) \times \mathbf{e}_+(-\mathbf{k}) = 0,$$

$$\mathbf{e}_+(\mathbf{k}) \times \mathbf{e}_+(-\mathbf{k}) = -\mathbf{e}_-(\mathbf{k}) \times \mathbf{e}_-(-\mathbf{k}) = -i\frac{\mathbf{k}}{|\mathbf{k}|}, \tag{12-103}$$

where the subscripts $\pm$ refer to right and left circular polarization. Therefore, the only possible final state of the two photons is

$$\frac{1}{\sqrt{2}}(\omega_{++} - \omega_{--}), \tag{12-104}$$

in agreement with the conclusions at the end of the previous section. The matrix element (12–102) must be summed over the final polarization states according to Eq. (12–104). Using $\mathbf{k}^2 = m^2$, this gives

$$(f|M|i) = \sum_{\mathrm{pol}} M = \frac{\sqrt{8\pi}}{m^2} \alpha f(0).$$

In the nonrelativistic approximation the n^1S states have a wave function $f(\mathbf{r})$ such that

$$f(0) = \frac{1}{\sqrt{\pi a^3 n^3}} = \frac{(\alpha m)^{3/2}}{\sqrt{8\pi n^3}}, \tag{12-105}$$

since the ground state has radius

$$a = 2a_0 = \frac{2}{\alpha m}.$$

The matrix element is therefore

$$(f|M|i) = \sqrt{\frac{\alpha^5}{mn^3}}. \qquad (12\text{--}106)$$

The annihilation probability per unit time follows from the general expression (8–40),

$$\Gamma = \frac{1}{2\pi} \int |(f|M|i)|^2 \delta(k_1 + k_2 - p - q) d^3k_1 d^3k_2$$

$$= \frac{1}{2\pi} |(f|M|i)|^2 \omega^2 \frac{d\omega}{dE_f} d\Omega$$

$$= |(f|M|i)|^2 m^2.$$

The last result is obtained for the center-of-momentum system when the motion of the electrons forming the positronium atom is neglected, $E_f = 2\omega = 2m$. The result thus obtained must still be divided by 2, since the equivalence of the two photons was ignored in the integrations, so that all final states were counted twice. Combined with Eq. (12–106), this gives the lifetime of an n^1S state:

$$\tau = \frac{1}{\Gamma} = \frac{2n^3}{\alpha^5 m}. \qquad (12\text{--}107)$$

This annihilation can occur only with the emission of two photons in the state (12–104).

The result (12–107) can be obtained much more simply from the nonrelativistic limit of the free-pair annihilation into two photons. Indeed, Eq. (12–52) yields, with

$$\rho = |f(0)|^2$$

and use of Eq. (12–105),

$$P = r_0^2 \pi \frac{(\alpha m)^3}{8\pi n^3}.$$

This expression must still be multiplied by a factor of four, because the free-pair annihilation probability was averaged over the four possible relative spin directions of the incident particles, whereas positronium annihilates from one definite state. (The three triplet states do not contribute, because of the selection rules.) The result is therefore

$$P = \frac{\alpha^5 m}{2n^3},$$

in agreement with Eq. (12–107). However, this derivation seems to us less satisfactory than the previous one, which also gives much more insight into the process.*

The lifetime for two-photon annihilation of the 1^1S_0 state, which is the ground state of positronium, becomes

$$\tau_2 = \frac{2}{\alpha^5 m} = 1.25 \times 10^{-10} \text{ sec}. \tag{12–108}$$

The methods used to discuss the two-photon annihilation can also be used for the three-photon case, but the detailed analysis would be considerably more complicated. We shall be content, therefore, with the short derivation just outlined.

The nonrelativistic limit of the free-pair annihilation into three photons is given in Eq. (12–67). The negaton density ρ at the position of the positon is again given by $|f(0)|^2$ for S-states. We must take account, however, of the selection rules which permit three-photon decay only from *triplet* states when L is even. The average over the three possible initial substates therefore necessitates a replacement of the factor $\frac{1}{4}$ of the free-pair average by $\frac{1}{3}$. Equation (12–67) gives

$$\frac{1}{\tau} = P = \frac{4}{3} \cdot \frac{4}{3} (\pi^2 - 9)\alpha r_0^2 \frac{(\alpha m)^3}{8\pi n^3} = \frac{2}{9} \cdot \frac{\pi^2 - 9}{\pi} \cdot \frac{\alpha^6 m}{n^3}. \tag{12–109}$$

The corresponding lifetime for three-photon decay of the 1^3S_1 state is†

$$\tau_3 = \frac{9\pi}{2(\pi^2 - 9)} \cdot \frac{1}{\alpha^6 m} = 1.386 \times 10^{-7} \text{ sec}. \tag{12–110}$$

This lifetime is 1120 times longer than τ_2, Eq. (12–108).

The annihilation of positronium from states with $L \neq 0$ will differ from the corresponding S-state annihilation by $(|\mathbf{p}|/m)^{2L}$ in order of magnitude, since this will be the relative magnitude of the leading term in an expansion for small momenta. States with $L \neq 0$ will therefore always decay by radiative transitions to lower states ($\tau \sim 10^{-8}$) rather than annihilate.

Many measurements have verified the theoretical results of this section within experimental error.‡

---

* The short derivation was first given by J. A. Wheeler and by J. Pirenne, *loc. cit.*
† This value was first obtained correctly by A. Ore and J. L. Powell, *Phys. Rev.* **75**, 1696 (1939).
‡ The reader is referred to a survey article by M. Deutsch, *Progress in Nuclear Physics*, Vol. 3 (Academic Press, New York, 1953), p. 131, which contains a rather complete list of references on positronium. The selection rules given there are unfortunately based on the incorrect work by Landau.

CHAPTER 13

THE PHOTON-PHOTON SYSTEM

Experiments for the observations of processes involving photon-photon collisions are extremely difficult to perform, because they are crossed-beam experiments which require the highest intensities and the most sensitive detection equipment. For this reason the photon-photon cross sections are at the present time still of little interest to the experimental physicist. However, in addition to their fundamental theoretical significance, these processes play an important role in various approximation methods which are used to estimate related observable processes.*

13–1 Photon-photon scattering as part of a diagram. The lowest-order diagrams which describe the scattering of two photons by each other are closed loops with four corners. There are six different diagrams which contribute to this process. Three of them are drawn in Fig. 13–1; the other three differ from these only by the arrow direction in the closed loop and, therefore, give just a factor of 2. The fourth-order diagrams, which consist of two unconnected second-order photon self-energy diagrams, give no physical contribution for free photons.

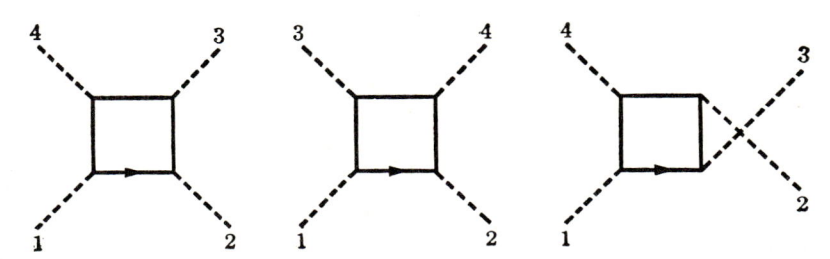

FIG. 13–1. The three fundamental diagrams for photon-photon scattering.

The diagrams of Fig. 13–1 may also occur as part of a larger diagram, so that the momenta k_1, k_2, k_3, and k_4 belong to *virtual* photons and do not satisfy $k_i^2 = 0$. This case is of particular importance when one or two of the photon lines are associated with an external field, as will be discussed in Section 15–8. It is therefore convenient to evaluate the photon-photon scattering diagrams first in the general case of a diagram part with $k_i^2 \neq 0$.

---

* For example, in the Weizsäcker-Williams method the field of a fast-moving charged particle is approximated by a suitable photon distribution [E. J. Williams, *Kgl. Dansk. Videnskab. Selskab* **13**, No. 4 (1935); C. F. v. Weizsäcker, *Z. Physik* **88**, 612 (1934)]. Pair production in the collision of two charged particles can thereby be reduced to pair production in photon-photon collisions.

This diagram part is a tensor of fourth rank $\Pi_{\mu\nu\lambda\sigma}(k_1,k_2,k_3,k_4)$, in complete analogy to the second-rank photon self-energy part $\Pi_{\mu\nu}$. For this reason, it is often called the fourth-rank *vacuum polarization tensor*.

A number of properties of this tensor are at once apparent. (1) It is a function of three four-vectors only, since the four k_i must satisfy*

$$\sum_{i=1}^{4} k_i = 0, \tag{13-1}$$

whether the photons are real or virtual. (2) When $\Pi_{\mu\nu\lambda\sigma}$ is multiplied by the photon operators a^μ, a^ν, $\cdots$ and a gauge transformation is carried out, the expression must remain invariant. Therefore,

$$k_1{}^\mu \Pi_{\mu\nu\lambda\sigma}(k_1 k_2 k_3 k_4) = 0,$$
$$k_2{}^\nu \Pi_{\mu\nu\lambda\sigma}(k_1 k_2 k_3 k_4) = 0, \tag{13-2}$$
$$\text{etc.}$$

The subscripts μ, ν, λ, and σ are associated with the photons 1, 2, 3, and 4. The tensor is symmetrical under simultaneous permutations of the indices and the arguments. (3) We recognize the diagrams of Fig. 13–1 as primitively divergent in the sense of Section 10–1, where it was proved that the associated integral is logarithmically divergent in the four-momentum p of the intermediate electron states. Let us enumerate the various methods available for dealing with this divergence:

(a) There is first the possibility of defining the value of the infinite integral by the requirement of gauge invariance. Equations (13–2) then become defining equations.

(b) As we shall see below, the infinite part of the integral which is not gauge-invariant can be interpreted as a *direct interaction* of the four photons. A corresponding term in the interaction Lagrangian, $\lambda(A_\mu A^\mu)^2$, would give rise to just such a term. It is therefore possible to combine the infinite terms and eliminate them by a renormalization of the interaction constant $\lambda \to \lambda_0 = \lambda - \delta\lambda$. This procedure can actually be carried out to all orders in λ and in α. One can then require gauge invariance for the renormalized theory, i.e., put the arbitrary finite constant λ_0 equal to zero. (*Cf.* also Section 9–2.)

(c) Finally, it is possible to use the method of regulators (Section 10–9) introducing n_e auxiliary fields with masses M_i and coupling constants $c_i e_i$. The result is evaluated in the limit as the auxiliary masses approach infinity.

---

* For reasons of symmetry it is more convenient to treat all four (virtual) photons of the polarization tensor as either ingoing or outgoing photons. We adopt the latter sign convention.

13-1] PHOTON-PHOTON SCATTERING AS PART OF A DIAGRAM

Consistent with our development of the theory (*cf.* especially Chapter 10), we shall adopt method (a). However, we shall now show that even this procedure is unnecessary—at least in lowest order—since the coefficient of the undetermined integral vanishes identically. For this purpose we must consider the form of $\Pi_{\mu\nu\lambda\sigma}$ in more detail.

Following Table 8–2, the integral for the polarization tensor can be written down in momentum space at once,

$$\Pi_{\mu\nu\lambda\sigma}(k_1 k_2 k_3 k_4) = T_{\mu\nu\lambda\sigma}(k_1 k_2 k_3 k_4) + T_{\mu\nu\sigma\lambda}(k_1 k_2 k_4 k_3) + T_{\mu\lambda\nu\sigma}(k_1 k_3 k_2 k_4), \quad (13\text{-}3)$$

$$T_{\mu\nu\lambda\sigma}(k_1 k_2 k_3 k_4) = -2 \frac{e^4}{(2\pi)^6} \int d^4p \, \text{Tr} \left\{ \gamma_\mu \frac{i\boldsymbol{p} - m}{p^2 + m^2} \gamma_\nu \right.$$

$$\left. \times \frac{i(\boldsymbol{p} - \boldsymbol{k}_2) - m}{(p - k_2)^2 + m^2} \gamma_\lambda \frac{i(\boldsymbol{p} - \boldsymbol{k}_2 - \boldsymbol{k}_3) - m}{(p - k_2 - k_3)^2 + m^2} \gamma_\sigma \frac{i(\boldsymbol{p} + \boldsymbol{k}_1) - m}{(p + k_1)^2 + m^2} \right\}. \quad (13\text{-}4)$$

The three terms in Eq. (13-3) represent the three diagrams of Fig. 13-1, and the factor 2 in Eq. (13-4) includes the other three diagrams, which differ only in the arrow direction but give the same integrals.

The introduction of three auxiliary variables, x, y, and z, permits a combination of the four denominators according to Eq. (A5-2),

$$(p^2 + m^2)z + [(p - k_2)^2 + m^2](y - z) + [(p - k_2 - k_3)^2 + m^2](x - y)$$

$$+ [(p + k_1)^2 + m^2](1 - x) = (p + K)^2 + a^2,$$

$$K = k_1(1 - x) - k_2(x - z) + k_3(x - y),$$

$$a^2 = m^2 + x(1 - x)k_1^2 + (x - z)(1 - x + z)k_2^2$$

$$+ (x - y)(1 - x + y)k_3^2 + 2[(1 - x)(x - z)k_1 \cdot k_2$$

$$+ (x - y)(1 + x - z)k_2 \cdot k_3 + (1 - x)(x - y)k_3 \cdot k_1]. \quad (13\text{-}5)$$

The tensor $T_{\mu\nu\lambda\sigma}$ can now be written

$$T_{\mu\nu\lambda\sigma} = -\frac{\alpha^2}{2\pi^4} \cdot 3! \int_0^1 dx \int_0^x dy \int_0^y dz \int \frac{X_{\mu\nu\lambda\sigma}}{(p^2 + a^2)^4} d^4p. \quad (13\text{-}6)$$

This equation is obtained after a shift of origin,

$$p \to p - K.$$

The tensor $X_{\mu\nu\lambda\sigma}$ therefore represents the trace in the numerator of Eq. (13-4) after this shift has been carried out. We remember that integrations over intermediate states must be carried out in a symmetrical way (*cf.* Section A5-2). The only surviving terms in $X_{\mu\nu\lambda\sigma}$ will thus be of fourth, second, and zeroth order in p:

$$X_{\mu\nu\lambda\sigma} = X_{\mu\nu\lambda\sigma}^{(4)} + X_{\mu\nu\lambda\sigma}^{(2)} + X_{\mu\nu\lambda\sigma}^{(0)}.$$

Only the integral involving $X_{\mu\nu\lambda\sigma}{}^{(4)}$ will be divergent. Since

$$X_{\mu\nu\lambda\sigma}{}^{(4)} = \text{Tr}\,(\gamma_\mu \not p\, \gamma_\nu \not p\, \gamma_\lambda \not p\, \gamma_\sigma \not p),$$

and by symmetrical integration,

$$\int \frac{p_\alpha p_\beta p_\gamma p_\delta}{(p^2+a^2)^4}\, d^4p = \frac{1}{24}(g_{\alpha\beta}g_{\gamma\delta} + g_{\alpha\gamma}g_{\beta\delta} + g_{\alpha\delta}g_{\beta\gamma}) \int \frac{(p^2)^2 d^4p}{(p^2+a^2)^4},$$

the divergent integral can be separated as follows:

$$\int \frac{(p^2)^2 d^4p}{(p^2+a^2)^4} = \int \frac{(p^2)^2 d^4p}{(p^2+m^2)^4} - 4\int_0^1 du \int d^4p\, \frac{(a^2-m^2)(p^2)^2}{[p^2+m^2 u+a^2(1-u)]^5}.$$

The first integral is an undetermined *constant* and therefore corresponds to a direct interaction between the four photons as mentioned above. The remaining term is finite, yielding

$$-i\pi^2 \int_0^1 \frac{a^2-m^2}{m^2 u + a^2(1-u)} = -i\pi^2 \ln \frac{a^2}{m^2}.$$

The coefficient of the infinite term contains the factor

$$F = (g_{\alpha\beta}g_{\gamma\delta} + g_{\alpha\gamma}g_{\beta\delta} + g_{\alpha\delta}g_{\beta\gamma})\text{Tr}\,(\gamma_\mu\gamma^\alpha\gamma_\nu\gamma^\beta\gamma_\lambda\gamma^\gamma\gamma_\sigma\gamma^\delta$$
$$+ \gamma_\mu\gamma^\alpha\gamma_\lambda\gamma^\beta\gamma_\nu\gamma^\gamma\gamma_\sigma\gamma^\delta + \gamma_\mu\gamma^\alpha\gamma_\nu\gamma^\beta\gamma_\sigma\gamma^\gamma\gamma_\lambda\gamma^\delta), \quad (13\text{-}7)$$

when we sum over all three diagrams. The term $g_{\alpha\beta}g_{\gamma\delta}$ times the trace in this expression is easily evaluated and gives

$$4\,\text{Tr}\,(\gamma_\mu\gamma_\nu\gamma_\lambda\gamma_\sigma + \gamma_\mu\gamma_\nu\gamma_\sigma\gamma_\lambda + \gamma_\mu\gamma_\lambda\gamma_\nu\gamma_\sigma) = 16(g_{\mu\nu}g_{\lambda\sigma} + g_{\mu\lambda}g_{\nu\sigma} + g_{\mu\sigma}g_{\nu\lambda}).$$

The last term, $g_{\alpha\delta}g_{\beta\gamma}$, times the trace gives exactly the same result, whereas the middle term, $g_{\alpha\gamma}g_{\beta\delta}$, yields

$$-32(g_{\mu\nu}g_{\lambda\sigma} + g_{\mu\lambda}g_{\nu\sigma} + g_{\mu\sigma}g_{\nu\lambda}).$$

It follows that
$$F = 0, \qquad (13\text{-}8)$$

so that the divergent integral has a coefficient which vanishes identically. This means that, when the logarithmically divergent integral $\Pi_{\mu\nu\lambda\sigma}$ is evaluated with a cutoff, the result will be finite and cutoff-independent.*

Since $\Pi_{\mu\nu\lambda\sigma}$ is finite, it must be gauge-invariant; this can be verified explicitly.

---

* The fact that $\Pi_{\mu\nu\lambda\sigma}$ is finite in lowest order seems to have been known to several people [*cf.* F. J. Dyson, *Phys. Rev.* **75**, 1736 (1949), p. 1747], but no proof in the literature is known to us. In particular, the only published paper in which $\Pi_{\mu\nu\lambda\sigma}$ is discussed in detail [R. Karplus and M. Neuman, *Phys. Rev.* **80**, 380 (1950)] uses regulators to eliminate the divergence, a procedure completely superfluous in view of Eq. (13-8).

From here on the evaluation of $\Pi_{\mu\nu\lambda\sigma}$ is straightforward, but exceedingly long and laborious. For this reason it is very important to make full use of all the symmetry properties and interrelations available. We shall present here a very brief outline of such a calculation.*

From Eqs. (13–3), (13–4), and (13–6) it follows that $\Pi_{\mu\nu\lambda\sigma}$ is of the form

$$\Pi_{\mu\nu\lambda\sigma} = \sum_{i,j,l,m} A^{ijlm} k_\mu{}^i k_\nu{}^j k_\lambda{}^l k_\sigma{}^m + \sum_{l,m} B_1{}^{lm} g_{\mu\nu} k_\lambda{}^l k_\sigma{}^m$$

$$+ \sum_{j,m} B_2{}^{jm} g_{\mu\lambda} k_\nu{}^j k_\sigma{}^m + C_1 g_{\mu\nu} g_{\lambda\sigma} + C_2 g_{\mu\lambda} g_{\nu\sigma} + C_3 g_{\mu\sigma} g_{\nu\lambda}, \quad (13\text{–}9)$$

where
$$i = 2, 3, 4, \qquad l = 1, 2, 4,$$
$$j = 1, 3, 4, \qquad m = 1, 2, 3.$$

The scalar functions A^{ijlm}, $B_1{}^{lm}$, $B_2{}^{jm}$, C_1, C_2, and C_3 depend on the scalar products of the four four-vectors† k^i ($i = 1, 2, 3, 4$) which satisfy Eq. (13–1). They are obtained after integration over p, x, y, and z.

The first considerable simplification arises from the gauge invariance of $\Pi_{\mu\nu\lambda\sigma}$. Since it satisfies Eqs. (13–2), the coefficients B and C are uniquely determined when the A's are known.

To show this we prove first that, because of Eqs. (13–2), all the coefficients B and C vanish identically when all the coefficients A vanish. It then follows that the B's and C's are uniquely determined by the A's, for the difference of two tensors which have the same A's but different B's and C's must vanish identically.

One can further show that the 81 coefficients A^{ijlm} can all be obtained by permutation of indices and by symmetry relations from 14 of the 81 coefficients $A_1{}^{ijlm}$. The latter are the contributions of the first diagram to the A^{ijlm}; they are the coefficients of $k^i k^j k^l k^m$ in Eq. (13–6).

The final result is more conveniently expressed by the five tensors $g_{\mu\nu\lambda\sigma}{}^{(n)}$ ($n = 1, 2, 3, 4, 5$). These tensors are defined by a consideration of the invariant

$$a^\mu(k_1) a^\nu(k_2) a^\lambda(k_3) a^\sigma(k_4) \Pi_{\mu\nu\lambda\sigma}(k_1 k_2 k_3 k_4).$$

Since this function is gauge-invariant, it can be expressed in terms of the four field strengths

$$f_{\mu\nu}(i) \equiv f_{\mu\nu}(k^i) = k_\mu{}^i a_\nu(k^i) - k_\nu{}^i a_\mu(k^i), \quad (i = 1, 2, 3, 4).$$

---

* The remaining part of this section is based on the work by Karplus and Neuman, loc. cit.

† Latin subscripts or superscripts distinguish between the four photons, whereas the Greek indices designate the covariant or contravariant components of the four-vectors, as usual.

Indeed, Karplus and Neuman show that the following five invariants are necessary and sufficient for this purpose:

$$f_{\alpha\beta}(1)f^{\beta\alpha}(2)f_{\gamma\delta}(3)f^{\delta\gamma}(4) = 4a^\mu(1)a^\nu(2)a^\lambda(3)a^\sigma(4)g_{\mu\nu\lambda\sigma}{}^{(1)}(1234),$$

$$f_{\alpha\beta}(1)f^{\beta\gamma}(2)f_{\gamma\delta}(3)f^{\delta\alpha}(4) = a^\mu(1)a^\nu(2)a^\lambda(3)a^\sigma(4)g_{\mu\nu\lambda\sigma}{}^{(2)}(1234),$$

$$f_{\alpha\beta}(1)f^{\beta\alpha}(2)k_1{}^\gamma f_{\gamma\delta}(3)f^{\delta\epsilon}(4)k_\epsilon{}^1 = -2a^\mu(1)a^\nu(2)a^\lambda(3)a^\sigma(4)g_{\mu\nu\lambda\sigma}{}^{(3)}(1234),$$

$$f_{\alpha\beta}(1)f^{\beta\alpha}(2)k_2{}^\gamma f_{\gamma\delta}(3)f^{\delta\epsilon}(4)k_\epsilon{}^1 = -2a^\mu(1)a^\nu(2)a^\lambda(3)a^\sigma(4)g_{\mu\nu\lambda\sigma}{}^{(4)}(1234),$$

$$k_2{}^\alpha f_{\alpha\beta}(4)f^{\beta\gamma}(1)[f_{\gamma\delta}(2)f^{\delta\epsilon}(3) - f_{\gamma\delta}(3)f^{\delta\epsilon}(2)]k_\epsilon{}^1$$

$$= a^\mu(1)a^\nu(2)a^\lambda(3)a^\sigma(4)g_{\mu\nu\lambda\sigma}{}^{(5)}(1234). \quad (13\text{--}10)$$

Of course, all permutations of 1, 2, 3, 4 in each $g_{\mu\nu\lambda\sigma}{}^{(i)}$ can occur, as they are determined by (10–9). The fourth-rank polarization tensor can then be expressed in terms of the five tensors $g_{\mu\nu\lambda\sigma}{}^{(n)}$ and the A^{ijlm}:

$$\Pi_{\mu\nu\lambda\sigma} = \sum_{24\text{ perm}} \left\{ \frac{1}{8}[A^{2143}(1234)g_{\mu\nu\lambda\sigma}{}^{(1)}(1234) + A^{2341}g_{\mu\nu\lambda\sigma}{}^{(2)}(1234)] \right.$$

$$+ \frac{1}{2k_3 \cdot k_4}[A^{2111}(1234)g_{\mu\nu\lambda\sigma}{}^{(3)}(1234) + A^{2121}(1234)g_{\mu\nu\lambda\sigma}{}^{(4)}(1234)]$$

$$\left. + \frac{1}{3k_2 \cdot k_4} A^{2311}(1234)g_{\mu\nu\lambda\sigma}{}^{(5)}(1234) \right\}. \quad (13\text{--}11)$$

The functions A^{ijlm} have never been evaluated in generality. A special case will be discussed in the following section.

When the "scattering of light by light" is encountered as part of a diagram, the appropriate insertion in momentum space is*

$$\Pi_{\mu\nu\lambda\sigma}\delta(k_1 + k_2 + k_3 + k_4). \quad (13\text{--}12)$$

13–2 Photon-photon scattering cross sections.
The scattering of two photons into a final state of two photons follows from the fourth-rank polarization tensor as the special case where $k_i^2 = 0$ ($i = 1, 2, 3, 4$). If one further chooses the center-of-momentum system where the two incident photons k_1 and k_2 are scattered into the two photons k_3 and k_4,

---

* This function times the Fourier transform of the four photon operators $a^\mu(x_1)a^\nu(x_2)a^\lambda(x_3)a^\sigma(x_4)$ is the integrand of the fourth-order term (in the S-operator expansion) associated with this process. Our $\Pi_{\mu\nu\lambda\sigma}$ differs therefore from the tensor $G_{\mu\nu\lambda\sigma}$ of Karplus and Neuman by a numerical factor [cf. Eq. (13–15)].

these four-vectors satisfy the relations*

$$\omega_1 = \omega_2 = \omega_3 = \omega_4 = \omega,$$

$$\mathbf{k}_1 = -\mathbf{k}_2, \quad \mathbf{k}_3 = -\mathbf{k}_4, \quad |\mathbf{k}_1| = |\mathbf{k}_3| = \omega.$$

Therefore, the system can be described by the three parameters

$$\alpha = -\frac{k_1 \cdot k_2}{m^2} = -\frac{k_3 \cdot k_4}{m^2} = \frac{\omega^2}{m^2},$$

$$\beta = \frac{k_1 \cdot k_3}{m^2} = \frac{k_2 \cdot k_4}{m^2} = -\frac{\omega^2}{m^2}\sin^2\frac{\theta}{2}, \qquad (13\text{--}13)$$

$$\gamma = \frac{k_1 \cdot k_4}{m^2} = \frac{k_2 \cdot k_3}{m^2} = -\frac{\omega^2}{m^2}\cos^2\frac{\theta}{2},$$

$$\alpha + \beta + \gamma = 0.$$

Nevertheless, the actual evaluation of $\Pi_{\mu\nu\lambda\sigma}$, Eq. (13–11), is very tedious. It was carried out in detail by Karplus and Neuman,† for the various possible polarization cases. We shall be content here to discuss briefly the unpolarized case.

The cross section, according to Section 8–6, is

$$\sigma = (2\pi)^2 \frac{\omega^2}{|k_1 \cdot k_2|} \mathsf{S}_f \bar{\mathsf{S}}_i \delta(k_3 + k_4 - k_1 - k_2)|(f|M|i)|^2.$$

The matrix element $(f|M|i)$ can be expressed in terms of $\Pi_{\mu\nu\lambda\sigma}$ of the previous section,

$$(f|M|i) = \frac{1}{(2\omega)^2} e^\mu(k_1) e^\nu(k_2) e^\lambda(k_3) e^\sigma(k_4) \Pi_{\mu\nu\lambda\sigma}(k_1 k_2 k_3 k_4). \qquad (13\text{--}14)$$

It is convenient to introduce the tensor $G_{\mu\nu\lambda\sigma}(k_1 k_2 k_3 k_4)$,

$$\Pi_{\mu\nu\lambda\sigma} = -\frac{i\alpha^2}{2\pi^2} G_{\mu\nu\lambda\sigma}. \qquad (13\text{--}15)$$

The denominator $|k_1 \cdot k_2| = 2\omega^2$ in the CM-system and the integration over the δ-function gives

$$\int \delta(k_3 + k_4 - k_1 - k_2) d^3k_3 d^3k_4 = \omega^2 d\Omega \frac{d\omega}{dE_f} = \frac{\omega^2}{2} d\Omega,$$

---

* All k_i were chosen as outgoing photons in the previous section. The momenta k_1 and k_2 in this section therefore differ by a minus sign from the k_1 and k_2 used previously.

† R. Karplus and M. Neuman, *Phys. Rev.* **83**, 776 (1951).

where $d\Omega$ is the solid angle into which one of the final photons scatters. The differential cross section follows from these relations as

$$\frac{d\sigma}{d\Omega} = \frac{\alpha^2 r_0^2}{64\pi^2} \left(\frac{m}{\omega}\right)^2 |e^\mu(1)e^\nu(2)e^\lambda(3)e^\sigma(4)G_{\mu\nu\lambda\sigma}(1234)|^2 \qquad (13\text{-}16)$$

for definite polarizations of the four photons. When we are dealing with unpolarized photons, we must average over the initial four polarization directions and sum over the final ones. This yields*

$$\frac{d\sigma}{d\Omega} = \frac{\alpha^2 r_0^2}{64\pi^2} \left(\frac{m}{\omega}\right)^2 \cdot \frac{1}{4}\sum_{\text{pol}} |e^\mu(1)e^\nu(2)e^\lambda(3)e^\sigma(4)G_{\mu\nu\lambda\sigma}(1234)|^2. \qquad (13\text{-}17)$$

A simple closed expression for this cross section can be obtained in the "nonrelativistic" limit only when $\omega \ll m$. In this case† the tensor $\Pi_{\mu\nu\lambda\sigma}$ is proportional to ω^4, which can be seen as follows. The functions A^{ijlm}, in an expansion in ω/m, reduce to their leading terms, which are constants, whereas the tensors $g_{\mu\nu\lambda\sigma}^{(i)}$ are proportional to ω^4 (for $i = 1, 2$) and to ω^6 (for $i = 3, 4, 5$). This means that the $g_{\mu\nu\lambda\sigma}^{(i)}$ with $i = 3, 4,$ and 5 can be neglected. The remaining two terms in the matrix element can be expressed in terms of the two invariants which can be constructed from the antisymmetrical tensor $f_{\mu\nu}$:

$$I_1 = \tfrac{1}{2} f_{\mu\nu} f^{\mu\nu} \quad \text{and} \quad I_2 = \tfrac{1}{8} \epsilon_{\mu\nu\lambda\sigma} f^{\mu\nu} f^{\lambda\sigma},$$

where $\epsilon_{\mu\nu\lambda\sigma}$ is the completely antisymmetric unit tensor of fourth rank, that is, $\epsilon_{\mu\nu\lambda\sigma} = \pm 1$ when $\mu\nu\lambda\sigma$ is an even or odd permutation of 0123, respectively, and $\epsilon_{\mu\nu\lambda\sigma} = 0$ otherwise. With the help of the two three-vectors,

$$\mathbf{E} = (f_{10}, f_{20}, f_{30}) \quad \text{and} \quad \mathbf{H} = (f_{23}, f_{31}, f_{12}),$$

the electric and the magnetic field strengths, these invariants can be written

$$I_1 = \mathbf{H}^2 - \mathbf{E}^2 \quad \text{and} \quad I_2 = \mathbf{H} \cdot \mathbf{E}.$$

The invariant of the fourth degree in the fields,

$$I_3 = \tfrac{1}{4} f_{\mu\nu} f^{\nu\lambda} f_{\lambda\sigma} f^{\sigma\mu},$$

---

* The abbreviation $\langle |M| \rangle_{\text{av}}^2$ of Karplus and Neuman (*loc. cit.*) differs from our $\tfrac{1}{4} \sum_{\text{pol}} |\cdots|^2$ by a factor $\tfrac{1}{16}$.

† H. Euler, *Ann. Physik* **26**, 398 (1936).

can be expressed in terms of the other two,

$$I_3 = \tfrac{1}{2}I_1{}^2 + I_2{}^2,$$

as can easily be verified.

The corresponding S-matrix element is most conveniently expressed in x-space

$$\begin{aligned}S_{fi}{}^{(4)} &= -\frac{i\alpha^2}{45}\frac{1}{m^4}\int[5I_1{}^2(x) - 14I_3(x)]d^4x \\ &= \frac{2i\alpha^2}{45}\frac{1}{m^4}\int[I_1{}^2(x) + 7I_2{}^2(x)]d^4x \\ &= \frac{2i\alpha^2}{45}\frac{1}{m^4}\int[(\mathbf{H}^2 - \mathbf{E}^2)^2 + 7(\mathbf{H}\cdot\mathbf{E})^2]d^4x.\end{aligned}\qquad(13\text{–}18)$$

Since $\Pi_{\mu\nu\lambda\sigma}$ increases like ω^4, the differential cross section increases like ω^6. It consists of a factor ω^6 and an angular factor. We find in the low-energy limit*

$$\frac{d\sigma}{d\Omega} = \frac{139}{8100}\left(\frac{\alpha}{2\pi}\right)^2 r_0{}^2\left(\frac{\omega}{m}\right)^6 (3 + \cos^2\theta)^2, \quad \text{(CM-system)}. \qquad (13\text{–}19)$$

In this limit the total cross section is easily found to be

$$\sigma = \frac{973}{10125}\frac{\alpha^2}{\pi^2}r_0{}^2\left(\frac{\omega}{m}\right)^6.$$

A factor $\tfrac{1}{2}$ was inserted to take account of the equivalence of the two final photons.

In the "extreme relativistic" limit ($\omega \gg m$) the angular distribution has a different energy dependence for different angles and does not factor into an angular function and an energy function.

In general, the differential cross section can be expressed in terms of three transcendental complex functions, $B(u)$, $T(u)$, and $I(u,v)$, in which the variables u and v are any of the three parameters α, β, and γ [Eq. (13–13)]. Karplus and Neuman evaluated these functions numerically for the cases

$$\theta = 0: \quad \alpha = -\gamma = \left(\frac{\omega}{m}\right)^2, \quad \beta = 0,$$

$$\theta = \frac{\pi}{2}: \quad \alpha = -2\beta = -2\gamma = \left(\frac{\omega}{m}\right)^2,$$

(13–20)

---

* H. Euler, *loc. cit.*; R. Karplus and M. Neuman, *loc. cit.*

and obtained the real and imaginary parts of the scattering amplitudes for various polarization cases. The resulting differential cross sections per unit solid angle,

$$\frac{d\sigma(\omega,0)}{d\Omega} \quad \text{and} \quad \frac{d\sigma(\omega,\pi/2)}{d\Omega}, \qquad (13\text{--}21)$$

as functions of ω and for unpolarized photons, have the following characteristic behavior.

For both cross sections the NR limit is an excellent approximation almost up to the pair production threshold $\omega = m$. Above this energy the imaginary part of the scattering amplitude is different from zero (see below), so that the derivative of (13–21) with respect to ω has a jump at that point. The forward scattering cross section has a broad maximum above $\omega = m$,

$$\left(\frac{d\sigma(\omega,0)}{d\Omega}\right)_{\max} = 4.1 \times 10^{-31} \text{ cm}^2/\text{sterad at } \omega = 3.5m, \qquad (13\text{--}22)$$

and for $\omega > 10m$ it is practically equal to its high-energy limit, which decreases like*

$$\frac{d\sigma(\omega,0)}{d\Omega} \sim \left(\frac{\alpha}{\pi}\right)^2 r_0^2 \left(\frac{m}{\omega}\right)^2 \left(\ln\frac{\omega}{m}\right)^4, \quad \text{(high-energy limit)}. \qquad (13\text{--}23)$$

The right-angle scattering above $\omega = m$ goes through its maximum,

$$\left(\frac{d\sigma(\omega,\pi/2)}{d\Omega}\right)_{\max} = 2.8 \times 10^{-31} \text{ cm}^2/\text{sterad at } \omega = 1.4m, \qquad (13\text{--}24)$$

and for $\omega > 3m$ decreases like

$$\frac{d\sigma(\omega,\pi/2)}{d\Omega} \sim \left(\frac{\alpha}{\pi}\right)^2 r_0^2 \left(\frac{m}{\omega}\right)^2, \quad \text{(high-energy limit)}. \qquad (13\text{--}25)$$

The above equations can be used as a crude estimate of the total scattering cross section. A more accurate value of this quantity is not known. However, as is evident from the above values, the photon-photon scattering cross section is at the present time still out of reach for the experimental physicist.

We want to return, now, to the diagrams for photon-photon scattering (Fig. 13–1). Apparently, this process can be regarded as a process of pair production in a photon-photon collision, followed by free-pair annihilation into two photons. The intermediate pair state may be virtual or real. In

---

* This energy dependence was first obtained by A. Achieser, *Physik. Z. Sowjetunion* **11**, 263 (1937).

the latter case, sufficient energy must be available in the initial state ($\omega \geq m$ in the CM-system) to produce the pair under conservation of both momentum and energy.

The separation into virtual and real intermediate states is very basic and is neatly expressed in the general formalism. Indeed, quite independent of the iteration solution, it was shown in the theory of the S-operator (Chapter 7) that there is a separation into the principal part and the contribution from the poles in the integration over intermediate energies. This can be seen, for example, in Eq. (7–34)$_+$, which is connected with the S-matrix by Eqs. (7–36)$_+$ and (7–69)$_+$. A pole therefore corresponds to real intermediate states and is always present when the energy is above the associated threshold energy. But let us return to the problem under discussion.

When two photons collide, they will not necessarily scatter, but may annihilate each other, thereby producing a negaton-positon pair. This *absorption process* for photons is actually much more probable than the scattering process, as we shall see in Section 13–3. However, as is shown in the Appendix, Section A7–1, the unitarity of the S-matrix leads to a relation between the *total* cross section, i.e., the sum of scattering and absorption cross sections, and the forward scattering amplitude

$$a(\omega,0) = a_1(\omega,0) + ia_2(\omega,0). \tag{13–26}$$

This relation is [*cf.* Eq. (A7–7)],

$$\sigma(\omega) = \frac{4\pi}{\omega} a_2(\omega,0). \tag{13–27}$$

It applies in our case where, in the center-of-momentum system, the scattering amplitude follows from Eq. (13–17),

$$a(\omega,0) = \frac{\alpha^2}{2\pi\omega} e^\mu e^\nu e^\lambda e^\sigma G_{\mu\nu\lambda\sigma}, \tag{13–28}$$

for definite polarizations of the four photons. Since the pair production cross section is of order $(137)^2$ larger than the photon-photon scattering cross section, the total cross section $\sigma(\omega)$ in Eq. (13–27) may be approximated very well by the former [*cf.* Eq. (13–40)]. This fact permits the calculation of forward photon-photon scattering from the photon-photon pair production cross section. This is useful, of course, because the latter is much easier to obtain. Starting from this cross section (for definite polarizations), $a_2(\omega,0)$ is given by (13–27). We can now make use of the analytic character of $a(\omega,0)$ and obtain $a_1(\omega,0)$ by analytic continuation from $a_2(\omega,0)$. This method is explained in the Appendix, Section A7–4. It furnishes an important cross check on the calculation by Karplus and Neuman.

In conclusion of this section, a few historical remarks are in order. The "scattering of light by light" was first discussed by Euler and Kockel* and was soon recognized as a particularly interesting phenomenon, since it is characteristic of quantized electrodynamics and contradicts the classical notions of electromagnetic radiation. The latter is described by the unquantized Maxwell equations, which are linear in the electromagnetic field and thus do not lead to a scattering phenomenon between light waves. This quantum-mechanical effect can be simulated, however, within the framework of a classical theory, if one assumes an effective nonlinear interaction between electromagnetic waves. For this purpose, one can start with a classical field theory with a Lagrangian

$$L = -\tfrac{1}{2}\partial_\mu a_\nu \partial^\mu a^\nu$$

[cf. Eq. (2–22)] for the free fields, to which an interaction Lagrangian is added:

$$L_1 = c_1[f_{\mu\nu}(x)f^{\mu\nu}(x)]^2 + c_2 f_{\mu\nu}(x)f^{\nu\lambda}(x)f_{\lambda\sigma}(x)f^{\sigma\mu}(x). \qquad (13\text{–}29)$$

L and L_1 will give rise to linear and cubic terms, respectively, in the field equations. The constants c_1 and c_2 can be so chosen that the low-energy limit of photon-photon scattering, Eq. (13–18), is reproduced in lowest order. The two invariants correspond to the two tensors $g_{\mu\nu\lambda\sigma}^{(1)}$ and $g_{\mu\nu\lambda\sigma}^{(2)}$ [cf. Eq. (13–10)]. In fact, it is not difficult to show that the constants c_1 and c_2 must be

$$c_1 = \frac{5}{180}\frac{\alpha^2}{m^4}, \quad c_2 = -\frac{7}{90}\frac{\alpha^2}{m^4}. \qquad (13\text{–}30)$$

They follow at once from $S^{(4)}$, the fourth-order S-operator, whose matrix element yields the low-energy limit of Eq. (13–15). These aspects of photon-photon scattering can be found in further detail in an interesting paper by Weisskopf.†

It is clear, however, that an effective Lagrangian, as in Eq. (13–29), can be constructed only in the low-energy limit, when the photon energy is too small to produce real pairs, since otherwise the pair field cannot be eliminated. On the other hand, the nonlinear corrections to the inherently linear classical Maxwell theory are extremely small, so that the basic principle of superposition is, in general, not violated to an *observable* degree. The scattering of photons by a Coulomb field is the only observable case at the present time (cf. Section 15–8).

---

* H. Euler and B. Kockel, *Naturwiss.* **23**, 246 (1935).
† V. Weisskopf, *Kgl. Danske Videnskab. Selskab* **14**, No. 6 (1936).

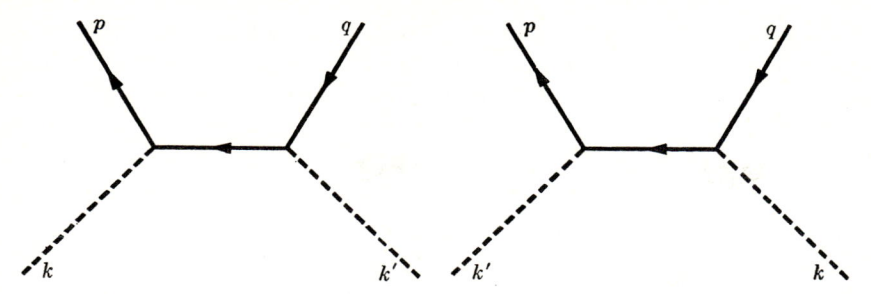

FIG. 13-2. Lowest-order diagrams for pair production in photon-photon collisions.

13-3 Pair production in photon-photon collisions. This process is exactly the inverse process to two-photon annihilation of free negaton-positon pairs (Section 12-4). Indeed, the transition probabilities are related by the principle of detailed balance, a special case of the principle of reciprocity, Eq. (7-119). Another method for calculating the cross section of this process from previous results is evident from the diagrams of Fig. 13-2. It consists in the use of the substitution law, Section 8-5, to convert either the Compton scattering or the two-photon pair annihilation diagrams into the pair creation diagrams of Fig. 13-2. The various methods yield the same result, of course.

With the assumption of unpolarized electrons, we find in the usual way

$$\sum_{\text{spins}} |(f|M|i)|^2 = \left(\frac{\alpha}{2\pi}\right)^2 \frac{X}{\epsilon_+ \epsilon_- \omega \omega'}. \qquad (13\text{-}31)$$

The trace X differs from that for pair annihilation, Eq. (12-41), only in the substitution

$$p \to -q, \quad k_1 \to -k', \quad k_2 \to -k. \qquad (13\text{-}32)$$

As is easily verified, this substitution leaves X invariant in form; this is to be expected from the principle of detailed balance.

The differential cross section follows from Eqs. (8-49) and (13-31):

$$d\sigma = (2\pi)^2 \frac{\omega \omega'}{|k \cdot k'|} \int d^3p \, d^3q \, \delta(p + q - k - k') \sum_{\text{spins}} |(f|M|i)|^2$$

$$= r_0^2 \frac{m^2}{|k \cdot k'|} \frac{|\mathbf{p}|}{\epsilon_+} \frac{d\epsilon_-}{dE_f} X \, d\Omega_-$$

$$= r_0^2 \frac{m^2}{|k \cdot k'|} \frac{\beta_-^2 \epsilon_- X}{E\beta_- - |\mathbf{P}| \cos \theta_-} d\Omega_-. \qquad (13\text{-}33)$$

This result simplifies considerably in the center-of-momentum system of the photons, where

$$\mathbf{P} = \mathbf{k} + \mathbf{k}' = \mathbf{p} + \mathbf{q} = 0,$$
$$\omega = \omega' = \epsilon_- = \epsilon_+, \quad (13\text{–}34)$$
$$\beta = \frac{|\mathbf{p}|}{\omega} = \frac{1}{\omega}\sqrt{\omega^2 - m^2}.$$

We find

$$\frac{d\sigma}{d\Omega} = \frac{1}{4} r_0^2 \left(\frac{m}{\omega}\right)^2 \beta X, \quad \text{(CM-system)}. \quad (13\text{–}35)$$

The solid angle $d\Omega$ can refer to either electron. The trace X is given by Eq. (12–41) with the simplifications (13–34), so that

$$\kappa_1 \to \omega^2(1 - \beta \cos\theta), \quad \kappa_2 \to \omega^2(1 + \beta \cos\theta), \quad (13\text{–}36)$$

where θ is the scattering angle. The resulting special cases of polarizations parallel and perpendicular to each other were first discussed by Breit and Wheeler* and can easily be integrated over the solid angle. The corresponding total cross sections are given by these authors. In the following we restrict ourselves to unpolarized photons. The sum over the polarizations in X has been carried out in Eq. (12–42). The averaging process requires a factor $\frac{1}{4}$. Thus, using Eq. (13–36), we obtain, after simple rearrangements,

$$d\sigma = \frac{r_0^2 \pi}{2} \beta \left(\frac{m}{\omega}\right)^2 \frac{1 - \beta^4 \cos^4\theta + 2(m/\omega)^2 \beta^2 \sin^2\theta}{(1 - \beta^2 \cos^2\theta)^2} \sin\theta\, d\theta,$$
$$\text{(CM-system)}. \quad (13\text{–}37)$$

This pair production cross section by unpolarized photons is, of course, symmetrical with respect to $\theta = \pi/2$. It has a threshold at $\omega = m$, where it reduces to

$$d\sigma = \frac{r_0^2 \pi}{2} \sqrt{1 - \left(\frac{m}{\omega}\right)^2} \sin\theta\, d\theta, \quad \text{(NR)}. \quad (13\text{–}38)$$

As expected, this is an isotropic distribution. At very high energies the electrons are produced at very small angles with respect to the line of incidence. We find

$$d\sigma = r_0^2 \pi \left(\frac{m}{\omega}\right)^2 \frac{1 + \cos^2\theta}{\sin^2\theta} \sin\theta\, d\theta, \quad \left(\theta \gtrsim \frac{m}{\omega}\right) \quad \text{(ER)} \quad (13\text{–}39)$$

and

$$d\sigma = r_0^2 \pi \sin\theta\, d\theta, \quad (\theta \sim 0 \text{ or } \pi).$$

---

* G. Breit and J. A. Wheeler, *Phys. Rev.* **46**, 1087 (1934).

The total cross section for unpolarized photons is obtained by integration of Eq. (13–37), and can be expressed* in terms of either β or ω:

$$\sigma = \frac{1}{2} r_0^2 \pi (1 - \beta^2) \left[(3 - \beta^4) \ln \frac{1+\beta}{1-\beta} - 2\beta(2 - \beta^2) \right]$$

$$= r_0^2 \pi \left(\frac{m}{\omega}\right)^2 \left\{ \left[2\left(1 + \left(\frac{m}{\omega}\right)^2\right) - \left(\frac{m}{\omega}\right)^4 \right] \cosh^{-1} \frac{\omega}{m} \right.$$

$$\left. - \left(1 + \left(\frac{m}{\omega}\right)^2\right) \sqrt{1 - \left(\frac{m}{\omega}\right)^2} \right\}. \qquad (13\text{–}40)$$

In the limits, this expression yields

$$\sigma = r_0^2 \pi \beta, \quad (\text{NR}) \qquad (13\text{–}41)$$

and

$$\sigma = r_0^2 \pi \left(\frac{m}{\omega}\right)^2 \left(\ln \frac{2\omega}{m} - 1 \right), \quad (\text{ER}). \qquad (13\text{–}42)$$

The cross section therefore increases slowly from $\omega = m$ on, reaches a maximum, and then decreases according to Eq. (13–42). Though σ is of the same order of magnitude as many easily observable cross sections, the producible photon intensities are much too small to make this process observable. However, the theoretical importance of this effect and its usefulness for various approximate calculations should not be underestimated.

---

* This result agrees with the formula found by Breit and Wheeler (*loc. cit.*) when their awkward notation is translated. The apparent difference of a factor of 2 arises from their definition of σ as the effective cross section for a photon beam incident on a photon gas. Our σ corresponds to the collision of two photon beams. Equation (13–40) also agrees with the result of Karplus and Neuman (*loc. cit.*), who obtain it from the imaginary part of the photon-photon forward scattering amplitude, provided one corrects their deplorable misprints.

CHAPTER 14

THEORY OF THE EXTERNAL FIELD

The theory which we have developed up to this point is completely general in the sense that it describes all the known physical processes which are caused by the interaction of electrons with photons. In particular, it contains as limiting cases the properties of classical charge and current distributions. The transition to the classical limiting case is carried out essentially by going from the operator relations to the relations between expectation values of these operators. In this transition the operators a_μ and ψ behave quite differently. While the expectation value $\langle a_\mu \rangle$ of a_μ corresponds to the classical electromagnetic vector potential, there is no quantity in the classical picture which corresponds to $\langle \psi \rangle$. Only bilinear expressions of ψ operators, such as $\langle \bar{\psi} \gamma_\mu \psi \rangle$, have a classical interpretation.

In many applications of great practical importance it appears useful to separate the total radiation field operators into two parts, one which is treated classically, and another which describes the quantum effects. The classical part is due to charge and current distributions which are not affected appreciably by the radiative processes under discussion. For this part we need no equation of motion, since its values are assumed to be given space-time functions. Examples of this situation will be found in Chapter 15, where applications will be discussed.

We emphasize that the method of separating the field into a classical part and a quantized part is not a new theory, and it is contained in the complete theory so far developed as a certain approximation. There is only one point where an extension of the theory is needed. Charges occur not only on electrons but also on different kinds of particles such as protons or mesons. With regard to particles of spin $\frac{1}{2}$ the theory can in some measure still be applied in the form so far developed by merely introducing other kinds of spinor fields with suitable commutation relations. It is also possible to develop the theory for charged particles with integral spin, such as the π-meson. Unfortunately, this theory exceeds the scope of the present work.

In most cases, the properties of the particles which are the sources of the external field are irrelevant and need not be specified. And for a large class of physically important problems the external field is a static field in a particular reference system.

In Section 14-1 we give the definition of the external field approximation and derive the basic equations in two mathematically different but physically equivalent forms. In Section 14-2 we introduce the bound interaction picture and establish its relation to the basic equations of the external field approximation. This is followed by a general discussion of commutation rules for the matter field in the presence of an external field (Section 14–3). In Section 14–4 we discuss explicit expressions for the propagation function. The following Section 14–5 is devoted to the calculation of the S-matrix in external field problems. Very little can be said regarding the S-matrix, in which the external field is included rigorously. Instead, most applications are confined to the Born approximation, which is defined as a successive approximation in powers of the external field. In Section 14–6 we make a few remarks concerning the problem of renormalization in the presence of external fields. Finally, Section 14–7 contains a derivation of the basic formulas needed for the applications to cross sections and energy-level shifts.

14–1 The external field approximation. A classical Maxwell field and current density distribution are described by two four-vector fields $\varphi_\mu(x)$ and $s_\mu(x)$. These are c-numbers (not operators) and are related by the equations

$$\partial^\lambda \partial_\lambda \varphi_\mu(x) = -s_\mu(x). \tag{14-1}$$

The current density satisfies the continuity equation

$$\partial^\mu s_\mu(x) = 0, \tag{14-2}$$

which is consistent with the Lorentz condition for the potentials

$$\partial^\mu \varphi_\mu(x) = 0. \tag{14-3}$$

These equations can be obtained from a quantized theory by identifying the expectation values of the field and current operators with the classical quantities. They are always a good approximation when the fluctuation of these quantities around the expectation values can be considered as negligible compared with the expectation values themselves.

We wish to find the modification of the basic equations for the radiation and matter fields a_μ and ψ, which describe the interaction of this system with such a classical field. The latter is called *the external field*. It is considered as a given field subject only to the relations (14–1), (14–2), and (14–3). In other words, the reaction of the quantized fields a_μ and ψ on the external field is neglected.

There are two different but, as we shall see, essentially equivalent ways of introducing the interaction of the photon-electron field with the external field.

In the first method, we may consider the effect of the external field φ_μ on the system by replacing in the interaction operator $H(\tau)$ the field operators a_μ by $a_\mu + \varphi_\mu$. The new interaction operator is then

$$H_a(\tau) = -\int_{\sigma(\tau)} j^\mu(x)(a_\mu(x) + \varphi_\mu(x))d\sigma. \tag{14-4a}$$

The field equations and the subsidiary condition are unchanged:

$$\partial^\lambda \partial_\lambda a_\mu(x) = 0, \quad (\partial + m)\psi(x) = 0, \tag{14-5a}$$

$$\{\partial^\lambda a_\lambda(x) - \int D(x - x') j^\lambda(x') d\sigma_\lambda'\}\omega = 0, \tag{14-6a}$$

$$\partial^\lambda \varphi_\lambda(x) = 0.$$

The Schroedinger equation for the state vector is

$$i\dot\omega(\tau) = H_a(\tau)\omega(\tau). \tag{14-7a}$$

In the second method, we consider the effect of the external current density operator on the system by substituting in $H(\tau)$ for j_μ the total current $j_\mu + s_\mu$. The basic equations which correspond to the last four are then

$$H_b(\tau) = -\int_{\sigma(\tau)} (j^\mu(x) + s^\mu(x)) a_\mu(x) d\sigma, \tag{14-4b}$$

$$\partial^\lambda \partial_\lambda a_\mu(x) = 0, \quad (\partial + m)\psi(x) = 0, \tag{14-5b}$$

$$\{\partial^\lambda a_\lambda(x) - \int_\sigma D(x - x')(j^\lambda(x') + s^\lambda(x')) d\sigma_\lambda'\}\omega = 0, \tag{14-6b}$$

$$i\dot\omega(\tau) = H_b(\tau)\omega(\tau). \tag{14-7b}$$

We shall now show that these two methods of describing the interaction with an external field are essentially equivalent.* To this end, we carry out a τ-dependent canonical transformation, which transforms the system (a) into the system (b).

Let $\omega_a(\tau)$ be the state vector satisfying Eq. (14-7a). We define a new state vector $\omega_b(\tau)$ by putting

$$\omega_b(\tau) = e^{i\Sigma(\tau)} \omega_a(\tau),$$

where the Hermitian operator $\Sigma(\tau)$ is to be suitably chosen. The transformed state vector $\omega_b(\tau)$ satisfies a new Schroedinger equation,

$$i\dot\omega_b(\tau) = G(\tau)\omega_b(\tau), \tag{14-8}$$

where

$$G(\tau) = e^{i\Sigma(\tau)} H_a(\tau) e^{-i\Sigma(\tau)} - i e^{i\Sigma(\tau)} \left(\frac{d}{d\tau} e^{-i\Sigma(\tau)}\right). \tag{14-9}$$

---

* J. Schwinger, *Phys. Rev.* **76**, 790 (1949), especially Section 2, p. 803.

We evaluate the exponentials as a power series and obtain for the first few terms

$$G(\tau) = H_a(\tau) - \frac{i}{1!}[H_a(\tau), \Sigma] - \dot{\Sigma} + \frac{i}{2!}[\dot{\Sigma}, \Sigma] + \cdots. \quad (14\text{--}10)$$

There are several possibilities for Σ, which correspond to the different solutions of Eq. (14–1). As is well known, the solution of this equation is determined only if we also give the boundary conditions for the functions φ_μ. The solution which corresponds to the physically realizable conditions are the so-called *retarded potentials* which can be characterized by the boundary conditions $\varphi_\mu = 0$ and $\partial^\mu \varphi_\mu = 0$ for $\tau = -\infty$. The operator Σ which corresponds to this solution can be given in two alternative forms,

$$\Sigma(\tau) = \int_{-\infty}^{\sigma(\tau)} s_\mu(x) a^\mu(x) d^4x = \int_{\sigma(\tau)} (\varphi_\mu \partial^\lambda a^\mu - a^\mu \partial^\lambda \varphi_\mu) d\sigma_\lambda, \quad (14\text{--}11)$$

which are easily transformed into each other by applying Gauss' theorem for the region bounded by the two surfaces $\sigma(\tau)$ and $\sigma(-\infty)$.

The nonvanishing terms in Eq. (14–10) are now readily evaluated:

$$-i[H_a, \Sigma] = \int_\sigma j_\mu(x) \varphi^\mu(x) d\sigma, \quad (14\text{--}12)$$

$$-\dot{\Sigma} = -\int_\sigma s_\mu(x) a^\mu(x) d\sigma, \quad (14\text{--}13)$$

$$\frac{i}{2}[\dot{\Sigma}, \Sigma] = \frac{1}{2} \int s_\mu(x) \varphi^\mu(x) d\sigma. \quad (14\text{--}14)$$

The last expression describes an interaction of the external current with its own field. It causes no radiative processes and may therefore be omitted. This is consistent with the external field approximation. With Eqs. (14–10) and (14–4a), $G(\tau)$ becomes

$$G(\tau) = -\int_\sigma (j_\mu(x) + s_\mu(x)) a^\mu(x) d\sigma, \quad (14\text{--}15)$$

which is identical with (14–4b).

The new subsidiary condition involves the operator

$$e^{i\Sigma} \partial^\lambda a_\lambda(x) e^{-i\Sigma} = \partial^\lambda a_\lambda(x) + i[\Sigma, \partial^\lambda a_\lambda(x)] + \cdots$$

$$= \partial^\lambda a_\lambda(x) - \int_\sigma D(x - x') s^\mu(x') d\sigma_\mu'.$$

This means that Eq. (14–6a) is transformed into Eq. (14–6b).

Since the commutation rules are not affected by the canonical transformation, we have therefore established that the two descriptions (a) and (b) are dynamically equivalent. In conformity with tradition, we shall adopt the system (a) as the basic set of equations for the external field approximation.

To our knowledge the limits of validity of these equations have never been carefully investigated. It is expected that the approximation is always valid provided the external current density is that of a large number of elementary particles such as a macroscopic charge-current distribution. Under this condition it would take a comparably large number of elementary transition processes to observe an appreciable effect on the external field.

The approximation is valid even if the number of particles which produce the external current is small, provided their mass is very much larger than the electron mass. This is the case for the field of atomic nuclei, but the limits of validity are here apparently much more critical. For instance, the inclusion of the interaction which causes the hyperfine splitting requires a partial extension of the external field approximation. In this case, the dynamical properties of the sources must be partially included in the system. The reaction of the radiation and matter field on the spin degrees of freedom of the source forms an essential ingredient of this interaction.

14–2 The bound interaction picture. For many purposes it is convenient to cast the basic equations into a different form which is, in a sense, intermediate between the interaction and the Heisenberg picture. In this *bound interaction picture* we introduce the state vector $\boldsymbol{\omega}$, which changes with τ only because of the interaction of the matter field with the radiation field.* This can be achieved by subjecting the system in the interaction picture to a τ-dependent canonical transformation which transforms the state vector according to

$$\omega(\tau) = V(\tau)\boldsymbol{\omega}. \tag{14–16}$$

If we assume $V(\tau)$ to satisfy the equation

$$i\dot{V}(\tau) = -\int_{\sigma(\tau)} j^\mu(x)\varphi_\mu(x)d\sigma V(\tau), \tag{14–17}$$

then this new state vector satisfies the Schroedinger equation

$$i\dot{\boldsymbol{\omega}} = -\int_{\sigma(\tau)} \mathbf{j}_\mu(x)a^\mu(x)d\sigma\boldsymbol{\omega}. \tag{14–18}$$

---

* W. H. Furry, *Phys. Rev.* **81**, 115 (1951), especially Section II.

14-2] THE BOUND INTERACTION PICTURE

The transformed current operator which appears in this equation is defined by

$$\mathbf{j}_\mu(x) = V^{-1}(\tau) j_\mu(x) V(\tau), \quad x \in \sigma(\tau), \tag{14-19}$$

which may be expressed in terms of the transformed matter field operators

$$\boldsymbol{\psi}(x) = V^{-1}(\tau)\psi(x)V(\tau), \tag{14-20}$$

$$\mathbf{j}_\mu(x) = -ie\overline{\boldsymbol{\psi}}(x)\gamma_\mu\boldsymbol{\psi}(x). \tag{14-21}$$

Since the transformation operator $V(\tau)$ depends explicitly on τ [*cf.* Eq. (14–17)], the operators $\boldsymbol{\psi}$ satisfy a new differential equation. It is easily obtained with the help of the relation

$$\partial_\mu \boldsymbol{\psi} = V^{-1}\partial_\mu \psi V + in_\mu V^{-1}[\psi, H_e]V, \tag{14-22}$$

where

$$H_e = -\int_{\sigma(\tau)} j_\nu(x)\varphi^\nu(x)d\sigma. \tag{14-23}$$

For the commutator, we obtain

$$[\psi(x), H_e] = en_\mu \gamma^\mu \varphi_\nu \gamma^\nu \psi(x) = en\gamma_\mu \varphi^\mu(x)\psi(x).$$

The new field equation for $\boldsymbol{\psi}$ is now obtained by substituting this expression into Eq. (14–22), contracting with γ^μ, and using Eq. (14–5a); the result is*

$$(\partial + ie\gamma_\mu\varphi^\mu + m)\boldsymbol{\psi}(x) = 0$$

and similarly,

$$\overline{\boldsymbol{\psi}}(x)(\overleftarrow{\partial} - ie\gamma_\mu\varphi^\mu - m) = 0. \tag{14-24}$$

In the following it will be sufficient to consider the first of these equations. It can be written in more compact form by introducing the notation

$$d^\mu = \partial^\mu + ie\varphi^\mu, \quad \boldsymbol{d} = d_\mu\gamma^\mu,$$

so that

$$(\boldsymbol{d} + m)\boldsymbol{\psi} = 0. \tag{14-25}$$

The radiation field operators $a_\mu(x)$ are not affected by the transformation (14–16), since $V(\tau)$ commutes with $a_\mu(x)$. It follows that the transformed

---

* The sign of the φ^μ term in these equations differs from the conventional notation because of our definition of the Fourier transform of ψ and the definition of e as a positive number. Its correctness can be checked by observing that for the *negaton* wave function $u(p)e^{ipx}$ we find

$$(ip + ie\varphi + m)u = 0$$

corresponding to the kinetic momentum $p^\mu + e\varphi^\mu$. The positon wave function ve^{-ipx} then gives the kinetic momentum $p^\mu - e\varphi^\mu$.

subsidiary condition is simply

$$[\partial^\lambda a_\lambda(x) - \int_{\sigma(\tau)} D(x - x')j^\mu(x')d\sigma_\mu']\omega = 0. \quad (14\text{--}26)$$

We collect the basic equations in the bound interaction picture for convenient reference:

$$i\dot{\omega} = -\int_{\sigma(\tau)} j_\mu(x)a^\mu(x)d\sigma\omega, \quad (14\text{--}27)$$

$$\partial^\lambda \partial_\lambda a_\mu(x) = 0, \quad (14\text{--}28)$$

$$(d + m)\psi = 0, \quad (14\text{--}29)$$

$$[\partial^\lambda a_\lambda(x) - \int D(x - x')j_\mu(x')d\sigma'^\mu]\omega = 0. \quad (14\text{--}30)$$

In this form, the external field φ_μ does not appear explicitly in the Schroedinger equation (14–27), but is contained in the field equations for the operator ψ, Eq. (14–29).

Equation (14–27) can be taken as the starting point for the evaluation of the S-matrix by the iteration method, using a procedure very similar to that of Chapter 8. Correspondingly, we shall encounter the same difficulties which the iteration solution presented in Chapters 9 and 10. In addition to these, the external field introduces two further problems. They are related to the propagation function of the matter field and to the definition of the matter vacuum. We shall discuss these two points in the following section.

14–3 Commutation rules. When we calculate the S-matrix elements with the operators ψ of the bound interaction picture, we obtain propagation functions which depend explicitly on the external field. These can be derived from the commutation rules of the ψ-operators.

We recall that the commutation rules on space-like surfaces are unaffected by the presence of the external field, as follows from an adaptation of the discussion given in Section 4–2 regarding the commutation rules of the Heisenberg fields.

For arbitrary x and x' we can still write

$$\{\psi(x), \overline{\psi}(x')\} = iS(x,x'). \quad (14\text{--}31)$$

But in the present case the function $S(x,x')$ will not be a function of the difference $x - x'$ only, as was the case for the free fields ψ without external fields. The spinor matrix $S(x,x')$ depends separately on both four-vectors x,x', that is, on eight variables in all. Because of Eq. (14–29), it satisfies

the differential equation
$$(\mathbf{d} + m)S(x,x') = 0. \tag{14-32}$$

For any space-like interval $[(x - x')^2 > 0]$, the boundary condition is fulfilled:

$$S(x,x') = \gamma^\mu \delta_\mu(x - x') = \partial D(x - x'), \quad [\text{valid for } (x - x')^2 > 0]. \tag{14-33}$$

In the last equation, $\delta_\mu(x - x')$ signifies the surface δ-function introduced in Section 1–12.

The evaluation of the function $S(x,x')$ is difficult, in general, and explicit expressions have been obtained only in certain very special cases.*

The solution of Eq. (14-32) can be reduced to the solution of a partial differential equation for a scalar function. To this end we express the matrix $S(x,x')$ in terms of another matrix, $\Delta(x,x')$, by writing

$$S(x,x') = (\mathbf{d} - m)\Delta(x,x'). \tag{14-34}$$

Because of the relations

$$\mathbf{dd} = d_\mu d^\mu - \frac{e}{2} \varphi_{\mu\nu} \sigma^{\mu\nu},$$

$$\varphi_{\mu\nu} = \partial_\mu \varphi_\nu - \partial_\nu \varphi_\mu, \tag{14-35}$$

$$\sigma^{\mu\nu} = \frac{1}{2i} (\gamma^\mu \gamma^\nu - \gamma^\nu \gamma^\mu),$$

we find that $S(x,x')$ is a solution of Eq. (14-32) provided $\Delta(x,x')$ satisfies the equation

$$(d_\mu d^\mu - m^2 + eM)\Delta(x,x') = 0, \tag{14-36}$$

where

$$M = -\tfrac{1}{2}\varphi_{\mu\nu}\sigma^{\mu\nu}.$$

Contrary to the field-free case ($\varphi_{\mu\nu} = 0$), Eq. (14-36) still contains a matrix operator, namely, M, and therefore $\Delta(x,x')$ is also a matrix in spinor space. However, we can express this matrix in a relatively simply way in terms of a scalar function $K(x,x')$ which is assumed to satisfy

$$(d_\mu d^\mu - m^2)K(x,x') = 0 \tag{14-37}$$

and the boundary condition

$$\partial_\mu K(x,x') = \delta_\mu(x - x'), \quad \text{for } (x - x')^2 > 0. \tag{14-38}$$

---

* J. Schwinger, *Phys. Rev.* **82**, 664 (1951), discusses the case of a constant field and a plane wave. G. Géhéniau and M. Demeur, *Physica* **17**, 71 (1951), give an integral representation for the function $S(x,x')$ in the case of a constant magnetic field. G. Géhéniau and F. Villars, *Helv. Phys. Acta* **23**, 178 (1950), have used approximate expressions for these functions.

When we differentiate Eq. (14–37) with respect to m^2, we obtain

$$(d_\mu d^\mu - m^2)\frac{dK}{d(m^2)} = K. \tag{14-39}$$

This relation can be generalized. Let us define the functions

$$K^{(\nu)} = \frac{dK^{(\nu-1)}}{d(m^2)},$$

$$K^{(0)} = K,$$

such that

$$(d_\mu d^\mu - m^2)K^{(\nu)} = \nu K^{(\nu-1)}. \tag{14-40}$$

With this relation we can express the solution of Eq. (14–36) in terms of the functions $K^{(\nu)}$ in the form

$$\Delta(x,x') = \sum_{\nu=0}^{\infty} \frac{(-1)^\nu}{\nu!} (eM)^\nu K^{(\nu)}. \tag{14-41}$$

which may be written formally as

$$\Delta(x,x') = e^{-eM[d/d(m^2)]} K(x,x'). \tag{14-42}$$

The problem is thus reduced to finding the scalar function $K(x,x')$. This would require an explicit solution of Eq. (14–37), which is difficult in general. We can, however, write an approximate solution in the form of an integral representation. Indeed, neglecting quadratic terms in the external field, we write Eq. (14–37) as

$$(\partial_\mu \partial^\mu - m^2)K(x,x') = -2ie\varphi^\mu(x)\partial_\mu K(x,x'). \tag{14-43}$$

In the absence of the external field, the solution of this equation is one of the homogeneous Δ-functions (Section A–1). We write, therefore,

$$K(x,x') = \Delta(x - x') + K_1(x,x'), \tag{14-44}$$

where $K_1(x,x')$ is assumed to be proportional to the fields and satisfies

$$(\partial_\mu \partial^\mu - m^2)K_1(x,x') = -2ie\varphi^\mu(x)\partial_\mu \Delta(x - x'). \tag{14-45}$$

Since the right side is a known function, this equation can be integrated with the Green's function method. If $G(x - x')$ is any solution of the inhomogeneous equation

$$(\partial_\mu \partial^\mu - m^2)G(x - x') = -\delta(x - x'), \tag{14-46}$$

i.e., any of the inhomogeneous Δ-functions, we find that

$$K_1(x,x') = +2ie \int G(x - x')\varphi^\mu(x'')\partial_\mu'' \Delta(x'' - x')dx'' \tag{14-47}$$

is a solution of Eq. (14–45). The solution is determined by imposing a suitable boundary condition which fixes the choice of the correct Green's function.

Combining this result with Eq. (14–41), we obtain the solution for $\Delta(x,x')$ correct to linear terms in the field

$$\Delta(x,x') = \Delta(x - x') + K_1(x,x') - eM \frac{d}{d(m^2)} \Delta(x - x'). \quad (14\text{–}48)$$

This same method can be used to find the inhomogeneous $S(x,x')$ function which, for instance, enters into the contraction symbol for the ψ-field. This function satisfies the equation

$$(\boldsymbol{d} + m)S(x,x') = -\delta(x - x') \quad (14\text{–}49)$$

instead of Eq. (14–37). It can be expressed in terms of an inhomogeneous K-function which satisfies

$$(d_\mu d^\mu - m^2)K(x,x') = -\delta(x - x') \quad (14\text{–}50)$$

instead of Eq. (14–37). The result is of the form

$$S(x,x') = (\boldsymbol{d} - m)e^{-eM\,[d/d(m^2)]} K(x,x'). \quad (14\text{–}51)$$

We conclude this section with a few remarks concerning the definition of the vacuum state. The definition of the vacuum in the presence of an external field requires some discussion because it is not unambiguous.

We recall briefly the vacuum definition in the absence of an external field. In Section 3–10 we introduced the separation of the matter field operator ψ into the negaton absorption and positon creation operators $\psi^{(-)}$ and $\psi^{(+)}$. We defined then the vacuum state vector ω_0 by the conditions

$$\psi^{(-)}\omega_0 = 0, \quad \bar{\psi}^{(-)}\omega_0 = 0.$$

These conditions express the absence of both kinds of particles in the state represented by ω_0.

In the present situation, the separation into the operators $\psi^{(-)}$ and $\psi^{(+)}$ is not so clear-cut. In the field-free case the frequency spectrum of the plane wave solutions of the Dirac equation consists of two continuous regions, which are separated by a gap of magnitude $2m$. The operator $\psi^{(-)}$ consists of those superpositions of plane waves with a time dependence in the form $e^{-i\omega x^0}$ ($\omega > 0$). On the other hand, the operator $\psi^{(+)}$ contains the time dependence in the form $e^{i\omega x^0}$ ($\omega > 0$). In the presence of external fields we may have solutions which correspond to bound states, such as the states in an attractive Coulomb potential. For these states the frequencies are smaller than m in magnitude. For a sufficiently strong external field, these bound-state levels may penetrate deeply into the gap of width $2m$, so that the separation into states with frequency dependence $e^{\mp i\omega x^0}$ becomes ambiguous. This situation is aggravated by the fact that the external potential φ_μ is determined only up to a gauge transformation. Thus, we could add a constant scalar potential V to the potential φ_0 which would change the position of all frequencies by this amount. The definition of the vacuum then becomes ambiguous.

This difficulty need not concern us too much here. It is simply due to the fact that the external fields which are so strong that they displace the frequencies of the monochromatic solutions well into the gap of width $2m$ cannot be realized within the external field approximation. The breakdown of this approximation would show itself in the fact that pairs could be produced by such a field which effectively neutralize the charges which produce the external field. This situation is closely related to the well-known paradox of Klein,[*] according to which potential barriers of height greater than $2m$ become easily penetrable by electrons. This paradox, which was often discussed in the early days of positon theory, is actually no paradox at all. It is an instance of the limit of the external field approximation.

For all cases of practical importance it is still possible to define the vacuum in the external field approximation by adjusting the gauge in such a way that the gap of order $2m$ in the frequency spectrum of the monochromatic solutions is located at the zero of the frequency scale. We can then define the vacuum state vector ω_0:

$$\psi^{(-)}\omega_0 = 0, \quad \bar{\psi}^{(-)}\omega_0 = 0, \tag{14-52}$$

where ψ is the electron field operator in the bound interaction picture.

14–4 The electron propagation function. Among the inhomogeneous S-functions discussed in the preceding section, the S_c-function (or propagation function) is of particular importance because, as we shall see in the following section, it occurs in the evaluation of the S-matrix. The propagation function S_c can be expressed in terms of the functions $S_\pm(x,x')$ into which $S(x,x')$ is separated, in analogy to the separation $S(x - x')$ by Eqs. (A1–7) and (A1–29):

$$S = S_+ + S_-.$$

In terms of these functions, the propagation function is

$$S_c(x,x') = \theta(x - x')S_+(x,x') - \theta(x' - x)S_-(x,x'), \tag{14-53}$$

where $\theta(x)$ is the discontinuous function defined by Eq. (A1–13). The propagation function S_c is an inhomogeneous S-function. This means that

---
[*] O. Klein, Z. Physik. **53**, 157 (1929).

it satisfies the inhomogeneous wave equation (14–49):

$$(\mathbf{d} + m)S_c(x,x') = -\delta(x - x'). \tag{14-54}$$

For practical applications it is useful to obtain our expression for $S(x,x')$ in terms of a complete set of the solutions $u_a(x)$ and $v_b(x)$ of Eq. (14–25), which are periodic in the time. Such solutions exist only for static potentials φ_μ. They are of the form

$$\begin{aligned} u_a(x) &= u_a(\mathbf{x})e^{-iE_a t}, \quad (E_a > 0), \\ v_b(x) &= v_b(\mathbf{x})e^{+iE_b t}, \quad (E_b > 0), \end{aligned} \quad (t = x^0). \tag{14-55}$$

As in the case of the free particle, we have denoted by u the "positive-energy" solutions, which represent negatons, and by v the "negative-energy" solutions, which represent positons. These solutions—which we employ now for the external field case—differ from the free-particle solutions because the functions $u_a(\mathbf{x})$, $v_b(\mathbf{x})$ satisfy the equations

$$Hu_a(\mathbf{x}) = E_a u_a(\mathbf{x}), \quad Hv_b(\mathbf{x}) = -E_b v_b(\mathbf{x}), \tag{14-56}$$

where

$$H = i\gamma^0 \boldsymbol{\gamma} \cdot \nabla - e\gamma^0 \gamma_\mu \varphi^\mu + i\gamma^0 m. \tag{14-57}$$

These equations are easily obtained from (14–25) after substitution of Eq. (14–55) and multiplication by $i\gamma^0$. We repeat that solutions of the type (14–55) are possible only for essentially static potentials. We can therefore assume that $\varphi^\mu = \varphi^\mu(\mathbf{x})$ is a function of the three-vector $\mathbf{x}$ alone.

The energy levels E_a and E_b may be either discrete or continuous. For simplicity, we write the following formulas for the discrete case. By a suitable change in notation (sums replaced by integrals, etc.) they can be applied also to the case of continuous or mixed energy levels.

The spinor operator H introduced here is the relativistic Hamiltonian of the one-particle theory with the characteristic solutions of both signs in the energy. It has the property that for any two spinors w_a and w_b, the following relation holds:

$$\int \overline{w_b(\mathbf{x})} i\gamma^0 H w_a(\mathbf{x}) d^3 x = \int \overline{H w_b(\mathbf{x})} i\gamma^0 w_a(\mathbf{x}) d^3 x. \tag{14-58}$$

When we apply this equation to the spinors $u_a(\mathbf{x})$ and $v_b(\mathbf{x})$, we obtain the orthogonality relations

$$\int \bar{u}_r(\mathbf{x}) i\gamma^0 u_s(\mathbf{x}) d^3 x = \delta_{rs}, \quad \int \bar{v}_r(\mathbf{x}) i\gamma^0 v_s(\mathbf{x}) d^3 x = \delta_{rs},$$

$$\int \bar{u}_r(\mathbf{x}) i\gamma^0 v_s(\mathbf{x}) d^3 x = 0, \quad \int \bar{v}_r(\mathbf{x}) i\gamma^0 u_s(\mathbf{x}) d^3 x = 0. \tag{14-59}$$

The normalization is, of course, arbitrary. We have chosen it in such a way that the operator $\psi(x)$ defined by

$$\psi(x) = \sum_r [u_r(\mathbf{x})a_r e^{-iE_r t} + v_r(\mathbf{x})b_r^* e^{iE_r t}] \tag{14-60}$$

satisfies the commutation rules (14–31), provided the a_r and b_s satisfy the usual anticommutation rules

$$\{a_r, a_s^*\} = \{b_r, b_s^*\} = \delta_{rs},$$
$$\{a_r, a_s\} = \{b_r, b_s\} = \{a_r, b_s\} = 0. \tag{14-61}$$

When we substitute Eq. (14–60) into Eq. (14–31) and use Eq. (14–61), we obtain the following explicit representation of $S(x,x')$:

$$S(x,x') = S_+(x,x') + S_-(x,x'),$$

with

$$S_+(x,x') = \sum_r u_r(\mathbf{x})\bar{u}_r(\mathbf{x}')e^{-iE_r(t-t')},$$

$$S_-(x,x') = \sum_s v_s(\mathbf{x})\bar{v}_s(\mathbf{x}')e^{+iE_s(t-t')}. \tag{14-62}$$

The property of the S-function expressed in Eq. (14–33) is equivalent to the completeness relation for the spinors u_r and v_s:

$$\sum_r (u_r(\mathbf{x})\bar{u}_r(\mathbf{x}') + v_r(\mathbf{x})\bar{v}_r(\mathbf{x}')) = i\gamma^0 \delta(\mathbf{x} - \mathbf{x}'). \tag{14-63}$$

From the expressions $S_\pm$ given in Eq. (14–62) and from the definition (14–53) of the S_c-function, we obtain now the following explicit form of the propagation function:†

$$S_c(x,x') = \begin{cases} \sum_r u_r(\mathbf{x})\bar{u}_r(\mathbf{x}')e^{-iE_r(t-t')}, & (\text{for } t > t'), \\ -\sum_r v_r(\mathbf{x})\bar{v}_r(\mathbf{x}')e^{iE_r(t-t')}, & (\text{for } t < t'). \end{cases} \tag{14-64}$$

The Fourier transform of $S_c(x,x')$ can be defined in analogy to $S_c(x - x')$:

$$S_c(x,x') = \frac{1}{(2\pi)^4} \int S_c(p,p') e^{ipx - ip'x'} d^4p\, d^4p'. \tag{14-65}$$

Operating on this expression with $\boldsymbol{d} + m$ from the left, we obtain

$$(\boldsymbol{d} + m)S_c(x,x') = \frac{1}{(2\pi)^4} \int [(i\boldsymbol{p} + m)\delta(p - q) - iV(p - q)]$$
$$\times S_c(q,p') e^{ipx - ip'x'} d^4q\, d^4p\, d^4p'. \tag{14-66}$$

---

† R. P. Feynman, *Phys. Rev.* **76**, 749 (1949).

In this last expression we have introduced the potential energy function

$$V(p) = V_\mu(p)\gamma^\mu, \qquad (14\text{-}67)$$

$$V_\mu(p) = \frac{-e}{(2\pi)^{3/2}} \varphi_\mu(\mathbf{p})\delta(p^0),$$

which is related to the external potential $\varphi^\mu(\mathbf{x})$ by

$$\varphi^\mu(\mathbf{x}) = \frac{1}{(2\pi)^{3/2}} \int \varphi^\mu(\mathbf{p}) e^{i\mathbf{p}\cdot\mathbf{x}} d^3p. \qquad (14\text{-}68)$$

The expression in square brackets which occurs on the right side of Eq. (14–66) may be looked upon as the matrix element of an integral operator in momentum space,

$$(i p + m)\delta(p - q) - iV(p - q) = (p|ip - iV + m|q).$$

A similar interpretation can be given to $S_c(p,p')$:

$$S_c(p,p') = (p|S_c|p').$$

In operator form, Eq. (14–54) is equivalent to the relation

$$(ip - iV + m)S_c = -1 \quad \text{or} \quad S_c = -(ip - iV + m)^{-1}. \qquad (14\text{-}69)$$

In this form, the propagation function S_c is an obvious generalization of the propagation function for no external field, when S_c reduces to $-(ip + m)^{-1}$. Just as in that case, we must make suitable provisions for the path of integration, which consist in assuming a mass with a small negative imaginary part $m - i\mu$ ($\mu > 0$) and in taking the limit $\mu \to +0$ at the end of the calculation.

The advantages of Eq. (14–69) become evident when the field is weak, so that S_c can be expanded:

$$S_c = -(ip - iV + m)^{-1} = -(ip + m)^{-1} - (ip + m)^{-1}iV(ip + m)^{-1}$$
$$- (ip + m)^{-1}iV(ip + m)^{-1}iV(ip + m)^{-1} - \cdots. \qquad (14\text{-}70)$$

When this expansion is not valid, or when terms of a certain order in the external field need to be separated, the following types of identities are useful:

$$(ip - iV + m)^{-1} = (ip + m)^{-1} + (ip + m)^{-1}iV(ip - iV + m)^{-1}$$
$$= (ip + m)^{-1} + (ip - iV + m)^{-1}iV(ip + m)^{-1}$$
$$= (ip + m)^{-1} + (ip + m)^{-1}iV(ip + m)^{-1}$$
$$+ (ip + m)^{-1}iV(ip - iV + m)^{-1}iV(ip + m)^{-1}. \qquad (14\text{-}71)$$

A typical application of this form of S_c will be discussed in Section 15–4.

For some applications of the propagation function it is convenient to have it in a form which corresponds to the nonrelativistic limit and to the case of the one-particle theory. The second assumption means that we consider only that part of the S-operator which refers to one kind of particles (negatons only or positons only). Such an abbreviated theory will be valid whenever one can disregard the contribution from processes of pair creation and pair annihilation. Since these processes involve an energy expense of at least $2m$, they can almost always be neglected in the nonrelativistic region.

If, in addition to the above assumption, it is also valid to disregard the interaction between orbital and spin angular momentum, then the wave function for one particle can be given by an ordinary scalar Schroedinger function $\psi(\mathbf{x})$.* In the following we shall refer to the above-mentioned three approximations simply as "the nonrelativistic limit."

The main problem in the discussion of the nonrelativistic limit is the reduction of the spinor wave functions $u(\mathbf{x})$ and $v(\mathbf{x})$ to the scalar functions $\psi(\mathbf{x})$. This can be accomplished by separating the spinors into *large* and *small* components.† We shall consider $u(x)$ in detail. The separation of $v(\mathbf{x})$ then follows by analogy.

Let us write
$$u = u_l + u_s,$$
where
$$u_l = \tfrac{1}{2}(1 + i\gamma^0)u, \quad u_s = \tfrac{1}{2}(1 - i\gamma^0)u.$$

Corresponding to this decomposition, we can split the equation

$$Hu = Eu \tag{14-72}$$

into a pair of equations for the spinors u_l and u_s. To this end, we write

$$H = K + V + i\gamma^0 m,$$
where
$$K = i\gamma^0 \boldsymbol{\gamma} \cdot (\nabla + ie\boldsymbol{\varphi}) = i\gamma^0 \boldsymbol{\gamma} \cdot \mathbf{d},$$

$$V = -e\varphi^0.$$

---

* The $\psi(\mathbf{x})$ introduced here should not be confused with the ψ which describes the operator of the electron field. The latter is a *spinor operator* and the former a *scalar wave function* (c-number). This overlap in notation is unfortunate. It is unavoidable, since it is firmly established by a long-standing tradition.

† The usual method of separation is almost always carried out in a special representation of the γ-matrices. See, for instance, W. Pauli, *Handbuch d. Physik* **24**, p. 236. However, as is shown in the text, this special choice of the representation is not necessary.

We now multiply Eq. (14–72) from the left by $i\gamma^0$, and use

$$i\gamma^0(K + V + i\gamma^0 m) = (-K + V + i\gamma^0 m)i\gamma^0.$$

The equation thus obtained from (14–72) can be added to or subtracted from Eq. (14–72), and we obtain the following system of two equations:

$$Ku_s = (E - V - m)u_l, \quad Ku_l = (E - V + m)u_s. \quad (14\text{–}73)$$

In this form it is easy to recognize that the terms "large" and "small" components are justified. Indeed, since the definition of E includes the rest energy m, the quantity $E - m$ is much smaller than $E + m$, in the nonrelativistic limit. For instance, for a Coulomb field this ratio is of order $\alpha^2 Z^2$ for the bound states. It follows that the components u_s are much smaller than the components u_l. In fact, using the second of Eqs. (14–73) and neglecting V compared with m, we can write

$$u_s \simeq \frac{K}{2m} u_l.$$

In the nonrelativistic limit it is permitted to neglect the small components. This implies that we everywhere replace u by u_l. Since $i\gamma^0 u_l = u_l$, we can also omit the factors $i\gamma^0$.

The spinors thus obtained are still four-component spinors, but we can reduce them to two-component spinors. This is possible because $\frac{1}{2}(1 + i\gamma^0)$ is a projection operator of rank two. By adapting the coordinate system to the two-dimensional subspace of the u_l, the spinors can be reduced to two components. The wave function can then be written as a product of a space function $\psi(\mathbf{x})$ and a two-component spin function:

$$u_l(\mathbf{x}) = \psi(\mathbf{x})\chi.$$

The spin function χ is so normalized that $\bar{\chi}\chi = 1$. We have a further relation for the two linearly independent spin functions χ_r ($r = 1, 2$), namely,

$$\sum_{r=1}^{2} \chi_r \bar{\chi}_r = 1.$$

It is now possible to express all quadratic expressions in $\bar{u}$ and u in terms of the scalar function $\psi(\mathbf{x})$ alone. For instance, the nonrelativistic form of the propagation function is now easily obtained as

$$S_c^{\text{NR}}(x,x') = \begin{cases} \sum_n \psi_n(\mathbf{x})\psi_n^*(\mathbf{x}')e^{-iE_n(t-t')}, & t > t', \\ 0, & t < t', \end{cases} \quad (\text{NR}). \quad (14\text{–}74)$$

The Fourier transform of this function follows from Eq. (14–65):

$$S_c^{NR}(p,p') = \frac{1}{(2\pi)^4} \int_{t>t'} \sum_n \psi_n(\mathbf{x})\psi_n^*(\mathbf{x}')e^{-iE_n(t-t')}e^{-ipx+ip'x'}d^4x d^4x'$$

$$= \frac{1}{2\pi} \int_{t>t'} dt dt' \sum_n \psi_n(\mathbf{p})\psi_n^*(\mathbf{p}')e^{-iE_n(t-t')}e^{iEt-iE't'},$$

where

$$\psi_n(\mathbf{p}) = \frac{1}{(2\pi)^{3/2}} \int \psi_n(\mathbf{x})e^{-i\mathbf{p}\cdot\mathbf{x}}d^3x.$$

We make use of

$$\int_0^\infty e^{i\alpha t}dt = \pi\delta(\alpha) + iP\frac{1}{\alpha}$$

and find, after a change of variables $t \to t - t'$,

$$S_c^{NR}(p,p') = \delta(E - E') \sum_n \left[\pi\delta(E - E_n)\psi_n(\mathbf{p})\psi_n^*(\mathbf{p}') + i\frac{\psi_n(\mathbf{p})\psi_n^*(\mathbf{p}')}{E - E_n}\right],$$

(NR). (14–75)

14–5 The S-matrix in the external field approximation. In this section we shall develop the methods for the evaluation of the S-matrix in the external field approximation. These methods are based on the iteration solution developed in Chapter 8, suitably modified to include the external field. It would be more desirable to base the calculations on the more general solution discussed in Chapter 7. But to our knowledge such a calculation has never been carried out successfully.

We may base the iteration solution on the bound interaction picture discussed in Section 14–2. The formal development is exactly the same as that at the beginning of Chapter 8. We obtain the S-operator in the form of Eqs. (8–7) and (8–8). The interaction operator $H(\tau)$ is given by

$$H(\tau) = ie \int_{\sigma(\tau)} \overline{\Psi}(x)\mathbf{a}(x)\Psi(x)d\sigma, \qquad (14\text{--}76)$$

where Ψ is the operator for the electron field in the bound interaction picture. It satisfies Eq. (14–29).

We can therefore develop the theory of the Feynman-Dyson diagrams in a manner completely analogous to the work of Section 8–2. The only difference is the appearance of the electron field operators $\Psi(x)$ in the presence of an external field instead of the free electron field operators $\psi(x)$. Consistently, the electron propagation function $S_c(x - x')$ is replaced by $S_c(x,x')$ discussed in the previous section. Otherwise Table 8–1 can be used.

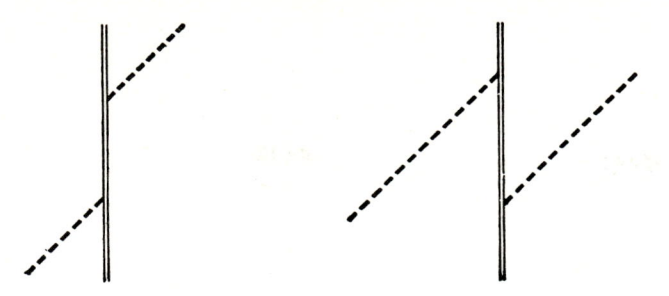

FIG. 14-1. Negaton-negaton scattering in an external field.

To express this change in the diagrams, we indicate the electron paths by double lines. For instance, the diagrams depicted in Figs. 8-2 and 8-3 would be modified, in the presence of an external field, as shown in Figs. 14-1 and 14-2. The analytical expressions for these two types of processes are given by Eqs. (8-10) and (8-11) provided one substitutes in these expressions $\boldsymbol{\psi}(x)$ for $\psi(x)$ and $S_c(x,x')$ for $S_c(x - x')$.

Unfortunately, $S_c(x,x')$ is not known in closed functional form (cf. the discussion in Section 14-3), but can be written only in operator form, Eq. (14-69). In general, useful results can be obtained only when expansions of the type in Eq. (14-70) can be employed.

It is therefore necessary to resort to an approximation method which we call the *Born approximation*.

The term *Born approximation* is frequently used for a more special class of problems encountered in wave-mechanical scattering theory.* Our more general definition in the present context is justified because the approximation thereby defined reduces to the latter definition for that restricted class of problems.

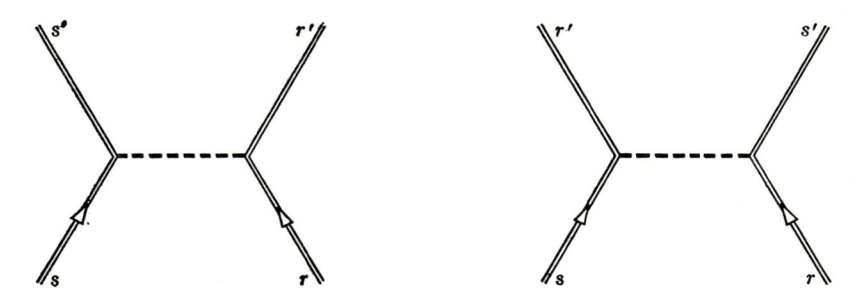

FIG. 14-2. The two second order diagrams for Compton scattering in the presence of an external field.

---

* Cf., for instance, N. F. Mott and H. W. S. Massey, *The Theory of Atomic Collisions*, Oxford, Clarendon Press (1949).

The Born approximation is obtained if we develop the iteration solution in powers of the coupling to the external field. This could be done, for instance, by developing the propagation functions $S_c(x,x')$ according to the scheme indicated in Eqs. (14–34) and (14–42), or in Eq. (14–69). It is formally more elegant to start with the system of equations labeled (a) in Section 14–1. The interaction operator in this case is

$$H(\tau) = H_r(\tau) + H_e(\tau), \qquad (14\text{–}77)$$

$$H_r(\tau) = -\int_{\sigma(\tau)} j_\mu(x) a^\mu(x) d\sigma, \quad H_e(\tau) = -\int_{\sigma(\tau)} j_\mu(x) \varphi^\mu(x) d\sigma. \qquad (14\text{–}78)$$

The operator $H_r(\tau)$ describes the interaction of the electron with the radiation field, while $H_e(\tau)$ contains the interaction with the external field.

The iteration solution is identical in form with Eqs. (8–7) and (8–8). But instead of Eq. (14–76) for $H(\tau)$ we must now substitute Eq. (14–77). Each term $S^{(n)}$ is a polynomial of order n in the external field. We can therefore write

$$S^{(n)} = \sum_{\nu=0}^{n} S^{(n\nu)}, \qquad (14\text{–}79)$$

where $S^{(n\nu)}$ contains all the terms of νth power in the external field φ_μ which contribute to $S^{(n)}$.*

The terms $S^{(n0)}$ describe the system in the absence of an external field. The sum of these terms is identical with the S-matrix discussed in Chapter 8. The terms $S^{(n1)}$ contain the external field in the first power only (first Born approximation). For most applications this approximation is sufficiently accurate. The lowest-order terms, $S^{(11)}$ and $S^{(21)}$, are the first Born approximations without radiative corrections, and $S^{(21)}$ describes to a remarkable degree of accuracy most of the radiative transitions in external fields usually encountered in atomic physics. The term $S^{(31)}$ contains the first-order radiative correction to the effects described by $S^{(11)}$ which cause some very small but, under favorable conditions, observable effects.

Examination of the term $S^{(n\nu)}$ reveals that this expression can be obtained from $S^{(n)}$, Eq. (8–8), by replacing ν operators $a_\mu(x)$ by ν external

---

* The question of the limit of validity and the convergence of the Born approximation has been the subject of several recent papers. In the case of the wave-mechanical scattering theory these questions were investigated by R. Jost and A. Pais, *Phys. Rev.* **82**, 840 (1951) and by W. Kohn, *Rev. Mod. Phys.* **26**, 292 (1954).

By applying the Fredholm theory of integral equations to these problems the limit of validity of the iteration solution can be established. The Fredholm theory is also applicable to the case of a quantized electron field in an external field, provided the effect of the radiation field is neglected [M. Neuman, *Phys. Rev.* **85**, 129 (1952); A. Salam and P. T. Matthews, *Phys. Rev.* **90**, 690 (1953)].

field functions $\varphi_\mu(x)$ in all possible ways and summing over all the expressions so obtained. The current operators are not affected by this operation.

The result of these substitutions may be most easily described in diagram form by introducing a new element into the diagrams, indicated by a corner marked by ×. It represents the external field acting at the corner so marked. If we use the diagrams in the momentum space, we need the Fourier components $\varphi_\mu(q)$ of the external fields, defined by

$$\varphi_\mu(x) = \frac{1}{(2\pi)^{3/2}} \int d^4q \varphi_\mu(q) e^{iqx}. \quad (14\text{–}80)$$

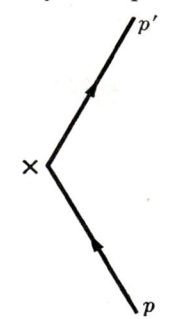

FIG. 14–3. The external field acting at a corner ×.

Table 8–2, which describes the analytical expression associated with the various elements in momentum space, can now be completed by adding one more rule: every external field action (Fig. 14–3) corresponds to a factor $\varphi_\mu(q)$.

The example of Fig. 14–3 describes the scattering of an electron in an external field. The lowest order radiative corrections to this process are shown in Fig. 15–3.

14–6 Renormalization. The method of calculation given in the preceding section for the S-matrix will, of course, contain all the divergent expressions which we have studied in Chapter 9. It is now necessary to show that these divergences can be removed by the method of renormalization. To this end we need to generalize the proof of Chapter 10 to include the case of an external field.

As far as the internal lines are concerned, everything remains exactly as in Chapter 10. The classification of the divergences, the concept of reducible and irreducible diagrams, the mass and charge renormalization, and finally, the renormalization of the electron wave functions are exactly the same as explained in that chapter.

The new feature comes with the renormalization of the external field. For the consistency of the method it is essential to include this in the general proof. One way of obtaining this renormalization factor is to consider for a moment the theory which includes the external field as a quantized field associated with the external charge-current distribution. In physical terms, this means that we enlarge the physical system by including the external part of the system with the original system. This procedure has the advantage that we may now apply all the results of the analysis of Chapter 10. There we have shown that the radiative corrections to the photon field operators results in a factor $\sqrt{\bar{Z}}$ for these operators. When this is translated back into the external field approximation, we find that the renormalized external field $\varphi_{\mu 0}$ is defined in accordance with Eq. (10–67) by

$$\varphi_\mu' = \sqrt{Z}\, \varphi_{\mu 0}. \tag{14–81}$$

Let us now consider the matrix element $M^{(\mu\nu)}$ of a radiative process of nth order containing ν external field corners. It is represented by an irreducible diagram of n corners. Further, let $M'^{(n\nu)}(e)$ be the matrix element for the same process including all the radiative corrections. It is divergent, of course. We remove the divergence due to the self-energy of the electron by renormalization of the mass, as described in Section 10–5. As we shall show, the remaining divergences can all be removed by renormalization of the charge:

$$M'^{(n\nu)}(e) = M_0^{(n\nu)}(e_0), \tag{14–82}$$

where M_0 is the matrix element obtained from M' by omitting all the divergent parts of the insertions into the irreducible diagram which corresponds to M.

To prove this equality, we need the relations

$$n = E_i + \tfrac{1}{2} E_e = 2P_i + P_e + \nu. \tag{14–83}$$

The first of these has already been used in Section 10–1. The second is a generalization which takes into account that each corner is either an external field corner or the source (or sink) of a photon line.

We write symbolically for the matrix element (cf. Section 10–8),

$$M'^{(n\nu)}(e) \sim e^n (\Gamma)^n (S')^{E_i} (D')^{P_i} (\bar{\psi}\psi)^{\frac{1}{2}E_e} (a')^{P_e} (\varphi)^\nu. \tag{14–84}$$

Each factor in this expression occurs with the same exponent as the corresponding factor in the expression for M. When we substitute into (14–84) the relations (10–48) to (10–52), (10–65), and (14–81), we obtain

$$M'^{(n\nu)}(e) = Z^{[-(n/2)+P_i+\frac{1}{2}P_e+\frac{1}{2}\nu]} Z'^{(-n+E_i+\frac{1}{2}E_e)} M_0^{(n\nu)}(e_0).$$

It follows, therefore, from Eq. (14–83) that all the Z-factors just cancel, and we have thus proved Eq. (14–82).

The consistency of the renormalization procedure is thereby also established for the external field problems.

14–7 Cross sections and energy levels. The considerations of Section 8–6 must here be repeated for the case of external fields. Although the general formula (8–49) is still valid with only minor modifications, several essential differences exist in its derivation for external fields.

14-7] CROSS SECTIONS AND ENERGY LEVELS

Since almost all the applications to scattering in external fields (Chapter 15) deal with the Coulomb field, we want to consider *static fields* in particular. Such a field has a Fourier transform of the form

$$\varphi_\mu(x) = n_\mu \frac{1}{(2\pi)^{3/2}} \int V(q)\delta(q^0)e^{iq\cdot x}d^4q. \tag{14-85}$$

The factor $(2\pi)^{-3/2}$ is chosen to conform with the Fourier expansions of the quantized fields, Eqs. (8-16) and (8-17). The S-matrix element for a given process can now be written down by the methods of Sections 8-2 and 8-3, and with the same numerical factor (8-20) as before. The matrix element $(f|S|i)$ contains, as usual, a δ-function of energy-momentum conservation. This δ-function contains the three-momentum transfer to the external field, but there is no energy transfer, since the field is static. This situation is borne out by the function $\delta(q^0)$ in Eq. (14-85). It follows that the integrations over d^4q at all corners of the diagram where the external field acts will eliminate the δ-function of conservation of three-momentum, but a function

$$\delta(E_f - E_i)$$

will survive. E_i and E_f here indicate the sum of the energies of the initial and final photons and electrons. In contradistinction to Eq. (8-29), we define the matrix $(f|M|i)$ for static fields by

$$(f|S|i) = (f|R|i) = \delta(E_f - E_i)(f|M|i). \tag{14-86}$$

The integration over the d^4q which must be carried out in the matrix element accounts for the action of the potential at all possible momentum transfers; i.e., it accounts for the spatial distribution of the potential. The spatial extent of the potential is the volume in which the interaction takes place. Therefore, there is no meaning to a transition probability per unit space-time volume as in Section 8-6, but instead there is only a transition probability per unit time,

$$\Gamma = \lim_{T \to \infty} \frac{|(f|S|i)|^2}{T}, \tag{14-87}$$

where T is the time during which the interaction takes place. The remaining argument proceeds as before:

$$\Gamma = |(f|M|i)|^2 \lim_{T \to \infty} \frac{1}{T} \cdot \left| \frac{1}{2\pi} \int_{-T/2}^{+T/2} e^{-i(E_f - E_i)t}dt \right|^2$$

$$= |(f|M|i)|^2 \frac{1}{4\pi^2} 2\pi\delta(E_f - E_i).$$

The transition probability per unit time is therefore

$$\Gamma = \frac{1}{2\pi} \delta(E_f - E_i)|(f|M|i)|^2 \qquad (14\text{--}88)$$

or, more generally,

$$\Gamma = \frac{1}{2\pi} S_f \bar{S}_i \delta(E_f - E_i)|(f|M|i)|^2, \qquad (14\text{--}89)$$

since the initial and final states may contain unresolved substates which must be averaged and summed appropriately.

The result (14–89) is to be compared with Eq. (8–40), where $(f|M|i)$ is defined differently and where we have a δ-function over the four-momenta. Both formulas are independent of the interaction volume V.

To show how V disappears in the formula for the cross section, we change the normalization in $(2\pi)^3$ to that in V, as in Section 8–6, and obtain for g incident particles,

$$\Gamma = \frac{(2\pi)^{3g-1}}{V^g} S_f \bar{S}_i \delta(E_f - E_i)|(f|M|i)|^2.$$

The interesting case is that of one particle incident on the field φ_μ, $g = 1$. The incident flux I is determined by the velocity β of the electron ($\beta = 1$ for a photon):

$$I = \frac{\beta}{V}. \qquad (14\text{--}90)$$

The cross section is therefore

$$\sigma = \frac{(2\pi)^2}{\beta} S_f \bar{S}_i \delta(E_f - E_i)|(f|M|i)|^2. \qquad (14\text{--}91)$$

This equation replaces the similar equation (8–49) for those scattering problems which involve one particle incident on a static field.

If more than one particle is incident, similar considerations show that V will always cancel and suitable factors $(2\pi)^{3g-1}$ will appear.

In addition to these formulas for scattering problems, methods must be developed for treating other types of problems which arise characteristically in the presence of external fields. We are referring here to essentially static effects which have their origin in the presence of the radiation field and are absent when radiative corrections are neglected. Examples are the "anomalous" magnetic moment of the electron and the shift of energy levels, as first observed by Lamb and Retherford.

For this purpose, it is essential to be able to separate the infinities before taking the required expectation values. In fact, the natural method of calculation, namely, the use of the bound interaction picture, does not allow us to do so, because the covariant form is lost in this picture. We must therefore separate all divergences first in the free interaction picture and only afterwards transform to the bound interaction picture.

This method proceeds as follows. We first transform from our usual free interaction picture, characterized by the state vector ω, to a different picture ω', defined by

$$\omega = \Omega\omega', \qquad (14\text{–}92)$$

where Ω is required to satisfy

$$i\dot{\Omega} = H\Omega \qquad (14\text{–}93)$$

and is therefore identical with the wave operator of Section **7–3**. This transformation yields, when applied to

$$i\dot{\omega} = (H_e + H)\omega,$$

a time dependence of ω' determined by the equation

$$i\dot{\omega}' = H_e'\omega', \quad H_e' = \Omega^* H_e \Omega. \qquad (14\text{–}94)$$

The ω' picture represents the *physical* electron in interaction with the external field. When the radiation field of the electron is ignored, i.e., when H' in

$$H_e' = H_e + H'$$

is neglected, we return to the bare electron in an external field, which is used, for example, in the Dirac theory of hydrogen-like atoms. In this particular case, H' will give rise to a radiative correction to the Dirac energy levels, the correction to the level E_n being

$$\Delta E_n = (n|H'|n). \qquad (14\text{–}95)$$

The essential point now is that for a one-particle problem the S-operator in the ω' picture is effectively the same as in the free interaction picture.* To see this, we consider the S-operator in the two pictures:

$$\omega(\infty) = S_F\omega(-\infty), \quad \omega'(\infty) = S'\omega'(-\infty). \qquad (14\text{–}96)$$

The transformation (14–92) therefore gives the relation

$$S' = \Omega^*(\infty)S_F\Omega(-\infty). \qquad (14\text{–}97)$$

We are now free to choose the initial conditions on Ω in Eq. (14–93). When we choose $\Omega(-\infty) = 1$ we have, according to Eq. (7–44)$_+$, the wave operator Ω_+ which satisfies $\Omega_+(\infty) = S$ [*cf.* Eq. (7–69)$_+$]. Therefore,

$$S' = S^* S_F. \qquad (14\text{–}98)_+$$

---

* Although this method was first used by Schwinger, Dyson seems to have been the first to emphasize this important point.

On the other hand, when we choose $\Omega(\infty) = 1$, we deal with Ω_- according to Eq. (7-44)$_-$. We can then use Eqs. (7-69)$_-$ and (7-70), and find $\Omega_-(-\infty) = S_- = S^*$, so that

$$S' = S_F S^*. \qquad (14\text{-}98)_-$$

Consider now the state vector $\omega_1'(\tau)$ which describes a one-particle system. The S-operator gives rise to no physical effects, since the resulting self-energy effects are all removed by renormalization. Therefore,

$$S\omega_1' = \omega_1',$$

We can thus use either Eq. (14-98)$_+$ or Eq. (14-98)$_-$ depending on whether the final or the initial state is a one-particle state, and we find, respectively

$$(\omega_1'(\infty), S'\omega'(-\infty)) = (\omega_1'(\infty), S^* S_F \omega'(-\infty)) = (\omega_1(\infty), S_F \omega'(-\infty)) \quad (14\text{-}99)_+$$

and

$$(\omega'(\infty), S'\omega_1'(-\infty)) = (\omega'(\infty), S_F S^* \omega_1'(-\infty)) = (\omega'(\infty), S_F \omega_1(-\infty)). \quad (14\text{-}99)_-$$

When both the initial and final states are one-particle states, we have, as a special case of either choice, $\Omega = \Omega_+$ or $\Omega = \Omega_-$,

$$(\omega_1'(\infty), S'\omega_1'(-\infty)) = (\omega_1'(\infty), S_F \omega_1'(-\infty)). \qquad (14\text{-}100)$$

This establishes that the expectation values of S' and of S_F are identical provided at least one of the two states, initial or final, is a one-particle state. The infinities can thus be removed in the usual way, since we are effectively working in the free interaction picture.

After this has been accomplished, we transform into the bound interaction picture, following Section 14-2:

$$S_B = V^{-1} S_F V.$$

Equations (14-95) and (14-100) then yield for the mth radiative correction to the level E_n,

$$-i \int_{-T}^{T} \Delta E_n dt = (\boldsymbol{\omega}_n, S_B^{(m)} \boldsymbol{\omega}_n).$$

Using Eq. (14-86), this can be written

$$\Delta E_n = -\frac{1}{2\pi i} (n | M_B^{(m)} | n). \qquad (14\text{-}101)$$

In first approximation, the bound particle wave functions are the Dirac-Coulomb wave functions. Since the radiative corrections are small, this approximation is sufficiently accurate. The transformation V then effectively involves only the plane wave expansion of the Coulomb wave functions.

CHAPTER 15

EXTERNAL FIELD PROBLEMS

Applications of the theory of the external field are numerous and of greatest importance. Many photon-electron processes of radiation theory are enhanced considerably when the incident photon is replaced by an external field. Of particular interest is the nuclear Coulomb field, since it is extremely strong. For example, Compton scattering is to be compared with Coulomb scattering, double Compton scattering with bremsstrahlung, pair production in photon-electron collisions with pair production by an electron in a Coulomb field. Similarly, the photon-photon processes come within the realm of observability: photon-photon scattering becomes Delbrück scattering, and pair production in photon-photon collisions becomes pair production by a photon in a Coulomb field.

The most accurate experiments, however, are those that can be carried out on bound systems. The theory of the external field permits us to extend our considerations from the fleeting positronium to the easily observable atomic spectra. In fact, the extremely precise measurements of atomic energy levels puts quantum electrodynamics to its severest test.

We shall first consider Coulomb scattering and its radiative corrections (Sections 15–1 and 15–2). These calculations lead to the discovery of the anomalous magnetic moment of the electron (Section 15–3). Then bound states are discussed. Radiative corrections to the Dirac energy levels are calculated (Section 15–4) and the well-known electric dipole transition probability is derived from our general theory (Section 15–5). The study of electrons in a Coulomb field is concluded with a section on bremsstrahlung (Section 15–6). The last two sections of this chapter are devoted to the interaction of photons with nuclear Coulomb fields. First, the inelastic process of pair production by photons (Section 15–7) is derived from bremsstrahlung by the substitution law, and then the elastic scattering of photons (Delbrück and Rayleigh scattering) is discussed in some detail (Section 15–8). This process enables us to obtain further insight into the physical meaning of the external field approximation.

15–1 Coulomb scattering. The most important approximation to the elastic scattering of electrons (negatons or positons) by atoms consists in the neglect of the negaton cloud surrounding the nucleus. The atom is then considered as an infinitely heavy positive point charge of magnitude Ze. The system under consideration is thereby simplified to the motion of an electron in a Coulomb field. We shall discuss at the end of this section how one can improve on this approximation.

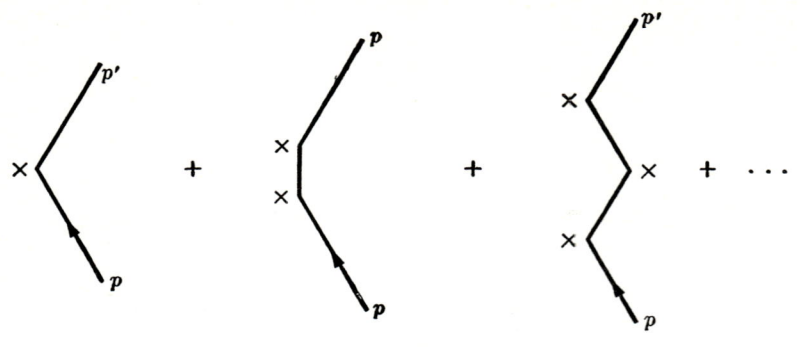

FIG. 15-1. Coulomb scattering.

The theory of the external field considered in the previous chapter can now be applied for this special case. We are dealing with a double expansion, one in α which describes the radiation field of the electron, and one in αZ which corresponds to a power series expansion in the external field. We shall refer to *Coulomb scattering* as the scattering of an electron in a Coulomb field to all orders in this field, but to zero order in the radiation field. The effect of the presence of virtual (or real) photons emitted and reabsorbed by the scattered electron, briefly called *radiative corrections*, will be dealt with in the following section.

It appears advantageous to evaluate the Coulomb scattering cross section to lowest order in αZ first in a purely formal way, and to postpone the discussion of the meaning and validity of such a calculation.

In terms of diagrams, the various orders in the Coulomb field can be depicted as in Fig. 15-1. The first diagram is evaluated as follows. The cross section for a static field is given by Eq. (14-91),

$$\sigma = (2\pi)^2 \frac{1}{\beta} S_f \bar{S}_i \delta(\epsilon_f - \epsilon_i) |(f|M|i)|^2, \qquad (15\text{-}1)$$

where β is the incident velocity. The matrix element, according to Eq. (14-86), for the case of a negaton is

$$(p'|M^{(11)}|p)\delta(\epsilon' - \epsilon) = \frac{e}{\sqrt{2\pi}} \sqrt{\frac{m^2}{\epsilon'\epsilon}} \int \bar{u}(p')\phi(q)\mathbf{n}u(p)\delta(p' - p - q)d^4q. \qquad (15\text{-}2)$$

For the Coulomb field of positive charge Ze,

$$\varphi_\mu(x) = n_\mu \frac{Ze}{4\pi r} = \frac{n_\mu}{(2\pi)^{3/2}} \int \phi(\mathbf{q})\delta(q^0)e^{iq\cdot x}d^4q, \qquad (15\text{-}3)$$

where n_μ is a time-like unit vector, $n^2 = -1$, we find

$$\phi(\mathbf{q}) = \frac{Ze}{(2\pi)^{3/2}} \frac{1}{|\mathbf{q}|^2}. \qquad (15\text{-}4)$$

Insertion of this expression into Eq. (15–2) yields

$$(p'|M^{(11)}|p) = \frac{Ze^2}{(2\pi)^2} \cdot \frac{m}{\epsilon} \frac{1}{|\mathbf{p}' - \mathbf{p}|^2} \bar{u}(\mathbf{p}')nu(\mathbf{p}). \tag{15-5}$$

For the scattering cross section of a polarized negaton beam into a solid angle $d\Omega$ and a definite polarization, Eqs. (15–2) and (15–5) yield, with $\epsilon = \gamma m$,

$$d\sigma^{(11)} = \frac{Z^2 e^4}{(2\pi)^2 \beta\gamma^2} \frac{1}{\int} \frac{|\mathbf{p}'|\epsilon' d\epsilon'}{|\mathbf{p}' - \mathbf{p}|^4} |\bar{u}(\mathbf{p}')nu(\mathbf{p})|^2 \delta(\epsilon' - \epsilon) d\Omega$$

$$= \frac{(Zr_0)^2}{4} \frac{|\bar{u}(\mathbf{p}')nu(\mathbf{p})|^2}{(\beta\gamma)^4 \sin^4 (\theta/2)} d\Omega. \tag{15-6}$$

For positons, $|\bar{u}(\mathbf{p}')nu(\mathbf{p})|$ is to be replaced by $|\bar{v}(\mathbf{p})nv(\mathbf{p}')|$. When the electron spin directions are not observed, we must average over the initial state and sum over the final spin state. Equation (A2–87) gives for negatons scattered through an angle θ,

$$\tfrac{1}{2} \sum_{\text{spins}} |\bar{u}(\mathbf{p}')nu(\mathbf{p})|^2 = -2(n\Lambda_-(p)n\Lambda_-(p'))$$

$$= \frac{1}{2m^2} (2(n \cdot p)^2 + p \cdot p' + m^2)$$

$$= \gamma^2 \left(1 - \beta^2 \sin^2 \frac{\theta}{2}\right).$$

Therefore, the differential cross section for unpolarized electrons is

$$\frac{d\sigma^{(11)}}{d\Omega} = \frac{Z^2 r_0^2}{4} \cdot \frac{1 - \beta^2 \sin^2 (\theta/2)}{(\beta^2\gamma)^2 \sin^4 (\theta/2)}. \tag{15-7}$$

This result differs from its NR-limit, the well-known *Rutherford cross section*:

$$\frac{d\sigma^{(11)}}{d\Omega} = \frac{Z^2 r_0^2}{4} \cdot \frac{1}{\beta^4 \sin^4 (\theta/2)}, \quad (\text{NR}), \tag{15-8}$$

by the spin correction $-\beta^2 \sin^2 (\theta/2)$, and by the relativistic factor $1 - \beta^2$. The cross sections (15–7) and (15–8) approach infinity for small momentum transfers (small θ). This situation originates in the infinite range of the Coulomb potential, which causes contributions from arbitrary large impact parameters.

The fact that we have derived the Rutherford cross section in this way is very remarkable. It is known to be an exact formula in the nonrelativistic limit, but appears here as a first approximation. Do all the other diagrams in Fig. 15–1 give no contribution? When we try to verify this contention by evaluating the second diagram, we find that it leads to a *divergent* integral. This difficulty seems to make the situation even more puzzling.

The answer to this problem has been most clearly stated by Dalitz.* It is based on the fundamental observation that the assumption of a pure Coulomb field is incompatible with the iteration solution for an external field; i.e., the S-matrix in the bound interaction picture (Section 14–5) cannot be expanded in powers of the external field. Indeed, in the NR-limit a far-distant particle cannot be described by a plane wave $\sim e^{ip \cdot x}$, as is assumed if one works with the double expansion [Eq. (14–79)], but must be expressed by a "Coulomb wave," i.e., a distorted plane wave of the form

$$e^{ip \cdot x + i\eta \ln\left[2|\mathbf{p}|r \sin^2\left(\frac{\theta}{2}\right)\right]} \tag{15-9}$$

with

$$\eta = \alpha Z/\beta. \tag{15-10}$$

If the infinite range of the Coulomb field prohibits the use of the double iteration solution, but requires the use of the bound interaction picture, it is understandable that the second and, in fact, all higher diagrams in Fig. 15–1 give divergent integrals, but it makes the result (15–8) even more remarkable.

The solution is found in a study of a *screened* Coulomb field, of the form

$$\frac{Ze}{4\pi} \frac{e^{-\lambda r}}{r}, \tag{15-11}$$

for which the double iteration solution exists. It then appears, as was first shown explicitly by Dalitz (*loc. cit.*) that in the NR-limit the expansion in αZ reduces to its lowest term times a *phase factor* which diverges in the limit $\lambda \to 0$. It follows that the *absolute value* of the matrix element is well defined in this limit and that the cross section is exact. Of course, if Coulomb scattering occurs coherently with other effects, e.g., nuclear scattering, our solution gives incorrect results, since the relative phases become important.

For relativistic velocities, the higher terms in αZ contribute (in addition to the phase factor) finite terms of which the lowest (second Born approximation) was first calculated correctly by McKinley and Feshbach.†

---

* R. H. Dalitz, *Proc. Roy. Soc.* (London) **A206**, 509 (1951).

† W. A. McKinley and H. Feshbach, *Phys. Rev.* **74**, 1759 (1948). We do not want to enumerate the many calculations and references in which this formula appears incorrectly, but quote only two important ones: N. F. Mott and H. S. W. Massey, *The Theory of Atomic Collisions* (Oxford, 1949), and H. A. Bethe and J. Ashkin in Vol. 1 of E. Segrè's *Experimental Nuclear Physics* (John Wiley and Sons, Inc., New York, 1953, p. 279). The second Born approximation given in Eq. (15–12) was improved with a third Born approximation by H. Mitter and P. Urban, *Acta Phys. Austriaca* **7**, 311 (1953).

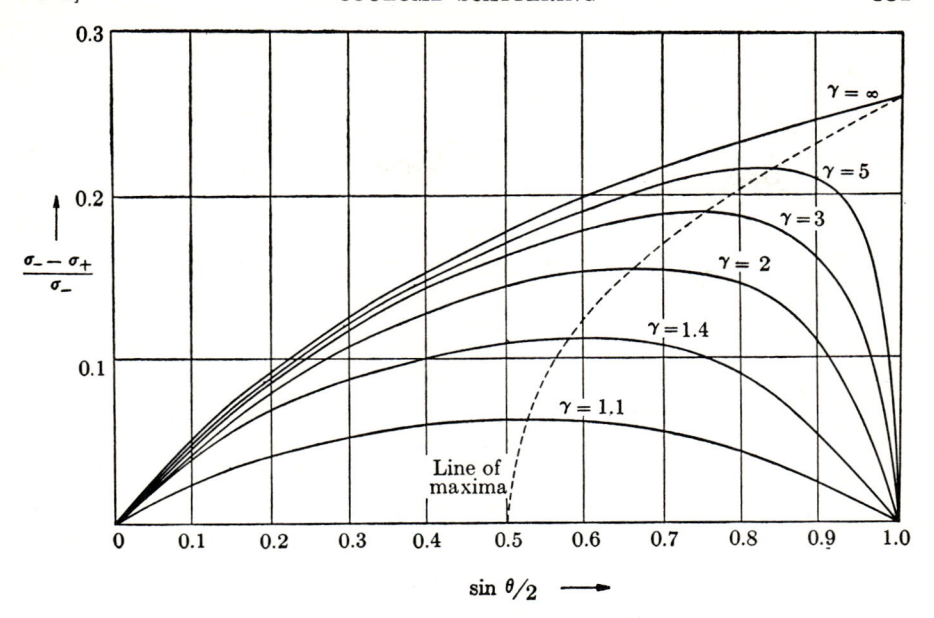

FIG. 15-2. Negaton-positon difference in Coulomb scattering on Al.

$$\frac{d\sigma^{(11+22)}}{d\Omega} = \frac{Z^2 r_0^2}{4} \frac{1}{(\beta^2\gamma)^2 \sin^4(\theta/2)}$$

$$\times \left[1 - \beta^2 \sin^2\frac{\theta}{2} + \alpha Z \beta \pi \sin\frac{\theta}{2}\left(1 - \sin\frac{\theta}{2}\right)\right]. \quad (15\text{--}12)$$

For positons, the sign of Z must be reversed. Of course, neither Eq. (15-7) nor Eq. (15-12) is a good approximation when αZ becomes too large. For medium or heavy nuclei it is necessary to carry out more accurate calculations using Coulomb wave functions.* The characteristic result of such a computation is a considerable increase above the cross section (15-7) for negatons; positons show less scattering than this first approximation indicates. The relative negaton-positon difference in the cross sections on Al is shown in Fig. 15-2 as a function of angle and energy, as calculated† from Eq. (15-12). The experiments on Coulomb scattering confirm these results, although further experiments on heavy nuclei and at large angles seem desirable because of several conflicting measurements reported in the

---

* W. A. McKinley and H. Feshbach, *loc. cit.*; H. Feshbach, *Phys. Rev.* **88**, 295 (1952). See also N. F. Mott and H. S. W. Massey, *loc. cit.*, for earlier work.
† These curves were prepared some time ago by Dr. B. C. Carlson and one of us (F. R.).

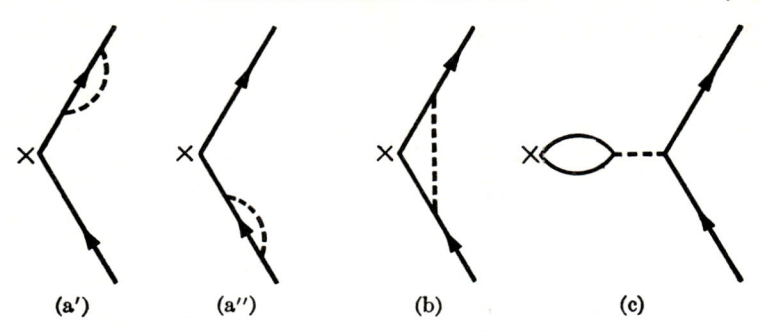

FIG. 15-3. Radiative corrections to Coulomb scattering.

literature. The reader is referred to a review article by Corson and Hanson,* where also the effect of the finite nuclear charge distribution is discussed.

15-2 Radiative corrections to Coulomb scattering. The first radiative correction to the first term in the external field expansion (first diagram of Fig. 15-1) takes into account the presence of one virtual photon (Fig. 15-3).

The diagrams (a) of Fig. 15-3 contribute only to the mass renormalization and the wave-function renormalization, since the radiative correction occurs in the external lines. Diagram (b) gives the bulk of the observable effect and contributes to the spurious charge renormalization, since it involves a vertex part. Diagram (c) contributes a true charge renormalization and an (observable) vacuum polarization effect, since it involves a photon self-energy part. In the calculation of these infinite diagrams the action of the external field (indicated by a cross in a diagram) can be treated formally in exactly the same way as an external photon line. This has been explained in Sections 14-5 and 14-6. In this section we shall first discuss the scattering in an arbitrary static external potential $\varphi^\mu(x)$ and we shall specialize to the Coulomb potential only at the end of the calculations. We put

$$\varphi^\mu(x) = \frac{1}{(2\pi)^{3/2}} \int \varphi^\mu(q) \delta(q^0) e^{iq \cdot x} d^4q \qquad (15\text{-}13)$$

and find for the lowest order scattering of a negaton

$$(p'|M^{(11)}|p) = \frac{e}{\sqrt{2\pi}} \frac{m}{\sqrt{\epsilon'\epsilon}} \bar{u}(\mathbf{p}')\gamma_\mu \varphi^\mu(\mathbf{q}) u(\mathbf{p}), \qquad (15\text{-}14)$$

with the definition for the momentum transfer to the external field

$$\mathbf{q} = \mathbf{p} - \mathbf{p}'. \qquad (15\text{-}15)$$

---

* D. R. Corson and A. O. Hanson in Vol. 3 of *Annual Review of Nuclear Science* (Annual Reviews, Inc., Stanford, Cal., 1953).

The radiative corrections drawn in Fig. 15–3 give the matrix element*

$$(p'|M^{(31)}|p) = \frac{e}{\sqrt{2\pi}} \frac{m}{\sqrt{\epsilon'\epsilon}} \bar{u}(\mathbf{p}')Q_\mu(p',p)\varphi^\mu(\mathbf{q})u(\mathbf{p}),$$

$$Q_\mu(p',p) = \Sigma(p')S_c(p')\gamma_\mu + \gamma_\mu S_c(p)\Sigma(p) + \Lambda_\mu(p',p) - \gamma_\mu D_c(q)\Pi(q).$$
(15–16)

The insertions Σ, Π, and Λ_μ, discussed in Chapter 9, simplify considerably, since they act on free electron wave functions. Thus, from (9–16), we find

$$\bar{u}(p')\Sigma(p')S_c(p')$$
$$= \bar{u}(p')\left[AS_c(p') + \frac{(i\not{p}' + m)(i\not{p}' - m)}{p'^2 + m^2}B + (i\not{p}' + m)^2\Sigma_f(p')S_c(p')\right].$$

When the δm term of mass renormalization is taken into account, the first term will exactly cancel [Eq. (9–28)]. The second term is undetermined but was found to be $-\frac{1}{2}B$ in Section 9–4. The last term vanishes because of $\bar{u}(p')(i\not{p}' + m) = 0$. The result is therefore

$$\bar{u}(p')\Sigma(p')S_c(p') = -\tfrac{1}{2}B\bar{u}(p').$$

In the same way, we find

$$S_c(p)\Sigma(p)u(p) = -\tfrac{1}{2}Bu(p).$$

The vertex part was separated into its infinite and finite terms in Eq. (9–87):

$$\Lambda_\mu(p',p) = L\gamma_\mu + \Lambda_{\mu f}(p',p).$$

Finally, Eq. (9–63a) yields

$$D_c(q)\Pi(q^2) = -C + q^2\Pi_f(q^2).$$

Combining these results, $Q(p',p)$ simplifies to

$$Q_\mu(p',p) = (L - B)\gamma_\mu + C\gamma_\mu + \Lambda_{\mu f}(p',p) - \gamma_\mu q^2\Pi_f(q^2). \quad (15\text{–}17)$$

This result expresses the introductory remarks to this section in a quantitative way. The spurious charge renormalization L and the wave-function renormalization B cancel exactly, as was shown to second order in Eq. (9–92) and to all orders in Eq. (10–28). The constant C is removed by a charge renormalization (Sections 9–5 and 10–6). The remaining terms are all observable.

After all renormalizations have been carried out, the matrix element (15–16) becomes

$$(p'|M^{(31)}|p) = \frac{e}{\sqrt{2\pi}} \frac{m}{\sqrt{\epsilon'\epsilon}} \bar{u}(\mathbf{p}')\varphi^\mu(q)[\Lambda_{\mu f}(p',p) - \gamma_\mu q^2\Pi_f(q^2)]u(\mathbf{p}). \quad (15\text{–}18)$$

---

* For the sign of the last term in Eq. (15–16) note the factor $(-1)^l$ in Eq. (8–20).

The functions $\Lambda_{\mu f}(p',p)$ and $\Pi_f(q^2)$ are given in Eqs. (9–91) and (9–66). The problem is thus reduced to the evaluation of these integrals.

Since the exact evaluation of $\Lambda_{\mu f}$ is quite complicated and contributes little to the physical insight, we shall restrict the following calculation to the case of small momentum transfers:

$$q^2 = \mathbf{q}^2 \ll m^2.$$

Unless the angles are very small, the following results will therefore be valid only for nonrelativistic velocities. We shall be content with the leading term in this expansion in powers of q^2/m^2; $\Pi_f(q)$ is therefore given by Eq. (9–68),

$$\Pi_f(0) = -\frac{\alpha}{15\pi m^2}. \tag{15-19}$$

In the calculation of $\Lambda_{\mu f}(p',p)$ we must take care of the infrared divergences which we want to treat by insertion of a small photon mass λ.* As one can easily see, the replacement of k^2 by $k^2 + \lambda^2$ in $D_c(k^2)$ affects Eq. (9–91) simply in the terms $m^2 x^2$. These terms become replaced by $m^2 x^2 + \lambda^2(1 - x)$. We find, using the Lorentz condition for the external field, $n^\mu q_\mu = 0$,

$$\Lambda_{\mu f}(p',p) = -\frac{\alpha}{2\pi}(\gamma_\mu I + \sigma_{\mu\nu}q^\nu J), \tag{15-20}$$

$$I = \int_0^1 dx \int_0^x dy \left\{ \frac{q^2(1-x+y)(1-y)}{m^2 x^2 + \lambda^2(1-x)} + \int_0^1 dz \frac{q^2 y(x-y)}{m^2 x^2 + \lambda^2(1-x)} \right.$$
$$\left. - 2m^2\left(1 - x - \frac{1}{2}x^2\right)\frac{q^2 y(x-y)}{[m^2 x^2 + \lambda^2(1-x)]^2} \right\},$$

$$J = \int_0^1 dx \int_0^x dy \frac{-mx(1-x)}{m^2 x^2 + \lambda^2(1-x)}.$$

It should be pointed out here that the term containing J is due to the anomalous magnetic moment of the electron, as will be shown in the following section. The effect of this term can easily be traced through the following calculations.

The integrals I and J are to be evaluated in the limit $\lambda \to 0$. Only I contains infrared divergences. We put $y = xu$ and see that these divergences arise from the lowest power of x in the numerator of the first and

---

* In this and the following sections the photon mass is denoted by λ to indicate its use as a convenient cutoff procedure rather than as the mass of a neutral vector meson which was previously denoted by κ (cf. Section 6–5). All terms which do not contribute in the limit $\lambda \to 0$ will be conveniently omitted.

third term in I,

$$I' = \int_0^1 x\,dx \int_0^1 du \left\{ \frac{q^2}{m^2 x^2 + \lambda^2(1-x)} - 2m^2 \frac{uq^2 x^2(1-u)}{[m^2 x^2 + \lambda^2(1-x)]^2} \right\}$$

$$= \frac{q^2}{m^2}\left(\frac{2}{3}\ln\frac{m}{\lambda} + \frac{1}{6}\right).$$

The remaining terms, I'', of I are finite and yield, after a simple calculation, $I'' = -\frac{5}{12}(q^2/m^2)$. Thus,

$$I = \left(\frac{2}{3}\ln\frac{m}{\lambda} - \frac{1}{4}\right)\frac{q^2}{m^2}.$$

The integral J converges and is

$$J = -\frac{1}{2m}.$$

Therefore,

$$\Lambda_{\mu f}(p',p) = -\frac{\alpha}{3\pi}\left[\gamma_\mu\left(\ln\frac{m}{\lambda} - \frac{3}{8}\right)\frac{q^2}{2m} - \frac{3}{4m}\sigma_{\mu\nu}q^\nu\right]. \quad (15\text{-}21)$$

We can now insert the results (15–19) and (15–21) into Eq. (15–18):

$$(p'|M^{(31)}|p) = -\frac{e}{\sqrt{2\pi}}\frac{m}{\sqrt{\epsilon'\epsilon}}\frac{\alpha}{3\pi}\left[\frac{q^2}{m^2}\left(\ln\frac{m}{\lambda} - \frac{3}{8} - \frac{1}{5}\right)\right.$$
$$\left. \times \bar{u}(\mathbf{p}')\gamma_\mu\varphi^\mu(\mathbf{q})u(\mathbf{p}) - \frac{3}{4im}\bar{u}(\mathbf{p}')\gamma_\mu\varphi^\mu \mathbf{q}u(\mathbf{p})\right]. \quad (15\text{-}22)$$

The term $-\frac{1}{5}$ is due to the vacuum polarization term (15–19).

As was encountered before repeatedly (*cf.*, for instance, Section 11–3) the infrared divergence arises whenever there is a possibility of emitting additional soft photons. These are created as the electron is accelerated in the external field. To this order of radiative corrections, only the emission of a single soft photon is relevant. The associated diagrams are the bremsstrahlung diagrams of Fig. 15–4. Only very soft photon emission

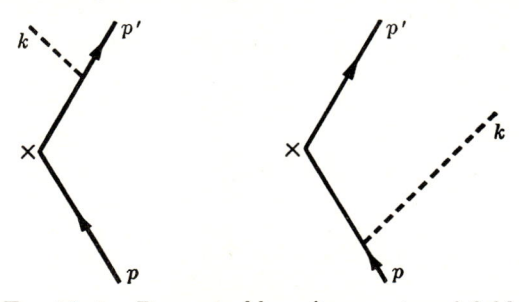

FIG. 15–4. Bremsstrahlung in an external field.

of maximum energy $\Delta\epsilon$ will interest us here. The complete bremsstrahlung process will be discussed in Section 15–6.

Let us not forget, however, that we have chosen as a cutoff method the use of photons of finite mass. This method must be carried through consistently. It implies that we must calculate the emission probability for photons of mass λ and of momentum $|\mathbf{k}|$ between 0 and $|\mathbf{k}_{\max}|$. In the limit $\lambda \to 0$, $|\mathbf{k}_{\max}|$ can then be identified with $\Delta\epsilon$.

The diagrams Fig. 15–4, for small k, give

$$(p'k|M^{(21)}|p) = \frac{ie^2}{(2\pi)^2} \frac{m}{\sqrt{\epsilon'\epsilon}} \frac{1}{\sqrt{2\omega}}$$
$$\times \bar{u}(\mathbf{p}')[eS_c(p'+k)\gamma_\mu\varphi^\mu + \gamma_\mu\varphi^\mu S_c(p-k)e]u(\mathbf{p}). \quad (15\text{–}23)$$

Here

$$\omega = \sqrt{\mathbf{k}^2 + \lambda^2} > |\mathbf{k}|$$

is the energy of the photon of finite mass (neutral vector meson). The square brackets can be further simplified for small k. In fact,

$$S_c(p-k)eu(\mathbf{p}) = \frac{i p - i k - m}{(p-k)^2 + m^2} eu(\mathbf{p})$$

$$= \frac{-e(i p + m) + 2ip \cdot e}{-2p \cdot k} u(\mathbf{p})$$

$$= -i \frac{p \cdot e}{p \cdot k} u(\mathbf{p}),$$

so that

$$\bar{u}(\mathbf{p}')[eS_c(p'+k)n + nS_c(p-k)e]u(\mathbf{p}) = \bar{u}(\mathbf{p}')nu(\mathbf{p})i\left(\frac{p' \cdot e}{p' \cdot k} - \frac{p \cdot e}{p \cdot k}\right),$$

and Eq. (15–23) simplifies to

$$(p'k|M^{(21)}|p) = -(p'|M^{(11)}|p)\frac{e}{(2\pi)^{3/2}}\frac{1}{\sqrt{2\omega}}\left(\frac{p' \cdot e}{p' \cdot k} - \frac{p \cdot e}{p \cdot k}\right), \quad (15\text{–}24)$$

where we used Eq. (15–14). This expression can be even further simplified, since we are working in the nonrelativistic region where

$$p' \cdot k = p \cdot k = -m\omega.$$

We obtain

$$(p'k|M^{(21)}|p) = -(p'|M^{(11)}|p)\frac{e}{(2\pi)^{3/2}}\frac{1}{\sqrt{2\omega}}\frac{1}{m\omega}\mathbf{q}\cdot\mathbf{e}, \quad (15\text{–}25)$$

which is valid for transverse polarizations.

Bremsstrahlung involves an additional photon in the final state when compared with Coulomb scattering, so that the respective matrix elements do not give rise to cross terms. The transition probability of interest is the sum of the transition probabilities for Coulomb scattering and low-energy bremsstrahlung. The latter must be summed over the polarizations and over all photon momenta from 0 to $\Delta\epsilon$. Since we are actually dealing with a vector meson which has two transverse and one longitudinal direction of **e**, we have to sum over these three directions. Following Section 6–5, Eq. (6–130), we know that the emission probability for longitudinal photons of the same momentum is* $\lambda^2/(\mathbf{k}^2 + \lambda^2)$ times the emission probability for transverse photons; thus, indicating the summation over the photon variables with $\mathsf{S}_f^{(k)}$,

$$\mathsf{S}_f^{(k)} |(p'k|M^{(21)}|p)|^2 = |(p'|M^{(11)}|p)|^2 \frac{e^2}{(2\pi)^3} \int \frac{d^3k}{2\omega^3} \cdot \frac{1}{3} \frac{\mathbf{q}^2}{m^2} \left(2 + \frac{\lambda^2}{\omega^2} \right).$$

The integrals, with $\omega^2 = \mathbf{k}^2 + \lambda^2$, in the limit $\lambda \to 0$, are

$$\int \frac{d^3k}{\omega^3} = 4\pi \int_0^{\Delta\epsilon} \frac{\mathbf{k}^2 d|\mathbf{k}|}{(\mathbf{k}^2 + \lambda^2)^{3/2}} = 4\pi \left(\ln \frac{2\Delta\epsilon}{\lambda} - 1 \right),$$

$$\lambda^2 \int \frac{d^3k}{\omega^5} = 4\pi\lambda^2 \int_0^{\Delta\epsilon} \frac{\mathbf{k}^2 d|\mathbf{k}|}{(\mathbf{k}^2 + \lambda^2)^{5/2}} = 4\pi \cdot \frac{1}{3}.$$

Therefore, Eq. (15–25) yields†

$$\mathsf{S}_f^{(k)} |(p'k|M^{(21)}|p)|^2 = |(p'|M^{(11)}|p)|^2 \frac{2\alpha}{3\pi} \frac{q^2}{m^2} \left(\ln \frac{2\Delta\epsilon}{\lambda} - \frac{5}{6} \right). \quad (15\text{–}26)$$

We can now return to Eqs. (15–14) and (15–22). The square of the absolute value of $(p'|M^{(11)} + M^{(31)}|p)$ yields, when summed over the final

---

* The longitudinal photons are not always treated correctly in the literature. The summation over the transverse polarizations gives

$$\sum_{\text{trans}} |\mathbf{A} \cdot \mathbf{e}|^2 = |\mathbf{A}|^2 - \left(\frac{\mathbf{A} \cdot \mathbf{k}}{|\mathbf{k}|} \right)^2,$$

which reduces to $\frac{2}{3}|\mathbf{A}|^2$ when averaged over the angles. It is incorrect to replace $|\mathbf{k}|^2$ in the denominator by ω^2, invoking $|\mathbf{k}| = \omega$, and to put $\omega^2 = \mathbf{k}^2 + \lambda^2$ in the later parts of the calculation. However, no mistake will thereby be introduced, if, at the same time, one ignores the contributions from the longitudinal photons, since the correct addition of the effect of the longitudinal photons will give, with $\omega^2 = \mathbf{k}^2 + \lambda^2$,

$$|\mathbf{A}|^2 - \left(\frac{\mathbf{A} \cdot \mathbf{k}}{|\mathbf{k}|} \right)^2 + \frac{\lambda^2}{\omega^2} \left(\frac{\mathbf{A} \cdot \mathbf{k}}{|\mathbf{k}|} \right)^2 = |\mathbf{A}|^2 - \left(\frac{\mathbf{A} \cdot \mathbf{k}}{\omega} \right)^2,$$

which is just the result obtained by the above incorrect procedure.

† The omission of the contributions from the longitudinal photons, which would replace $-\frac{5}{6}$ in this equation by -1, caused considerable confusion in the first attempts to derive these radiative corrections. *Cf.* R. P. Feynman, *Phys. Rev.* **76**, 769 (1949), footnote 13.

spins and averaged over the initial spins [cf. Eq. (A2–87)],

$$-\frac{e^2}{4\pi}\frac{m^2}{\epsilon'\epsilon}\varphi^\mu(\mathbf{q})\varphi^\nu(\mathbf{q})\left\{\left[1-\frac{2\alpha}{3\pi}\frac{\mathbf{q}^2}{m^2}\left(\ln\frac{m}{\lambda}-\frac{3}{8}-\frac{1}{5}\right)\right]\text{Tr}\left[\gamma_\mu\Lambda_-(p)\gamma_\nu\Lambda_-(p')\right]\right.$$

$$\left.+\frac{2\alpha}{3\pi}\cdot\frac{3}{4im}\cdot\text{Tr}\left[\gamma_\mu q\Lambda_-(p)\gamma_\nu\Lambda_-(p')\right]\right\}.$$

The highest term in powers of α has been omitted because it can be discussed consistently only in conjunction with the radiative corrections $M^{(51)}$.

It is now convenient to specialize the external field to a Coulomb field:

$$\varphi^\mu(\mathbf{q})=n^\mu\phi(\mathbf{q}),$$

where $\phi(\mathbf{q})$ is given by Eq. (15–4). The nonrelativistic limit gives

$$\text{Tr}\left[n\Lambda_-(p)n\Lambda_-(p')\right]=-2,$$

$$\text{Tr}\left[nq\Lambda_-(p)n\Lambda_-(p')\right]=\frac{i}{m}q^2.$$

We combine these results and find

$$\tfrac{1}{2}\sum_{\text{spins}}|(p'|M^{(11)}+M^{(31)}|p)|^2$$

$$=2\alpha\frac{m^2}{\epsilon'\epsilon}|\phi(\mathbf{q})|^2\left[1-\frac{2\alpha}{3\pi}\frac{\mathbf{q}^2}{m^2}\left(\ln\frac{m}{\lambda}-\frac{3}{8}-\frac{1}{5}+\frac{3}{8}\right)\right]. \quad (15\text{–}27)$$

The last $\frac{3}{8}$ arises from the term containing the second trace, which is the magnetic moment term.

Finally, Eq. (15–27) and the spin sum over Eq. (15–26) must be combined. The dependence on the photon mass thereby exactly cancels:

$$\tfrac{1}{2}\sum_{\text{spins}}|(p'|M^{(11)}+M^{(31)}|p)|^2+\tfrac{1}{2}\sum_{\text{spins}}S_f^{(k)}|(p'k|M^{(21)}|p)|^2$$

$$=2\alpha\frac{m^2}{\epsilon'\epsilon}|\phi(\mathbf{q})|^2\left[1-\frac{2\alpha}{3\pi}\frac{\mathbf{q}^2}{m^2}\left(\ln\frac{m}{2\Delta\epsilon}+\frac{5}{6}-\frac{1}{5}\right)\right]. \quad (15\text{–}28)$$

The elimination of the infrared divergence is thereby completed. The momentum transfer is

$$\mathbf{q}^2=|\mathbf{p}'-\mathbf{p}|^2=4m^2\beta^2\sin^2\frac{\theta}{2}$$

and the factor multiplying the square bracket in Eq. (15–28) is exactly

$$\sum_{\text{spins}}|(p'|M^{(11)}|p)|^2.$$

The scattering cross section, including first radiative corrections, can therefore be written*

$$\frac{d\sigma}{d\Omega} = \frac{d\sigma^{(11)}}{d\Omega}(1-\delta), \qquad (15\text{-}29)$$

$$\delta = \frac{8\alpha}{3\pi}\beta^2 \sin^2\frac{\theta}{2}\left(\ln\frac{m}{2\Delta\epsilon} + \frac{19}{30}\right), \quad (\text{NR}). \qquad (15\text{-}30)$$

Our result (15-30) is valid only for $\beta \ll 1$. For Coulomb scattering $\sigma^{(11)}$ was discussed in the preceding section.

To derive the radiative corrections in the relativistic case, one must evaluate the integrals in $\Lambda_{\mu f}$ and Π_f exactly (Eq. 15-18). This tedious calculation was first carried out by Schwinger.† Of particular interest is the extreme relativistic limit, which was found to be

$$\delta = \frac{4\alpha}{\pi}\left[\left(\ln\frac{\epsilon}{\Delta\epsilon} - \frac{13}{12}\right)\left(\ln\frac{2\epsilon \sin(\theta/2)}{m} - \frac{1}{2}\right) + \frac{17}{72} + \phi(\theta)\right], \quad (\text{ER}),$$
$$(15\text{-}31)$$

where

$$\phi(\theta) = \frac{1}{2}\sin\frac{\theta}{2}\int_{\cos(\theta/2)}^{1}\left[\frac{\ln[(1+x)/2]}{1-x} - \frac{\ln[(1-x)/2]}{1+x}\right]\frac{dx}{\sqrt{x^2 - \cos^2(\theta/2)}}$$

$$\sim \frac{1-c}{\sqrt{2c(1+c)}}\left(\ln\frac{1}{2(1-c)} + \frac{1-c}{2} + 1\right), \quad \left(c = \cos\frac{\theta}{2}\right).$$

The function $\phi(\theta)$ must, in general, be evaluated numerically, but is approximated well by the last formula, whose positive error increases from 0 at $\theta = 0$ to 8.6% at $\theta = \pi/2$. Exact values are

$$\phi(0) = 0, \quad \phi\left(\frac{\pi}{2}\right) = 0.292, \quad \phi(\pi) = \frac{\pi^2}{24}.$$

---

* To be consistent in the order of approximation, it is actually necessary to include the second Born approximation, Eq. (15-12):

$$\frac{d\sigma}{d\Omega} = \frac{d\sigma^{(11+22)}}{d\Omega}(1-\delta).$$

The cross term of order $\alpha^2 Z$, of course, has no significance and should be dropped.

† J. Schwinger, *Phys. Rev.* **75**, 898 (1949), and **76**, 790 (1949). These calculations contain two errors associated with the joining of the two transition probabilities [our Eqs. (15-26) and (15-27)] which eliminates the infrared divergence. Since these two errors exactly cancel each other, Schwinger's final result is correct. *Cf.* L. R. B. Elton and H. H. Robertson, *Proc. Phys. Soc.* **A65**, 145 (1952).

The radiative corrections (15–30) and (15–31) of the elastic scattering of electrons (there is no difference between negatons and positons in this approximation) depend on the experimental energy resolution $\Delta\epsilon/\epsilon$, where $\Delta\epsilon \ll \epsilon - m$. The better this resolution is, the more soft-photon bremsstrahlung can be excluded, and the *smaller* is the scattering cross section. In the limit $\Delta\epsilon \to 0$ this *elastic* cross section does not diverge, as might be concluded erroneously from these equations, but approaches zero (*cf.* Section 16–1). There is no perfectly elastic scattering. Formally, this cancellation comes about when in Eq. (15–29) further corrections are considered which include multiple soft-photon emission. The smaller $\Delta\epsilon/\epsilon$, the larger the number of bremsstrahlung photons which must be included together with higher radiative corrections. (*Cf.* again the discussion on the infrared divergence in Section 16–1.) A scattering experiment of relatively "poor" resolution (for instance $\Delta\epsilon/\epsilon \sim 1\%$!) will mainly measure one-photon bremsstrahlung. Such a measurement was carried out by Lyman, Hanson, and Scott* and satisfactory agreement with theory was found within the limits of the experimental accuracy.

In the above experiments it is essential to measure the energy of the outgoing electron, so as to ensure that the scattering is elastic. If, on the other hand, one measures only the scattering angle and disregards the final electron energy ϵ', the observed cross section will include inelastic collisions in which part of the energy has been lost by radiation. We now want to consider this case, but restrict ourselves again to the nonrelativistic limit.

Apparently, one simply has to add to the elastic cross section given by Eqs. (15–29) and (15–30), the bremsstrahlung cross section for the emission of a photon of energy ω, integrated from $\omega = \Delta\epsilon$ to $\omega = T = \epsilon - m$, where T is the maximum possible energy loss of the incident electron. The final electron momentum $\mathbf{p}''$ will now be different from $\mathbf{p}'$ in the elastic scattering case,

$$|\mathbf{p}''| < |\mathbf{p}'| = |\mathbf{p}|.$$

However, the photon momentum $\mathbf{k}$, in the NR case, satisfies

$$|k| = \omega = \frac{\mathbf{p}^2}{2m} - \frac{\mathbf{p}''^2}{2m} \ll |\mathbf{p}|,$$

so that the electron recoil due to photon emission can be neglected and $\mathbf{p}''$ can be taken parallel to $\mathbf{p}'$. It follows that $\mathbf{p}''$ differs from $\mathbf{p}'$ only in magnitude:

$$\mathbf{p}'' = x\mathbf{p}'. \tag{15–32}$$

---

* E. M. Lyman, A. O. Hanson, and M. B. Scott, *Phys. Rev.* **84**, 626 (1951).

15-2] RADIATIVE CORRECTIONS TO COULOMB SCATTERING

The cross section for the emission of a photon of energy between ω and $\omega + d\omega$ and arbitrary polarization follows from Eq. (15-25):

$$\frac{d\sigma_B}{d\Omega} = \frac{(2\pi)^2}{\beta}\frac{1}{2}\sum_{\text{spins}}|(p''|M^{(11)}|p)|^2|\mathbf{p}''|\epsilon'' \frac{e^2}{(2\pi)^3}\frac{1}{2\omega^3}\frac{1}{3}\frac{q^2}{m^2}2d^3k.$$

Here we do not need to use photons of finite mass, since the radiated energy has the lower limit $\Delta\epsilon$. When we insert the value of $M^{(11)}$ for the case of a Coulomb field, Eq. (15-5), and note that

$$q^2 = |\mathbf{p} - \mathbf{p}''|^2 = m^2\beta^2(1 + x^2 - 2x\cos\theta),$$

we find

$$\frac{d\sigma_B}{d\Omega} = Z^2 r_0^2 \frac{8\alpha}{3\pi} \frac{1}{\beta^2(1+x^2-2x\cos\theta)} \cdot \frac{d\omega}{\omega}.$$

The relation between ω and x follows from the conservation relations

$$\frac{\omega}{m} = \frac{1}{2}\beta^2(1-x^2), \qquad \frac{d\omega}{\omega} = -\frac{2x\,dx}{1-x^2}.$$

These relations are to be inserted into the bremsstrahlung cross section and the integration from $x = 0$ to $x = x_{\max}$ is to be carried out:

$$\frac{d\sigma_B}{d\Omega} = \frac{Z^2 r_0^2}{4}\frac{1}{\beta^4 \sin^4(\theta/2)}\frac{8\alpha}{3\pi} 4\beta^2 \sin^4\frac{\theta}{2} I,$$

(15-33)

$$I = 2\int_0^{x_{\max}} \frac{x\,dx}{(1-x^2)(1-2x\cos\theta+x^2)}.$$

The maximum value of x is determined by the minimum value of ω, which is $\Delta\epsilon$:

$$x_{\max} = \sqrt{1 - \frac{\omega_{\min}}{T}} = \sqrt{1 - \frac{\Delta\epsilon}{T}} \simeq 1 - \frac{\Delta\epsilon}{2T}. \qquad (15\text{-}34)$$

Since this value is very close to 1, we can evaluate I in this limit and find

$$\frac{1}{2}I = \int_0^{x_{\max}} \frac{dx}{(1-x^2)(1-2x\cos\theta+x^2)} - I_0$$

$$= \frac{1}{8\sin^2(\theta/2)}\ln\frac{2T}{\Delta\epsilon}$$

$$+ \int_0^1 \frac{dx}{1-x}\left[\frac{1}{(1+x)(1-2x\cos\theta+x^2)} - \frac{1}{8\sin^2(\theta/2)}\right] - I_0,$$

with
$$I_0 = \int_0^1 \frac{dx}{1 - 2x\cos\theta + x^2} = \frac{\pi - \theta}{2\sin\theta}.$$

The other integral is also elementary:

$$2\int_0^1 \frac{dx}{1-x}\left[\frac{1}{(1+x)(1-2x\cos\theta+x^2)} - \frac{1}{8\sin^2(\theta/2)}\right]$$

$$= \frac{\cot\theta}{\sin\theta}\ln\sin\frac{\theta}{2} + \frac{\ln 2}{4\sin^2(\theta/2)} + I_0.$$

It follows, therefore, from Eq. (15–33) that the cross section for the emission of photons of energy $\Delta\epsilon$ to T in a Coulomb field is

$$\frac{d\sigma_B}{d\Omega} = \frac{Z^2 r_0^2}{4} \frac{1}{\beta^4 \sin^4(\theta/2)} \cdot \frac{8\alpha}{3\pi}\beta^2 \sin^2\frac{\theta}{2}\left[\ln\frac{4T}{\Delta\epsilon} - (\pi-\theta)\tan\frac{\theta}{2}\right.$$

$$\left. + \frac{\cos\theta}{\cos^2(\theta/2)}\ln\sin\frac{\theta}{2}\right], \quad \text{(NR)}. \quad (15\text{–}35)$$

We combine this result with the elastic scattering cross section (15–29) and (15–30):*

$$\frac{d\sigma}{d\Omega} = \frac{Z^2 r_0^2}{4} \frac{1}{\beta^4 \sin^4(\theta/2)} \cdot \left[1 - \frac{8\alpha}{3\pi}\beta^2 \sin^2\frac{\theta}{2}\left(\ln\frac{1}{4\beta^2} + \frac{19}{30}\right.\right.$$

$$\left.\left. + (\pi-\theta)\tan\frac{\theta}{2} - \frac{\cos\theta}{\cos^2(\theta/2)}\ln\sin\frac{\theta}{2}\right)\right], \quad \text{(NR)}. \quad (15\text{–}36)$$

This is the cross section for scattering of an electron of velocity β by an angle θ, irrespective of the associated radiation loss in the nonrelativistic limit. We notice that the correction is of the order

$$\alpha\beta^2 \sin^2\frac{\theta}{2}\ln\beta^2$$

and therefore vanishes in the nonrelativistic limit. Radiative corrections to scattering irrespective of radiation loss can be observed only at relativistic energies.

15–3 The magnetic moment of the electron. It is well known that the electron which satisfies the wave equation of Dirac has a magnetic dipole moment of magnitude $\mu_0 = e/2m$.† In our field theoretical formulation of

---

* This formula was also first obtained by Schwinger, *loc. cit.*
† *Cf.* any book on quantum mechanics.

quantum electrodynamics this property of the electron is implicitly contained in the interaction operator

$$H(\tau) = ie \int_{\sigma(\tau)} \bar{\psi}\gamma_\mu\psi\varphi^\mu d\sigma \qquad (15\text{-}37)$$

for the matter field ψ which interacts with an external electromagnetic field φ^μ. In the nonrelativistic limit a magnetic moment μ would give rise to an interaction operator

$$H'(\tau) = -\mu_0 \int \bar{\psi}(x)\boldsymbol{\sigma}\psi(x)\cdot\mathbf{H}(x)d^3x, \qquad (15\text{-}38)$$

where

$$\boldsymbol{\sigma} = (\sigma_{23},\sigma_{31},\sigma_{12}), \quad \mathbf{H} = (\varphi_{23},\varphi_{31},\varphi_{12}),$$

$$\sigma_{\mu\nu} = \frac{1}{2i}(\gamma_\mu\gamma_\nu - \gamma_\nu\gamma_\mu), \quad \varphi_{\mu\nu} = \partial_\mu\varphi_\nu - \partial_\nu\varphi_\mu.$$

We can exhibit explicitly the presence of this term in the expression (15-37) for $H(\tau)$ by decomposing the current (according to a method due to Gordon*) into an *orbital* and a *polarization* current. This decomposition is based on the identity

$$j_\mu = -ie\bar{\psi}\gamma_\mu\psi = j_{1\mu} + j_{2\mu},$$

$$j_{1\mu} = \frac{ie}{2m}(\bar{\psi}\partial_\mu\psi - \overline{\partial_\mu\psi}\,\psi), \qquad (15\text{-}39)$$

$$j_{2\mu} = \frac{e}{2m}\partial^\nu(\bar{\psi}\sigma_{\mu\nu}\psi),$$

which is easily derived with the help of the field equations

$$(\partial + m)\psi = 0, \quad \bar{\psi}(\overleftarrow{\partial} - m) = 0.$$

Correspondingly, the interaction operator may be separated into two parts,

$$H(\tau) = H_1(\tau) + H_2(\tau).$$

One can easily see that H_1 vanishes in the nonrelativistic limit when φ^μ is a static field. In fact, the Fourier transform of $j_{1\mu}$ is then proportional to

$$(p' - p)\bar{\psi}(p')\psi(p),$$

since $p_0' = p_0$, so that $j_{1\mu} = 0$.

---

* W. Gordon, Z. *Physik* **50**, 630 (1928).

The second term,

$$H_2(\tau) = -\int_{\sigma(\tau)} j_{2\mu}\varphi^\mu d\sigma,$$

contains the magnetic interaction (15–38) in the nonrelativistic limit. To show this, we transform $H_2(\tau)$ by partial integration into

$$H_2(\tau) = -\tfrac{1}{2}\mu_0 \int \bar{\psi}\sigma_{\mu\nu}\psi\varphi^{\mu\nu}d\sigma, \qquad (15\text{--}40)$$

where we have omitted the surface terms. This omission is always justified if the external field vanishes at large space-like distances, as we shall explicitly assume. In the nonrelativistic limit the components in (15–40) which involve the electric field $\varphi^{0\mu}$ can be neglected, since they contain only products of large with small components of ψ, and the expression (15–40) reduces to (15–38):

$$H_2(\tau) \rightarrow H'(\tau).$$

This relation justifies the statement that the magnetic moment of the electron in this theory has, without radiative corrections, the value μ_0.

The effect of radiative corrections on the magnetic moment of the electron can be seen most easily by considering the interaction (15–37) in the form $H_1 + H_2$, where H_2 is given by Eq. (15–40). When this interaction is used to construct an S-operator, the iteration solution will be

$$S = 1 - i\int H(\tau)d\tau + \cdots = 1 + \frac{i}{2}\mu_0 \int \bar{\psi}(x)\sigma_{\mu\nu}\psi(x)\varphi^{\mu\nu}(x)d^4x + \cdots . \qquad (15\text{--}41)$$

The S-matrix element of $S^{(11)}$ in momentum space is

$$(p'|S^{(11)}|p) = -\frac{i}{2}\mu_0 \frac{1}{\sqrt{2\pi}} \bar{u}(\mathbf{p}')\sigma_{\mu\nu}u(\mathbf{p})\varphi^{\mu\nu}(q)\delta(\epsilon' - \epsilon), \qquad (15\text{--}42)$$

where

$$\varphi^{\mu\nu}(q) = i(q^\mu \varphi^\nu(q) - q^\nu \varphi^\mu(q)).$$

The corresponding matrix element $(p'|M^{(11)}|p)$ is therefore

$$(p'|M^{(11)}|p) = \frac{1}{\sqrt{2\pi}}\mu_0 \bar{u}(\mathbf{p}')\varphi^\mu \sigma_{\mu\nu}q^\nu u(\mathbf{p}). \qquad (15\text{--}43)$$

Here we made use of the antisymmetry of $\sigma_{\mu\nu}$.

Consider now the nonrelativistic limit of the matrix element $M^{(31)}$ given by Eqs. (15–18) and (15–20) for the case of an external static vector potential with Fourier transform $\varphi^\mu(q)$, where φ^μ is a space-like vector. The terms of the form

$$\bar{u}(\mathbf{p}')\gamma_\mu u(\mathbf{p})\varphi^\mu(q)$$

do not contribute in the NR-limit, since they connect large with small components of u. We are left with

$$(p'|M^{(31)}|p) = \frac{e}{\sqrt{2\pi}}\bar{u}(\mathbf{p}')\left(\frac{\alpha}{2\pi}\frac{1}{2m}\varphi^\mu\sigma_{\mu\nu}q^\nu\right)u(\mathbf{p}). \qquad (15\text{--}44)$$

Comparison with Eq. (15–43) shows that the matrix elements of $M^{(31)}$ and $M^{(11)}$ are of exactly the same form. But whereas $M^{(11)}$ corresponds to a magnetic moment μ_0, $M^{(31)}$ corresponds to a magnetic moment

$$\mu^{(31)} = \frac{\alpha}{2\pi}\mu_0. \qquad (15\text{--}45)$$

It follows that the contribution $\mu^{(31)}$ of the radiative corrections to the magnetic moment can be combined with $\mu^{(11)} = \mu_0$ to give*

$$\mu = \left(1 + \frac{\alpha}{2\pi}\right)\mu_0. \qquad (15\text{--}46)$$

Clearly, higher radiative corrections will continue this power series in α. Indeed, the next order correction which was calculated by Karplus and Kroll† contributes $\mu^{(51)} = -0.328(\alpha^2/\pi^2)\mu_0$, so that to this order**

$$\mu = \left(1 + \frac{\alpha}{2\pi} - 0.328\frac{\alpha^2}{\pi^2}\right)\mu_0 = 1.0011596\,\mu_0. \qquad (15\text{--}47)$$

This result is slightly more accurate than the best measurement to date,‡

$$\mu = (1.0011612 \pm 0.0000024)\,\mu_0, \qquad (15\text{--}48)$$

which is therefore in excellent agreement with the theoretical predictions.

15–4 Energy levels in hydrogen-like atoms. The earliest and most decisive experiment for the effect of radiative corrections in quantum electrodynamics was the accurate determination of certain energy levels in hydrogen-like atoms, especially in hydrogen and deuterium. These experiments were first carried out by Lamb and Retherford§ and were successively im-

---

* This "anomalous" magnetic moment of the electron was calculated by J. Schwinger, *Phys. Rev.* **73**, 416 (1948), and **76**, 790 (1949).

† R. Karplus and N. M. Kroll, *Phys. Rev.* **77**, 536 (1950); correction by C. M. Sommerfield, *Phys. Rev.* **107**, 328 (1957) and A. Petermann, *Nucl. Phys.* **5**, 677 (1958).

‡ S. Koenig, A. G. Pradell, and P. Kusch, *Phys. Rev.* **88**, 191 (1952); P. Franken and S. Liebes, *Phys. Rev.* **104**, 1197 (1956). The value quoted is the most recent, A. A. Schupp, R. W. Pidd, and H. R. Crane, *Bull. Am. Phys. Soc.* **4**, 250 (1959).

§ W. E. Lamb, Jr., and R. C. Retherford, *Phys. Rev.* **72**, 241 (1947); this paper was followed by a series of papers which described their extensive and beautiful work: *Phys. Rev.* **75**, 1325 (1949), **79**, 549 (1950), **81**, 222 (1951), **85**, 259 (1952), **86**, 1014 (1952); the last two papers in this series are by S. Triebwasser, E. S. Dayhoff, and W. E. Lamb, Jr., *Phys. Rev.* **89**, 98 and 106 (1953).

** See Supplement Section S 5–1 for more accurate results.

proved by them until they arrived at the very accurate values of†

$$\Delta E_H = 1057.77 \pm 0.10 \text{ Mc/sec},$$
$$\Delta E_D = 1059.00 \pm 0.10 \text{ Mc/sec} \qquad (15\text{--}49)$$

for the $2\ {}^2s_{1/2} - 2\ {}^2p_{1/2}$ level separation in hydrogen and deuterium respectively. This level separation is zero in the Dirac theory when radiative corrections are neglected. The latter shift both the $2\ {}^2s_{1/2}$ and the $2\ {}^2p_{1/2}$. The effect is often referred to as the "Lamb shift."

Because of the importance of these experiments, we shall discuss the corresponding calculations in some detail. For this purpose it is, of course, essential to work in the bound interaction picture. The electron field operators then satisfy Eq. (14–24) for an external Coulomb field which yields the usual energy levels in hydrogen-like atoms.* The lowest-order radiative corrections ΔE to such a level E can be calculated by the method of Section 14–7, Eq. (14–101):

$$\Delta E = -\frac{1}{2\pi i}(E|M|E). \quad (15\text{--}50)$$

The diagrams for these corrections in the bound interaction picture are shown in Fig. 15–5. They represent the *fluctuation* effect (due to virtual emission and reabsorption of photons) and the *polarization* effect (due to vacuum polarization in the presence of an external field).

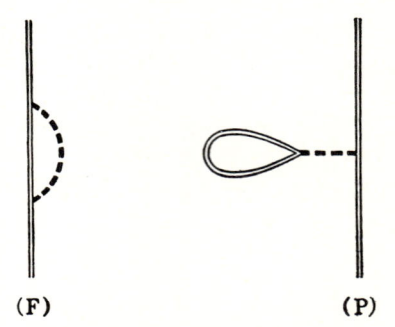

FIG. 15–5. Diagrams for the lowest order radiative corrections to the motion of an electron in an external field.

Consider the fluctuation diagram. It is exactly the lowest-order self-energy diagram in this picture. The corresponding S-matrix element for a negaton can easily be written down in terms of the operator notation of Section 14–4. Using Eq. (15–50),

$$\Delta E_F = \frac{ie^2}{(2\pi)^4}\int \bar{u}_0 \gamma_\mu (i\boldsymbol{p} - i\boldsymbol{k} - iV + m)^{-1}\gamma^\mu u_0 \frac{d^4k}{k^2}. \quad (15\text{--}51)$$

The subscript zero refers to the energy level E_0 whose corrections are sought. The integral operator V is defined in terms of the external field φ^μ by Eq. (14–67). In momentum space,

$$Vu = -\frac{e}{(2\pi)^{3/2}}\int \varphi(\mathbf{p} - \mathbf{p}')u(\mathbf{p}')d^3p'. \quad (15\text{--}52)$$

---

* Cf. any book on relativistic quantum mechanics, e.g. L. I. Schiff, *Quantum Mechanics*, 2d ed., McGraw-Hill Book Co., New York, 1955.

† See Supplement Section S 5–3 for more recent results.

15-4] ENERGY LEVELS IN HYDROGEN-LIKE ATOMS

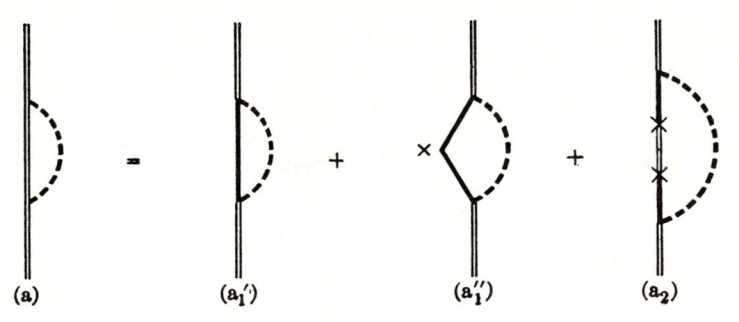

FIG. 15–6. Decomposition of the fluctuation diagram.

The expression (15–51) for ΔE_F includes the self-energy of the free electron, which, of course, must be subtracted in order to obtain the *observable* level shift. In other words, the level shift is the *energy difference* between the self-energy of the bound and the free electron. Only this difference has physical meaning. We shall see soon how this free negaton self-energy can be subtracted in a most convenient way.

The difficulty of the evaluation of (15–51) hinges entirely on the expression for the electron propagation function. Since no closed expression is known, approximation methods must be used. To this end, we use the identity (14–71):

$$(i p - i k - i V + m)^{-1}$$
$$= (i p - i k + m)^{-1} + (i p - i k + m)^{-1} i V (i p - i k + m)^{-1}$$
$$+ (i p - i k + m)^{-1} i V (i p - i k - i V + m)^{-1} i V (i p - i k + m)^{-1}.$$

The corresponding separation of the matrix element (15–51) can be expressed in terms of diagrams (Fig. 15–6). The external double lines indicate Coulomb wave functions; the internal double lines indicate Coulomb propagation functions.

The separation of the fluctuation term into the three parts depicted in Fig. 15–6 is exact. However, in order to make further progress it is now necessary to resort to approximation methods. To this end, let us first estimate the order of magnitude of the effect in question. The diagram a_1' is the expectation value of the free electron self-energy part. Although this expectation value is to be taken with Coulomb wave functions, to be sure, we surmise that it will not contribute observable terms, but will be relevant in the cancellation of infinite terms. We shall therefore estimate the order of magnitude of the next diagram, (a_1'').

This diagram gives a factor α from the emission and reabsorption of the virtual photon; it gives a factor αZ from the single action of the Coulomb field in the intermediate state; and, finally, it gives a factor a^{-3}, where a

is the atomic radius, which is the order of magnitude of the normalization of the Coulomb wave functions. For hydrogen-like atoms,

$$a = \frac{n}{Z} a_0 = \frac{n}{\alpha Z m},$$

where n is the principal quantum number and a_0 is the Bohr radius. It follows that the diagram (a_1'') gives an energy correction of order of magnitude $\alpha(\alpha Z)^4 m$ times the corresponding integral. This integral would, in general, be expected to be of order one. However, we know that the vertex part of which we are here taking the expectation value contains an infrared divergence in its observable part (the *total* vertex part is free of such a divergence!). This divergence is logarithmic and will thus give rise to a logarithm in the evaluation of this integral. We expect the divergence to be compensated by terms which must necessarily arise from diagram (a_2), thereby replacing the infinite argument by a finite one, but the functional form of a logarithmic factor cannot be expected to disappear. A very similar situation arose in the case of radiative corrections to Coulomb scattering, which is indeed the same calculation, but is carried out solely in the continuous part of the spectrum. In the low-energy limit this logarithm was found to be of order v^2, the square of the electron velocity. In our case, the negaton's velocity is of order αZ. We estimate, therefore, that diagram (a_1'') will contain terms of order $\alpha(\alpha Z)^4 \ln \alpha Z$ and $\alpha(\alpha Z)^4$.

It is now essential to recognize that the three-vector of momentum in the atom is of the same order αZ as the ratio of two successive Born approximations. This means that there is a competition between relativistic corrections and higher Born approximations. The separation of Fig. 15–6 is therefore very convenient, since it separates the first Born approximation in intermediate states from higher ones. The diagrams (a_1) and (a_2) are often referred to as the *one-potential part* and the *many-potential part*. From the above argument follows that we can evaluate (a_2) in the nonrelativistic limit, whereas (a_1) must be calculated relativistically.

From the physical point of view it is clear that we are dealing here with virtual transitions from the initial bound state of energy E_0 to all discrete and continuum levels and back again to E_0. Therefore, the intermediate states which have high enough momentum $(p \gg \alpha Z m)$ to make a relativistic calculation necessary, will be so far in the continuum that their binding energy [order $(\alpha Z)^2 m$] can be neglected. It follows that for high excitations the Coulomb field needs to be taken into account only to first approximation in the intermediate state, but that the calculation must be carried out relativistically; on the other hand, for low excitations the Coulomb field in intermediate states is essential, but a nonrelativistic approximation is sufficient. The separation into one-potential and many-potential parts is therefore also a separation into high-energy and low-energy intermediate states. Although there is no sharp separation between them, we can imagine some characteristic energy of order αZm to be the lower limit of the high-energy part, $\sim m$, and the upper limit of the low-energy part, $\sim(\alpha Z)^2 m$.

Since it follows from our estimate above that the term containing the logarithm is the leading term and that this term must be common to the high-energy and the low-energy parts (because of the cancellation of the infrared divergence), a satisfactory first estimate of ΔE can be obtained by calculating only ΔE_2, the nonrelativistic many-potential part. The upper cutoff of the logarithmically divergent result, $O(\alpha Zm)$ for ΔE_2, can now be chosen to be of order m, thus including approximately the relativistic calculation ΔE_1 which must have the same functional form. In this way, the first Lamb-shift calculation was carried out by Bethe* with the result $\Delta E_H = 1040$ Mc/sec. This is in excellent agreement with the experimental result [see Eq. (15–49)], considering the approximate nature of the calculation.

In the following pages, we shall calculate the radiative corrections of order $\alpha(\alpha Z)^4 \ln (\alpha Z)$ and $\alpha(\alpha Z)^4$ to an arbitrary bound state of a hydrogen-like atom with principal quantum number n_0, orbital angular momentum quantum number l, and energy E_0. This calculation will be carried out in two steps. In the first step ΔE_1 will be obtained by showing that it is exactly the expectation value of the operator associated with the Coulomb scattering diagram of Fig. 15–3. In the second step we shall evaluate ΔE_2 which, as we shall see, has contributions only from diagram (a_2), Fig. 15–6, within the order of interest.

We now turn to the calculation of the one-potential part, ΔE_1. It consists in the first Born approximation of the fluctuation and the polarization diagrams, Fig. 15–5:

$$\Delta E_1 = \Delta E_{F1} + \Delta E_{P1}.$$

The diagrams for ΔE_{F1} are (a_1') and (a_1'') in Fig. 15–6, from which the free electron self-energy must be subtracted. This can easily be accomplished by expanding the external lines of (a_1') in Born approximation. We therefore assume the bound electron to be an electron at rest or a free electron of momentum $|\mathbf{p}| = \alpha Zm$ in zeroth approximation, so that the Coulomb function can be approximated by

$$u = u_f + (i\mathbf{p} + m)^{-1} iVu. \qquad (15\text{–}53)$$

The function u_f must, of course, be properly normalized to one electron per atomic volume. The diagram (a_1') can be separated according to this

---

* H. A. Bethe, *Phys. Rev.* **72**, 339 (1947).

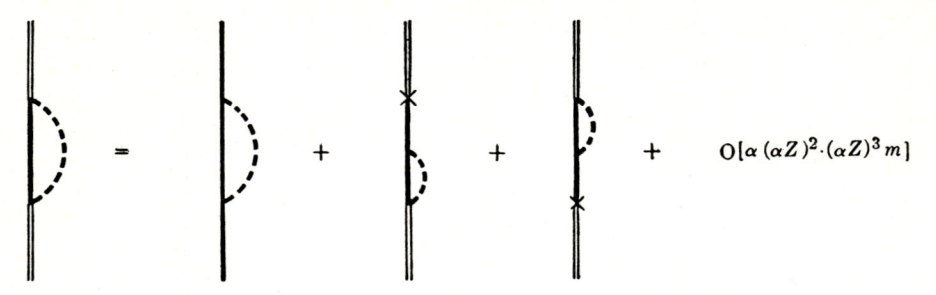

FIG. 15-7. Approximate decomposition of diagram (a_1').

approximation, as shown in Fig. 15-7. The terms of order $\alpha(\alpha Z)^5 m$ are negligible in our approximation. Actually, it has been shown that this approximation gives ΔE_{F1} correct to order $\alpha(\alpha Z)^6 m$.

The first diagram in this separation is just the free electron self-energy which must be subtracted, so that ΔE_{F1} consists of the other two diagrams and diagram (a_1'') of Fig. 15-6. These diagrams are exactly the same as diagrams (a) and (b) in Fig. 15-3 for the radiative corrections to Coulomb scattering. They differ only in the external lines which refer to the states of the continuum and the discrete spectrum, respectively. In the scattering case, the Coulomb functions for the continuum were approximated by plane waves, i.e., free-particle states. The results obtained there, Eq. (15-22), are valid for free initial and final states irrespective of the momentum. In particular, they will be valid for momenta of order $\alpha Z m$ as they occur in bound states. In fact, we can take the integrand obtained there and simply evaluate it for Coulomb functions. This corresponds to an expansion of the Coulomb functions in plane waves. The error introduced thereby will be of the order of the binding energy $(\alpha Z)^2 m$ and is therefore negligible.

We can make exactly the same arguments about the polarization diagram, Fig. 15-5 (P). The first Born approximation in intermediate states yields diagram (c) of Fig. 15-3, except for the external lines. By the same argument as above we can take over the integrand of the scattering calculation and calculate its expectation value for Coulomb functions. The polarization diagram does not contribute to ΔE_2, the many-potential part, within our approximation, since the second Born approximation in intermediate states vanishes, because it is a closed loop of three corners, and the third Born approximation is of order $\alpha(\alpha Z)^5$.

The method developed in Section 14-7 for the radiative corrections to energy levels can now be recalled. This method needs to be used for the one-potential part only, since only this part contains infinities. These must be removed in the free interaction picture and then a transformation must be carried out to the bound interaction picture to first order in the potential in intermediate states. To this end, we can remove the infinities in the covariant first Born approximation and evaluate the result for the momentum distribution of the state of energy E_0. We are thus again led to the expectation value of the expression (15–22). The previous discussion shows that the result is exactly the one-potential part of the diagrams of Fig. 15–5. To the order of this calculation we are left with diagram (a_2) of Fig. 15–6 as the only contribution to the many-potential part to be evaluated in the nonrelativistic limit.

The one-potential part follows therefore from Eq. (15–22):

$$\Delta E_1 = -\frac{ie}{(2\pi)^{3/2}}\left[\frac{\alpha}{3\pi}\left(\ln\frac{m}{\lambda} - \frac{3}{8} - \frac{1}{5}\right)\int \bar{u}_0(\mathbf{p}')\frac{\mathbf{q}^2}{m^2}\varphi(\mathbf{q})u_0(\mathbf{p})d^3p'd^3p \right.$$

$$\left. + \frac{\alpha}{4\pi}\int \bar{u}_0(\mathbf{p}')\varphi(\mathbf{q})\frac{i\mathbf{q}}{m}u_0(\mathbf{p})d^3p'd^3p\right]. \quad (15\text{–}54)$$

This result can be written somewhat more conveniently in x-space, putting $\partial_\mu = (\partial/\partial x^\gamma)$:

$$\Delta E_1 = \frac{i\alpha}{3\pi}\left(\ln\frac{m}{\lambda} - \frac{3}{8} - \frac{1}{5}\right)\int \bar{u}_0(\mathbf{x})\left(\frac{\nabla^2}{m^2}e\varphi(\mathbf{x})\right)u_0(\mathbf{x})d^3x$$

$$+ \frac{i\alpha}{4\pi}\int \bar{u}_0(\mathbf{x})\left(\boldsymbol{\gamma}\cdot\frac{\nabla}{m}e\varphi(\mathbf{x})\right)u_0(\mathbf{x})d^3x$$

$$= \frac{\alpha}{3\pi}\left(\ln\frac{m}{\lambda} - \frac{3}{8} - \frac{1}{5}\right)\left(0\left|\frac{\nabla^2}{m^2}V\right|0\right) - \frac{i\alpha}{4\pi}\left(0\left|\boldsymbol{\gamma}\cdot\frac{\nabla}{m}V\right|0\right), \quad (15\text{–}55)$$

where V is the potential energy operator defined in Eq. (15–52) but written in the x-space representation. For an attractive Coulomb field

$$V = V^0 = -\frac{\alpha Z}{r}.$$

We next consider the many-potential term, diagram (a_2) in Fig. 15–6. Only the fluctuation term contributes to ΔE_2. We find

$$\Delta E_2 = \frac{e^2}{(2\pi)^4}\sum_{\text{pol}}\int \bar{u}_0 e\frac{i\mathbf{p} - i\mathbf{k} - m}{(p-k)^2 + m^2}iV(i\mathbf{p} - i\mathbf{k} - iV + m)^{-1}$$

$$\times iV\frac{i\mathbf{p} - i\mathbf{k} - m}{(p-k)^2 + m^2}eu_0\frac{d^4k}{k^2 + \lambda^2}.$$

Here we have assumed a small photon mass λ and a corresponding summation over polarizations, including the longitudinal photons. Since this procedure is a cutoff procedure, it is important to carry it out in exactly the same way as in ΔE_1. Only then will the joining of the two infrared divergent parts, ΔE_1 and ΔE_2, yield the correct result.*

The evaluation of ΔE_2 is considerably simplified by taking into account all the permissible approximations to order $\alpha(\alpha Z)^4 \ln \alpha Z$ and $\alpha(\alpha Z)^4$. First we notice that due to the $k^2 + \lambda^2$ term in the denominator the photon momentum will contribute mainly when it is of the same order of magnitude as the photon energy ω, that is, $O[(\alpha Z)^2 m]$, the order of energy-level differences. Thus, ik in the numerator of the free propagation functions is of order $(\alpha Z)^2$ compared with m and can be neglected. The negaton's momentum is of order $\alpha Z m$, its energy is $m - O[(\alpha Z)^2 m]$. In the denominator of the free-particle propagation function, $p^2 + m^2$ vanishes† and k^2 is of order $(\alpha Z)^4 m^2$, whereas the leading term,

$$-2p \cdot k = 2m\omega + O[(\alpha Z)^3 m^2],$$

is of order $(\alpha Z)^2 m^2$. The free propagation functions can thus be written

$$\frac{i p - m}{2m\omega}.$$

Since we must evaluate ΔE_2 in the nonrelativistic limit only, we can use the nonrelativistic Coulomb propagation function. That means that we replace it by the sum over the large components of the negaton functions $u_l \bar{u}_l$ and ignore the positon states which involve transitions of order $(\alpha Z)^2 m / 2m$ smaller. We can thus replace S_c by S_c^{NR}, given in Eq. (14–75), though we have to retain the spin functions χ because of the spinor matrices in the other factors; dropping the subscript l, we have

$$S_c(p - k, p' - k)$$
$$\to 2\pi \delta(E_0 - E') \sum_n \delta_+(E_n - E_0 + \omega) u_n(\mathbf{p} - \mathbf{k}) \bar{u}_n(\mathbf{p}' - \mathbf{k}),$$

$$\delta_+(E) = \frac{1}{2} \delta(E) + \frac{i}{2\pi} P \frac{1}{E}.$$

Here we have used the *total* energy of the state u_0, $m - E_0$, and the total energy of the intermediate states u_n, $m - E_n - \omega$. The above factor can be simplified, since $\mathbf{k}$ in the argument of u_n can be ignored, being of order αZ compared with $\mathbf{p}$. This corresponds to the dipole approximation where in x-space the factor $e^{-i\mathbf{k} \cdot \mathbf{x}}$ is replaced by unity.

---

* *Cf.* the footnote on p. 337.

† Note that u_0 is the plane wave expansion of the Coulomb wave function, each plane wave being regarded as a free particle in this approximation.

With these simplifications, ΔE_2 can be written in operator notation:

$$\Delta E_2 = \frac{\alpha}{2\pi^2} \sum_{\text{pol}} \int \bar{u}_0 e \frac{i\not{p} - m}{2m\omega} iV \sum_n \delta_+(E_n - E_0 + \omega)$$
$$\times u_n \bar{u}_n iV \frac{i\not{p} - m}{2m\omega} eu_0 \frac{d^4k}{k^2 + \lambda^2}.$$

Following the same arguments that led to Eq. (15–26), we can sum over the polarizations:

$$\Delta E_2 = \frac{\alpha}{2\pi^2} \frac{1}{12m^2} \sum_n \int d^3k \int \frac{\delta_+(E_n - E_0 + \omega)(2 + \lambda^2/\omega^2)d\omega}{\omega^2(|\mathbf{k}|^2 - \omega^2 + \lambda^2)}$$
$$\times |(0|\boldsymbol{\gamma}(i\not{p} - m)iV|n)|^2,$$

where we used the notation

$$(0|M|n) = \bar{u}_0 M u_n = \int \bar{u}_0(\mathbf{p}')M(\mathbf{p}',\mathbf{p})u_n(\mathbf{p})d^3p'd^3p.$$

This expectation value can be simplified:

$$(0|\boldsymbol{\gamma}(i\not{p} - m)iV|n) = (0|2i\not{p}iV - (i\not{p} + m)\boldsymbol{\gamma}iV|n)$$
$$= (0| - 2\mathbf{p}V - iV\boldsymbol{\gamma}(i\not{p} + m)|n)$$
$$= (0|2[V,\mathbf{p}] + iV(i\not{p} - m)\boldsymbol{\gamma}|n)$$
$$= (0|2[V,\mathbf{p}] - (p^2 + m^2)\boldsymbol{\gamma}|n).$$

The last term vanishes between large components, whereas the first term can be simplified by use of the Hamiltonian

$$H = \frac{p^2}{2m} + V.$$

Inserting this operator, we find

$$-i(0|2[V,\mathbf{p}]|n) = 2(0|[V,\mathbf{p}]|n) = 2(0|[H,\mathbf{p}]|n) = 2(E_0 - E_n)(0|\mathbf{p}|n).$$

Therefore, after a trivial integration over angles,*

$$\Delta E_2 = \frac{4\alpha}{3\pi m^2} \sum_n \int_0^\infty |\mathbf{k}|^2 d|\mathbf{k}| \int_{-\infty}^\infty \left(1 + \frac{\lambda^2}{2\omega^2}\right)$$
$$\times \frac{(E_0 - E_n)^2 |(0|\mathbf{p}|n)|^2 \delta_+(E_n - E_0 + \omega)}{\omega^2(\mathbf{k}^2 - \omega^2 + \lambda^2)} d\omega.$$

---

* The remaining integration is identical with the corresponding part of the work by M. Baranger, H. A. Bethe, and R. P. Feynman, *Phys. Rev.* **92**, 482 (1953). However, the inclusion of both parts of the δ_+-function in our calculation gives a simple justification for their integration method in the ω-plane.

The integration over ω is to be carried out in the usual way by displacing the poles in the complex ω-plane. Both λ and m receive small negative imaginary parts. The δ_+-function has the integral representation

$$\delta_+(E_n - E_0 + \omega) = \frac{1}{2\pi}\int_0^\infty e^{-i(E_n-E_0+\omega)t}dt,$$

so that the contour is to be closed by a large half-circle in the negative imaginary part of the ω-plane. Therefore, only the pole at

$$\omega = \omega_0 = \sqrt{\mathbf{k}^2 + \lambda^2}$$

will contribute. The pole at $\omega = 0$ does not contribute, since it arose from the denominators $p^2 + m^2 + k^2 - 2p\cdot k$, which contain a small negative imaginary part $(-2p\cdot k \to 2m\omega)$, corresponding to a small displacement of the pole at the origin into the upper half plane.

The result is

$$\Delta E_2 = -i\frac{8\alpha}{3m^2}\sum_n (E_n - E_0)^2|(n|\mathbf{p}|0)|^2 \int_0^\infty \frac{\mathbf{k}^2 d|\mathbf{k}|}{2\omega_0^3}$$

$$\times \left(1 + \frac{\lambda^2}{2\omega_0^2}\right)\delta_+(E_n - E_0 + \omega_0).$$

The remaining integration is elementary. It yields a complex value for ΔE_2 and therefore also for ΔE. The imaginary part describes the natural *line width* of the state of energy E_0 and will be discussed in detail in Section 16–3. It does not affect the line *shift* and it can therefore be ignored in the present calculation. The real part, in the limit $\lambda \to 0$, gives

$$\Delta E_2 = \frac{2\alpha}{3\pi m^2}\sum_n |(n|\mathbf{p}|0)|^2 (E_n - E_0)\left(\ln\frac{\lambda}{2|E_n - E_0|} + \frac{5}{6}\right). \quad (15\text{–}56)$$

This result for the many-potential part must now be combined with the one-potential part (Eq. 15–55). To this end we note the well-known relation *

$$\sum_n |(n|\mathbf{p}|0)|^2 (E_n - E_0) = \tfrac{1}{2}(0|\nabla^2 V|0). \quad (15\text{–}57)$$

---

* This equation can easily be verified:

$$\sum_n |(n|\mathbf{p}|0)|^2 (E_n - E_0) = \tfrac{1}{2}\sum_n \{(0|\mathbf{p}|n)\cdot(n|[H,\mathbf{p}]|0) + (0|[\mathbf{p},H]|n)(n|\mathbf{p}|0)\}$$

$$= \tfrac{1}{2}(0|[\mathbf{p},[H,\mathbf{p}]]|0)$$

$$= \tfrac{1}{2}(0|[\mathbf{p},[V,\mathbf{p}]]|0)$$

$$= \tfrac{1}{2}(0|\nabla^2 V|0).$$

Equation (15–56) can therefore be written

$$\Delta E_2 = \frac{\alpha}{3\pi}\left(\ln\frac{\lambda}{2k_0} + \frac{5}{6}\right)\left(0\left|\frac{\nabla^2}{m^2}V\right|0\right). \tag{15–58}$$

The energy k_0 introduced here is an *average excitation* energy. In writing down its definition, which follows by comparison of Eqs. (15–58) and (15–56) using Eq. (15–57), we want to indicate explicitly the quantum numbers n_0 and l associated with E_0.

$$\ln\frac{k_0(n_0,l)}{Z^2\mathrm{Ry}} = \frac{\sum\limits_n |(n|\mathbf{p}|n_0l)|^2 (E_n - E_0)\ln(|E_n - E_0|/\mathrm{Ry})}{\sum\limits_n |(n|\mathbf{p}|n_00)|^2 (E_n - E_0)}. \tag{15–59}$$

The Rydberg unit of energy, Ry, is conveniently chosen as the reference energy.

In the form (15–58) the similarity between this calculation and the corresponding calculation for the continuum, Eq. (15–26), is particularly striking. In fact, the expectation value of Eq. (15–24) between the appropriate Coulomb functions would have given exactly the same result.

Combining Eqs. (15–58) and (15–55), we find for the total radiative correction to the level E_0 exact to order $\alpha(\alpha Z)^4$ inclusive,

$$\Delta E = \Delta E_1 + \Delta E_2$$

$$= \frac{\alpha}{3\pi}\left(\ln\frac{m}{2k_0} - \frac{3}{8} - \frac{1}{5} + \frac{5}{6}\right)\left(0\left|\frac{\nabla^2}{m^2}V\right|0\right) - \frac{i\alpha}{4\pi}\left(0\left|\boldsymbol{\gamma}\cdot\frac{\nabla}{m}V\right|0\right). \tag{15–60}$$

The last term is the effect of the anomalous magnetic moment which was discussed in the previous section. This can easily be seen by writing it in the form

$$-i\frac{\alpha}{4\pi}\left(0\left|\boldsymbol{\gamma}\cdot\frac{\nabla}{m}V\right|0\right) = \frac{\alpha}{2\pi}\cdot\frac{e}{2m}\cdot(0|\sigma_{i0}f^{i0}|0).$$

This term can be separated into two parts which contribute in states with orbital angular momentum $l = 0$ and $l \neq 0$, respectively. The matrices $\boldsymbol{\gamma}$ will give nonvanishing expectation values only between large and small components (*cf.* the last part of Section 14–4). Since we are interested in the nonrelativistic limit, we can introduce the large components in first approximation, *cf.* Eq. (14–73):

$$u_s = \frac{1}{2m}i\gamma^0\boldsymbol{\gamma}\cdot\nabla u_l = -\frac{1}{2m}\boldsymbol{\gamma}\cdot\nabla u_l,$$

so that we find with $V = V^0 = -V_0$ and $i\gamma^0 u_l = u_l$,

$$\begin{aligned}
-i2m\bar{u}\boldsymbol{\gamma}\cdot\boldsymbol{\nabla}Vu &= 2m\bar{u}\gamma^0\boldsymbol{\gamma}\cdot\boldsymbol{\nabla}Vu \\
&= 2m(\bar{u}_l i\gamma^0\boldsymbol{\gamma}\cdot\boldsymbol{\nabla}Vu_s + \bar{u}_s i\gamma^0\boldsymbol{\gamma}\cdot\boldsymbol{\nabla}Vu_l) \\
&= -\bar{u}_l\boldsymbol{\gamma}\cdot\boldsymbol{\nabla}V\boldsymbol{\gamma}\cdot\boldsymbol{\nabla}u_l - (\boldsymbol{\nabla}\bar{u}_l)\cdot\boldsymbol{\gamma}\boldsymbol{\gamma}\cdot\boldsymbol{\nabla}Vu_l \\
&= -\bar{u}_l\boldsymbol{\gamma}\cdot\boldsymbol{\nabla}V\boldsymbol{\gamma}\cdot\boldsymbol{\nabla}u_l + \bar{u}_l\boldsymbol{\gamma}\cdot\boldsymbol{\nabla}(\boldsymbol{\gamma}\cdot\boldsymbol{\nabla}Vu_l),
\end{aligned}$$

by partial integration. The divergence vanishes, since it can be transformed into a surface integral at large distances.

We now use the identity (A2–53):

$$\boldsymbol{\gamma}\cdot\mathbf{A}\boldsymbol{\gamma}\cdot\mathbf{B} = \mathbf{A}\cdot\mathbf{B} + i\boldsymbol{\sigma}\cdot\mathbf{A}\times\mathbf{B}$$

and find with $\boldsymbol{\nabla}\times\boldsymbol{\nabla}V = 0$,

$$\begin{aligned}
-i2m\bar{u}\boldsymbol{\gamma}\cdot\boldsymbol{\nabla}Vu &= -\bar{u}_l(\boldsymbol{\nabla}V\cdot\boldsymbol{\nabla} + i\boldsymbol{\sigma}\cdot\boldsymbol{\nabla}V\times\boldsymbol{\nabla})u_l \\
&\quad + \bar{u}_l[\nabla^2 V + (\boldsymbol{\nabla}V)\cdot\boldsymbol{\nabla} - i\boldsymbol{\sigma}\cdot\boldsymbol{\nabla}V\times\boldsymbol{\nabla}]u_l \\
&= \bar{u}_l\nabla^2 V u_l - 2i\bar{u}_l\boldsymbol{\sigma}\cdot\boldsymbol{\nabla}V\times\boldsymbol{\nabla}u_l.
\end{aligned}$$

This result can be written in a nicer form by introducing the orbital angular momentum operator

$$\mathbf{L} = \mathbf{r}\times\mathbf{p} = -i\mathbf{r}\times\boldsymbol{\nabla}$$

and by observing that the Coulomb potential V is spherically symmetrical,

$$\boldsymbol{\nabla}V = \mathbf{r}\frac{1}{r}\frac{dV}{dr}.$$

We then find

$$-i2m\bar{u}\boldsymbol{\gamma}\cdot\boldsymbol{\nabla}Vu = \bar{u}_l\nabla^2 V u_l + 2\bar{u}_l\frac{1}{r}\frac{dV}{dr}\boldsymbol{\sigma}\cdot\mathbf{L}u_l. \tag{15–61}$$

We conclude from this separation that the anomalous magnetic moment also contributes to the level shift in s-states. Its contribution can be inserted in Eq. (15–60), with the result

$$\Delta E = \frac{\alpha}{3\pi}\left(\ln\frac{m}{2k_0} - \frac{1}{5} + \frac{5}{6}\right)\left(0\left|\frac{\nabla^2}{m^2}V\right|0\right) \quad \text{for } s\text{-states.} \tag{15–62}$$

For states with $l \neq 0$, we note that in the NR-limit

$$\left(0\left|\frac{\nabla^2}{m^2}V\right|0\right) = 0, \quad (l \neq 0).$$

The one-potential part therefore contributes only the magnetic moment term, and the many-potential part (15–58) contributes only the term ln (Ry/k_0), since the coefficient of ln $(\lambda/2 \text{ Ry})$ vanishes. Using Eq. (15–61), we find*

$$\Delta E = \frac{\alpha}{3\pi}\left[\ln\frac{Z^2\text{Ry}}{k_0(n_0,l)}\left(n_0 0 \left|\frac{\nabla^2}{m^2}V\right| n_0 0\right) + \frac{3}{4m^2}\left(n_0 l\left|\boldsymbol{\sigma}\cdot\mathbf{L}\frac{1}{r}\frac{dV}{dr}\right| n_0 l\right)\right].$$

(15–63)

The expectation values are easily evaluated for Schroedinger wave functions. Since

$$V = -e\varphi^0(r) = -\frac{\alpha Z}{r}, \quad \nabla^2 V = \alpha Z 4\pi\delta(\mathbf{r}),$$

we obtain

$$(n_0 0|\nabla^2 V|n_0 0) = 4\pi\alpha Z|\psi_{n_0,0}(0)|^2 = 4\pi\alpha Z\frac{1}{\pi a^3},$$

$$a = \frac{n_0}{Z}a_0 = \frac{n_0}{\alpha Z m}.$$

This result can be expressed in Rydberg units,

$$1 \text{ Ry} = \tfrac{1}{2}\alpha^2 m,$$

so that

$$\left(n_0 0\left|\frac{\nabla^2}{m^2}V\right|n_0 0\right) = \frac{8Z^4}{n_0^3}\alpha^2 \text{ Ry}. \qquad (15\text{–}64)$$

Similarly, the expectation value of the spin-orbit coupling term is an elementary problem of quantum mechanics. We find for the Coulomb field,

$$\frac{1}{m^2}\left(n_0 l\left|\boldsymbol{\sigma}\cdot\mathbf{L}\frac{1}{r}\frac{dV}{dr}\right|n_0 l\right) = c_{lj}l(l+1)\frac{1}{m^2}\alpha Z\int_0^\infty |\psi_{n_0,l}(\mathbf{r})|^2\frac{dr}{r}d\Omega$$

$$= \frac{4Z^4}{n^3}\alpha^2 \text{ Ry}\frac{c_{lj}}{2l+1},$$

where

$$c_{lj} = \begin{cases} \dfrac{1}{l+1}, & \text{for } j = l + \dfrac{1}{2}, \\ -\dfrac{1}{l}, & \text{for } j = l - \dfrac{1}{2}. \end{cases} \qquad (15\text{–}65)$$

---

* Note that the definition of $k_0(n_0,l)$, Eq. (15–59) contains in its denominator the matrix element between the *s-state* associated with n_0 and the intermediate state. The first term in Eq. (15–63) is therefore this same term.

The two expressions (15–62) and (15–63) can therefore be written, omitting the subscript zero on n,

$$\Delta E = \frac{8Z^4}{n^3} \frac{\alpha^3}{3\pi} \left[\ln \frac{m}{2k_0(n,0)} + \frac{19}{30} \right] \text{Ry}, \quad (l = 0), \tag{15-66}$$

$$\Delta E = \frac{8Z^4}{n^3} \frac{\alpha^3}{3\pi} \left[\ln \frac{Z^2 \text{Ry}}{k_0(n,l)} + \frac{3}{8} \frac{c_{lj}}{2l+1} \right] \text{Ry}, \quad (l \neq 0).$$

The average excitation energy k_0 must be obtained numerically. Bethe, Brown, and Stehn* found the following values for the $2s$ and $2p$ states in hydrogen,**

$$\ln \frac{k_0}{\text{Ry}} = \begin{cases} 2.8121 \pm 0.0004, & \text{for } 2s, \\ -0.0300 \pm 0.0002, & \text{for } 2p. \end{cases} \tag{15-67}$$

Using the best values for the fundamental constants, one obtains for the $2\,^2s_{1/2} - 2\,^2p_{1/2}$ separation in H the value†

$$\Delta E_H = 1052.10 \pm 0.08 \text{ Mc/sec}. \tag{15-68}$$

This value superseded the older nonrelativistic calculation by Bethe (*loc. cit.*) and was obtained by several independent calculations which all led to the result (15–66).‡ One can see that ΔE_H arises primarily from the radiative correction to the $2\,^2s_{1/2}$ level, which is shifted up by about 1039 Mc/sec, of which the magnetic moment contributes about 51 Mc/sec and the polarization of the vacuum§ about -27 Mc/sec. The spin-orbit interaction shifts the $2\,^2p_{3/2}$ level up and the $2\,^2p_{1/2}$ level down, as is to be expected, the latter shift being twice the former. The $2\,^2p_{1/2}$ shift consists of -17 Mc/sec from the spin-orbit interaction and $+4$ Mc/sec from the average excitation energy. The magnetic moment therefore contributes about 68 Mc/sec to ΔE_H. (See Table 15–1.)

When the theoretical result (15–68) is compared with the measured value (15–49), we notice that it falls short by 5.7 Mc/sec. This discrepancy is of

---

* H. A. Bethe, L. M. Brown, and J. R. Stehn, *Phys. Rev.* **77**, 370 (1950).

** The results in the remainder of this Section have been obtained to higher accuracy; see Supplement Section S 5–3 and S 5–2.

† E. E. Salpeter, *Phys. Rev.* **89**, 92 (1953).

‡ N. M. Kroll and W. E. Lamb, *Phys. Rev.* **75**, 388 (1949); J. B. French and V. F. Weisskopf, *Phys. Rev.* **75**, 1240 (1949); R. P. Feynman, *Phys. Rev.* **74**, 1430 (1949), and correction in *Phys. Rev.* **76**, 769 (1949), footnote 13; J. Schwinger, *Phys. Rev.* **76**, 790 (1949); H. Fukuda, Y. Miyamoto, and S. Tomonaga, *Progr. Theor. Phys.* (Japan) **4**, 47 and 121 (1948).

§ The effect of the vacuum polarization was first calculated by R. Serber, *Phys. Rev.* **48**, 49 (1935), and A. E. Uehling, *Phys. Rev.* **48**, 55 (1935).

Table 15-1

Contributions (in Mc/sec) to the $2s_{1/2}$ and $2p_{1/2}$ level corrections in hydrogen

Type of correction	Order	Effect	$2s_{1/2}$	$2p_{1/2}$	$2s_{1/2} - 2p_{1/2}$
2nd-order radiative corrections	$\alpha(\alpha Z)^4 \ln(\alpha Z)$ and $\alpha(\alpha Z)^4$	Magnetic moment, Vacuum polarization, Other	$\left.\begin{array}{r}50.86\\-27.13\\1015.48\end{array}\right\}1039.21$	$\left.\begin{array}{r}-16.96\\0\\4.07\end{array}\right\}-12.89$	$\left.\begin{array}{r}67.82\\-27.13\\1011.41\end{array}\right\}1052.10\pm.1$
	$\alpha(\alpha Z)^5$	Relativistic corrections to above	7.14	0	7.14
4th-order radiative corrections	$\alpha^2(\alpha Z)^4$	Magnetic moment, Vacuum polarization, Other	$\left.\begin{array}{r}-0.078\\-0.24\\0.24\pm.1\end{array}\right\}-0.08\pm.1$	$\left.\begin{array}{r}0.025\\0\\0\end{array}\right\}0.025$	$\left.\begin{array}{r}0.10\\-0.24\\+0.24\pm.1\end{array}\right\}-0.10\pm.1$
		Corrections to fine structure	0.379	-0.017	0.396
Mass corrections	$\alpha \dfrac{m}{M}$ (FS)				-1.175
	$\dfrac{m}{M}$ (rad. corr.)	Corrections to above radiative corrections			-1.571
Nucleon-electron direct interaction as known from neutron-negaton interaction			0.025	0	0.025
Total					$1057.99 \pm .2$

the order of magnitude of the terms neglected in the above calculation, $O[\alpha(\alpha Z)^5]$. A more accurate calculation is therefore needed.

Recently, such a calculation was carried out by two independent groups.* They both obtain the following $\alpha(\alpha Z)^5$ corrections to the results (15–66) for s-levels:

$$\frac{8Z^4}{n^3}\frac{\alpha^3}{3\pi} \cdot 3\pi\alpha Z \left(1 + \frac{11}{128} - \frac{1}{2}\ln 2 + \frac{5}{192}\right) \text{Ry.} \qquad (15\text{–}69)$$

The last term inside the parentheses is due to the polarization term. The $2^2 s_{\frac{1}{2}}$ level in H is raised by 7.14 Mc/sec due to these corrections.

Further corrections have been discussed by Salpeter (*loc. cit.*). They consist first in the next order (fourth order) radiative corrections, which give a total of -0.10 ± 0.1 Mc/sec.† Secondly, there is the correction due to the finite proton mass M which yields further -1.18 Mc/sec.‡

In this way we obtain a final value of

$$\Delta E_H = 1057.99 \pm 0.2 \text{ Mc/sec.} \qquad (15\text{–}70)$$

The various contributions to this result are listed in Table 15–1.

Within the experimental error and the accuracy of the theoretical estimate, the agreement between the measured and the computed $n = 2$ level shifts in the hydrogen atom is therefore quite satisfactory. A similar analysis for deuterium is more complicated because of the small effects due to the structure of the deuterium nucleus, which are very difficult to estimate. One finds §

$$\Delta E_D = 1059.34 \pm 0.2 \text{ Mc/sec,} \qquad (15\text{–}71)$$

which agrees, but not quite as well as in hydrogen, with the measured value (15–49). However, the difference,

$$\Delta E_D - \Delta E_H = 1.35 \pm 0.04 \text{ Mc/sec,} \qquad (15\text{–}72)$$

---

* M. Baranger, H. A. Bethe, and R. P. Feynman, *loc. cit.*, first reported by M. Baranger, *Phys. Rev.* **84,** 866 (1951); R. Karplus, A. Klein, and J. Schwinger, *Phys. Rev.* **86,** 288 (1952).

† They consist of the fourth-order magnetic moment (*cf.* Section 15–3), the fourth-order vacuum polarization [M. Baranger, F. J. Dyson, E. E. Salpeter, *Phys. Rev.* **88,** 680 (1952)] and many small terms from various diagrams [R. Bersohn, J. Weneser, and N. M. Kroll, *Phys. Rev.* **86,** 596 (1952)]. The numerical results of these authors are given in Table 15–1.

‡ The mass correction to fine structure was calculated by E. E. Salpeter, *Phys. Rev.* **87,** 328 (1952).

§ This value contains the various small effects discussed by Salpeter (*loc. cit.*) as well as the corrected fourth-order magnetic moment term and other recent corrections. *Cf.* also H. A. Bethe and E. E. Salpeter, *Quantum Mechanics of One- and Two-Electron Atoms*, Springer Verlag, Berlin, and Academic Press, New York, 1957.

is in very good agreement with the experimental value

$$\Delta E_D - \Delta E_H = 1.23 \pm 0.15 \text{ Mc/sec.} \tag{15-73}$$

Measurements in other hydrogen-like atoms, especially in He^+, have been carried out, but are so far not accurate enough to be of special interest.

We conclude this discussion with the remark that the small value of 0.6 Mc/sec, when compared with the total energy of the $n = 2$ state of 8×10^8 Mc/sec, shows that the proton-electron interaction has an over-all deviation from the Coulomb field of not more than 1 in 10^9.

Of course, radiative corrections are not restricted to pure Coulomb effects. The interaction with the nuclear magnetic dipole moment in conjunction with the Coulomb interaction gives rise to *hyperfine structure*. The radiative corrections to this effect are of considerable interest. The calculations proceed along lines similar to the above level shift calculation, except that the external field is not only the Coulomb potential $\varphi^0(x)$, but also the vector potential $\varphi_i(x)$ produced by the nuclear magnetic moment. The result of this calculation was obtained by two different methods.* It yields a correction factor to the Fermi formula for the hyperfine separation $\Delta \nu$ of s-levels. For the hydrogen ground state this factor is

$$1 + \frac{\alpha}{2\pi} - \frac{0.328 \alpha^2}{\pi^2} - \alpha^2 \left(\frac{5}{2} - \ln 2 \right). \tag{15-74}$$

The second and third term will be recognized as due to the anomalous magnetic moment of second and fourth order, respectively (*cf.* Section 15-3).

The main importance of this calculation, which agrees very well with experiments, lies in the use of this good agreement for the determination of a very accurate value of the fine-structure constant α. The best value obtained in this way is, at the present time,†

$$\alpha^{-1} = 137.039 \pm 0.002. \tag{15-75}$$

15-5 Radiative transitions between bound states. The problem to be discussed in this section is historically one of the oldest and most elementary applications of radiation theory. It is the problem of the spontaneous transitions between stationary states of an electron with the simultaneous emission of radiation. Precisely this problem eventually led to the formulation of quantum electrodynamics. Whereas ordinary wave mechanics with its classical external field is quite successful in leading to correct expressions for scattering and absorption of radiation, it completely fails

---

* R. Karplus and A. Klein, *Phys. Rev.* **85**, 972 (1952); N. M. Kroll and F. Pollock, *Phys. Rev.* **86**, 876 (1952).

† H. A. Bethe and E. E. Salpeter (*loc. cit.*), p. 353.

when applied to the problem of spontaneous emission of radiation. Before there existed a theory of the quantized radiation field, the only way to obtain the correct expression for the probability of spontaneous emission was via the principle of detailed balance or the correspondence principle.

The result which is usually given for the total transition probability per unit time of a transition from a state a to a state b is†

$$\Gamma_{\text{tot}} = \tfrac{4}{3}\alpha\omega^3 |\mathbf{x}_{ba}|^2, \qquad (15\text{-}76)$$

where the matrix element $\mathbf{x}_{ba}$ is

$$\mathbf{x}_{ba} = \int \psi_b^*(\mathbf{x}) \mathbf{x} e^{-i\mathbf{k}\cdot\mathbf{x}} \psi_a(\mathbf{x}) d^3x. \qquad (15\text{-}77)$$

The scalar Schroedinger wave functions $\psi(\mathbf{x})$ were introduced in Section 14–4.

For optical dipole radiation the exponential in Eq. (15–77) can be disregarded and we find approximately

$$\mathbf{x}_{ba} = \int \psi_b^*(\mathbf{x}) \mathbf{x} \psi_a(\mathbf{x}) d^3x. \qquad (15\text{-}78)$$

We shall now see how these well-known results are directly obtained from the general S-matrix theory in quantum electrodynamics. To this end, we work in the bound interaction picture (cf. Section 14–2), and we shall assume that we have a complete set of *stationary* states described by the relativistic spinor functions $u_r(x)$ and $v_s(x)$:

$$\begin{aligned} u_r(x) &= u_r(\mathbf{x}) e^{-iE_r t}, \quad (E_r > 0), \\ v_s(x) &= v_s(\mathbf{x}) e^{+iE_s t}, \quad (E_s > 0). \end{aligned} \qquad (15\text{-}79)$$

As was explained in Section 14–4, the existence of such a complete set implies that the external field is static. The functions $u_r(\mathbf{x})$ and $v_s(\mathbf{x})$ are the solutions of the stationary state equations (14–56):

$$H u_r = E_r u_r, \quad H v_s = -E_s v_s,$$

with H given by Eq. (14–57).

The probability for the transition from the initial state u_a to the final state u_b is given by Eq. (14–89) with

$$E_i = E_a, \quad E_f = E_b + \omega,$$

$$\Gamma = \frac{1}{2\pi} S_f |(f|M|i)|^2 \delta(E_f - E_i). \qquad (15\text{-}80)$$

---

† See any textbook on quantum mechanics. We assume for simplicity that the two states are nondegenerate.

The matrix element is obtained from the first term in the iteration solution:

$$(f|M|i) = \frac{e}{\sqrt{2\pi}} \frac{1}{\sqrt{2\omega}} \int \bar{u}_b(\mathbf{p}')e u_a(\mathbf{p})\delta(\mathbf{p}' + \mathbf{k} - \mathbf{p})d^3p'd^3p$$

$$= \frac{e}{\sqrt{2\pi}} \frac{1}{\sqrt{2\omega}} \int \bar{u}_b(\mathbf{x})e u_a(\mathbf{x})e^{-i\mathbf{k}\cdot\mathbf{x}}d^3x. \quad (15\text{–}81)$$

The reduction to the nonrelativistic case permits the replacement of the exponential by 1. The decomposition

$$u = u_l + u_s, \quad \bar{u} = \bar{u}_l + \bar{u}_s$$

then permits the replacement of u_a and u_b by their large components. Using

$$i\gamma^0 u_l = u_l$$

and

$$\boldsymbol{\gamma} = i\gamma^0[H,\mathbf{x}]$$

$$(15\text{–}82)$$

we find, dropping the subscript l,

$$\bar{u}_b e u_a = \bar{u}_b[H, \mathbf{e}\cdot\mathbf{x}]u_a = (E_b - E_a)\bar{u}_b \mathbf{e}\cdot\mathbf{x} u_a.$$

Therefore, the sum over the final polarization states yields

$$\sum_{\text{pol}} |(f|M|i)|^2 = \frac{e^2}{4\pi\omega}\omega^2 \cdot \frac{2}{3}\left|\int \bar{u}_b \mathbf{x} u_a d^3x\right|^2$$

$$= \tfrac{2}{3}\alpha\omega |\mathbf{x}_{ba}|^2, \quad (15\text{–}83)$$

where $\mathbf{x}_{ba}$ is given by Eq. (15–78). In the last equation we used the decomposition of u_l,

$$u_l(\mathbf{x}) = \psi(\mathbf{x})\chi, \quad (15\text{–}84)$$

with the spin function χ normalized to unity,

$$\bar{\chi}\chi = 1.$$

The integration over the density of final states gives a factor $4\pi\omega^2$ and eliminates the δ-function of energy conservation. Therefore, from Eqs. (15–80) and (15–83),

$$\Gamma = \frac{1}{2\pi} \cdot 4\pi\omega^2 \cdot \frac{2}{3}\alpha\omega |\mathbf{x}_{ba}|^2$$

$$= \tfrac{4}{3}\alpha\omega^3 |\mathbf{x}_{ba}|^2.$$

This is exactly the expression (15–76) with $\mathbf{x}_{ba}$ given by Eq. (15–78). The deduction of the electric dipole transition probability from the general radiation theory is thereby completed.

At this point the elementary radiation theory can be further developed in the way just indicated. In particular, the theories of Raman scattering, dispersion, multipole radiation, etc., can be developed along familiar lines. These topics are discussed elsewhere in detail* and will therefore be omitted here. The contribution of these problems to the basic understanding of quantum electrodynamics is primarily of historical interest.

15–6 Bremsstrahlung. The process of radiative transitions between bound state levels has a counterpart in the continuous spectrum, namely, the change in momentum of an electron scattering in an external field with the emission of a photon. This *bremsstrahlung* or *deceleration radiation* with the emission of a single photon is a well-defined process only within certain limits: the simultaneous emission of very soft photons—too soft to be observable within the accuracy of the energy determination of the incident and outgoing electron—can never be excluded. In fact, this radiation is always present, even in the so-called elastic scattering (*cf.* Section 16–1). Therefore, it will be impossible to make a clean physical distinction between bremsstrahlung and radiationless scattering when the emitted photon is very soft. This problem was discussed in connection with Coulomb scattering and its radiative correction in the first two sections of this chapter; we do not need to elaborate on it further. The relation of this problem to the infrared divergence is discussed in Section 16–1.

We shall restrict ourselves, therefore, to the emission of one not-too-soft photon k by an electron of momentum p which is scattered in an external field. The final electron momentum is p'. The external field is assumed to be the Coulomb field of a nucleus of charge Z,

$$\phi = \frac{Ze}{4\pi r}. \tag{15-85}$$

The general problem is extremely difficult. It requires the bound interaction picture (a singularly inappropriate name in this case!) and the use of the continuum state Coulomb wave functions. This problem has never been solved, except for the nonrelativistic limit and a very few special cases where numerical calculations were carried out for definite energies.

Fortunately, most of the physically interesting bremsstrahlung occurs at high energies, where some approximation methods have been developed. The simplest method is the Born approximation, where one effectively expands in the coupling to the external field αZ, and which is valid for high enough energy. The iteration solution in the interaction picture is then most suitable. In particular, in a first Born approximation the matrix elements $S^{(21)}$ (*cf.* Chapter 14) can easily be read from the diagrams of Fig. 15–4. Following Chapter 8 and Section 14–7, we have for a negaton

---

* See, e.g., W. Heitler, *The Quantum Theory of Radiation*, Third Edition, Oxford, Clarendon Press, 1954, especially §§17–20, where further references can be found.

$$(p'k|M^{(21)}|p) = \frac{ie^2}{(2\pi)^2} \frac{m}{\sqrt{\epsilon'\epsilon 2\omega}} \int \bar{u}(\mathbf{p}') \left[e \frac{i(\not{p}' + \not{k}) - m}{(p' + k)^2 + m^2} \not{n} \right.$$

$$\left. + \not{n} \frac{i(\not{p} - \not{k}) - m}{(p - k)^2 + m^2} e \right] u(\mathbf{p})\phi(\mathbf{q})\delta(\mathbf{p}' + \mathbf{q} + \mathbf{k} - \mathbf{p})d^3q.$$

The Fourier transform of the Coulomb potential ϕ is given in Eq. (15-4), and the relation

$$\varphi^\mu = n^\mu \phi, \quad (n^2 = -1)$$

is used. Since we are in the interaction picture, $M^{(21)}$ can be simplified considerably:

$$(p'k|M^{(21)}|p) = +i\alpha \frac{\sqrt{\alpha Z^2}}{2\pi^2} \frac{m}{\sqrt{\epsilon'\epsilon\omega}} \frac{1}{|\mathbf{q}|^2} \bar{u}(\mathbf{p}')Q(p',p)u(\mathbf{p}),$$

$$Q(p',p) = e \frac{i(\not{p}' + \not{k}) - m}{2p'\cdot k} \not{n} + \not{n} \frac{i(\not{p} - \not{k}) - m}{-2p\cdot k} e, \quad (15\text{-}86)$$

$$\mathbf{q} = \mathbf{p} - \mathbf{p}' - \mathbf{k}.$$

The cross section follows from Eq. (14-91)

$$d\sigma = \frac{(2\pi)^2}{\beta} \mathsf{S}_f\bar{\mathsf{S}}_i \delta(\epsilon' + \omega - \epsilon)|(p'k|M^{(21)}|p)|^2$$

$$= \frac{(2\pi)^2}{\beta} \frac{\alpha Z^2}{4\pi^4} r_0^2 \int \frac{m^4}{\epsilon'\epsilon\omega} \frac{1}{|\mathbf{q}|^4} \frac{1}{(2m)^2} \frac{1}{2} X d^3p'd^3k\delta(\epsilon' + \omega - \epsilon), \quad (15\text{-}87)$$

where X is the trace resulting from the summation over the initial and final spin directions,

$$X = (2m)^2 \operatorname{Tr}[Q\Lambda_-(p)\overline{Q}\Lambda_-(p')]. \quad (15\text{-}88)$$

The factor $\frac{1}{2}$ in front of X [Eq. (15-87)] arises from the spin average in the initial state.

The expression for Q can be further simplified, since u satisfies

$$(i\not{p} + m)u = 0,$$

$$\frac{1}{p'\cdot k} e(i\not{p}' + i\not{k} - m)\not{n} + \not{n}(i\not{p} - i\not{k} - m)e \frac{1}{-p\cdot k}$$

$$= \frac{1}{p'\cdot k}(2ip'\cdot en + ei\not{k}\not{n}) + (\not{n}i\not{k}e - 2ip\cdot en)\frac{1}{p\cdot k}.$$

The computation of X is now straightforward. It can be carried out, for example, as follows. We first commute $\Lambda_-(p)$ to the right and note that

$$(m + i\not{p})(m - i\not{p}') = m^2 - m(i\not{p}' - i\not{p}) + \not{p}\not{p}' = (i\not{p} + m)(i\not{p} - i\not{p}'),$$

so that

$$X = \text{Tr}\left[\left(\frac{2ip'\cdot en + eikn}{2p'\cdot k} + \frac{nike - 2ip\cdot en}{2p\cdot k}\right)(i\not{p} - m)\right.$$

$$\left.\left(\frac{2ip'\cdot en + nike}{2p'\cdot k} + \frac{eikn - 2ip\cdot en}{2p\cdot k}\right)(i\not{p}' - m)\right]$$

$$= X_1 + X_2,$$

$$X_1 = \text{Tr}\,[Q\bar{Q}i\not{p}(i\not{p} - i\not{p}')],$$

$$X_2 = \text{Tr}\left\{(i\not{p}' - m)Q\left[\frac{1}{p'\cdot k}(2p'\cdot e\epsilon + \epsilon ke + np\cdot ke - nkp\cdot e)\right.\right.$$

$$\left.\left. - \frac{1}{p\cdot k}(p\cdot ekn - ep\cdot kn - ek\epsilon + 2p\cdot e\epsilon)\right]\right\}.$$

It is now convenient to choose a special gauge. This is no loss of generality, since the theory is gauge-invariant. We choose $e^0 = 0$; the following equations can then be used advantageously:

$$k^2 = \mathbf{k}^2 = 0, \qquad kn = -2\omega - nk,$$
$$p^2 = p'^2 = -m^2, \qquad pn = -2\epsilon - np,$$
$$e^2 = 1, \qquad ke = -ek,$$
$$n^2 = -1, \qquad ne = -en.$$

With these relations, the first trace is easily found to be

$$X_1 = 4(m^2 + p'\cdot p)\left[\left(\frac{p'\cdot e}{p'\cdot k} - \frac{p\cdot e}{p\cdot k}\right)^2 - \frac{\omega^2}{p'\cdot k\,p\cdot k}\right]. \tag{15-89}$$

In a similar way, but after a somewhat longer calculation, we find

$$X_2 = 4 + \frac{2}{(p'\cdot k)^2}[2(p'\cdot e)^2(2\epsilon\epsilon' + \eta) - 2p\cdot ep'\cdot ep'\cdot k - p'\cdot k\eta]$$

$$+ \frac{2}{(p\cdot k)^2}[2(p\cdot e)^2(2\epsilon\epsilon' - \eta') + 2p\cdot ep'\cdot ep\cdot k - p\cdot k\eta']$$

$$- \frac{4}{p\cdot k\,p'\cdot k}\{p\cdot ep'\cdot e(\eta - \eta' + 4\epsilon\epsilon' - 2\omega^2)$$

$$- p\cdot k[\epsilon'\omega - (p'\cdot e)^2] - p'\cdot k[\epsilon\omega + (p\cdot e)^2]\}, \tag{15-90}$$

where

$$\eta = 2\epsilon\omega + p\cdot k = \epsilon\omega(1 + \beta\cos\theta),$$
$$\eta' = 2\epsilon'\omega + p'\cdot k = \epsilon'\omega(1 + \beta'\cos\theta'), \tag{15-91}$$

and θ and θ' are the angles between the direction of the emitted photon and the ingoing and outgoing negaton momentum.

The two traces X_1 and X_2 [Eqs. (15–89) and (15–90)] can now be combined to yield

$$X = 2\left[\left(\frac{p'\cdot e}{p'\cdot k}\right)^2(4\epsilon^2 - \mathbf{q}^2) + \left(\frac{p\cdot e}{p\cdot k}\right)^2(4\epsilon'^2 - \mathbf{q}^2)\right.$$

$$\left. - 2\frac{p'\cdot e\, p\cdot e}{p'\cdot k\, p\cdot k}(4\epsilon'\epsilon - \mathbf{q}^2) + \left(2 + \frac{\omega^2\mathbf{q}^2}{p'\cdot k\, p\cdot k} - \frac{p\cdot k}{p'\cdot k} - \frac{p'\cdot k}{p\cdot k}\right)\right]. \quad (15\text{–}92)$$

The cross section of interest is the differential cross section for scattering of the negaton into a solid angle $d\Omega'$ with the emission of a photon of energy ω into the solid angle $d\Omega_k$. The relative directions are conveniently expressed in polar coordinates with the polar axis in the direction of $\mathbf{k}$; thus, the direction of the outgoing negaton is determined by θ' and φ', whereas the ingoing negaton direction is determined by θ and φ. Inserting X into Eq. (15–87), we obtain

$$d\sigma = \frac{\alpha Z^2}{(2\pi)^2} r_0^2 \cdot \frac{m^2}{|\mathbf{q}|^4} \cdot \frac{|\mathbf{p}'|}{|\mathbf{p}|} \cdot \frac{d\omega}{\omega} \cdot d\Omega' d\Omega_k$$

$$\times \left[\left(2\epsilon\omega \frac{p'\cdot e}{p'\cdot k} - 2\epsilon'\omega \frac{p\cdot e}{p\cdot k}\right)^2 - \omega^2\mathbf{q}^2\left(\frac{p'\cdot e}{p'\cdot k} - \frac{p\cdot e}{p\cdot k}\right)^2\right.$$

$$\left. + \omega^2\left(2 + \frac{\omega^2\mathbf{q}^2}{p'\cdot k\, p\cdot k} - \frac{p\cdot k}{p'\cdot k} - \frac{p'\cdot k}{p\cdot k}\right)\right]. \quad (15\text{–}93)$$

This result can be expressed in terms of the angles θ and θ', the relative azimuth $\phi = \varphi' - \varphi$, and the angle ψ between the polarization direction $\mathbf{e}$

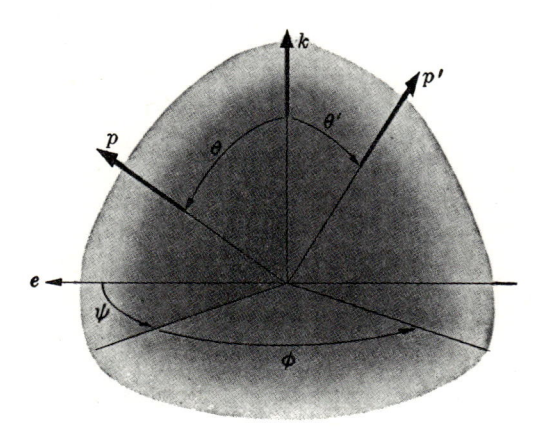

FIG. 15–8. Relative directions in bremsstrahlung.

and the **p-k** plane. These angles are drawn in Fig. 15–8. The cross section in terms of these angles is

$$d\sigma = \frac{\alpha Z^2}{(2\pi)^2} r_0^2 \cdot \frac{m^2}{|\mathbf{q}|^4} \frac{|\mathbf{p}'|}{|\mathbf{p}|} \frac{d\omega}{\omega} d\Omega' d\Omega_k$$

$$\times \left[\left(2\epsilon \frac{\beta' \sin \theta' \cos (\phi + \psi)}{1 - \beta' \cos \theta'} - 2\epsilon' \frac{\beta \sin \theta \cos \psi}{1 - \beta \cos \theta} \right)^2 \right.$$

$$- \mathbf{q}^2 \left(\frac{\beta' \sin \theta' \cos (\phi + \psi)}{1 - \beta' \cos \theta'} - \frac{\beta \sin \theta \cos \psi}{1 - \beta \cos \theta} \right)^2$$

$$\left. + \omega^2 \frac{(\gamma'/\gamma)\beta'^2 \sin^2 \theta' + (\gamma/\gamma')\beta^2 \sin^2 \theta - 2\beta'\beta \sin \theta' \sin \theta \cos \phi}{(1 - \beta' \cos \theta')(1 - \beta \cos \theta)} \right].$$

(15–94)

In this first Born approximation it appears that the same $d\sigma$ is valid for both positons and negatons.

The result (15–94) was first given by May.* The differential cross section can be integrated over the direction of the outgoing electron, $d\Omega'$. For this purpose, it is sufficient to calculate the two cases with polarization **e** parallel and perpendicular to the **p-k** plane. The integration can then be done analytically, but the result is very complicated.† A further summation over the polarization directions of the outgoing photons yields the angular distribution for the emission of a photon of energy ω into solid angle $d\Omega_k$ in the direction θ, φ, irrespective of polarization or direction of the outgoing electron. This result was also obtained by Gluckstern and Hull.† Finally, we can integrate over $d\Omega_k$ and obtain the total cross section for the emission of a photon of energy ω into an energy interval $d\omega$. We find the following frequency distribution‡

---

* M. M. May, *Phys. Rev.* **84**, 265 (1951). See also R. L. Gluckstern, M. H. Hull, and G. Breit, *Phys. Rev.* **90**, 1026 (1953); this paper contains the first published derivation of this cross section including polarization.

† R. L. Gluckstern and M. H. Hull, *Phys. Rev.* **90**, 1030 (1953).

‡ H. A. Bethe and W. Heitler, *Proc. Roy. Soc.* **A146**, 83 (1934). Their work is primarily concerned with bremsstrahlung as an energy loss of electrons penetrating through matter.

$$d\sigma(\omega,\epsilon) = \alpha Z^2 r_0^2 \frac{|\mathbf{p}'|}{|\mathbf{p}|} \frac{d\omega}{\omega} \left\{ \frac{4}{3} - \frac{2}{(\beta'\beta)^2} \left(\frac{\gamma}{\gamma'} + \frac{\gamma'}{\gamma} - \frac{2}{\gamma'\gamma} \right) \right.$$

$$+ \left(l \frac{\gamma'}{\gamma^2 - 1} + l' \frac{\gamma}{\gamma'^2 - 1} - ll' \right) + L \left[\frac{8}{3} \gamma'\gamma + \frac{\omega^2}{m^2} \left(\frac{1}{(\beta'\beta)^2} + 1 \right) \right]$$

$$\left. + \frac{1}{2} \frac{\omega}{m} \left(l \left(1 + \frac{\gamma'}{\beta^2 \gamma} \right) - l' \left(1 + \frac{\gamma}{\beta'^2 \gamma'} \right) + 2 \frac{\omega/m}{\beta^2 \gamma \beta'^2 \gamma'} \right) \right] \right\}, \quad (15\text{-}95)$$

where

$$|\mathbf{p}| = \beta\epsilon = \beta\gamma m, \quad |\mathbf{p}'| = \beta'\epsilon' = \beta'\gamma' m, \quad \gamma = \gamma' + \omega/m$$

and

$$l = \frac{1}{\beta\gamma} \ln \frac{1+\beta}{1-\beta}, \quad l' = \frac{1}{\beta'\gamma'} \ln \frac{1+\beta'}{1-\beta'},$$

$$L = \frac{2}{\beta\gamma\beta'\gamma'} \ln \frac{\gamma\gamma'(1+\beta\beta') - 1}{\omega/m}.$$

The same result can be obtained, of course, by first summing $d\sigma$ [Eqs. (15-93) or (15-94)] over the polarizations, and then integrating over $d\Omega' d\Omega_k$. In the latter form of $d\sigma$, this summation can be carried out by averaging over ψ and multiplying by 2. Using

$$\sum_{\text{pol}} \cos^2 (\phi + \psi) = \sum_{\text{pol}} \cos^2 \psi = 1$$

$$\sum_{\text{pol}} \cos (\phi + \psi) \cos \psi = \cos \phi$$

we find

$$d\sigma = \frac{\alpha Z^2}{(2\pi)^2} r_0^2 \frac{m^2}{|\mathbf{q}|^4} \frac{|\mathbf{p}'|}{|\mathbf{p}|} \frac{d\omega}{\omega} d\Omega' d\Omega_k$$

$$\times \left[\frac{\beta'^2 \sin^2 \theta'}{(1 - \beta' \cos \theta')^2} (4\epsilon^2 - \mathbf{q}^2) + \frac{\beta^2 \sin^2 \theta}{(1 - \beta \cos \theta)^2} (4\epsilon'^2 - \mathbf{q}^2) \right.$$

$$- 2 \frac{\beta'\beta \sin \theta' \sin \theta \cos \phi}{(1 - \beta' \cos \theta')(1 - \beta \cos \theta)} (4\epsilon'\epsilon - \mathbf{q}^2 + 2\omega^2)$$

$$\left. + 2\omega^2 \frac{(\gamma'/\gamma)\beta'^2 \sin^2 \theta' + (\gamma/\gamma')\beta^2 \sin^2 \theta}{(1 - \beta' \cos \theta')(1 - \beta \cos \theta)} \right]. \quad (15\text{-}96)$$

This is the famous Bethe-Heitler formula.* Its integration leads again to the cross section (15–95). We refer to the quoted original papers and to the work by Heitler† for an extensive discussion of these formulas and their physical implications. We shall be content to mention the nonrelativistic and the extreme relativistic limit of these results.

In the nonrelativistic limit the incident electron is assumed to have a momentum $|\mathbf{p}| \ll m$, so that also $|\mathbf{p}'| \ll m$:

$$\omega = \frac{\mathbf{p}^2}{2m} - \frac{\mathbf{p}'^2}{2m} \ll |\mathbf{p}|$$

and

$$\mathbf{q} = \mathbf{p} - \mathbf{p}' - \mathbf{k} \doteq \mathbf{p} - \mathbf{p}'.$$

Equation (15–94) yields

$$d\sigma = \frac{\alpha Z^2}{\pi^2} r_0^2 \frac{m^4}{|\mathbf{p} - \mathbf{p}'|^4} \frac{\beta'}{\beta} \frac{d\omega}{\omega} d\Omega' d\Omega_k$$

$$\times (\beta' \sin \theta' \cos(\phi + \psi) - \beta \sin \theta \cos \psi)^2, \quad \text{(NR)} \quad (15\text{–}97)$$

or, when summed over the polarizations,

$$d\sigma = \frac{\alpha Z^2}{\pi^2} r_0^2 \frac{m^4}{|\mathbf{p} - \mathbf{p}'|^4} \frac{\beta'}{\beta} \frac{d\omega}{\omega} d\Omega' d\Omega_k$$

$$\times (\beta'^2 \sin^2 \theta' + \beta^2 \sin^2 \theta - 2\beta'\beta \sin \theta' \sin \theta \cos \phi), \quad \text{(NR)}. \quad (15\text{–}98)$$

The total cross section follows upon integration over $d\Omega' d\Omega_k$ or, as the limit of Eq. 15–95),

$$d\sigma = \frac{16}{3} \alpha Z^2 r_0^2 \frac{d\omega}{\omega} \frac{1}{\beta^2} \ln \frac{\beta + \beta'}{\beta - \beta'}, \quad \text{(NR)}. \quad (15\text{–}99)$$

The infrared divergence in the limit $\omega \to 0$ is evident in all these formulas. The breakdown of the Born approximation for low incident energies is not responsible for it. Indeed, one can show that the exact nonrelativistic

---

* This formula was obtained independently by F. Sauter, *Ann. Physik* **20**, 404 (1934), Bethe and Heitler, *loc. cit.*, and G. Racah.

† W. Heitler, *The Quantum Theory of Radiation*, 3rd edition, Oxford, Clarendon Press, 1954.

solution* obtained by use of Coulomb wave functions can be very well approximated by multiplying Eqs. (15–97) to (15–99) by the "Sommerfeld factor"†

$$f(\beta',\beta) = \frac{\beta}{\beta'} \frac{1 - e^{-2\pi\alpha Z/\beta}}{1 - e^{-2\pi\alpha Z/\beta'}}. \qquad (15\text{–}100)$$

This factor changes nothing on the infrared divergence.

In the extreme relativistic case, it is easily seen from Eq. (15–94) that most of the radiation is emitted into small angles and that the momentum transfer to the Coulomb field is almost always small. The whole effect takes place with an average angle of the order of $\theta_0 = 1/\gamma$. The angular distribution in the extreme relativistic limit after integration over $d\Omega'$ was obtained by Hough.‡ The total cross section follows from the general formula (15–95),

$$d\sigma = 4\alpha Z^2 r_0^2 \frac{d\omega}{\omega}\left(1 + \left(\frac{\gamma'}{\gamma}\right)^2 - \frac{2}{3}\frac{\gamma'}{\gamma}\right)\left(\ln\frac{2m\gamma\gamma'}{\omega} - \frac{1}{2}\right), \quad \text{(ER)}. \quad (15\text{–}101)$$

This and other results which can be obtained from the general equations (15–93) to (15–95) in the extreme relativistic limit, in which bremsstrahlung is of greatest practical importance, suffer from two significant defects. One arises from the above-mentioned small momentum transfers, which correspond to large impact parameters. As a result, the incident electron is to a large extent screened from the nucleus by the atomic electrons. Thus, the pure Coulomb field should be replaced by a screened Coulomb field and the whole calculation should be repeated. However, since the screening effect is of greatest importance at very high energies, a relatively simple modification of the results for a pure Coulomb field can correct matters to a sufficiently good approximation.§ For intermediate energies numerical integrations are necessary. A complete discussion of the results for screening can be found in the article by Bethe and Ashkin.¶

---

* A. Sommerfeld, *Ann. Physik* **11**, 257 (1931); see also *Atombau und Spectrallinien*, Vol. II, Braunschweig, 1939.

† This factor was derived by G. Elwert, *Ann. Physik* **34**, 178 (1939), and is an improvement over the factor obtained by F. Sauter, *Ann. Physik* **18**, 486 (1933); see also P. Kirkpatrick and L. Wiedemann, *Phys. Rev.* **67**, 321 (1945), for a more accurate study and criticism of previous work.

‡ P. V. C. Hough, *Phys. Rev.* **74**, 80 (1948); see also M. Stearns, *Phys. Rev.* **76**, 836 (1949).

§ H. A. Bethe, *Proc. Camb. Phil. Soc.* **30**, 524 (1934), and H. A. Bethe and W. Heitler, *loc. cit.*

¶ H. A. Bethe and J. Ashkin, Vol. I, p. 166, of E. Segrè's *Experimental Nuclear Physics*, John Wiley and Sons, Inc., New York, 1953, especially p. 259.

The other defect is more serious and is more difficult to correct. It concerns the use of the first Born approximation.* Clearly, when αZ is not much smaller than unity, this approximation will be rather poor. This occurs for heavy elements, where it is known that the Bethe-Heitler formulas disagree with experiments, being about 10% too large.

Very recently this problem has found a satisfactory solution at high energies.† To this end an approximate Coulomb wave function‡ is used which agrees with the exact solution to order $(\alpha Z)^2/l$, where l is the associated orbital angular momentum. The resulting cross section is then correct to order $(\alpha Z)^2(\ln \epsilon)/\epsilon$, i.e., for energies $\epsilon \gtrsim 50$ Mev. In this energy region the results are in excellent agreement with experiment.

Finally, a small correction must be applied for the bremsstrahlung of the incident electron by the Z negatons surrounding the nucleus. The bremsstrahlung cross section in the field of an electron (Section 12–3) differs from the corresponding cross section in the field of the nucleus only by the factor Z^2 [cf. Eq. (12–29)] at high energies. The Z atomic negatons can thus be easily included in this energy region by replacing Z^2 by $Z(Z+1)$ in the above results. This effect in combination with screening was discussed by Wheeler and Lamb.§

We conclude this section with a few remarks about processes which are related to bremsstrahlung. For this purpose we consider the diagram for bremsstrahlung and its inverse in the bound interaction picture (Fig. 15–9).

The first diagram in this figure can represent photon emission from an atom (Section 15–5), radiative capture, or bremsstrahlung, depending on the initial and final states, which can be either in the discrete spectrum or in the continuum. Similarly, the second diagram can represent photon absorption by an atom causing excitation (final state is a bound state) or ionization (final state in the continuum). The latter process is usually referred to as the *photoelectric process*.

The calculation of the photoelectric cross section is made difficult because of the necessary use of Coulomb wave functions in the initial state. The final state can be approximated well by a plane wave when its energy is sufficiently high above the ionization potential. One can show that about 80% of the cross section is contributed by the K-electrons because of their considerably larger binding energy. As a result, it is *not* sufficient to use nonrelativistic Coulomb wave functions in the initial state for elements

---

* There are also, of course, the radiative corrections to bremsstrahlung, but they are of little interest theoretically, and are much too small to be observable.

† H. A. Bethe and L. C. Maximon, *Phys. Rev.* **93**, 768 (1954), and H. Davies, H. A. Bethe, and L. C. Maximon, *Phys. Rev.* **93**, 788 (1954).

‡ This wave function is a slight modification of a wave function proposed by W. H. Furry, *Phys. Rev.* **46**, 391 (1934).

§ *Cf.* the references given at the end of Chapter 11.

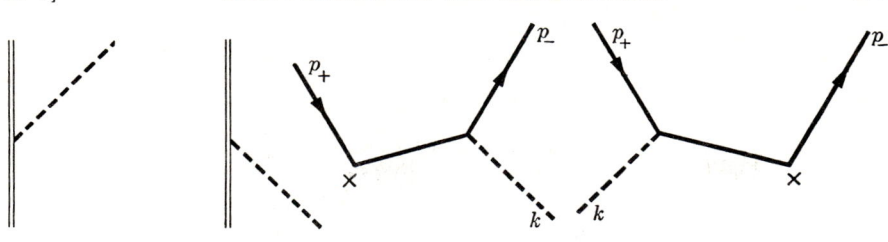

Fig. 15-9. Photon emission and absorption by an electron in an external field.

Fig. 15-10. Pair production by a photon in an external field.

of high Z. For these elements, the final state must also be represented by a *relativistic* Coulomb wave when the final energy is close to the absorption edge. It follows that an exact calculation is difficult; however, good approximations can be obtained in most cases of interest. The reader is referred to Heitler's book (§21) and to the Bethe-Ashkin article* for further details on this process.

15-7 Pair production and annihilation. The process of pair production by a photon in the field of a nucleus is very closely related to the process of bremsstrahlung discussed in the previous section. A comparison of Figs. (15-4) and (15-10) shows that the corresponding diagrams are related by the substitution law. In fact, this case seems to have been the first instance of the application of this law. Going from bremsstrahlung to pair production, we see that the outgoing photon becomes an ingoing photon and the ingoing negaton becomes an outgoing positon. In symbols,

$$k \to -k, \quad p \to -p_+. \tag{15-102}$$

We shall, furthermore, use the notation p_- instead of p' for the outgoing negaton. The conservation law becomes

$$k = p_+ + p_- + q \tag{15-103}$$

and indicates a threshold energy of $\omega = 2m$.

The substitution (15-102) can be inserted into the trace X, Eq. (15-92), and yields the trace $-X$ for pair production; the additional minus sign arises from the replacement of $\Lambda_-(p)$ by $-\Lambda_+(p_+)$.

The differential cross section differs from the corresponding expression (15-87) for bremsstrahlung only in the velocity of the incident particle,

---

* H. A. Bethe and J. Ashkin, "Penetration of Particles through Matter," published in E. Segrè, *Experimental Nuclear Physics*, Vol. I, John Wiley and Sons, Inc., New York, 1953.

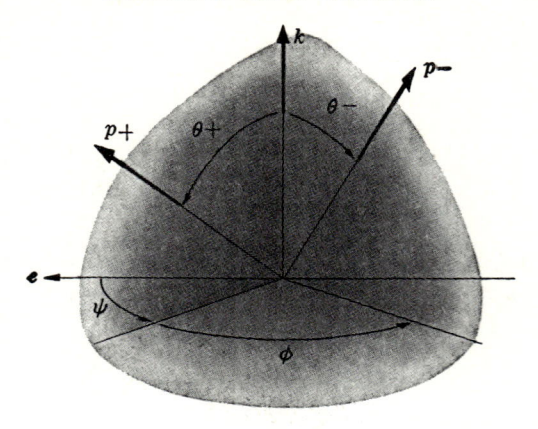

FIG. 15–11. Relative directions in pair production.

the density of final states, and a factor of 2, since the average over the spins of negaton p is to be replaced by a sum over the spins of positon p_+. Therefore,

$$d\sigma = (2\pi)^2 \frac{\alpha Z^2}{4\pi^4} r_0^2 \int \frac{m^4}{\epsilon_+ \epsilon_- \omega} \frac{1}{|\mathbf{q}|^4} \frac{1}{(2m)^2} X d^3 p_+ d^3 p_- \delta(\epsilon_+ + \epsilon_- - \omega).$$

Inserting for the trace X its explicit expression, we find for the differential cross section for pair production by a linearly polarized photon,

$$d\sigma = -\frac{\alpha Z^2}{(2\pi)^2} r_0^2 \frac{m^2}{|\mathbf{q}|^4} \frac{|\mathbf{p}_-| |\mathbf{p}_+| d\epsilon_+}{\omega} d\Omega_+ d\Omega_-$$

$$\times 2 \left[\left(\frac{p_- \cdot e}{p_- \cdot k}\right)^2 (4\epsilon_+^2 - \mathbf{q}^2) + \left(\frac{p_+ \cdot e}{p_+ \cdot k}\right)^2 (4\epsilon_-^2 - \mathbf{q}^2) + 2 \frac{p_- \cdot e\, p_+ \cdot e}{p_- \cdot k\, p_+ \cdot k} \right.$$

$$\left. \times (4\epsilon_+ \epsilon_- + \mathbf{q}^2) + \left(2 - \frac{\mathbf{q}^2 \omega^2}{p_- \cdot k\, p_+ \cdot k} + \frac{p_+ \cdot k}{p_- \cdot k} + \frac{p_- \cdot k}{p_+ \cdot k}\right) \right]$$

$$= -\frac{\alpha Z^2}{(2\pi)^2} r_0^2 \frac{m^2}{|\mathbf{q}|^2} \frac{|\mathbf{p}_-| |\mathbf{p}_+| d\epsilon_+}{\omega} d\Omega_+ d\Omega_-$$

$$\times 2 \left[\left(2\epsilon_+ \frac{p_- \cdot e}{p_- \cdot k} + 2\epsilon_- \frac{p_+ \cdot e}{p_+ \cdot k}\right)^2 - \mathbf{q}^2 \left(\frac{p_- \cdot e}{p_- \cdot k} - \frac{p_+ \cdot e}{p_+ \cdot k}\right)^2 \right.$$

$$\left. + \left(2 - \frac{\mathbf{q}^2 \omega^2}{p_+ \cdot k\, p_- \cdot k} + \frac{p_+ \cdot k}{p_- \cdot k} + \frac{p_- \cdot k}{p_+ \cdot k}\right) \right]. \tag{15-104}$$

In analogy to Eq. (15–93) for bremsstrahlung, which has been written in terms of the angles (Eq. 15–94), we can write this cross section in terms of the angles θ_+ and θ_- of $\mathbf{p}_+$ and $\mathbf{p}_-$ with respect to $\mathbf{k}$, their relative azimuth φ, and the angle ψ between the $\mathbf{k}$-$\mathbf{e}$ and the $\mathbf{k}$-$\mathbf{p}_+$ plane (cf. Fig. 15–11).

$$d\sigma = -\frac{\alpha Z^2}{(2\pi)^2} r_0^2 \frac{m^2}{|\mathbf{q}|^4} \frac{|\mathbf{p}_-||\mathbf{p}_+|d\epsilon_+}{\omega^3} d\Omega_+ d\Omega_-$$

$$\times 2\left[\left(2\epsilon_+ \frac{\beta_- \sin\theta_- \cos(\phi+\psi)}{1-\beta_-\cos\theta_-} + 2\epsilon_- \frac{\beta_+ \sin\theta_+ \cos\psi}{1-\beta_+\cos\theta_+}\right)^2\right.$$

$$- \mathbf{q}^2 \left(\frac{\beta_- \sin\theta_- \cos(\phi+\psi)}{1-\beta_-\cos\theta_-} - \frac{\beta_+ \sin\theta_+ \cos\psi}{1-\beta_+\cos\theta_+}\right)^2$$

$$\left. - \omega^2 \frac{|\mathbf{p}_+|^2 \sin^2\theta_+ + |\mathbf{p}_-|^2 \sin^2\theta_- + 2|\mathbf{p}_+||\mathbf{p}_-|\sin\theta_+ \sin\theta_- \cos\phi}{\epsilon_+\epsilon_-(1-\beta_+\cos\theta_+)(1-\beta_-\cos\theta_-)}\right].$$

(15-105)

This result was first published by May.* The average over polarizations proceeds as in bremsstrahlung, i.e., we average over ψ. The result is the Bethe-Heitler formula for pair production (H. A. Bethe and W. Heitler, loc. cit.):

$$d\sigma = -\frac{\alpha Z^2}{(2\pi)^2} r_0^2 \frac{m^2}{|\mathbf{q}|^4} \frac{|\mathbf{p}_-||\mathbf{p}_+|d\epsilon_+}{\omega^3} d\Omega_+ d\Omega_-$$

$$\times \left[\frac{\beta_+^2 \sin^2\theta_+}{(1-\beta_+\cos\theta_+)^2}(4\epsilon_-^2 - \mathbf{q}^2) + \frac{\beta_-^2 \sin^2\theta_-}{(1-\beta_-\cos\theta_-)^2}(4\epsilon_+^2 - \mathbf{q}^2)\right.$$

$$+ \frac{2\beta_+\beta_- \sin\theta_+ \sin\theta_- \cos\phi}{(1-\beta_+\cos\theta_+)(1-\beta_-\cos\theta_-)}(4\epsilon_+\epsilon_- + \mathbf{q}^2 - 2\omega^2)$$

$$\left. - 2\omega^2 \frac{(\gamma_+^2 - 1)\sin^2\theta_+ + (\gamma_-^2 - 1)\sin^2\theta_-}{\gamma_+\gamma_-(1-\beta_+\cos\theta_+)(1-\beta_-\cos\theta_-)}\right]. \quad (15\text{-}106)$$

This cross section for *unpolarized* incident γ-rays gives a completely symmetrical distribution of the emitted negatons and positons, since $d\sigma/d\epsilon_+$ is invariant under the interchange of the subscripts $+$ and $-$. This result is, of course, not quite correct, especially for low electron velocities, because the positive nuclear charge will cause a charge asymmetry of the emitted pair. In this first Born approximation such effects are ignored, but they will appear when the electrons are represented by Coulomb wave functions rather than plane waves.

The integration over angles was also performed by Bethe and Heitler (*loc. cit.*) and is completely analogous to the bremsstrahlung case [*cf.* Eq.

---

* M. M. May, *loc. cit.*, in Section 15-6. His cross section, however, is too small by a factor of 2.

(15–95)]. It leads to the symmetrical energy distribution

$$d\sigma(\omega,\epsilon_+) = \alpha Z^2 r_0^2 \frac{|\mathbf{p}_+||\mathbf{p}_-|d\epsilon_+}{\omega^3}\left[-\frac{4}{3} - \frac{2}{(\beta_+\beta_-)^2}\left(\frac{\gamma_+}{\gamma_-} + \frac{\gamma_-}{\gamma_+} - \frac{2}{\gamma_+\gamma_-}\right)\right.$$

$$+\left(l_+\frac{\gamma_-}{\gamma_-^2-1} + l_-\frac{\gamma_+}{\gamma_+^2-1} - l_+l_-\right)$$

$$+ L\left\{-\frac{8}{3}\gamma_+\gamma_- + \left(\frac{\omega}{m}\right)^2\left(1+\frac{1}{(\beta_+\beta_-)^2}\right) + \frac{1}{2}\frac{\omega}{m}\left[l_+\left(1-\frac{\gamma_-}{\beta_+^2\gamma_+}\right)\right.\right.$$

$$\left.\left.\left. + l_-\left(1-\frac{\gamma_+}{\beta_-^2\gamma_-}\right) - 2\frac{\omega}{m}\frac{1}{\beta_+^2\gamma_+\beta_-^2\gamma_-}\right]\right\}\right], \quad (15\text{–}107)$$

where

$$l_\pm = \frac{1}{\beta_\pm\gamma_\pm}\ln\frac{1+\beta_\pm}{1-\beta_\pm}, \quad L = \frac{2}{\beta_+\gamma_+\beta_-\gamma_-}\ln\frac{\gamma_+\gamma_-(1+\beta_+\beta_-)+1}{\omega/m},$$

$$|\mathbf{p}_\pm| = \beta_\pm\epsilon_\pm = \beta_\pm\gamma_\pm m.$$

This energy distribution is conveniently plotted as a function of the ratio kinetic energy of the positons (or negatons) to the total kinetic energy available in the final state, $(\epsilon_+ - m)/(\omega - 2m)$. This ratio varies from 0 to 1, and $d\sigma(\omega,\epsilon_+)$ is symmetrical about $\frac{1}{2}$. Over most of the range it is a very flat distribution up to very high energies ω.

Finally, we can integrate over $d\epsilon_+$ and obtain the total cross section $\sigma(\omega)$ for pair production by unpolarized photons of energy ω. This result cannot be expressed in closed form in terms of tabulated functions. The most convenient form, using $\eta = 2m/\omega$, is

$$\sigma(\omega) = \alpha Z^2 r_0^2 \left\{2\eta^2[2C_2(\eta) - D_2(\eta)]\right.$$

$$\left. - \frac{2}{27}[(109 + 64\eta^2)E_2(\eta) - (67 + 6\eta^2)(1-\eta^2)F_2(\eta)]\right\}, \quad (\eta \leq 1),$$

$$(15\text{–}108)$$

where

$$C_2(\eta) = \int_1^{1/\eta} \frac{\cosh^{-1} x}{x}\cosh^{-1}\frac{1}{\eta x}\,dx, \quad (\eta \leq 1),$$

$$D_2(\eta) = \int_1^{1/\eta} \frac{\cosh^{-1}(1/\eta x)}{\sqrt{x^2-1}}\,dx, \quad (\eta \leq 1),$$

$$(15\text{–}109)$$

$$E_2(\eta) = F(\sqrt{1-\eta^2}) - E(\sqrt{1-\eta^2}), \quad (\eta \leq 1),$$

$$F_2(\eta) = F(\sqrt{1-\eta^2}), \quad (\eta \leq 1).$$

The functions F and E denote the complete elliptic integrals of the first and second kind, respectively.

The total cross section was first calculated by Racah,* and was given in essentially this form by Jost, Luttinger, and Slotnick.† Near threshold, i.e., for $\omega - 2m \ll m$, $\sigma(\omega)$ can be approximated by

$$\sigma(\omega) = \frac{\pi}{12} \alpha Z^2 r_0^2 \left(\frac{\omega}{m} - 2\right)^3, \quad (\omega - 2m \ll m). \qquad (15\text{-}110)$$

The total cross section can also be calculated by use of the Bohr-Peierls-Placzek relation [cf. Appendix A–7, Eq. (A7–7)],

$$\sigma(\omega) = \frac{4\pi}{\omega} a_2(\omega).$$

In this case, the scattering process associated with the absorptive process of pair production is Delbrück scattering, i.e., the scattering of light by a Coulomb field (cf. Section 15–8). On calculating the imaginary part of the forward scattering amplitude for this process, $a_2(\omega)$, one also obtains $\sigma(\omega)$ for pair production. The relation between these two processes is completely analogous to the relation between the scattering of photons by photons (Section 13–2) and the associated absorption process, the pair production in photon-photon collisions. In fact, the diagrams differ only in the replacement of two external photon lines by two external field actions. (Compare Fig. 13–1 with Fig. 15–13.)

This method was used by Jost, Luttinger, and Slotnick (loc. cit.). The same calculation also gave the nuclear recoil distribution in pair production. To this end, it is only necessary to express the total cross section as an integral over **q**, the recoil momentum. The integrand then yields the required distribution. The resulting formulas are quite involved and will not be given here.

Returning to Eqs. (15–107) and (15–108), we note that these rather complicated formulas simplify considerably in the extreme relativistic limit, i.e., when $\omega \gg 2m$. The energy distribution (15–107) becomes

$$d\sigma(\omega,\epsilon_+) = 4\alpha Z^2 r_0^2 \frac{d\epsilon_+}{\omega^3} \left(\epsilon_+^2 + \epsilon_-^2 + \frac{2}{3}\epsilon_+\epsilon_-\right)\left(\ln \frac{2\epsilon_+\epsilon_-}{m\omega} - \frac{1}{2}\right), \quad (\text{ER}), \qquad (15\text{-}111)$$

---

* G. Racah, *Nuovo Cimento* **13**, 69 (1936).
† R. Jost, J. M. Luttinger, and M. Slotnick, *Phys. Rev.* **80**, 189 (1950).

and the total cross section (15–108) yields

$$\sigma(\omega) = \alpha Z^2 r_0{}^2 \left(\frac{28}{9} \ln \frac{2\omega}{m} - \frac{218}{27}\right), \quad \text{(ER)}. \qquad (15\text{–}112)$$

All these results are subject to the same criticism as the corresponding cross sections for bremsstrahlung:* (1) the neglect of screening, (2) the neglect of the Coulomb field for the emerging pair (plane waves are assumed in this first Born approximation), and (3) the neglect of pair production by the atomic electrons.

Screening is again of importance when the energies are high enough to permit contributions from large impact parameters. Complete screening replaces the arguments in the logarithms of Eqs. (15–111) and (15–112) by a constant, $183\, Z^{-1/3}$, and changes the last term. The reader is referred to the excellent summary by Bethe and Ashkin (*loc. cit.*) for a detailed discussion of screening and many other aspects of pair production.

The corrections for the Coulomb effects beyond the first Born approximation are very difficult. In the "nonrelativistic limit" (i.e., near threshold), when the electron velocities are small, an almost exact calculation with Coulomb wave functions can be carried out.† The Sommerfeld factor as proposed by Sauter for bremsstrahlung (*cf.* Section 15–6),

$$f(\beta_+, \beta_-) = \frac{\eta_+ \eta_-}{(e^{\eta_+} - 1)(1 - e^{-\eta_-})}, \quad (\eta_\pm = 2\pi\alpha Z/\beta_\pm), \qquad (15\text{–}113)$$

when used as a multiplicative factor to the first Born approximation, affords a good approximation to the exact results in the NR limit.

In the extreme relativistic limit, the results obtained by Bethe and Maximon, and by Davies, Bethe, and Maximon (*loc. cit.*) for bremsstrahlung and extended by them to pair production are valid in the same approximation. They yield a reduction of the total cross section $\sigma(\omega)$ as calculated above, which is about 10% for Pb. This correction is almost independent of energy for $\omega \gtrsim 50m$.

The effect of the atomic electrons can easily be estimated. We have seen in Section 11–4 that for high energies the pair production cross section in photon-electron collisions when the electron is initially at rest is identical

---

* The radiative correction is again too small to be of interest. It was investigated in detail for *internal* pair production by R. H. Dalitz, *Proc. Roy. Soc.* **A206**, 521 (1951). In that case it is less than one percent, except in the forward and backward direction, where it is a few percent.

† Y. Nishina, S. Tomonaga, and S. Sakata, *Sci. Papers Inst. Phys. and Chem. Res.* (Japan) **24**, No. 17 (1934), supplement. It should be clear that the Born approximation is completely inapplicable near threshold, since it is an expansion in $\alpha Z/\beta$. Numerical calculations in this energy region were carried out by H. R. Hulme and J. C. Jaeger, *Proc. Roy. Soc.* **153**, 443 (1936).

with Eq. (15–112), except for a slightly different constant term, and except for the factor Z^2. In this ER limit the contribution of the Z electrons can thus be included in $\sigma(\omega)$ by replacing Z^2 by $Z(Z+1)$. One thereby obtains the total cross section for pair production in the field of the neutral *atom* rather than in the field of the nucleus. The combined effect of pair production in the field of the atomic electrons and of screening was computed by Wheeler and Lamb (*loc. cit.*).

The inverse process to pair production in the field of an atom is *pair annihilation* with the emission of a single quantum (*one-quantum annihilation*). This process can take place only in an external field: the incident positon annihilates with one of the atomic negatons giving the recoil momentum to the nucleus. In the bound interaction picture the two processes of pair creation and pair annihilation are shown in Fig. 15–12.

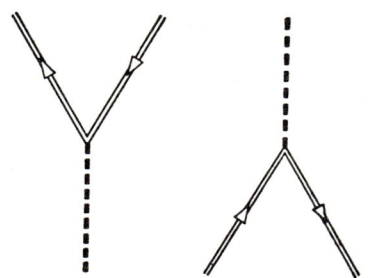

FIG. 15–12. Pair creation and annihilation.

Although the diagrams make obvious their relationship by detailed balance, this fact cannot be used advantageously in practice. Pair production is usually calculated in a first Born approximation, whereas in pair annihilation it is essential to use bound state Coulomb wave functions for the negaton. The relation of these processes is quite similar to the relation of bremsstrahlung to the photoelectric effect. In fact, one-quantum annihilation can be obtained from the photoelectric effect by the substitution law, provided the incident positon in the former and the outgoing negaton in the latter can be treated as free particles.

This pair-annihilation process is to be compared with the free-pair annihilation process discussed in Section 12–4. A positon penetrating matter will annihilate primarily through the latter process, which is of order $Z\pi r_0^2$; one-quantum annihilation is of the order of $(\alpha Z)^4 Z\pi r_0^2$ and contributes, therefore, only for heavy elements, where it is of relative order 10%. For Pb it actually contributes as much as 20%.

We shall not pursue pair annihilation further. The interested reader is referred to Heitler's book, where additional references can also be found.

15–8 Delbrück and Rayleigh scattering. The scattering of photons by an external electric field is called *Delbrück scattering*.* Since the effect is

---

* M. Delbrück, *Z. Physik* **84**, 144 (1933), first showed on theoretical grounds that such a process can occur. The effect was first observed in the laboratory by R. R. Wilson, *Phys. Rev.* **90**, 720 (1953). Exact calculations have been carried out so far only for forward scattering: F. Rohrlich and R. L. Gluckstern, *Phys. Rev.* **86**, 1 (1952).

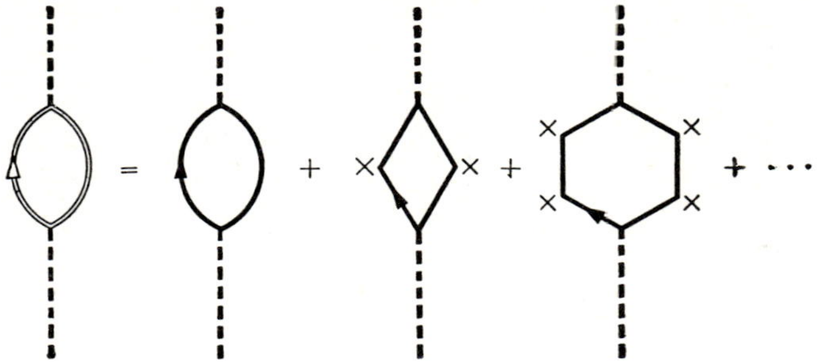

FIG. 15–13. Delbrück scattering.

very small, the strongest electric fields are necessary. Only the fields of atomic nuclei are sufficiently large to make the process observable. When the incident photon energy is very small compared with the rest energy of the nucleus, the nuclear recoil is negligible and the nuclear field can be assumed to be a static Coulomb field. The scattering in this case is elastic and is sometimes also referred to as *potential scattering of light*. The situation is similar to the case of Compton scattering, which yields the elastic Thomson scattering in the nonrelativistic limit (*cf.* Section 11–1).

This superficial similarity is actually very deep. In fact, *Delbrück scattering is a radiative correction to Compton scattering of the incident photon by the nucleus.* To prove this assertion, let us first consider the diagram for Delbrück scattering. It is shown in zero order of radiative corrections in Fig. 15–13. The incident photon creates a pair in the external field. This pair annihilates, emitting a photon of the same energy as the incident photon, provided the external field is static and is therefore unable to take up energy. In the same figure the diagrams of the iteration solution corresponding to an expansion in the external field are also given in a schematic way. Note that odd-cornered loops vanish because of Furry's theorem (Section 8–4).

Let us now assume that the nucleus can be described by a quantized electron field of mass M and coupling constant Ze.* The interaction of the ordinary electron field (mass m) and the "heavy electron" field with the photon field yields an S-matrix whose iteration solution can be described by diagrams, as in Chapters 8 to 13. In particular, the Compton scattering of a photon by a nucleus will be represented by the diagrams Fig. 15–14(a). The cross section for this process is given by the Klein-Nishina formula, Eq. (11–21), provided the nucleus is initially at rest, except that the classical

---

* This assumption implies spin $\frac{1}{2}$, a magnetic moment $Ze/2M$, and a point charge for the nucleus. Although these conditions are obviously not fulfilled, they will not cause errors in the nonrelativistic limit when the spin effects can be ignored. We shall be interested in this limit only.

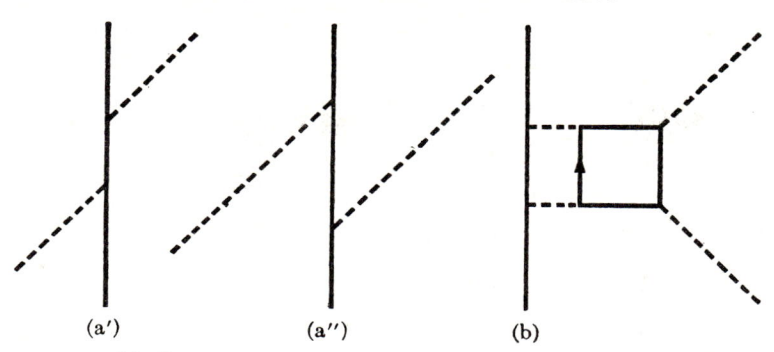

FIG. 15-14. (a) Compton scattering of a photon by a nucleus, and (b) one of the radiative corrections.

electron radius $r_0 = \alpha/m$ is to be replaced by the "classical nuclear radius"*

$$R_0 = \frac{\alpha Z^2}{M} = Z^2 \frac{m}{M} r_0.$$

In the nonrelativistic limit, $\omega \ll M$, this cross section approaches the nuclear Thomson scattering cross section [cf. Eq. (11-26)] for unpolarized photons,

$$\frac{d\sigma}{d\Omega} = \tfrac{1}{2} R_0^2 (1 + \cos^2 \theta), \quad \text{(NR)}. \tag{15-114}$$

If only the nuclear field and the photon field were present, the theorem at the end of Section 11-3 would hold and only this Thomson cross section would survive in the NR limit of Compton scattering *and all its radiative corrections*. However, the presence of the electron field of mass $m \ll M$ permits two different limits of $\omega \ll M$, namely, $\omega \ll m \ll M$ and $m \gtrsim \omega \ll M$. In the first case, the theorem is still correct, but in the second case, terms of order ω/m so far neglected will play an important role. In that case the electron field is *not* nonrelativistic, although the nuclear field is. Therefore, radiative corrections to Fig. 15-14(a) which involve the emission and reabsorption of the same virtual photon will not contribute, but those corrections which involve closed electron loops will be important. (Closed nuclear loops contribute always by a factor about m/M less.) The situation becomes clearer when one observes that the Coulomb field is due to virtual longitudinal photons whose momentum four-vector k^μ satisfies $k^2 > 0$. This leads to an *instantaneous* interaction which can only take place with another charged particle, i.e., a virtual nucleus or a virtual electron. The diagram in Fig. 15-14(b) represents the lowest nonvanishing radiative

---

* This radius R_0 is of the order $Z^2/2000$ times as small as the actual nuclear radius R.

correction to the diagram (a). (Other diagrams of the same order differ only in permutations of the closed loop corners.) The next order radiative correction involves a closed electron loop connected to the nucleon by *four* virtual photon lines. We see that these diagrams correspond exactly to the Delbrück scattering diagrams (Fig. 15–13) for the iteration solution in the external field. This concludes the arguments which prove that Delbrück scattering is actually a radiative correction to Compton scattering by the nucleus.*

The above discussion exhibits another interesting feature of Delbrück scattering, namely, that Delbrück scattering is nothing but the "scattering of light by light" when all but one incident and one outgoing photon is real and all the others are virtual. The only difference is that the coupling constant to all the virtual photons is Ze rather than e, so that diagram 15–14(b) will give a cross section which is $(e^2 Z^2)^2 \sim \alpha^2 Z^4$ times larger than the photon-photon scattering cross section discussed in Section 13–2. Equation (13–17) shows that the latter is of order $\alpha^2 r_0^2$. It follows that Delbrück scattering is of order $(\alpha Z)^4 r_0^2$.

In the discussion of photon-photon scattering it was found that this interesting process is unfortunately too small to be observable at the present time. Paradoxically, it now appears that this same process, when it occurs as a partly virtual process, is actually observable because of the factor $\alpha^2 Z^4 \sim (Z/12)^4$, which can be much larger than 1 for heavy nuclei. For example, for Pb we find $\alpha^2 Z^4 = (6.45)^4 \sim 1700$. This factor is sufficient to make the cross section observable.

The necessity for calculating Delbrück scattering for large Z makes the Born approximation, which is effectively an expansion in αZ, a poor approximation. Other, better approximations are much more difficult as will be briefly discussed below.

Let us, then, turn to the expansion indicated schematically in Fig. 15–13. The first diagram vanishes identically, since it represents the self-energy of a free photon. The second term, which is of second order in the external field, consists of six diagrams, the three diagrams depicted in Fig. 15–15 and their charge conjugates (with respect to the electron field). The latter differ from the diagrams shown only in the arrow direction of the closed loop; therefore they give simply a factor of 2, according to Furry's theorem. A comparison of these diagrams and those for photon-photon scattering shows the obvious relationship. In fact, the only essential difference between the matrix element calculated there and the matrix element of interest to us here is that for two of the photons the relation $k^2 = 0$ is not satisfied. The considerations of Section 13–1 are still valid, since this

---

* Note that arguments of this type can be used to construct the whole theory of the static external field. They add considerably to the physical insight into this theory.

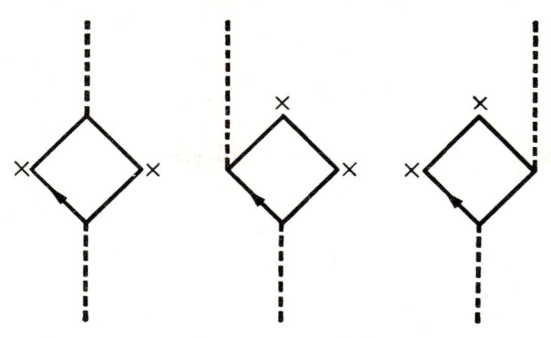

FIG. 15–15. Lowest order Delbrück scattering.

relation was not assumed there, but the calculations of Section 13–2 are no longer applicable. The matrix element for Delbrück scattering can be written, in general, in terms of the fourth-rank polarization tensor $\Pi_{\mu\nu\lambda\sigma}$ of Section 13–1,

$$(k'|M|k) = \frac{1}{2\omega} e^\mu e'^\nu \int n^\lambda(\mathbf{q}) n^\sigma(\mathbf{q}') \phi(\mathbf{q}) \phi(\mathbf{q}')$$

$$\Pi_{\mu\nu\lambda\sigma}(-kk'qq') \delta(\mathbf{k} - \mathbf{k}' - \mathbf{q} - \mathbf{q}') d^3q\, d^3q', \quad (15\text{--}115)$$

where $\phi(\mathbf{q})$ is the Fourier transform of the Coulomb potential, (15–4),

$$\phi(\mathbf{q}) = \frac{Ze}{(2\pi)^{3/2}} \frac{1}{|\mathbf{q}|^2}.$$

The cross section for unpolarized photons, using Eq. (13–15), can be written

$$d\sigma = (2\pi)^2 \frac{1}{2} \sum_{\text{pol}} |(k'|M|k)|^2 \delta(\omega' - \omega) d^3k'.$$

$$\frac{d\sigma(\omega,\theta)}{d\Omega} = \frac{(\alpha Z)^4}{32\pi^6} r_0^2 \sum_{\text{pol}} \left| e_u e_v' \int n_\lambda(\mathbf{q}) n_\sigma(\mathbf{k} - \mathbf{k}' - \mathbf{q}) G^{\mu\nu\lambda\sigma}(-\mathbf{k}\mathbf{k}'\mathbf{q}\mathbf{q}') \right.$$

$$\left. \times \frac{m\, d^3q}{|\mathbf{q}|^2 |\mathbf{k} - \mathbf{k}' - \mathbf{q}|^2} \right|^2. \quad (15\text{--}116)$$

This is the differential cross section for elastic Delbrück scattering. The evaluation of the integral can be carried out in principle along the lines of the general discussion in Section 13–1.

From that section we learn that the infinite term disappears when the contributions of all the diagrams are combined. We also see how the integration can be reduced to a minimum number of integrals. These integrals, however, are very complicated and cannot be integrated in terms of known functions. The expressions become very long and the work appears to be prohibitive.* Only in the special case of forward scattering ($\theta = 0$) does it seem possible to carry the exact integrations through analytically.†

However, for this particular case of forward scattering a much simpler method is available. This is the method of analytic continuation discussed in detail in Appendix A7. When two of the internal electron lines of the diagrams in Fig. 15-15 conserve energy, as can happen at photon energies $\omega > 2m$, the process can be thought of as pair production in the nuclear Coulomb field followed by pair annihilation. It follows that pair production is the absorptive process associated with Delbrück scattering. The Bohr-Peierls-Placzek relation then gives immediately [cf. Eq. (A7-7)]

$$a_2(\omega) = \frac{\omega}{4\pi}\sigma(\omega).$$

In this equation $a_2(\omega)$ is the imaginary part of the Delbrück forward scattering amplitude, and $\sigma(\omega)$ is the total pair production cross section. The assumption involved here, of course, is that the total cross section, i.e., the *sum* of $\sigma(\omega)$ and the total Delbrück scattering cross section, can be well approximated by $\sigma(\omega)$ alone. That this is satisfied follows from the fact that $\sigma(\omega)$ is of order $\alpha Z^2 r_0^2$, whereas the elastic Delbrück scattering cross section is only of order $(\alpha Z)^4 r_0^2$; therefore, even for large Z, $\sigma(\omega)$ is much larger (for Pb the ratio is about 380).

The knowledge of the total pair production cross section thus gives us $a_2(\omega)$. The real part of the forward scattering amplitude $a_1(\omega)$ can now be obtained by analytic continuation. One finds, with Eq. (A7-28),

$$a_1(\omega) = \frac{\omega^2}{2\pi^2} P \int_0^\infty \frac{\sigma(\omega')d\omega'}{\omega'^2 - \omega^2}.$$

This method is far simpler than the direct calculation using $\Pi_{\mu\nu\lambda\sigma}$. With the help of expression (15-108) for $\sigma(\omega)$, Rohrlich and Gluckstern‡ found by straightforward integration in terms of $\eta = 2m/\omega$,

$$a_1(\omega) = (\alpha Z)^2 r_0 \left\{ \frac{\eta}{\pi}[2C_1(\eta) - D_1(\eta)] + \frac{1}{27\pi\eta}[(109 + 64\eta^2)E_1(\eta) - (67 + 6\eta^2)(1 - \eta^2)F_1(\eta)] - \frac{\eta^2}{9} - \frac{9}{4} \right\} \quad (15\text{-}117)$$

and

$$a_2(\omega) = (\alpha Z)^2 r_0 \left\{ \frac{\eta}{\pi}[2C_2(\eta) - D_2(\eta)] - \frac{1}{27\pi\eta}[(109 + 64\eta^2)E_2(\eta) - (67 + 6\eta^2)(1 - \eta^2)F_2(\eta)] \right\}, \quad (15\text{-}118)$$

---

* Qualitative results on Delbrück scattering were first obtained by A. Achieser and I. Pomerantschuk, *Physik. Z. Sowjetunion* **11**, 478 (1937).

† F. Rohrlich and R. L. Gluckstern, *loc. cit.*

‡ *Loc. cit.* The same results were obtained by J. S. Toll, *Thesis*, Princeton University, 1952.

where
$$C_1(\eta) = \operatorname{Re} \int_0^{1/\eta} \frac{\sin^{-1} x}{x} \cosh^{-1} \frac{1}{\eta x} dx, \quad (\eta > 0),$$

$$D_1(\eta) = \operatorname{Re} \int_0^{1/\eta} \frac{\cosh^{-1}(1/\eta x)}{\sqrt{1-x^2}} dx, \quad (\eta > 0),$$

$$E_1(\eta) = \begin{cases} E(\eta), & (\eta \le 1), \\ \eta E\left(\frac{1}{\eta}\right) + \left(\frac{1}{\eta} - \eta\right) F\left(\frac{1}{\eta}\right), & (\eta \ge 1), \end{cases} \quad (15\text{-}119)$$

$$F_1(\eta) = \begin{cases} F(\eta), & (\eta \le 1), \\ \frac{1}{\eta} F\left(\frac{1}{\eta}\right), & (\eta \ge 1), \end{cases}$$

and C_2, D_2, E_2, and F_2 are given in Eqs. (15-109). Here, $F(\eta)$ and $E(\eta)$ again denote the complete elliptic integrals of the first and second kind, respectively.

For small energies ($\omega \ll m$), the real part is*

$$a_1(\omega) = \frac{73}{72} \cdot \frac{1}{32} \cdot (\alpha Z)^2 r_0 \left(\frac{\omega}{m}\right)^2, \quad (\omega \ll m). \quad (15\text{-}120)$$

This formula affords an excellent approximation up to energies ω close to $2m$, where $a_1(\omega)$ becomes more and more linear until it approaches its high-energy limit

$$a_1(\omega) = \frac{7}{18} (\alpha Z)^2 r_0 \left(\frac{\omega}{m}\right), \quad (\omega \gg 2m). \quad (15\text{-}121)$$

The imaginary part has the same threshold, of course, as the pair-production cross section; i.e., it vanishes for $\omega \le 2m$ and is [cf. Eq. (15-110)]

$$a_2(\omega) = \frac{1}{24} (\alpha Z)^2 r_0 \left(\frac{\omega}{m} - 2\right)^3, \quad (\omega - 2m \ll m). \quad (15\text{-}122)$$

It therefore increases faster than $a_1(\omega)$ and crosses $a_1(\omega)$ near $\omega \simeq 20m$. Its high energy limit is

---

* This energy dependence of $a_1(\omega)$ for small ω could have been predicted, because the cross section in the nonrelativistic limit should be purely classical and independent of $\hbar$. Indeed, inserting $\hbar$ and c, we find

$$a_1(\omega) \sim \left(\frac{Ze^2}{\hbar c}\right)^2 \left(\frac{e^2}{mc^2}\right) \left(\frac{\hbar \omega}{mc^2}\right)^2 = Z^4 \left(\frac{\omega}{c}\right)^2 \left(\frac{e^2}{mc^2}\right)^6,$$

which is independent of $\hbar$.

$$a_2(\omega) = \frac{7}{9\pi}(\alpha Z)^2 r_0 \left(\frac{\omega}{m}\right)\left(\ln\frac{2\omega}{m} - \frac{109}{42}\right), \quad (\omega \gg 2m), \quad (15\text{-}123)$$

in agreement with Eqs. (15–112) and (A7–7).

The physical difference between the real and the imaginary part of $a(\omega)$ lies in the energy conservation in intermediate states: the real part corresponds to virtual pairs, the imaginary part corresponds to real intermediate pairs. This separation is evident throughout the S-matrix formalism and has its origin in the δ_+-function of energy, as, for example, in Eq. $(7\text{-}34)_+$. The real and imaginary parts of $a(\omega)$ are therefore often referred to as the *dispersive and absorptive parts* of the amplitude, a designation which is evident from the connection between $a(\omega)$ and the index of refraction $n(\omega)$ [*cf.* Eq. (A7–6)].

We conclude from the above equations that the dispersive part is dominant not only below threshold where $a_2 = 0$, but also above threshold up to about 10 Mev. Above this energy the absorptive part dominates.

The differential cross section follows from the amplitudes,

$$\frac{d\sigma(\omega,0)}{d\Omega} = |a_1(\omega) + ia_2(\omega)|^2 = a_1^2(\omega) + a_2^2(\omega). \quad (15\text{-}124)$$

At high energies ($\omega \gg 2m$) these results, of course, are subject to the same corrections of screening and atomic electron participation as pair production. However, this energy region is not of great physical interest, since the important process is dispersive scattering, which is masked by the absorptive scattering at high energies.

So far we have considered forward scattering only in lowest order of αZ. An estimate of higher order corrections (Coulomb corrections) can be obtained again by dispersion techniques. Since the Coulomb corrections for low energy pair production are fairly well known, $a_2(\omega)$ follows trivially as above, and $a_1(\omega)$ is obtained from (A7 − 28). In this way it can be estimated that the Coulomb corrections to forward Delbrück scattering are at most of the order of ten percent.*

We now turn to the very difficult problem of computing Delbrück scattering for finite angles, $\theta > 0$. As a first attempt an approximation was applied which is valid only for high energies and small angles.† The method used in this approximation consists of a combination of an impact parameter method and the analytic continuation method. The semiclassical integral expansion of the scattering amplitude in terms of partial cross sections of definite impact parameters is well known. It is a Bessel transform. The partial cross sections are related by analytic con-

---

* F. Rohrlich, *Phys. Rev.* **108**, 169 (1957).

† H. A. Bethe and F. Rohrlich, *Phys. Rev.* **86**, 10 (1952).

tinuation in the same way as the amplitudes are. The dispersive partial cross sections can thus be obtained when the absorptive ones are known. To determine these, one must extend the Bohr-Peierls-Placzek relation to finite angles. This can be done for $\omega \gg m$ and angles $\theta \ll 1$. The partial cross sections for pair production can then be related to the absorptive partial cross sections. The problem is thereby reduced to a few integrals. For details the reader is referred to the original paper.

This first attempt unfortunately proved not very satisfactory and has been replaced by a much more sophisticated approximation twenty years later. This is the "impact factor formalism" of Cheng and Wu.*

It is physically obvious that the forward scattering of a photon by a Coulomb field does not depend on its polarization. The impact parameter method used by Bethe and Rohrlich leans heavily on the forward scattering results and does not take account of polarization effects. If $a^\perp = a_1^\perp + ia_2^\perp$ and $a^\| = a_1^\| + ia_2^\|$ are the complex amplitudes for scattering with incident linear polarization perpendicular and parallel to the scattering plane, then for unpolarized photons (15 − 124) is generalized to

$$\frac{d\sigma(\omega,\vartheta)}{d\Omega} = \frac{1}{2}[|a^\perp(\omega,\vartheta)|^2 + |a^\|(\omega,\vartheta)|^2] \qquad (15\text{–}125)$$

For forward scattering $a_i^\perp = a_i^\| = a_i$ ($i = 1, 2$), while for $\vartheta \neq 0$ $a_i^\perp$ and $a_i^\|$ can be quite different, even for small angles and high energies, as was shown by Cheng and Wu.

The idea of the high-energy scattering approximation by Cheng and Wu is the following: consider all Feynman diagrams for elastic two-body scattering, pick out those which contribute to the leading term in the high-energy limit, and attempt to sum them. This is indeed as formidable a task as it seems. However, one discovers eventually, that there exists a systematic scheme of calculation which greatly simplifies this work. It involves the use of certain diagrams (impact diagrams) and associated rules. These diagrams correspond to sums of Feynman diagrams. The theory is closely related to null plane quantum electrodynamics which is discussed very briefly in Supplement S 1-5.

For Delbrück scattering the impact factor method yields the high energy limit for momentum transfers $q = 2\omega \sin \vartheta/2$ in the range $m^2/\omega \ll q \ll \omega$ valid to *all* order of αZ. One finds

$$a^{\text{pol}} = \frac{i}{\pi}(\alpha Z)^2 r_0 \left(\frac{\omega}{m}\right) \frac{\sinh(\alpha Z\pi)}{\alpha Z\pi} f^{\text{pol}}\left(\frac{q}{m}, \alpha Z\right) \qquad (15\text{–}126)$$

---

* H. Cheng and T. T. Wu, *Phys. Rev.* **D5,** 3077 (1972). This paper is based on earlier work by these authors which is quoted there.

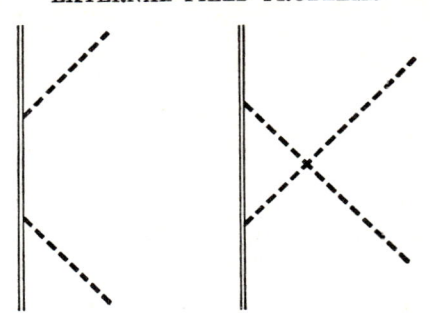

FIG. 15–16. Rayleigh scattering.

where "pol" stands for $\perp$ or $\|$. The factor f^{pol} is a complicated integral given by Cheng and Wu.* It was evaluated numerically by J. Kraus† for $Z = 92$. The cross section is found to fall off very steeply with increasing q. Also, the lowest order approximation in αZ can be too large by as much as a factor of the order of 10. The term Coulomb *corrections* is therefore rather misleading in this case.

While these results are extremely valuable, they do not answer the need of those experiments which are carried out in the energy range of a few Mev. A courageous calculation of the lowest order of $\pi\mu\gamma\lambda\sigma$ was carried out analytically,‡ but the resulting expression is too complicated for the evaluation of cross sections.

Numerical calculations have thus been carried out for a_2 using the known pair creation and annihilation cross sections,§ followed by the use of dispersion relations to find a_1.** However, the latter are open to criticism since the analyticity assumption underlying this use of the dispersion relations is not well established.

At this point it is well to remember that Delbrück scattering is actually a radiative correction and that it therefore occurs coherently only with the lowest order process, which is nuclear Thomson scattering in this case. The corresponding amplitudes must therefore be combined before the cross section is calculated. However, we have so far ignored another process which takes place coherently with Delbrück and nuclear Thomson scattering. This process is *Rayleigh scattering* or *electron resonance scattering:* the incident photon of energy ω is absorbed by a bound atomic electron which is excited to a higher discrete level or to the continuum. Subsequently, a photon of the same energy ω is emitted while the electron returns to its original state. The corresponding diagrams are given in Fig. 15–16.

---

* H. Cheng and T. T. Wu, *loc. cit.*
† J. Kraus, *Nuclear Phys.* **B89**, 133 (1975).
‡ V. Costantini et al., *Nuovo Cim.* **A2**, 733 (1971).
§ P. Kessler, *J. Phys. Rad.* **19**, 739 (1958).
** F. Ehlotzky and G. C. Sheppey, *Nuovo Cim.* **33**, 1185 (1964).

The calculation of this cross section is clearly very difficult except in the nonrelativistic limit, where it reduces to ordinary Thomson scattering times a suitable form factor which accounts for the Z atomic electrons in a statistical fashion.* When the energy is conserved in the intermediate state, Rayleigh scattering can be regarded as a photoelectric effect followed by radiative capture. It follows that at intermediate energies (momentum transfer $\sim \alpha Z m$) the K-electrons will again contribute most of the effect. Indeed, they contribute very nearly 80% to the amplitude.

Finally there is the possibility of another process which can interfere coherently with Delbrück scattering: nuclear resonance scattering. When all these effects are taken into account the cross section for the elastic scattering of unpolarized photons by heavy (i.e., non-recoiling) atoms has the form (15–125) with

$$a^{\text{pol}} = a_D^{\text{pol}} + a_T^{\text{pol}} + a_R^{\text{pol}} - a_N^{\text{pol}} \quad (15\text{–}127)$$

where D, T, R, and N stand for Delbrück, Thomson, Rayleigh, and nuclear resonance.

High energy experiments permit one to ignore all the other processes since Delbrück scattering (at small angles) by far dominates. The results are in very good agreement with the predictions by Cheng and Wu.†

At energies of a few Mev the situation is at present still not clear: discrepancies seem to exist between measurements and the present theoretical predictions which are in question as indicated above. In particular, while a_{2D} seems to be confirmed, the more interesting dispersive part, a_{1D}, seems to be smaller than predicted, at least at the higher regions of this energy range.‡

In conclusion of this section the process of *photon splitting* should be mentioned because it is diagrammatically closely related to Delbrück scattering: one of the external field points in each of the diagrams of Fig. 15–15 is replaced by an outgoing photon line. The theory of this process is unfortunately not yet fully developed for external electric fields,§ as well as for external magnetic fields.†† Cross sections are available only for special cases. The process is just beginning to be observed.**

---

* W. Franz, *Z. Physik* **98**, 314 (1936).

† G. Jarlskog et al., *Phys. Rev.* **D8**, 3813 (1973) used the energy range of 1 to 7 GeV.

‡ The measurements of S. Kahane and R. Moreh, *Phys. Rev.* **C9**, 2384 (1974) are for 7.9 MeV photons. Earlier work at lower energies are referred to in this paper.

§ M. Bolsterli, *Phys. Rev.* **4**, 367 (1954); A. P. Bukhvostov, *Sov. Phys. JETP* **16**, 467 (1963); D. Boccalletti et al., *Nuovo Cim.* **43**, 1115 (1966); Y. Shima, *Phys. Rev.* **142**, 944 (1966); V. Costantini et. al., *Nuovo Cim.* **2A**, 733 (1971).

†† S. Adler, *Ann. Phys.* **67**, 599 (1971). Z. Bialynicka-Birula and I. Bialynicki-Birula, *Phys. Rev.* **D2**, 2341 (1970).

** G. Jarlskog et al., *loc. cit.*

CHAPTER 16

SPECIAL PROBLEMS

In this last chapter we shall discuss a number of special problems for which the iteration solution of the S-matrix of a definite finite order is not directly applicable. Foremost among problems in this class is the infrared divergence which is here for the first time completely discussed from the modern point of view (Section 16–1). The problem of radiation damping, which is discussed in Section 16–2, can be most conveniently formulated by use of the Cayley transform of the S-operator and the Heitler integral equation. Closely related to this problem is the theory of the finite width of spectral lines (Section 16–3). In Section 16–4 we discuss the problem of the self-stress of the electron which, from the modern point of view, dissolves into a pseudoproblem. A section entitled "Outlook" (Section 16–5) concludes this chapter and the book.

16–1 The infrared divergences.* In this section we shall examine the divergences which arise for very long wavelengths. These were encountered again and again in the previous chapters. They were seen to arise in two different ways: the emission probability of soft photons becomes infinite with increasing wavelength, and the radiative corrections diverge when one integrates over the energy of the virtual photons down to the limit zero.

We shall see that the solution of this difficulty offers considerable physical insight into the mechanism of radiative processes. The questions which we raise at this point are the following:

(a) In all the problems which were considered in detail the infrared divergences exactly canceled. Can this be proved to hold in general, i.e., for any process and to all orders of the iteration solution?

(b) Since the emission of soft photons is the more likely the softer the photons, what meaning is there to elastic scattering, e.g., Coulomb scattering (Section 15–1)?

(c) In Section 15–2 it was shown that the corrections to Coulomb scattering depend on the experimental resolution of the electron energy, and decrease as this resolution increases. Is it possible to show what happens in the limit of perfect resolution?

We shall first answer question (a) and give a formal proof that when the appropriate terms in the iteration solution are combined, the infrared divergences completely disappear for any process and any order.

---

* The main results of this section can be found in J. M. Jauch and F. Rohrlich, *Helv. Phys. Acta* **27**, 613 (1954) and F. Rohrlich, *Phys. Rev.* **98**, 181 (1955).

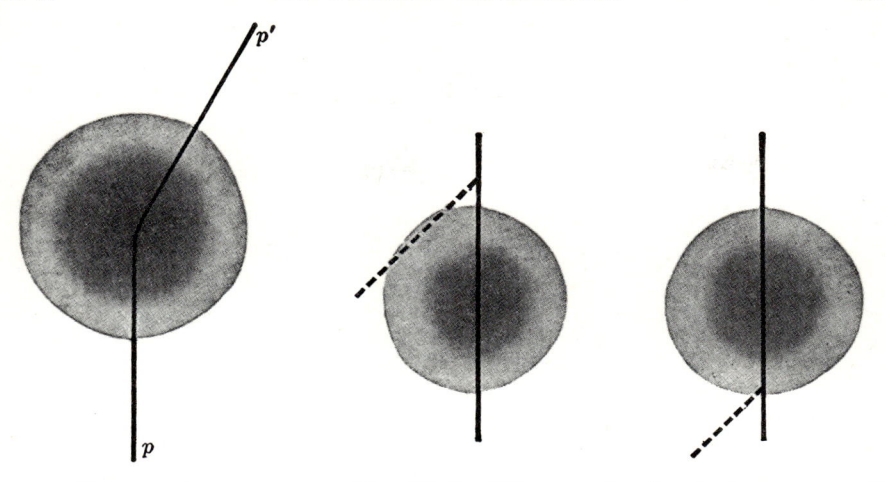

FIGURE 16-1

FIG. 16-2. Diagrams for the basic process with one additional photon.

Consider the basic process indicated in Fig. 16–1. The shaded area stands for an arbitrary diagram involving possibly external fields as well as internal and external photon lines. The generalization to processes with more than one electron path will be discussed later. The two external electron lines in Fig. 16–1 may refer to either negatons or positons. The following argument is obviously independent of this assumption. For negatons the S-matrix element is of the form

$$M = \bar{u}(\mathbf{p}')Q(p'p)u(\mathbf{p}), \qquad (16\text{-}1)$$

where $Q(p'p)$ is a constant times the spinor matrix associated with the shaded area of the basic process.

We wish to calculate the matrix element M_1 corresponding to the emission of one additional soft photon of momentum $k = (\omega, \mathbf{k})$ and polarization e, $(\omega \ll m)$. This calculation is the generalization of the evaluation of the diagrams in Fig. 15–4.

According to the general rules of Chapter 8, this matrix element has the form

$$M_1 = \frac{ie}{(2\pi)^{3/2}} \bar{u}(\mathbf{p}') \left[\frac{e(k)}{\sqrt{2\omega}} \frac{i(p'+k) - m}{(p'+k)^2 + m^2} Q(p'+k, p) \right.$$

$$\left. + Q(p', p-k) \frac{i(p-k) - m}{(p-k)^2 + m^2} \frac{e(k)}{\sqrt{2\omega}} \right] u(\mathbf{p}). \qquad (16\text{-}2)$$

In this calculation we have ignored the emission of photons from internal lines of M, since they do not give rise to infrared divergences.

The expression (16-2) can be considerably simplified if we use the fact that $k^0 = \omega \ll m$ and that the propagation functions operate on free-particle wave functions. Retaining only the terms of lowest order in ω/m, we obtain, after some rearrangement,

$$M_1 = \beta(k) M$$

with

$$\beta(k) = -\frac{e}{(2\pi)^{3/2}} \frac{1}{\sqrt{2\omega}} \left(\frac{p' \cdot e}{p' \cdot k} - \frac{p \cdot e}{p \cdot k} \right). \tag{16-3}$$

A special case of this result was obtained in Eq. (15-24). The total transition probability is modified by the factor

$$b = \frac{e^2}{(2\pi)^3} \sum_{\text{pol}} \int \frac{1}{2\omega} \left| \frac{p' \cdot e}{p' \cdot k} - \frac{p \cdot e}{p \cdot k} \right|^2 d^3k \tag{16-4}$$

due to the emission of one soft photon. In this form the infrared divergence is manifest as a logarithmic divergence of the k-space integral at low frequencies.

The evaluation of this integral in the general case cannot be carried out in terms of elementary functions.* However, it is evidently sufficient to give our proof in the special coordinate system in which the initial electron is at rest ($\mathbf{p} = 0$). In this system the integral (16-4) is elementary.

Since we are interested here in the divergent terms only, it is not necessary to introduce the photon mass explicitly, nor do we need to consider the associated longitudinal photons. These were shown to give only finite contributions in the derivation of Eq. (15-26).

The polarization sum in Eq. (16-4) becomes

$$\sum_{\text{pol}} |p' \cdot e|^2 = \mathbf{p}'^2 - \left(\frac{\mathbf{p}' \cdot \mathbf{k}}{\omega} \right)^2,$$

so that we have

$$b = \frac{\alpha}{4\pi^2} \int \frac{d^3k}{\omega^3} \left[\frac{\mathbf{p}'^2}{(\epsilon' - |\mathbf{p}'| \cos \theta')^2} - \frac{\mathbf{p}'^2 \cos^2 \theta'}{(\epsilon' - |\mathbf{p}'| \cos \theta')^2} \right]$$

$$= \frac{\alpha}{2\pi} \beta'^2 \int \frac{d\omega}{\omega} \int_{-1}^{1} \frac{1 - x^2}{(1 - \beta' x)^2} dx$$

$$= \frac{\alpha}{\pi} \left(\frac{1}{\beta'} \ln \frac{1 + \beta'}{1 - \beta'} - 2 \right) \int \frac{d\omega}{\omega}$$

$$= \frac{2\alpha}{\pi} \left(\frac{\tanh^{-1} \beta'}{\beta'} - 1 \right) \int \frac{d\omega}{\omega}. \tag{16-5}$$

---

* J. Schwinger, *Phys. Rev.* **76**, 790 (1949), p. 810 ff.

16–1] THE INFRARED DIVERGENCES

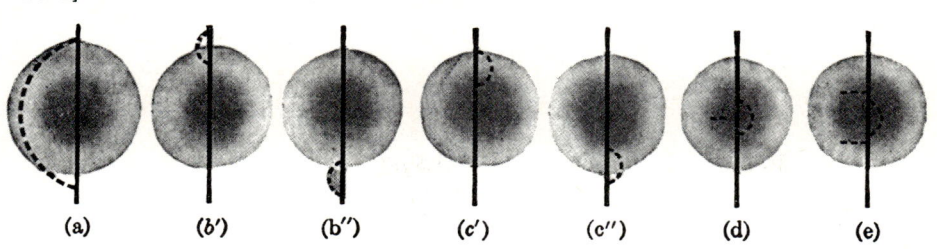

FIG. 16–3. Radiative corrections which give rise to infrared divergences.

This result is exactly the same as that obtained for the special case of double Compton scattering, when one of the outgoing photons is very soft [cf. Eq. (11–44)].

We now turn to the first radiative corrections of the basic process M shown in Fig. 16–1. These are obtained by inserting into M one internal photon line in all possible ways. If both ends of this photon line are attached to internal electron lines, no infrared divergences will arise, unless possibly when an electron self-energy part or a vertex part is thereby created. This is true even when only one end of the added photon line is attached to an internal line, whereas the other end is attached to one of the two external lines. The remaining radiative correction diagrams are shown schematically in Fig. 16–3. The meaning of diagrams (a) and (b) is clear. In the diagrams (c) we mean to indicate that the added photon line begins on an external line and ends on the *first* internal line of M, thereby creating a vertex part at the very edge of the diagram. In (d) and (e) the added photon line creates a vertex part and an electron self-energy part.

As we shall see below, the infrared divergent parts of the radiative corrections to M can all be written in the form

$$\rho M.$$

The corrections of order α to $|M|^2$ are obtained from the cross term between M and the correction terms $M^{(2)}$:

$$|M + M^{(2)}|^2 = |M|^2(1 + 2\rho + \cdots).$$

Our aim is to prove the compensation of the divergence in b by the term

$$r = 2\rho; \tag{16–6}$$

that is, we want to show that

$$b + r = \text{convergent integral}. \tag{16–7}$$

The diagram (a) gives* for small **k**,

$$M_{(a)}^{(2)} = \frac{ie^2}{(2\pi)^4} \int \bar{u}(\mathbf{p}')\gamma_\mu \frac{i(p'-k)-m}{(p'-k)^2+m^2} Q \frac{i(p-k)-m}{(p-k)^2+m^2} \gamma^\mu u(\mathbf{p}) \frac{d^4k}{k^2}$$

$$= \frac{i\alpha}{4\pi^3} \int \frac{-ip'_\mu}{p'\cdot k} \bar{u}(\mathbf{p}')Qu(\mathbf{p}) \frac{-ip^\mu}{p\cdot k} \frac{d^4k}{k^2}$$

$$= \rho_1 M,$$

with

$$\rho_1 = -\frac{i\alpha}{4\pi^3} \int \frac{p'\cdot p}{p'\cdot k\, p\cdot k} \cdot \frac{d^4k}{k^2}. \qquad (16\text{-}8)$$

We first carry out the integration over k^0 along the path C_{1R} in the complex plane (*cf.* Section A–1). Since all poles are on the real axis of k^0-space, we can rotate the path by $\pi/2$ and integrate along the imaginary axis. Replacing k^0 by ik^0, we have

$$\rho_1 = -\frac{i\alpha}{4\pi^3} p'\cdot p \int d^3k \int_{-\infty}^\infty \frac{idk^0}{(\omega^2+k^{02})(|\mathbf{p}'|\omega\cos\theta' - i\epsilon'k^0)(|\mathbf{p}|\omega\cos\theta - i\epsilon k^0)}$$

$$= -\frac{i\alpha}{4\pi^3} \cdot \frac{i\pi}{\epsilon'\epsilon} p'\cdot p \int \frac{d^3k}{\omega^3}\left(\frac{\beta'\cos\theta'}{1-\beta'^2\cos^2\theta'} - \frac{\beta\cos\theta}{1-\beta^2\cos^2\theta}\right)$$

$$\times \frac{1}{\beta'\cos\theta' - \beta\cos\theta}.$$

As in the evaluation of b, we can choose the reference frame defined by $\mathbf{p} = 0$, so that $p'\cdot p = -m\epsilon'$.

$$\rho_1 = -\frac{\alpha}{4\pi^2} \int \frac{d\omega}{\omega} 2\pi \int_{-1}^1 \frac{dx}{1-\beta'^2x^2} = -\frac{\alpha}{2\pi}\frac{1}{\beta'}\ln\frac{1+\beta'}{1-\beta'}\int\frac{d\omega}{\omega}.$$

From Eq. (16–6) follows

$$r_1 = -\frac{2\alpha\,\tanh^{-1}\beta'}{\pi\,\beta'}\int\frac{d\omega}{\omega}. \qquad (16\text{-}9)$$

We see that this term exactly cancels the first term in b, Eq. (16–5).

We return to the remaining diagrams (b) to (e) in Fig. 16–3. Their infrared contributions will be denoted by ρ_2. The value of ρ_2 need not be

---

* Since the main contributions to this integral came from the vicinity of $\omega^2 = |\mathbf{k}|^2 = (k^0)^2$, $|\mathbf{k}|^2 \ll m$ implies $k^0 \ll m$ throughout, as can be verified explicitly. Equation (16–8) can thus be integrated over k^0 from $-\infty$ to $+\infty$.

calculated by integration of the respective matrix elements, but can be obtained by the following argument.

We remember that $\Sigma(p)$ and $\Lambda_\mu(p',p)$ in second order were found to be free of infrared divergences (*cf.* Sections 9–4 and 9–6). However, both B and L contain such divergences. Since δm is not infrared divergent, we have the relations

$$-B^i + \Sigma_f{}^i(p)S_c{}^{-1}(p) = 0, \qquad (16\text{–}10)$$

$$\gamma_\mu L^i + \Lambda_{\mu f}{}^i = 0. \qquad (16\text{–}11)$$

The superscript i means "infrared divergence of." It was further proved that the constants B and L are equal [Eq. (9–92) and Eq. (10–28)], so that their infrared divergences are also equal,

$$B^i = L^i.$$

Since the Σ and Λ_μ in every diagram occur in such a way that the constants B and L always cancel, it follows from Eqs. (16–10) and (16–11) that

$$\gamma_\mu \Sigma_f{}^i S_c{}^{-1}(p) \text{ cancels } \Lambda_{\mu f}{}^i. \qquad (16\text{–}12)$$

By the latter statement is meant that $\Lambda_{\mu f}{}^i$ cancels with *half* the $\Sigma_f{}^i$ from either side of the electron path, as was shown repeatedly for the cancellation of B and L [*cf.*, for example, the evaluation of the diagrams (a) and (b) in Fig. 15–3]. Therefore, the sum of all the diagrams (b), (c), (d), and (e) of Fig. 16–3 would exactly cancel in their infrared contributions, were it not for the fact that the terms $\Sigma_f(p')$ and $\Sigma_f(p)$ in the external lines [diagrams (c)] vanish because of

$$\bar{u}(p')S_c{}^{-1}(p')\Sigma_f(p') = \Sigma_f(p)S^{-1}(p)u(p) = 0.$$

The corresponding parts of the divergences in (c) are therefore not compensated. The *missing* parts are

$$\frac{1}{2}S^{-1}(p')\Sigma_f{}^i(p') + \frac{1}{2}\Sigma_f{}^i(p)S^{-1}(p) = B^i = -\frac{\alpha}{\pi}\int\frac{dx}{x} = -\frac{\alpha}{\pi}\int\frac{d\omega}{\omega},$$

according to Eq. (9–25). The diagrams (b) to (e) therefore contribute

$$\rho_2 = \frac{\alpha}{\pi}\int\frac{d\omega}{\omega},$$

that is,

$$r_2 = \frac{2\alpha}{\pi}\int\frac{d\omega}{\omega}. \qquad (16\text{–}13)$$

These diagrams are seen to yield an infrared divergence which exactly compensates the second term in b, Eq. (16–5). We can combine Eqs. (16–9) and (16–13):

$$r = r_1 + r_2 = -\frac{2\alpha}{\pi}\left(\frac{\tanh^{-1}\beta'}{\beta'} - 1\right)\int \frac{d\omega}{\omega}, \qquad (16\text{–}14)$$

which, when compared with b, completes the proof of Eq. (16–7).*

The diagrams of Fig. 16–2 may now be regarded as the basic processes and the previous argument can be repeated, such that the emission of *two* low-energy photons is seen by this same argument to be compensated as well. The same is then true for an arbitrary number of soft photons.

Finally, it becomes evident that if a diagram contains more than one open electron path the above proof can easily be extended.

Closed loops do not cause difficulties in these proofs, because they never contribute to soft photon emission (*cf.* the last theorem of Section 8–4) and the associated radiative corrections exactly cancel in the sense of Eq. (16–12). Thus, closed loops contribute neither to b nor to r.

The results established by this proof answer the first question raised at the beginning of this section in the affirmative. They can be summarized as follows:

The infrared divergences, class (b), are entirely due to the unphysical description of certain processes by the iteration solution. The processes of soft photon emission and radiative corrections due to soft virtual photons are evaluated as two quite different processes in the iteration solution, and originate in S-matrix elements of different order in e. Physically, these processes should not be separated. Their separation causes the appearance of the infrared divergences, which is thus introduced in an artificial way, due to the particular mathematical treatment adopted. *The recombination of soft-photon emission and soft-photon corrections indeed eliminates these divergences completely for all processes and to all orders in e.*

We have thus come to the important conclusion that the model for the mechanism of radiative processes which follows so naturally from the iteration solution in the form of the Feynman-Dyson diagrams is actually a very

---

* The conjecture that this direct cancellation occurs in all orders has been expressed by many people. Special cases were discussed previously by H. A. Bethe and J. R. Oppenheimer, *Phys. Rev.* **70**, 451 (1946). E. Corinaldesi and R. Jost, *Helv. Phys. Acta* **21**, 183 (1948), discuss it for scalar particles and R. Jost, *Phys. Rev.* **72**, 815 (1947), for the Compton effect. This cancellation was explicitly verified for Compton scattering by M. R. Schafroth, *Helv. Phys. Acta* **23**, 542 (1950), and by L. M. Brown and R. P. Feynman, *Phys. Rev.* **85**, 231 (1952). A general proof was attempted by K. Baumann, *Acta Phys. Austr.* **7**, 248 (1953), which remained incomplete, because the infrared divergence of the self-energy was not calculated. The same remark applies to T. Kinoshita, *Progr. Theor. Phys.* **5**, 1045(L), (1950).

poor model. It breaks down completely for the description of soft-photon processes, as we have just seen.

To arrive at a better description for this case, we evidently have to obtain closed expressions which contain the effect of photons of various low energies and which are *independent of the number* of the participating photons. This number is not an observable, since the "free" and the virtual photons apparently become indistinguishable in the soft-photon limit. In fact, there seems to exist a not very sharp separation between photons which are observable individually, and the photons which are observable only collectively. The latter appear as an energy loss, $\Delta\epsilon$, whose details cannot be differentiated.

To study these conjectures we ask for the modification of a given process due to this collective soft-photon effect. We thereby expect to obtain a dependence on $\Delta\epsilon$, so that the whole collective effect *vanishes* as $\Delta\epsilon$ approaches zero. This calculation will be free of infrared divergences, as follows from the above proof. It was first carried out in the fundamental paper by Bloch and Nordsieck.*

Since we are interested in soft-photon processes only, virtual pair production and annihilation will be unimportant in the limit. We can thus simplify the calculation considerably by working with classical currents, rather than with the quantized electron field.

We return to the basic process (Fig. 16–1) and calculate the modification of the associated transition probability due to the collective soft-photon effect. The treatment by a classical current distribution

$$s_\mu(x) = \frac{1}{(2\pi)^{5/2}} \int s_\mu(k) e^{ikx} d^4k \qquad (16\text{--}15)$$

requires the choice

$$s_\mu(k) = \frac{ie}{(2\pi)^{3/2}} \cdot \left(\frac{p_\mu'}{p' \cdot k} - \frac{p_\mu}{p \cdot k}\right), \qquad (16\text{--}16)$$

in order to obtain the same single-photon emission probability amplitude $\beta(k)$ (Eq. 16–3) as is obtained with the quantized field. To see this, we note that the first-order contribution to the S-matrix for a system described by

---

* F. Bloch and A. Nordsieck, *Phys. Rev.* **52**, 54 (1937). See also W. Braunbeck and E. Weinmann, *Z. Physik* **110**, 360 (1938), and W. Pauli and M. Fierz, *Nuovo Cimento* **15**, 167 (1938). A review of the older work is found in the paper by G. Morpurgo, *Nuovo cimento*, Suppl. No. 2, p. 109 (1951). In these papers, which use a noncovariant calculation, the problem of the infrared divergence could not be separated from the divergence at high energy. In the covariant formalism this difficulty is now overcome by the renormalization of mass and charge. See R. J. Glauber, *Phys. Rev.* **84**, 395 (1951), and W. Thirring and B. Touschek, *Phil. Mag.* **42**, 244 (1951).

the interaction operator

$$H(\tau) = -\int_{\sigma(\tau)} s_\mu(x) a^\mu(x) d\sigma \qquad (16\text{--}17)$$

is the expression

$$S_1 = i\int s_\mu(x) a^\mu(x) d^4x. \qquad (16\text{--}18)$$

The matrix element for the emission of one photon with momentum k and polarization e is then given by

$$\beta(k) = \frac{i}{\sqrt{2\omega}} \frac{1}{(2\pi)^{3/2}} \int s_\mu(x) e^\mu e^{-ikx} d^4x,$$

$$\beta(k) = \frac{i}{\sqrt{2\omega}} s_\mu(k) e^\mu. \qquad (16\text{--}19)$$

This is indeed the same expression as (16–3), provided the Fourier component of the current is defined by (16–15) and (16–16).

We note in passing that the classical current with the Fourier components

$$s_\mu(k) = \frac{ie}{(2\pi)^{3/2}} \left(\frac{p_\mu'}{p' \cdot k} - \frac{p_\mu}{p \cdot k} \right)$$

has a simple physical interpretation which can be seen most easily when it is transformed into x-space. The identities

$$\int_{-\infty}^{0} \delta(x - v\tau) d\tau = \frac{i}{(2\pi)^4} \int \frac{d^4k}{k \cdot v} e^{ikx},$$

$$\int_{0}^{\infty} \delta(x - v\tau) d\tau = \frac{-i}{(2\pi)^4} \int \frac{d^4k}{k \cdot v} e^{ikx}$$

allow us to express the current in x-space (Eq. 16–15) in the form

$$s_\mu(x) = -e \left[\int_{-\infty}^{0} v_\mu \delta(x - v\tau) d\tau + \int_{0}^{\infty} v_\mu' \delta(x - v'\tau) d\tau \right]$$

with $\qquad (16\text{--}20)$

$$v_\mu = \frac{p_\mu}{m}, \quad v_\mu' = \frac{p_\mu'}{m}.$$

This is the current of a classical point charge $-e$ moving for $x^0 < 0$ with the constant four-velocity v_μ and for $x^0 > 0$ with the velocity v_μ'.

Before we proceed with the calculation, we consider a lemma which is of importance for the S-matrix of a classical current distribution.

16-1] THE INFRARED DIVERGENCES

Let $H(\tau)$ be an interaction operator which satisfies the condition

$$[H(\tau),H(\tau')] = iC(\tau,\tau'), \qquad (16\text{--}21)$$

where $C(\tau,\tau')$ is a c-number. The S-matrix for this system differs from the matrix

$$S' = e^{-i\Sigma}, \qquad (16\text{--}22)$$

$$\Sigma = \int_{-\infty}^{+\infty} H(\tau)d\tau \qquad (16\text{--}23)$$

only by a phase factor.

The significance of this lemma lies in the fact that the phase factor can be ignored and that the S-matrix can be calculated according to Eq. (16–22).

An alternative way of expressing this lemma is to state that the time-ordering operation in the iteration solution contributes only a phase factor.

To prove this lemma, we carry out the τ-dependent transformation

$$\omega'(\tau) = e^{i\Sigma(\tau)}\omega(\tau). \qquad (16\text{--}24)$$

The new state vector $\omega'(\tau)$ satisfies a transformed Schroedinger equation

$$i\dot{\omega}'(\tau) = H'(\tau)\omega'(\tau),$$

with $H'(\tau)$ given by

$$H'(\tau) = e^{i\Sigma(\tau)}\left(H(\tau) - i\frac{d}{d\tau}\right)e^{-i\Sigma(\tau)}.$$

This expression can be written in terms of iterated commutators:

$$H'(\tau) = H(\tau) + i[\Sigma(\tau),H(\tau)] - \dot{\Sigma} + \frac{i}{2}[\dot{\Sigma}(\tau),\Sigma(\tau)] + \cdots . \qquad (16\text{--}25)$$

If we choose $\Sigma(\tau)$ such that it satisfies

$$\dot{\Sigma}(\tau) = H(\tau), \quad \Sigma(-\infty) = I, \qquad (16\text{--}26)$$

or

$$\Sigma(\tau) = \int_{-\infty}^{\tau} H(\tau')d\tau', \qquad (16\text{--}27)$$

only the commutator in Eq. (16–25) is different from zero. Thus, the transformed interaction operator

$$H'(\tau) = -\tfrac{1}{2}\int_{-\infty}^{\tau} C(\tau',-\tau)d\tau' \qquad (16\text{--}28)$$

is a c-number.

The S-operator is defined in terms of the general transformation operator $V(\tau,\tau_0)$ which satisfies

$$\omega'(\tau) = V(\tau,\tau_0)\omega'(\tau_0).$$

This transformation can therefore be written

$$V(\tau,\tau_0) = e^{-i\int_{\tau_0}^{\tau}H'(\tau)d\tau}.$$

We can now express the S-operator in the form

$$S = e^{-i\Sigma(\infty)} V(\infty, -\infty) e^{i\Sigma(-\infty)}.$$

With Eq. (16–26) and the definition

$$\Sigma \equiv \Sigma(\infty) = \int_{-\infty}^{+\infty} H(\tau) d\tau,$$

we may write

$$S = e^{i\theta} e^{-i\Sigma}, \qquad (16\text{–}29)$$

where

$$\theta = \tfrac{1}{2} \int_{-\infty}^{+\infty} d\tau \int_{-\infty}^{\tau} C(\tau, \tau') d\tau', \qquad (16\text{–}30)$$

which proves the lemma.

The interaction operator (16–17) with the classical current density $s_\mu(x)$ satisfies the conditions of our lemma. We therefore write Eq. (16–22) for the S-operator with Σ given by

$$\Sigma = -\int d^4 x\, s_\mu(x) a^\mu(x). \qquad (16\text{–}31)$$

Let us now divide the k-space into finite but small cells Δ_r ($r = 1, \cdots$) and denote by $\mathbf{k}_r$ a representative value of the k-vectors in cell Δ_r. We then evaluate the matrix-element M_r which connects the photon vacuum state with a state representing precisely n_r photons in cell Δ_r.

It is clear that this matrix element M_r is different from zero only for terms in the expansion for S,

$$S = \sum_{m=0}^{\infty} \frac{(-i)^m}{m!} \Sigma^m, \qquad (16\text{–}32)$$

which contain at least n_r photon creation operators. Therefore only terms for which $m \geq n_r$ will contribute to this matrix element. For a definite value of $m = n_r + 2l$ these contributions can be calculated by use of the ordering theorem (*cf.* Appendix A4). Since Σ contains one emission and one absorption operator, the matrix element in question is obtained by contracting l pairs of photon operators.

In the present case, there is a considerable simplification due to the fact that the current is a *c*-number. It follows that each contraction results in the same term and is easily found to be

$$\rho_r \equiv \rho(\mathbf{k}_r) = \frac{\Delta_r}{2\omega_r} s_\mu^*(\mathbf{k}_r) s^\mu(\mathbf{k}_r). \qquad (16\text{–}33)$$

The l contractions therefore introduce a factor ρ_r^l and also a combinatorial factor

$$\frac{1}{2^l} \frac{m!}{n_r! l!},$$

which is the number of times l ordered pairs can be selected from m distinct objects. The result is

$$M_r = \frac{1}{n_r!}\left(\frac{-i}{\sqrt{2\omega_r}}\, s_\mu(\mathbf{k}_r)e^\mu(\mathbf{k}_r)\right)^{n_r} \sum_{l=0}^{\infty} (-1)^l \frac{\rho_r^l}{2^l l!}. \tag{16-34}$$

The matrix element for the simultaneous emission of n_1 photons into Δ_1, n_2 photons into Δ_2, etc., is the product

$$M = \prod_{r=1}^{\infty} M_r, \tag{16-35}$$

so that the total probability for the emission process is

$$P = \sum_{\text{pol}} \prod_{r=1}^{\infty} n_r! |M_r|^2 \Delta_r^{n_r}. \tag{16-36}$$

The factors $\Delta_r^{n_r}$ are the densities of the final states; the factor $n_r!$ takes account of the different possible sequences in which n_r photons can be emitted. In this expression we also sum over the unobserved polarization states of the final photons, which gives for each r:

$$\sum_{\text{pol}} |s_\mu e^\mu|^2 = |\mathbf{s}|^2 - \frac{1}{\omega^2}|\mathbf{s}\cdot\mathbf{k}|^2 = s^\mu s_\mu^*. \tag{16-37}$$

In the last equation we have used the conservation law for the current in the form $s_\mu k^\mu = 0$.

The transition probability becomes, with Eq. (16-33),

$$P = \prod_{r=1}^{\infty} e^{-\rho_r} \frac{\rho_r^{n_r}}{n_r!}. \tag{16-38}$$

This result shows that the final photons are emitted according to the Poisson distribution law into each volume element in k-space and that therefore the average number of photons emitted into Δ_r is just exactly ρ_r of Eq. (16-33). We also see that the average total number of emitted photons is

$$\bar{N} = \sum_r \rho_r = \sum_r \frac{\Delta_r}{2\omega_r} s_\mu^*(k_r)s^\mu(k_r). \tag{16-39}$$

If this expression is transformed into an integral by letting $\Delta_r \to 0$, and if the current (16-16) is substituted, we obtain precisely the integral b in Eq. (16-4), which was shown to be logarithmically divergent. It follows that the average number of photons emitted into all phase space is infinite. This divergence is now no longer a difficulty, since it appears as an infinity for the total number of photons emitted rather than as an infinite transition probability, as in the case of Eq. (16-4).

The correction factor $b(\Delta\epsilon)$ which is defined as the correction to the total transition probability due to photon emission with total energy loss $\leq \Delta\epsilon$ is

$$b(\Delta\epsilon) = {\sum_{n_r}}' \prod_{r=1}^{\infty} e^{-\rho_r} \frac{\rho_r^{n_r}}{n_r!}. \qquad (16\text{–}40)$$

The summation is restricted by the condition that the numbers n_r must satisfy the inequality

$$\sum_{r=1}^{\infty} n_r \omega_r \leq \Delta\epsilon. \qquad (16\text{–}41)$$

If the summation in Eq. (16–40) were not restricted by Eq. (16–41) it could easily be performed. We can restore the unrestricted summation by introducing a discontinuous function

$$I(n_r) = \begin{cases} 1, & \text{for } \sum_r n_r \omega_r \leq \Delta\epsilon, \\ 0, & \text{for } \sum_r n_r \omega_r > \Delta\epsilon. \end{cases} \qquad (16\text{–}42)$$

A useful explicit form for this function is

$$I(n_r) = \int_{-\Delta\epsilon}^{+\Delta\epsilon} \delta\left(\sum_r n_r \omega_r - x\right) dx$$

$$= \frac{1}{2\pi} \int_{-\Delta\epsilon}^{+\Delta\epsilon} dx \int_{-\infty}^{+\infty} d\sigma e^{\left(\sum_r n_r \omega_r - x\right) i\sigma}.$$

The restriction of the summation in Eq. (16–40) can now be dropped, and we obtain

$$b(\Delta\epsilon) = \frac{1}{2\pi} \int_{-\Delta\epsilon}^{+\Delta\epsilon} dx \int_{-\infty}^{+\infty} d\sigma F(\sigma) e^{-ix\sigma}$$

$$F(\sigma) = \sum_{n_r} \prod_r e^{-\rho_r} \frac{(\rho_r e^{i\omega_r \sigma})^{n_r}}{n_r!}. \qquad (16\text{–}43)$$

The product and sum can be interchanged. Each individual sum for any r is of the form

$$\sum_n \frac{(\rho e^{i\omega\sigma})^n}{n!} = e^{\rho e^{i\omega\sigma}}.$$

Therefore we obtain for F,

$$F(\sigma) = e^{\sum_r \rho_r (e^{i\omega_r \sigma} - 1)} = e^{G(\sigma)}. \qquad (16\text{–}44)$$

The sum in the exponential can be reduced to an integral in the limit $\Delta_r \to 0$:

$$G(\sigma) = \int \frac{d^3k}{2\omega} s^\mu(k) s_\mu{}^*(k)(e^{i\omega\sigma} - 1). \tag{16-45}$$

It is now easy to see how the infrared divergence disappears. For, when we substitute the expression (16-16) in Eq. (16-45), we obtain

$$G(\sigma) = \frac{\alpha}{(2\pi)^2} \int \frac{d^3k}{\omega} \left[\frac{p'^2}{(p'\cdot k)^2} + \frac{p^2}{(p\cdot k)^2} - \frac{2p'\cdot p}{(p'\cdot k)(p\cdot k)} \right] (e^{i\omega\sigma} - 1). \tag{16-46}$$

The last factor, $e^{i\omega\sigma} - 1$, approaches $i\omega\sigma$ for $\omega \to 0$ and eliminates the singularity at the origin. After the angular integration this expression becomes†

$$G(\sigma) = \alpha C \int_0^{\Delta\epsilon} \frac{d\omega}{\omega} (e^{i\omega\sigma} - 1), \tag{16-47}$$

where C is a positive numerical factor of order 1 which depends on p and p'. In the coordinate system in which the initial electron is at rest ($\mathbf{p} = 0$), C was evaluated to be [cf. Eq. (16-5)]

$$C = \frac{2}{\pi}\left(\frac{\tanh^{-1} \beta'}{\beta'} - 1 \right). \tag{16-48}$$

The final expression for the correction factor $b(\Delta\epsilon)$ is obtained by substituting $G(\sigma)$ into Eq. (16-44) and making use of Eq. (16-43). The integrations can be interchanged and the x-integration can be carried out first, so that

$$b(\Delta\epsilon) = \frac{1}{\pi}\int_{-\infty}^{\infty} \frac{d\sigma}{\sigma} \sin(\sigma\Delta\epsilon) \exp\left[\alpha C \int_0^{\sigma\Delta\epsilon} \frac{dy}{y}(e^{iy} - 1) \right]. \tag{16-49}$$

In this form it is obvious that $b = 0$ for $\Delta\epsilon = 0$. This means that the cross section for the basic process is zero when the energy loss due to soft-photon emission is zero. It follows that *every scattering process is accompanied by some radiative energy loss.*

When $\Delta\epsilon \neq 0$, we can introduce the new variable $x = \sigma\Delta\epsilon$ and obtain

$$b(\Delta\epsilon) = \frac{1}{\pi}\int_{-\infty}^{\infty} \frac{dx}{x} \sin x \exp\left[\alpha C \int_0^{x} \frac{dy}{y}(e^{iy} - 1) \right]. \tag{16-50}$$

This integral no longer depends on $\Delta\epsilon$. It differs from a constant only by its dependence on αC, which involves the initial and final state of the scat-

---

† Since the n_r are positive integers or zero, the restriction (16-41) does not allow larger values of ω than $\Delta\epsilon$.

tered electron. We write, therefore,

$$b(\Delta\epsilon) = 0, \quad (\Delta\epsilon = 0),$$

$$b(\Delta\epsilon) \equiv b(\alpha C), \quad (\Delta\epsilon \neq 0).$$

It does not seem possible to evaluate the integral (16–50) analytically; however, since C is of order 1 and α is small, we have in first approximation

$$b(\alpha C) \simeq 1. \tag{16–51}$$

Expanding $b(\alpha C)$ in powers of αC, we notice that

$$b'(0) \equiv \left[\frac{db(\alpha C)}{d(\alpha C)}\right]_{\alpha C = 0} = \frac{1}{\pi} \int_{-\infty}^{\infty} \frac{dx}{x} \sin x \int_{0}^{x} \frac{dy}{y} (e^{iy} - 1)$$

$$= \frac{2}{\pi} \int_{0}^{\infty} \frac{dx}{x} \sin x [Ci(x) - \ln \gamma x],$$

where

$$Ci(x) = \int_{x}^{\infty} \frac{dy}{y} \cos y$$

is the cosine integral and $\ln \gamma$ is Euler's constant.* The first term gives

$$\int_{0}^{\infty} \frac{dx}{x} \sin x\, Ci(x) = \int_{1}^{\infty} \frac{dy}{y} \int_{0}^{\infty} \frac{dx}{x} \sin x \cos xy$$

$$= \frac{1}{2} \int_{1}^{\infty} \frac{dy}{y} \int_{0}^{\infty} \frac{dx}{x} \{\sin [x(1 + y)] - \sin [x(1 - y)]\} = 0,$$

since both x-integrals give $\pi/2$. The second term of $b'(0)$ also vanishes,† so that $b'(0) = 0$.

The next coefficient in the expansion, $b''(0)$, can be evaluated analytically and is found to be $-\pi^2/12$. Therefore,

$$b(\alpha C) = 1 - \frac{\pi^2}{12} (\alpha C)^2 + \cdots. \tag{16–52}$$

Since C is of order 1, $b(\alpha C)$ differs from 1 by about 5×10^{-5}. But even for very high energies, when $C \gg 1$ (β' very close to 1), b does not decrease very much below 1. One finds numerically

$$b(\tfrac{1}{2}) = 0.85, \quad b(1) = 0.58.$$

---

* See, e.g., E. Jahnke and F. Emde, *Tables of Functions*, Dover, 1945, p. 1.
† See, e.g., A. Erdélyi, *Tables of Integral Transforms*, McGraw-Hill, 1945, formula 3, p. 76.

The last case corresponds to the energy

$$\epsilon = e^{\pi/(2\alpha)}m = e^{137(\pi/2)}m \sim 10^{100} \text{ Mev.}$$

We are now in a position to answer the remaining questions raised at the beginning of this section. The probability for every scattering process (in the broad sense of the word) is multiplied by a factor $b(\Delta\epsilon)$ which takes into account the soft-photon effects. These effects manifest themselves only collectively by a minimum energy loss $\Delta\epsilon$. It was shown that this energy loss is due to the emission of an undeterminable number of photons whose average is infinite. The factor b is a step function for small $\Delta\epsilon$, so that every transition probability vanishes unless it is accompanied by a finite, though small, energy loss. Therefore, when this energy loss is small enough, $\Delta\epsilon \ll m$, the usual calculation for elastic scattering (e.g., Coulomb scattering) which completely ignores soft-photon processes, will give the correct result within an error of 10^{-4}, since $b = 1$ in this case, Eq. (16–51), unless the energy is *extremely* high. In this sense only is the notion of elastic scattering a meaningful concept.

When the energy loss $\Delta\epsilon$ is not very small, a finite, *observable number* of photons can be held responsible for it. This number decreases to one as $\Delta\epsilon$ increases. Our previous calculation is then no longer valid and only a qualitative behavior of $b(\Delta\epsilon)$ can be inferred.

In order to illustrate this behavior, we recall the calculation of radiative corrections and energy loss in Coulomb scattering (Section 15–2). The corrections of order α were shown to depend on the minimum resolution $\Delta\epsilon$ for the emission of a *single* photon, through the factor

$$1 - \delta = 1 - O\left(\alpha\beta^2 \ln \frac{m}{\Delta\epsilon}\right)$$

[*cf.* Eq. (15–30)]. This result is valid so long as $\Delta\epsilon/m$ is not too small; otherwise the emission of two or more photons contributes. In the limit $\Delta\epsilon \to 0$ an infinite number of photons contributes. But as the number of contributing photons increases their total effect decreases, that is, b approaches 1, so that the cross section approaches the elastic cross section.

16–2 Radiation damping in collision processes. In all applications of the S-matrix to the various processes so far considered we have ignored radiation damping. This effect will give a correction to the previous results. These results are all based on the formulas (8–40) and (8–49) for the transition probabilities and cross sections, which were obtained without taking into account the fact that during the transition process the initial state ω_i decays and therefore the effective probability for a process $\omega_i \to \omega_f$ is decreased. In the *exact* solution for the S-matrix this effect would, of course, be automatically contained. In the iteration solution it appears in the higher-order terms. But in this form it is difficult to isolate the effect of radiation damping from the higher-order radiative corrections.

It is therefore desirable to have a formalism in which the damping effect appears in a natural way in the explicit expression for the S-matrix. This formalism is based on the integral equation of Heitler, which was derived in Section 7–5, Eq. (7–85). Instead of working with the iteration solution of S we can develop an iteration solution for the operator K defined in Eq. (7–47), and then calculate the S-operator in the form (7–83)

$$S = \frac{1 - (i/2)K}{1 + (i/2)K}. \tag{16-53}$$

The iteration solution for K is easily obtained from the operator

$$G_0(\tau) = e^{iH_0\tau} G_0 e^{-iH_0\tau}. \tag{16-54}$$

The matrix element $G_0(q,q')$ of the operator G_0 is the solution of the integral equation $(7\text{–}35)_0$. The operator $G_0(\tau)$ is the operator G_0 in the interaction picture. From the definition (7–47) of the operator K and Eq. (16–54) follows immediately

$$K = \int_{-\infty}^{+\infty} G_0(\tau) d\tau. \tag{16-55}$$

We can obtain the iteration solution for K by transforming the integral equation (7–35) first into one for the τ-dependent operator $G(\tau)$. To this end, we make use of the formula

$$P \frac{e^{-i\omega\tau}}{\omega} = \frac{i}{2} \int_{-\infty}^{+\infty} \epsilon(\tau' - \tau) e^{-i\omega\tau'} d\tau',$$

where the left side is to be understood as the principal value. With $\omega = E - E''$, Eq. $(7\text{–}35)_0$ becomes, in the interaction picture,

$$G_0(\tau) = H(\tau) \left[1 - \frac{i}{2} \int_{-\infty}^{+\infty} \epsilon(\tau - \tau') G_0(\tau') d\tau' \right]. \tag{16-56}$$

This equation is now easily iterated and we find, with Eq. (16–55),

$$K = \sum_{n=1}^{\infty} K_n, \tag{16-57}$$

$$K_n = \left(-\frac{i}{2}\right)^{n-1} \int_{-\infty}^{+\infty} \cdots \int_{-\infty}^{+\infty} d\tau_n \cdots d\tau_1 \epsilon(\tau_n, \tau_{n-1}) \cdots \epsilon(\tau_2, \tau_1)$$
$$\times H(\tau_n) \cdots H(\tau_1), \tag{16-58}$$

$$K_1 = \int_{-\infty}^{+\infty} H(\tau) d\tau.$$

Although this iteration solution is strikingly similar in appearance to the iteration solution for the S-operator [cf. Eqs. (8–7) and (8–8)], the algebraic structure is quite different. The occurrence of the ϵ-factors in Eq. (16–58) prevents us from extracting terms for a definite matrix element with the ordering procedure used for the S-operator.

It is possible, however, to reduce the evaluation of the operators K_n to the evaluation of the successive approximations S_m ($m \leq n$) for the S-operator.†

$$S = \sum_{m=0}^{\infty} S_m.$$

Indeed, by a suitable algebraic manipulation, one can derive a recursion relation for the operators K_n. The simplest way to obtain such relations is to start with Eq. (16–53) and substitute in it the respective power series developments for S and K. The resultant equation may be written as

$$\left(\sum_{m=0}^{\infty} S_m\right)\left(1 + \frac{i}{2}\sum_{n=1}^{\infty} K_n\right) = 1 - i\sum_{n=1}^{\infty} K_n. \tag{16-59}$$

By equating the same powers of e in this relation, we find the recursion relations

$$K_1 = iS_1,$$
$$K_n = i\left(S_n + \frac{i}{2}\sum_{\nu=1}^{n} S_{n-\nu}K_\nu\right). \tag{16-60}$$

It is possible to express the K_n in terms of the anti-Hermitian parts of S_n because all K_n are Hermitian. Thus, if we define

$$A_n = \frac{1}{2}i(S_n{}^* - S_n),$$

the following relations are a consequence of (16–60):‡

$$K_n = -A_n - \frac{1}{4}\sum_{\substack{\mu>0,\nu>0 \\ \mu+\nu<n-1}} A_{n-\mu-\nu}K_\mu K_\nu. \tag{16-61}$$

In the absence of external fields, $S_1 = 0$. In this case, the relations (16–61) give for $n \leq 5$

$$K_n = -A_n.$$

---

† Such reductions were given by N. Fukuda and T. Miyazima, *Progr. Theor. Phys.* **5**, 849 (1950); S. N. Gupta, *Proc. Cambridge Phil. Soc.* **47**, 454 (1951); J. Pirenne, *Phys. Rev.* **86**, 395 (1952).

‡ For the derivation, see J. Pirenne, *loc. cit.*, Eq. (34).

In these successive approximations for K_n there will occur infinities, just as in the expressions for S_n. These infinities must be eliminated by the method of mass and charge renormalization, as explained in Chapters 9 and 10. Only after this step do we have an expression for K which lends itself to physical interpretation.*

In order to calculate $S = 1 + R$ from the operator K, it is necessary to solve the Heitler integral equation (7–85)†

$$iR = K + \tfrac{1}{2}KR.$$

This equation is difficult to solve in general. In the cases where damping effects are important it is not sufficient to solve it by an iteration method. Fortunately, it can be shown that the damping effects are always negligibly small in quantum electrodynamics and are probably unobservable.‡

This fortuitous result makes it possible to calculate all transition processes in this theory with complete disregard of the damping effect. The damping theory is, however, of great theoretical importance and in certain meson theories with gradient couplings it is a very strong effect.§

16–3 The natural line width of stationary states. The problem of the natural width of spectral lines was first discussed in a fundamental paper by Weisskopf and Wigner.¶ This elementary theory gives the correct shape of spectral lines to a high degree of approximation.

However, when such a calculation is carried out consistently, it leads not only to a broadening of the line associated with the decay of the initial state but also to a shift of the line. This is precisely the electromagnetic level shift discussed in Section 15–4. In the older treatments on the subject this level shift is usually omitted with little comment.

We shall here outline a treatment of the level broadening which restores its intimate relation to the level shift. This treatment has several advantages. First, the calculation of the level broadening can be directly reduced to the calculation of the S-matrix in the bound interaction picture. It is

---

* Before the advent of the renormalization theory Heitler used a subtraction procedure which proved to be too drastic, inasmuch as the classical limit of this mutilated theory does not exist. See H. A. Bethe and J. R. Oppenheimer, *Phys. Rev.* **70**, 451 (1946).

† This kind of equation was first obtained and discussed for a special case by I. Waller, *Z. Physik* **88**, 436 (1934). In connection with damping theory in field theory, the equation was used by W. Heitler, *Proc. Cambridge Phil. Soc.* **37**, 291 (1941); A. H. Wilson, *ibid.*, 301 (1941); and E. Gora, *Z. Physik* **120**, 121 (1943).

‡ A. H. Wilson, *loc. cit.*

§ See W. Heitler and H. W. Peng, *Proc. Roy. Ir. Acad.* **7**, 101 (1943).

¶ V. Weisskopf and E. Wigner, *Z. Physik* **63**, 54 (1930); *ibid.* **65**, 18 (1930).

therefore possible to extend the powerful methods for the calculation of the S-operator also to this problem. Second, the calculation is covariant in character and the renormalization procedure is easily carried out, although no renormalization is necessary in lowest order. Finally, higher-order radiative corrections can be calculated in the same way as the corrections for the level shift.*

This method of obtaining the line broadening is based on the observation that the wave function associated with a decaying stationary state has a time dependence of the form

$$\omega \sim e^{-iE't},$$

where E' is a complex number:

$$E' = E - i\frac{\Gamma}{2} = E_0 + \Delta E'.$$

The real part E is the sum of the unperturbed energy level, E_0 say, and the level shift ΔE:

$$E = E_0 + \Delta E.$$

The imaginary part, $-i(\Gamma/2)$, gives the exponential decay of the stationary state:

$$e^{-iE't} = e^{-(\Gamma/2)t}e^{-iEt}.$$

This Γ is the line-width parameter which we wish to calculate. The average lifetime of the state is $\tau = 1/\Gamma$.

This imaginary part of E' arises in a natural way when one calculates radiative corrections to the Dirac energy levels. In fact, it arose in our calculation of the radiative level shift in hydrogen-like atoms, where we omitted it in the later parts of the calculations in order to concentrate on the level shift proper. This imaginary part of

$$\Delta E' = \Delta E - i\frac{\Gamma}{2}$$

originates in the δ-function part of the δ_+-function associated with the electron propagator S_c. Let us consider this term now. For this purpose, we turn to the equations which precede Eq. (15–56). We can drop the

---

* Radiative corrections to the line shape were studied by W. Heitler and S. T. Ma, *Proc. Roy. Ir. Acad.* **52**, 109 (1949); E. Arnous and S. Zienau, *Helv. Phys. Acta* **24**, 279 (1951); F. Low, *Phys. Rev.* **88**, 53 (1952).

small photon mass λ, since there is no infrared divergence in this part of the integral. Thus, we find, with $\omega_{0n} = E_0 - E_n$,

$$\Gamma = \frac{4\alpha}{3}\frac{1}{m^2} \sum_n (\omega_{0n})^2 |(n|\mathbf{p}|0)|^2 \int_0^\infty \frac{d|\mathbf{k}|}{|\mathbf{k}|} \delta(\omega_{n0} + |\mathbf{k}|),$$

$$\Gamma = \frac{4\alpha}{3}\frac{1}{m^2} \sum_{\substack{n \\ E_n < E_0}} \omega_{0n} |(n|\mathbf{p}|0)|^2. \tag{16-62}$$

This expression can be brought into a more familiar form by using the relation

$$i\omega_{n0}(n|\mathbf{x}|0) = \frac{1}{m}(n|\mathbf{p}|0),$$

$$\Gamma = \frac{4\alpha}{3} \sum_{\substack{n \\ E_n < E_0}} (\omega_{0n})^3 |(n|\mathbf{x}|0)|^2. \tag{16-63}$$

This is precisely the total transition probability for the transitions from the state E_0 to all other states with $E_n < E_0$.

The line shape for the emitted radiation in the transition $E_0 \rightarrow E_n$ is obtained by calculating the probability of finding the system in the state E_n with one photon present:

$$P_{0 \rightarrow n} = \frac{e^2}{(2\pi)^3} \left| \int \bar{u}_n(\mathbf{x}) e u_0(\mathbf{x}) d^3x \right|^2 \left| \frac{e^{[i(\omega - \omega_{0n}) - \frac{1}{2}\Gamma]t} - 1}{i(\omega - \omega_{0n}) - \frac{1}{2}\Gamma} \right|.$$

For times sufficiently large to make $\Gamma t \gg 1$, $P_{0 \rightarrow n}$ becomes proportional to the well-known resonance denominator including damping:

$$P_{0 \rightarrow n} \simeq \frac{1}{(\omega - \omega_{0n})^2 + \frac{1}{4}\Gamma^2}. \tag{16-64}$$

16–4 The self-stress of the electron. The problem of the self-stress in quantum electrodynamics is a pseudoproblem which the theory inherited from its classical ancestor, the Lorentz theory of the electron.* It survived at first because of the noncovariant formalism employed, and later because of the remaining ambiguities which the infinite self-energy integrals seem to offer. The consistent handling of these integrals, as it has been carried out throughout this work (cf. especially Chapters 9 and 10, and Appendix

---

* H. A. Lorentz, *The Theory of the Electron*, Teubner, 1916. A brief historical survey of the problems of electron theory can be found in *Developments in the Theory of Electrons*, by A. Pais, Institute for Advanced Study and Princeton University, 1948.

5) eliminates this problem completely. As we shall see, it can be considered, at best, as a check on the internal consistency of the formalism.

Consider the total momentum operator P^μ of the system. It can be written as the integral over the plane σ with normal vector $d\sigma^\nu = n^\nu d\sigma$:

$$P^\mu = \int \Theta^{\mu\nu} d\sigma_\nu,$$

where $\Theta^{\mu\nu}$ is the symmetrical momentum tensor of the system. We are concerned with the transformation properties of the expectation values of this operator for states which represent an electron of definite momentum **p** and energy ϵ. For simplicity, we assume that the electron moves in the 1-direction such that

$$|\mathbf{p}| = \langle P^1 \rangle, \quad \epsilon = \langle P^0 \rangle. \tag{16-65}$$

At this point we remind the reader that P^μ transforms like a four-vector provided $\partial_\mu \Theta^{\mu\nu} = 0$. In fact, this transformation property of P^μ is equivalent to the conservation law of energy and momentum.

We imagine now two observers O and O', and assume that the electron is moving with momentum **p** in the frame of O' and is at rest in O. The question is whether the values of **p** and ϵ calculated by the two observers are indeed connected by a Lorentz transformation.

The transformation law which connects the coordinates in the two reference systems, using $x^1 = x$, $x^0 = t$, is

$$x' = \gamma(x + \beta t), \quad t' = \gamma(t + \beta x). \tag{16-66}$$

The transformation law for the expectation values of the tensor components $\Theta^{\mu\nu}$ simplifies because, in the rest system,

$$\langle \Theta^{\mu\nu} \rangle = 0, \quad \text{for } \mu \neq \nu,$$

and $\tag{16-67}$

$$\langle \Theta^{\mu\nu}(\mathbf{x},t_1) \rangle = \langle \Theta^{\mu\nu}(\mathbf{x},t_2) \rangle.$$

We then find, by applying the transformation law for the tensor field $\Theta^{\mu\nu}(x)$,

$$\langle \Theta'^{00}(x') \rangle = \gamma^2 \langle \Theta^{00}(x) + \beta^2 \Theta^{11}(x) \rangle,$$

$$\langle \Theta'^{10}(x') \rangle = \gamma^2 \beta \langle \Theta^{00}(x) + \Theta^{11}(x) \rangle.$$

Consequently,

$$\epsilon = \int \langle \Theta'^{00}(x') \rangle d^3x' = \gamma^2 \int \langle \Theta^{00}(x) + \beta^2 \Theta^{11}(x) \rangle d^3x',$$

$$|\mathbf{p}| = \int \langle \Theta'^{10}(x') \rangle d^3x' = \beta\gamma^2 \int \langle \Theta^{00}(x) + \Theta^{11}(x) \rangle d^3x'.$$

The integrands on the right side are to be taken on the three-dimensional hyperplane which is obtained from the plane $t' = 0$ by the Lorentz transformation (16–66). Because of the second of equations (16–67), the integration can be extended to the hyperplane in x'-space determined by the equation $t = 0$. Since it follows from the first of Eqs. (16–66) that $x = (1/\gamma)x'$ for this plane, the three-dimensional volume element transforms according to

$$d^3x' = \frac{1}{\gamma} d^3x.$$

Thus, we find

$$\epsilon = \gamma \int_{t=0} \langle \Theta^{00}(x) + \beta^2 \Theta^{11}(x) \rangle d^3x,$$

$$|\mathbf{p}| = \beta\gamma \int_{t=0} \langle \Theta^{00}(x) + \Theta^{11}(x) \rangle d^3x.$$

(16–68)

Now the energy ϵ and the momentum $\mathbf{p}$ of the electron must, for all β, satisfy the relation

$$|\mathbf{p}| = \beta\epsilon.$$

Therefore, the expectation value of the quantity

$$S(0) = \int_{t=0} \Theta^{11}(x) d^3x \qquad (16\text{–}69)$$

must vanish. This quantity is called the *self-stress* operator.

Because of the rotational symmetry in the rest system,

$$\Theta_1{}^1 = \Theta_2{}^2 = \Theta_3{}^3 = \tfrac{1}{3}(\Theta_\mu{}^\mu - \Theta_0{}^0).$$

The symmetrical momentum tensor $\Theta^{\mu\nu}$ is defined by Eqs. (1–98), (1–99), and (1–97). For the interacting fields the Lagrangians are given in Eqs. (4–10) to (4–14). Inserting these into the definition for $\Theta^{\mu\nu}$, we find

$$\Theta^{\mu\nu} = -\tfrac{1}{4}[\overline{\Psi}(\gamma^\mu D^\nu + \gamma^\nu D^\mu)\Psi - \overline{\Psi}(\gamma^\mu \overleftarrow{D}^\nu + \gamma^\nu \overleftarrow{D}^\mu)\Psi] + \Theta_\text{rad}{}^{\mu\nu} \quad (16\text{–}70)$$

where

$$D^\mu = \partial^\mu + ieA^\mu, \quad \overleftarrow{D}^\mu = \overleftarrow{\partial}^\mu - ieA^\mu,$$

and $\Theta_\text{rad}{}^{\mu\nu}$ is the momentum tensor of the free radiation field. Since the trace of the latter vanishes, we find from the field equations (4–15) and (4–16)

$$\Theta_\mu{}^\mu = -\tfrac{1}{2}\overline{\Psi}(\gamma_\mu D^\mu - \gamma_\mu \overleftarrow{D}^\mu)\Psi = m\overline{\Psi}\Psi. \qquad (16\text{–}71)$$

The operator

$$H = -n_\mu P^\mu = -\int n_\mu \Theta^{\mu\nu} d\sigma_\nu$$

reduces to
$$H = P^0 = -\int \Theta^{00} d^3x = \int \Theta_0{}^0 d^3x \qquad (16\text{-}72)$$
for an electron at rest.

An explicit expression for H can easily be obtained from $\Theta^{\mu\nu}$, Eq. (16-70). It is of the form
$$H = m \int \overline{\Psi}\Psi d^3x + H',$$
where H' does not depend on m explicitly. [Cf. Eq. (14-57), which shows the same separation.] Combining this result with Eq. (16-71), we find
$$\int \Theta_\mu{}^\mu d^3x = m \frac{\partial H}{\partial m}, \qquad (16\text{-}73)$$
so that the self-stress operator for (16-69) can be written in the form
$$3S(0) = m \frac{\partial H}{\partial m} - H. \qquad (16\text{-}74)$$

This elegant equation was first suggested by Pais and Epstein.* Unfortunately, the rest of their work is incorrect and their arguments are misleading, as we shall see below.

The expectation value of H for a *physical* electron at rest is the experimental mass
$$m_0 = m + \delta m,$$
where the δm are the contributions to the rest energy which arise from the surrounding field [*cf.* Eq. (10-40)]. Therefore, the expectation value of the self-stress operator in this case is
$$3\langle S(0)\rangle = m \frac{\partial \delta m}{\partial m} - \delta m = m^2 \frac{\partial}{\partial m}\left(\frac{\delta m}{m}\right). \qquad (16\text{-}75)$$

We conclude that the necessary and sufficient condition for the vanishing of this expression is that
$$\delta m = \text{const. } m. \qquad (16\text{-}76)$$

If this equation were not fulfilled, the self-stress of an electron at rest would not vanish and there would be an inconsistency in the transformation properties of the momentum-energy four-vector of an electron in interaction with its own field. It would be equivalent to a lack of energy-momentum conservation.

---

* A. Pais and S. T. Epstein, *Rev. Mod. Phys.* **21**, 445 (1949).

Pais and Epstein (*loc. cit.*) use the second-order expression for δm [*cf.* Eqs. (9–17) and (9–28)] in the form of Eq. (9–21). They argue that the upper cutoff M is a constant independent of m, and that therefore Eq. (16–76) is not fulfilled because of the logarithmic term, leading to a nonvanishing self-stress. The simplest though not the most satisfactory argument against this calculation is to observe that the theory contains only one mass m and that the cutoff value M is necessarily proportional to m, so that the logarithm is independent of m.

The correctness of Eq. (16–76), of course, is basic to the mass renormalization procedure. In fact, the terms A in Σ were originally recognized as mass corrections on the basis of their transformation properties. (*Cf.* Section 9–4.)

The first correct treatment of the self-stress problem* consisted in a proof that the apparently contradictory transformation properties arise entirely from the ambiguities of the infinite integrals, so that any arbitrary cutoff procedure, so long as it is a covariant procedure, will eliminate this difficulty. Since the theory is formally Lorentz invariant, it is evident that only an incorrect treatment of ambiguous expressions can lead to an internal inconsistency regarding transformation properties. As a convenient relativistic cutoff procedure Rohrlich used the method of regulators (*cf.* Section 10–9). This method was originally proposed as a cutoff procedure for the S-matrix (*cf.* Pauli and Villars, *loc. cit.*, in Section 10–9), and in this form is not suitable for the self-stress problem. The reason for this situation lies in the *ad hoc* way in which these auxiliary fields are introduced in the S-matrix. Even though one eventually proceeds to the limit ($M_i \rightarrow \infty$, etc.), these fields affect the energy balance. A *consistent* treatment therefore requires *the introduction of these regulators at the beginning of the theory*, i.e., in the Lagrangian. (The same is true, of course, for any cutoff procedure.) This was carried out first by Rohrlich in connection with the self-stress, and later in greater detail and more completely by Gupta (*loc. cit.*, in Section 10–9). The regularization of the Lagrangian then leads to a regularization of the S-matrix which is identical with the one proposed by Pauli and Villars. However, it would be inconsistent to regularize the S-matrix only and not the Lagrangian, although it does not affect the calculation of cross sections and related problems associated with S-matrix elements only. It then follows that the failure of the Pauli-Villars regularization to yield a vanishing self-stress is completely inconsequential.† The consistent application of a cutoff, i.e., the introduction of it in the Lagrangian, leads to a vanishing self-stress and automatically guarantees that the conservation laws are fulfilled. It is easy to modify Eq. (16–75) for the case of regulators,

---

* F. Rohrlich, *Phys. Rev.* **77**, 357 (1950).

† Compare the argument by S. Borowitz and W. Kohn, *Phys. Rev.* **86**, 985 (1952), Appendix II, which may seem to be misleading concerning this point.

and we find
$$\langle S(0) \rangle = 0 \tag{16-77}$$

with exactly the same conditions on the regulators, which produce a finite electron self-energy [Eqs. (10-74)]. An explicit second-order calculation gives the same result.

Equation (16-75) was obtained for an electron at rest. If we want to be independent of the coordinate system, it is necessary to calculate the radiative corrections to $\Theta^{\mu\nu}$ explicitly. A second-order calculation of this type was carried out by Villars.* He also uses regulators as relativistic cutoff and shows—to second order—that the self-stress causes no difficulties.

A very different method was adopted by Borowitz and Kohn (*loc. cit.*). They calculated the second-order radiative corrections to $\Theta^{\mu\nu}$ without any cutoff and determined the value of certain infinite integrals by the requirement that energy-momentum conservation,

$$\partial_\mu \Theta^{\mu\nu} = 0, \tag{16-78}$$

must be satisfied throughout. This procedure is very similar to the method used in Section 9-5 for the photon self-energy. There the undetermined integral $D(0)$ (Eq. 9-50) must be chosen to have the value zero in order to preserve gauge invariance. The advantage of this method lies in its independence of any cutoff procedure. It shows that it is possible to choose the undetermined integrals in accordance with the invariance properties and conservation laws. Therefore, it proves something quite different from the previously discussed methods. In these it was shown that there is no problem when the theory is made finite and that no arbitrariness remains, since the result is cutoff independent. The cutoff procedure may thus be regarded as a guide for the consistent handling of the infinite integrals. This guide is much more general than the explicit use of Eq. (16-78), since the same guide is able to eliminate all ambiguities at the same time and is based only on the invariance properties of the cutoff itself.

16-5 Outlook. We hope that the reader who has followed us to this point will retain two major aspects of the strange situation in which the theory of photons and electrons finds itself at the present time.

With respect to the applications, i.e., the actual description of physical quantities, we have here one of the best-established physical theories. Whenever the theory is subjected to an experimental test we find the theoretical prediction in complete agreement with the experimental result. The accuracy in the agreement is limited only by the experimental error and the endurance and ingenuity of the computer.

---

* F. Villars, *Phys. Rev.* **79**, 122 (1950).

With respect to the fundamental concepts, on the contrary, we are not so fortunate. The theory is incomplete insofar as we are forced to introduce the charge of the electron and the masses of electron and photon as phenomenological quantities. The value of these quantities cannot be deduced from the theory but must be accepted as empirically given. This point of view enabled us to carry through the program of renormalization which was essential for the removal of the divergences in the iteration solution.

In the light of this situation, we see the following possibilities for future developments of this theory.

A number of improvements need to be made in the calculation of cross sections, energy levels, and lifetimes. These are essentially technical in character. Improved approximations, especially with respect to Coulomb wave functions and high-energy limits, are necessary. In this connection the modern high-speed electronic computing machine will no doubt be of increasing importance. These theoretical developments will to some extent go hand in hand with the improvement of experimental accuracies and techniques.

Quite independent of this computational technique are the questions of the theoretical framework and the basic concepts. We have here a quite paradoxical situation. No matter how profoundly our concepts and theoretical structure are changed in the future, the verifiable results will hardly be affected.

We can only speculate about the kind of changes we may expect in the future. The restriction to one type of interaction between two kinds of elementary particles is an enormous abstraction from the actual physical world. There exists an alarmingly increasing number of so-called elementary particles and they have many different kinds of interactions. We considered it quite justifiable to concentrate on photons and electrons and ignore the rest of the elementary particles because there are a large number of processes for which the existence of other particles is irrelevant. But the question always looms in the background whether, in so doing, we may not have omitted a significant element in the basic concepts. The failure of a realistic regulator theory is somewhat reassuring in this respect, but the question cannot be regarded as settled so long as infinite renormalization constants occur in the theory.

Other questions arise in connection with the theory of the S-operator. The only method known for calculating this operator is the iteration solution. This forces upon us a very unphysical picture of a charged particle. The radiation field accompanying the particle is first removed, leaving a *bare* electron, and then it is added again, photon by photon. This leads to the separation of every physical process into a "lowest-order" process and its radiative corrections. The essentially unphysical nature of this procedure is glaringly exemplified in the problem of the infrared divergences.

The solution of this problem presented here gives considerable insight into that question.

Finally, we have felt that the theoretical physicist of the future would be greatly helped by a deeper understanding of many mathematical problems. We have found "solutions" to equations, guided by physical intuition, without knowing whether such solutions actually exist in a strict mathematical sense. We have used iteration processes and power series without the slightest idea about the convergence of these infinite processes. It is extremely unfortunate that some of the important mathematical problems we encounter here are apparently beyond the scope of knowledge of our mathematical contemporaries. We see here a widening field of research for the future.

MATHEMATICAL APPENDIX

APPENDIX A1

THE INVARIANT FUNCTIONS

In the course of the development of the theory, we found it necessary to introduce various invariant functions of the four-vector x. For convenience, we summarize here the properties of these functions. We shall denote these functions* by $\Delta_i(x)$.

All $\Delta_i(x)$-functions which occur in the quantum theory of fields can be reduced to one standard integral representation of the form

$$\Delta_i = \frac{1}{(2\pi)^4} \int_{C_i} \frac{e^{ipx}}{p^2 + m^2} d^4p. \tag{A1-1}$$

The different types of functions are obtained by the appropriate choice of the path of integration, C_i, in the complex p^0-plane. The integration over the variable p^0 is to be carried out first. The poles at the points

$$p^0 = \pm \epsilon, \quad \epsilon = \sqrt{p_1^2 + p_2^2 + p_3^2 + m^2},$$

are avoided by the paths C_i. If the entire path C_i is located in a finite portion of the p^0-plane, we call the corresponding $\Delta_i(x)$ *homogeneous*. In this case $\Delta_i(x)$ satisfies the homogeneous wave equation

$$(\partial_\mu \partial^\mu - m^2) \Delta_i(x) = 0. \tag{A1-2}$$

On the other hand, if C_i passes through the point $p^0 = \infty$, the function $\Delta_i(x)$ satisfies the *in*homogeneous wave equation

$$(\partial_\mu \partial^\mu - m^2) \Delta_i(x) = -\delta(x), \tag{A1-3}$$

and $\Delta_i(x)$ is called *inhomogeneous*.

Since the case of photons ($m = 0$) is of special importance, we shall denote by $D_i(x)$ the limit $m \to 0$ of the $\Delta_i(x)$-functions.

---

* The first invariant Δ_i-function was introduced by P. Jordan and W. Pauli, *Z. Physik* **47**, 151 (1928).

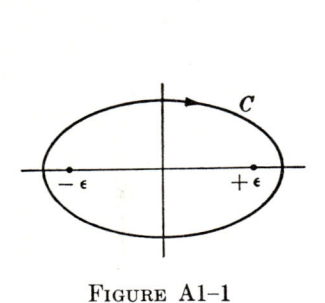

FIGURE A1–1

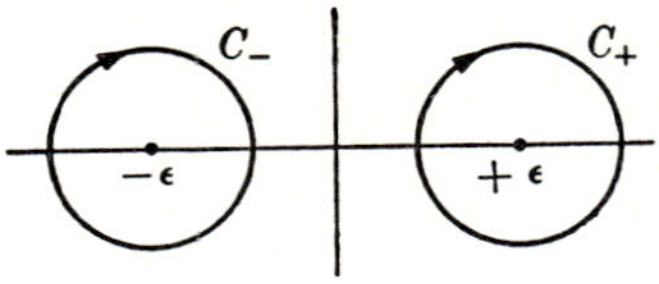

FIGURE A1–2

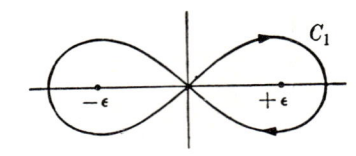

FIGURE A1–3

A1–1 The homogeneous delta-functions. The paths C, C_+, C_-, and C_1 shown in Figs. A1–1, A1–2, and A1–3 define the four functions $\Delta(x)$, $\Delta_+(x)$, $\Delta_-(x)$, and $i\Delta_1(x)$. Note the factor i in the definition of $\Delta_1(x)$ which was inserted in order to conform with the usual definition of this function.

A1–2 The inhomogeneous delta-functions. We define the five functions $\Delta_R(x)$, $\Delta_A(x)$, $\Delta_{1R}(x)$, $\Delta_{1A}(x)$, and $\Delta_P(x)$ by the five paths C_R, C_A, C_{1R}, C_{1A}, and C_P shown in Figs. A1–4, A1–5, and A1–6. The path C_P is identical with the Cauchy principal value of the integral.

The notation for the various Δ-functions is somewhat confused in the literature. In particular, the function Δ_P is often denoted by $\bar{\Delta}$ [J. Schwinger, *Phys. Rev.* **75**, 651 (1949)], the function Δ_{1R} by Δ_c [Stueckelberg and Rivier; see D. Rivier, *Helv. Phys. Acta* **22**, 265 (1949)], and also by $\frac{1}{2}i\Delta_F$ [F. J. Dyson, *Phys. Rev.* **75**, 486 (1949)]. Because of its special importance, we shall also denote the function Δ_{1R} by the somewhat simpler symbol Δ_c:

$$\Delta_{1R}(x) \equiv \Delta_c(x), \qquad (A1-4)$$

which stresses the causal nature of this function. Correspondingly, we put $D_{1R}(x) \equiv D_c(x)$.

By "causal nature" we refer to the boundary conditions associated with Eq. (A1–3), which give Δ_c. Clearly, the Δ-functions can also be defined by the differential equations (A1–2) and (A1–3) with appropriate boundary conditions. For example, the function Δ can be defined as the solution of Eq. (A1–2) with the following boundary conditions:

$$\Delta(x)|_{x^0=0} = 0,$$
$$\partial_\mu \Delta(x)|_{x^0=0} = \delta_\mu(x), \qquad (A1-5)$$

FIGURE A1-4

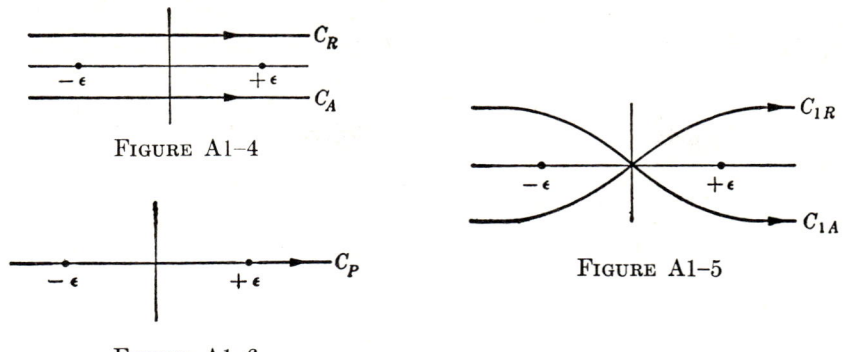

FIGURE A1-5

FIGURE A1-6

where $\delta_\mu(x)$ is the *surface δ-function* with the property that for any integrable function $f(x)$ on a space-like plane σ,

$$\int f(x)\delta_\mu(x)d\sigma^\mu = f(0). \tag{A1-6}$$

A1-3 Relations between the Δ-functions. From the definitions given in Sections A1-1 and A1-2, we can immediately obtain a number of relations which may be classified according to the following subheadings:

Linear relations

$$\Delta = \Delta_+ + \Delta_- = \Delta_R - \Delta_A,$$
$$i\Delta_1 = \Delta_+ - \Delta_- = \Delta_{1R} - \Delta_{1A}, \tag{A1-7}$$
$$\Delta_P = \tfrac{1}{2}(\Delta_R + \Delta_A) = \tfrac{1}{2}(\Delta_{1R} + \Delta_{1A}).$$

$$\Delta_{1R} + \Delta_- = \Delta_R, \quad \Delta_{1A} - \Delta_- = \Delta_A. \tag{A1-8}$$

$$\Delta_{1R} - \Delta_+ = \Delta_A, \quad \Delta_{1A} + \Delta_+ = \Delta_R. \tag{A1-9}$$

$$\Delta_P - \tfrac{1}{2}\Delta = \Delta_A, \quad \Delta_P + \tfrac{1}{2}\Delta = \Delta_R,$$
$$\tag{A1-10}$$
$$\Delta_P - \frac{i}{2}\Delta_1 = \Delta_{1A}, \quad \Delta_P + \frac{i}{2}\Delta_1 = \Delta_{1R}.$$

The last four relations show that Δ_{1R} and Δ_{1A} are related to $i\Delta_1$ as Δ_R, Δ_A are related to Δ. This justifies the notation used here.

Actually, only six of these linear relations are independent of each other. All Δ-functions can thus be linearly combined from three basic ones, for instance Δ_P, Δ_1, Δ.

Reality conditions. By taking the complex conjugate of the various integrals and expressing the result again in terms of the original integrals, we obtain the relations

$$\Delta_+^* = \Delta_-, \quad \Delta_R^* = \Delta_R,$$
$$\Delta_A^* = \Delta_A, \quad \Delta_{1R}^* = \Delta_{1A}. \tag{A1-11}$$

Symmetry properties. By replacing x by $-x$ and expressing the result again in terms of the original integrals, we find the symmetry relations

$$\Delta(-x) = -\Delta(x), \quad \Delta_1(-x) = \Delta_1(x), \quad \Delta_P(-x) = \Delta_P(x),$$
$$\Delta_R(-x) = \Delta_A(x), \quad \Delta_{1R}(-x) = \Delta_{1R}(x), \tag{A1-12}$$
$$\Delta_{1A}(-x) = \Delta_{1A}(x), \quad \Delta_+(-x) = -\Delta_-(x).$$

Relations involving step functions. We define the two step functions

$$\theta(x) = \begin{cases} 1 & x^0 > 0 \\ \frac{1}{2} & x^0 = 0, \\ 0 & x^0 < 0, \end{cases} \tag{A1-13}$$

$$\varepsilon(x) = 2\theta(x) - 1, \tag{A1-14}$$

which satisfy the relations (for all $x^0 \neq 0$)

$$\theta^2(x) = \theta(x), \quad \theta(x)\theta(-x) = 0,$$
$$\theta(x) + \theta(-x) = 1, \quad \varepsilon^2(x) = 1. \tag{A1-15}$$

Multiplication of a homogeneous Δ-function by one of the functions θ or ε changes it into an inhomogeneous function, and the following relations are obtained:

$$\Delta_R(x) = \theta(x)\Delta(x) = 2\theta(x)\Delta_P(x)$$
$$\Delta_A(x) = -\theta(-x)\Delta(x) = 2\theta(-x)\Delta_P(x),$$
$$\theta(x)\Delta_-(x) = \theta(x)\Delta_{1A}(x),$$
$$\theta(x)\Delta_+(x) = \theta(x)\Delta_{1R}(x), \tag{A1-16}$$
$$\Delta_{1R}(x) = \theta(x)\Delta_+(x) - \theta(-x)\Delta_-(x),$$
$$\Delta_{1A}(x) = \theta(x)\Delta_-(x) - \theta(-x)\Delta_+(x),$$
$$\theta(x)\Delta(x) - \theta(-x)\Delta(x) = 2\Delta_P(x),$$
$$\varepsilon(x)\Delta(x) = 2\Delta_P(x).$$

From these formulas follow the relations

$$\Delta_R(x) = 0 \text{ for } x^0 < 0, \quad \Delta_A(x) = 0 \text{ for } x^0 > 0. \tag{A1-17}$$

A1-4 Integral representations.

With the identities

$$f(z_0) = \frac{1}{2\pi i} \oint \frac{f(z)}{z - z_0} dz = \int_{-\infty}^{+\infty} f(z)\delta(z - z_0)dz \qquad (A1\text{-}18)$$

and

$$\delta(z^2 - z_0^2) = \frac{1}{2|z_0|}[\delta(z - z_0) + \delta(z + z_0)], \qquad (A1\text{-}19)$$

we can rewrite the four-dimensional integrals (A1–1) to give the following equivalent expressions for the homogeneous functions:

$$\Delta(x) = \frac{i}{(2\pi)^3} \int e^{ipx} \delta(p^2 + m^2) \varepsilon(p) d^4p, \qquad (A1\text{-}20)$$

$$\Delta_1(x) = \frac{1}{(2\pi)^3} \int e^{ipx} \delta(p^2 + m^2) d^4p, \qquad (A1\text{-}21)$$

$$\Delta_\pm(x) = \frac{\pm i}{2} \frac{1}{(2\pi)^3} \int (1 \pm \varepsilon(p)) e^{ipx} \delta(p^2 + m^2) d^4p. \qquad (A1\text{-}22)$$

The integrals over p^0 can be carried out and we then arrive at the following three-dimensional representations:

$$\Delta(x) = \frac{1}{(2\pi)^3} \int e^{i\mathbf{p}\cdot\mathbf{x}} \frac{\sin \epsilon x^0}{\epsilon} d^3p, \qquad (A1\text{-}23)$$

$$\Delta_1(x) = \frac{1}{(2\pi)^3} \int e^{i\mathbf{p}\cdot\mathbf{x}} \frac{\cos \epsilon x^0}{\epsilon} d^3p, \qquad (A1\text{-}24)$$

where $\epsilon = +\sqrt{\mathbf{p}^2 + m^2}$. Finally, the following one-dimensional integral representations are sometimes useful:*

$$\Delta_P = \frac{1}{8\pi^2} \int_{-\infty}^{+\infty} e^{i[\lambda\alpha + (m^2/4\alpha)]} d\alpha,$$

$$\Delta_1(x) = -\frac{1}{2\pi^2} \int_0^\infty \sin\left(\lambda\alpha + \frac{m^2}{4\alpha}\right) d\alpha, \qquad (A1\text{-}25)$$

$$\Delta_{1R}(x) = \frac{1}{4\pi^2} \int_0^\infty e^{-i[\lambda\alpha + (m^2/4\alpha)]} d\alpha.$$

---

* Cf. J. Schwinger, *Phys. Rev.* **75**, 651 (1949), Appendix. In this paper the function $\Delta(x)$ is defined with its sign opposite to ours.

A1–5 Explicit expressions. The integrals for the functions Δ_P, Δ_1 can be reduced to standard integral representations for cylinder functions, and the following explicit formulas are obtained:

$$\Delta_P(x) = \frac{1}{4\pi}\delta(\lambda) - \begin{cases} \dfrac{1}{8\pi}\dfrac{m}{\lambda^{1/2}} J_1(m\lambda^{1/2}), & \text{for } \lambda > 0, \\ 0, & \text{for } \lambda < 0, \end{cases} \qquad (A1\text{--}26)$$

$$\Delta_1(x) = \frac{m^2}{4\pi} \times \begin{cases} \dfrac{1}{m\lambda^{1/2}} N_1(m\lambda^{1/2}), & \lambda > 0 \\ \dfrac{2}{\pi m(-\lambda)^{1/2}} K_1[m(-\lambda)^{1/2}], & \lambda < 0, \end{cases} \qquad (A1\text{--}27)$$

where $\lambda = -x^2$. By expanding the cylinder functions involved here for small values of λ, we can explicitly exhibit the singularities of these functions at $\lambda = 0$.

$$\Delta_P(x) = \frac{1}{4\pi}\delta(\lambda) - \frac{1}{16\pi}m^2\theta(\lambda) + \cdots,$$

$$\Delta_1(x) = -\frac{1}{2\pi^2\lambda} + \frac{m^2}{4\pi^2}\left(\ln\frac{\gamma m|\lambda|^{1/2}}{2} - \frac{1}{2}\right) + \cdots, \qquad (A1\text{--}28)$$

$$\gamma = 1.781 \cdots.$$

A1–6 The S-functions. In the quantization of the electron waves we encounter the spinor functions S. We define for any path C_i the spinor function S_i by the expression

$$S_i(x) = (\partial - m)\Delta_i(x). \qquad (A1\text{--}29)$$

The spinor function is a 4×4 matrix, as well as a function of x, and satisfies the equation

$$(\partial + m)S_i(x) = \begin{cases} 0, & \text{for } C_i = C, C_+, C_-, C_1, \\ -\delta(x), & \text{for } C_i = C_R, C_A, C_{1R}, C_{1A}, C_P. \end{cases} \qquad (A1\text{--}30)$$

In accordance with Eq. (A1–4), we define

$$S_{1R}(x) \equiv S_c(x).$$

APPENDIX A2

THE GAMMA-MATRICES

A2–1 Various representations.* The γ-matrices are defined by the anticommutation rules

$$\{\gamma_\mu, \gamma_\nu\} = 2g_{\mu\nu} \qquad (A2\text{--}1)$$

and the requirement that they shall be an irreducible set. Two sets of matrices γ_μ, γ_μ' which are connected by an S-transformation

$$\gamma_\mu' = S\gamma_\mu S^{-1} \qquad (A2\text{--}2)$$

are said to be equivalent. It is at once evident that if γ_μ is an irreducible set satisfying (A2–1), then γ_μ' is also such a set. In this sense, we can always obtain families of equivalent representations by letting S go through all nonsingular matrices. The members of these families of matrices may be looked upon as representing the same transformations referred to different coordinate systems.

It is a remarkable feature of these matrices that the conditions (A2–1) define only one such family. In other words, we have the fundamental

THEOREM. *Let γ_μ, γ_μ' ($\mu = 0, \cdots 3$) be two irreducible sets of matrices, satisfying* (A2–1). *Then there exists a nonsingular matrix S such that*

$$\gamma_\mu' = S\gamma_\mu S^{-1}, \qquad (A2\text{--}3)$$

and S is, moreover, unique except for an arbitrary multiplicative factor.

This theorem is a special case of a general theorem on simple matrix algebras.† We shall prove it here by more elementary methods, but we shall also obtain some useful relations in the course of the proof.

We introduce the 16 matrices

$$\gamma_r = 1, \ \gamma_\mu, \ \gamma_\mu\gamma_\nu \ (\mu < \nu), \ \gamma_5\gamma_\mu, \ \gamma_5 \qquad (A2\text{--}4)$$

---

* Most of the content of this and the following section is contained in two papers by W. Pauli, *Zeeman Verhandelingen*, 1935, p. 31–43, and *Ann. Inst. Henri Poincaré* **6,** 109 (1936).

† See, for instance, H. Weyl, *The Classical Groups*, Princeton University Press, 1939, Chapter III, especially Section 3, p. 87 ff.

which are subdivided into the five sets Γ_t ($t = 1, \cdots, 5$):

$$\Gamma_1 = 1,$$
$$\Gamma_2 = \gamma_\mu, \quad (\mu = 0, \cdots, 3),$$
$$\Gamma_3 = \gamma_\mu \gamma_\nu, \quad (\mu < \nu), \tag{A2-5}$$
$$\Gamma_4 = \gamma_5 \gamma_\mu,$$
$$\Gamma_5 = \gamma_5 \equiv \gamma_0 \gamma_1 \gamma_2 \gamma_3.$$

Thus, the sets Γ_1 and Γ_5 contain one matrix each, Γ_2 and Γ_4 contain four matrices each, and Γ_3 contains six matrices. If γ^r is defined in a manner analogous to (A2-4) except that the γ_μ are replaced by $\gamma^\mu = g^{\mu\nu}\gamma_\nu$, we may write, because of (A2-1),

$$\gamma_r \gamma^r = \gamma^r \gamma_r = \xi_t 1, \quad \text{for } \gamma_r \in \Gamma_t, \quad \text{(not summed over } r\text{)}, \tag{A2-6}$$

where ξ_t is a sign factor depending only on the class t and is given by $+1$ for $t = 1, 2,$ and 5, and by -1 for $t = 3$ and 4.

Lemma. *The trace of all γ_r is zero except for $\gamma_1 = 1$:*

$$\text{Tr } \gamma_r = 0, \quad (r \neq 1). \tag{A2-7}$$

The proof is based on the observation that the trace of a product of two anticommuting matrices is zero:

$$\text{Tr } (AB) = 0, \quad \text{if } AB = -BA. \tag{A2-8}$$

It is therefore sufficient to write every γ_r as a product of two anticommuting matrices. This is already done for Γ_3 and Γ_4. For γ_5, we write $\gamma_5 = \gamma_0(\gamma_1\gamma_2\gamma_3) = -(\gamma_1\gamma_2\gamma_3)\gamma_0$, and for γ_μ we may write, with $\nu \neq \mu$,

$$\gamma_\mu = -\gamma^\nu(\gamma_\mu\gamma_\nu) = (\gamma_\mu\gamma_\nu)\gamma^\nu,$$

where we do not sum over ν. This proves the lemma.

From this lemma follows immediately that the sixteen matrices γ_r are linearly independent. Indeed, a relationship

$$\sum_{r=1}^{16} c_r \gamma_r = 0 \tag{A2-9}$$

would, after multiplication with γ^s and taking the trace, lead to

$$\sum_{r=1}^{16} c_r \text{ Tr } (\gamma_r \gamma^s) = \xi_s c_s n = 0, \tag{A2-10}$$

where n is the dimension of the matrices. Since n and ξ_s are $\neq 0$, it follows that $c_s = 0$ for all s.

Since the maximum number of linearly independent matrices of order n is n^2, it follows further that
$$n \geq 4. \qquad (A2\text{-}11)$$

It is easy to give an explicit representation of the order $n = 4$. We first construct the 2×2 spin matrices defined by

$$\sigma_1 = \begin{pmatrix} 0 & 1 \\ 1 & 0 \end{pmatrix}, \quad \sigma_2 = \begin{pmatrix} 0 & -i \\ i & 0 \end{pmatrix}, \quad \sigma_3 = \begin{pmatrix} 1 & 0 \\ 0 & -1 \end{pmatrix}, \qquad (A2\text{-}12)$$

satisfying the product rules
$$\sigma_k^2 = 1,$$
$$\sigma_k \sigma_l = -\sigma_l \sigma_k = i\sigma_m \qquad (A2\text{-}13)$$

(k, l, m = cyclic permutation of 1, 2, 3).

Then we construct the 4×4 matrices

$$\gamma_i = \begin{pmatrix} 0 & \sigma_i \\ \sigma_i & 0 \end{pmatrix} \; (i = 1, 2, 3), \quad \gamma_0 = i \begin{pmatrix} 1 & 0 \\ 0 & -1 \end{pmatrix}. \qquad (A2\text{-}14)$$

We verify by the use of (A2–13) that the γ_μ defined by (A2–14) actually satisfy the commutation rules (A2–1). We also know from (A2–11) that the representation is irreducible, since $n = 4$ is the smallest possible order of the matrices. The representation (A2–14) will be referred to as the *standard representation*.

Although (A2–14) is useful for the purpose of demonstrating the existence of at least one representation, it is never needed in actual calculations, since all the results of physical content are independent of the representation and can ultimately be expressed in terms of invariant traces of products of the γ's.

The proof of the main theorem is greatly facilitated by the powerful lemma of Schur* which, for our purpose, may be formulated as follows.

Let γ_r, γ_r' be two irreducible representations of degree n, n' ($n \leq n'$), and let S be a matrix with n' rows and n columns which connects the two representations by

$$\gamma_r' S = S \gamma_r. \qquad (A2\text{-}15)$$

Then S is either the null matrix (the matrix which consists only of zeros) or it is nonsingular. In the latter case, $n = n'$.

To prove this lemma, we start with the remark that the matrices γ_r, γ_r' induce linear transformations in the linear vector spaces P and P' of dimensions n and n', respectively. Since the representations are assumed irre-

---

* I. Schur, *Neue Begründung der Theorie der Gruppencharaktere*, Sitzungsber. Preuss. Akad., 1905, p. 406.

ducible, the only invariant subspaces are the two trivial ones: the space consisting of the null vector and the whole space.

The matrix S maps the space P into a subspace P_1 of P':

$$P \to P_1 = S(P) \subset P'. \tag{A2–16}$$

where P_1 is an invariant subspace of P', since, by (A2–15),

$$\gamma_r'(P_1) = \gamma_r'S(P) = S\gamma_r(P) = S(P) = P_1. \tag{A2–17}$$

Thus P_1 is either 0 or P'. In the former case, S is the null matrix, and in the latter $n = n'$ and S has an inverse. This concludes the proof of Schur's lemma.

As a special consequence of this lemma, we note that if a matrix S commutes with all γ_μ it is a multiple of the unit matrix I. Indeed, if S commutes with all γ_μ, so does $S - \lambda I$. Now choose λ such that

$$\text{Det } |S - \lambda I| = 0.$$

Then, according to Schur's lemma, the matrix $S - \lambda I$, being singular, must be zero. Thus,

$$S = \lambda I,$$

as asserted.

The last step in the proof of the main theorem stated at the beginning of this section now simply consists in constructing explicitly a matrix S connecting two arbitrary irreducible representations. This can be done in the following manner.

Let F be an arbitrary matrix of n' rows and n columns and define

$$S = \sum_{r=1}^{16} \gamma_r' F \gamma^r \xi_r. \tag{A2–18}$$

where ξ_r is the sign factor introduced in (A2–6). Multiply (A2–18) from the left by γ_s' (s fixed) and from the right by γ_s:

$$\gamma_s' S = \sum_{r=1}^{16} \gamma_s' \gamma_r' F \gamma^r \xi_r, \tag{A2–19}$$

$$S \gamma_s = \sum_{r=1}^{16} \gamma_r' F \gamma^r \xi_r \gamma_s. \tag{A2–20}$$

Now it follows from the anticommutation rules that

$$\gamma_s' \gamma_r' = \epsilon_{sr} \gamma_t', \tag{A2–21}$$

where ϵ_{sr} is ± 1 and is independent of the representation. Furthermore,

$$\gamma^t \xi_t = \gamma_t^{-1} = \epsilon_{sr} \gamma_r^{-1} \gamma_s^{-1} = \epsilon_{sr} \xi_r \gamma^r \gamma_s^{-1}, \quad \text{(no summations)}. \tag{A2–22}$$

In the first equation, (A2–19), we substitute (A2–21) and sum over t instead of r, and in the second, (A2–20), we substitute (A2–22) after changing the summation index to t. In this way, we obtain

$$\gamma_s'S = \sum_{t=1}^{16} \epsilon_{sr}\gamma_t' F\gamma^r \xi_r = S\gamma_s. \tag{A2–23}$$

Finally, since F was arbitrary and the γ_r are a linearly independent set, we can always choose F such that at least one matrix element of S does not vanish (for instance, by taking all but one matrix element of F equal to zero). The conditions of Schur's lemma are thus satisfied, leading to

$$\gamma_r' = S\gamma_r S^{-1} \tag{A2–24}$$

and, therefore, to Eq. (A2–3).

The above arguments also imply that S is determined up to a numerical factor. For, if S' were another connecting matrix, then $S^{-1}S'$ would commute with all γ_r and would thus be a multiple of the unit matrix. This concludes the proof of the main theorem.

As a final consequence of Eqs. (A2–18) and (A2–24), we note that if $\gamma_r = \gamma_r'$ then S is a multiple of the unit matrix $S = \lambda I$; thus

$$\sum_{r=1}^{16} \gamma_r F \gamma^r \xi_r = \lambda I. \tag{A2–25}$$

The numerical factor on the right side can be found from the trace

$$\sum_{r=1}^{16} \text{Tr}\,(\gamma_r F \gamma^r \xi_r) = \sum_{r=1}^{16} \text{Tr}\,(F\gamma^r \xi_r \gamma_r) = 16\,\text{Tr}\,F = 4\lambda.$$

Thus

$$\lambda = 4\,\text{Tr}\,F.$$

A2–2 The matrices A, B, and C. Let X be any matrix. We denote the Hermitian conjugate by X^+, the transpose by $X^\sim$, and the complex conjugate by X^*. Obviously, the three operations are related by

$$(X^*)^\sim = (X^\sim)^* = X^+. \tag{A2–26}$$

Since the commutation rules (A2–1) are invariant under these three operations, we may define three matrices A, B, C by the equations

$$-\gamma_\mu^+ = A\gamma_\mu A^{-1}, \quad \gamma_\mu^\sim = B\gamma_\mu B^{-1}, \quad \gamma_\mu^* = C\gamma_\mu C^{-1}. \tag{A2–27}$$

From these, we immediately obtain

$$\begin{aligned} A &= aA^+, & aa^* &= 1, \\ B^\sim &= bB, & b^2 &= 1, \\ C^* &= cC^{-1}, & c &> 0. \end{aligned} \tag{A2–28}$$

The assertion $c > 0$ in the third of Eqs. (A2–28) is not trivial and can be proved, for instance, in the following way. We observe first that $C^*C = cI$ is a multiple of the unit matrix, since it commutes with all the γ_μ. Next we find that under an S-transformation, $\gamma_\mu \to \gamma_\mu' = S\gamma_\mu S^{-1}$, the C transforms according to $C' = S^*CS^{-1}$. It follows that C^*C is invariant under these transformations. It is therefore sufficient to prove $c > 0$ for a particular representation, for instance the standard representation (A2–14). For this case, C is given by $C = \gamma_0\gamma_2$ and $C^* = \gamma_0\gamma_2$. Therefore $CC^* = I$, and $c > 0$ holds generally. There is no simple proof known to us which does not make use of a special representation.

The arbitrary factors in A and C can thus be chosen so that $a = c = 1$. On the other hand, a similar arbitrary factor in B does not affect b, which is subject only to the defining property (A2–28). We shall prove that $b = -1$. From (A2–28) and (A2–1), we obtain the general relation

$$(B\gamma_r)^\sim = b\xi_r B\gamma_r. \tag{A2–29}$$

Thus, $B\gamma_r$ is either symmetrical or antisymmetrical according to whether $b\xi_r$ is positive or negative. Furthermore, all the $B\gamma_r$ are linearly independent, just as the γ_r themselves. In the antisymmetrical class, there are either 10 or 6 matrices, depending on whether b is $+1$ or -1. Now the maximum number of linearly independent antisymmetrical 4×4 matrices is 6, which leaves us only with the possibility $b = -1$.†

We also note the relations

$$\Gamma_t^+ = \zeta_t A \Gamma_t A^{-1}, \quad \Gamma_t^\sim = \xi_t B \Gamma_t B^{-1},$$
$$\Gamma_t^* = C\Gamma_t C^{-1}, \quad \Gamma_t = \eta_t \gamma_5 \Gamma_t \gamma_5^{-1}. \tag{A2–30}$$

We have used the notation Γ_t to designate any of the γ_r contained in the set Γ_t. The three sign factors are collected in Table A2–1. The product

TABLE A2–1

The three sign factors ζ_t, ξ_t, and η_t

t	1	2	3	4	5
ζ_t	1	-1	-1	1	1
ξ_t	1	1	-1	-1	1
η_t	1	-1	1	-1	1

---

† W. Pauli, *Ann. Inst. Henri Poincaré* **6**, 109 (1936).

of any two of these factors can be replaced by the third one, because of the relation

$$\zeta_t \xi_t \eta_t = 1. \tag{A2-31}$$

Since the three operations are related by (A2–26), there exists a relation between the matrices A, B, C, and γ_5. The as yet undetermined factor in B can be chosen in such a way that this relation takes on the form

$$A = C^+ B \gamma_5. \tag{A2-32}$$

Some further useful relations are obtained by considering transformations of the γ_r's induced by *proper* Lorentz transformations:

$$\gamma_\mu' = a_\mu{}^\nu \gamma_\nu = S^{-1} \gamma_\mu S, \quad (a_\mu{}^\nu \text{ real}).$$

By considering the relations $\gamma^{+\prime} = \gamma'^+$, $\gamma^{*\prime} = \gamma'^*$, and $\gamma^{\sim\prime} = \gamma'^\sim$, we obtain, with the help of (A2–27) and (A2–28),

$$S^+ = A S^{-1} A^{-1}, \tag{A2-33}$$

$$S^* = C S C^{-1}, \tag{A2-34}$$

$$S^\sim = B\gamma_5 S^{-1} (B\gamma_5)^{-1}. \tag{A2-35}$$

These relations remain correct for space reflections for which $S = \gamma_0$, $S^{-1} = \gamma^0$. However, if we consider time reflections for which $S = \gamma_1 \gamma_2 \gamma_3 = -S^{-1}$, we obtain, instead of (A2–33), the relation

$$S^+ = -A S^{-1} A^{-1}, \tag{A2-36}$$

and instead of (A2–35),

$$S^\sim = -B\gamma_5 S^{-1} (B\gamma_5)^{-1}.$$

Equation (A2–34) remains unchanged.

A2–3 The amplitudes of the plane wave solutions.

The equation

$$(\partial + m)\psi = 0 \tag{A2-37}$$

has two solutions in the form of plane waves,

$$\psi = u e^{ip \cdot x}, \tag{A2-38}$$

$$\psi = v e^{-ip \cdot x}. \tag{A2-39}$$

The four-vector p satisfies

$$p^2 + m^2 = 0,$$

and the amplitudes u and v of the plane waves are functions of the vector p which satisfy the algebraic equations

$$(i\not{p} + m)u = 0, \qquad (A2-40)$$

$$(i\not{p} - m)v = 0. \qquad (A2-41)$$

It is convenient to introduce the notation

$$\Lambda_{\pm} = \frac{\pm i\not{p} + m}{2m}. \qquad (A2-42)$$

The operators $\Lambda_{\pm}$ satisfy the relations

$$\Lambda_{\pm}(p)\Lambda_{\mp}(p) = 0, \qquad (A2-43)$$

$$\Lambda_{+}(p) + \Lambda_{-}(p) = 1, \qquad (A2-44)$$

$$\Lambda_{\pm}^{2}(p) = \Lambda_{\pm}(p). \qquad (A2-45)$$

These relations show that the operators $\Lambda_{\pm}$ are two projection operators each of rank two, which decompose the four-dimensional spinor space S into two subspaces U, V. This means that any arbitrary spinor $x \in S$ may be decomposed in an unambiguous fashion into two spinors $u \in U$, $v \in V$:

$$x = u + v,$$

such that

$$\Lambda_{+}u = \Lambda_{-}v = 0. \qquad (A2-46)$$

Indeed, the two spinors are determined in terms of x by

$$u = \Lambda_{-}x, \quad v = \Lambda_{+}x. \qquad (A2-47)$$

From this relation and Eq. (A2–45) follows that $\Lambda_{-}u = u$, $\Lambda_{+}v = v$. Since

$$\Lambda_{\pm}^{+}A = A\Lambda_{\pm}, \qquad (A2-48)$$

where A is given by Eq. (A2–27), any two spinors u and v are orthogonal in the sense

$$\bar{u}v = 0, \quad \bar{v}u = 0, \quad (\bar{x} = x^{+}A). \qquad (A2-49)$$

This follows from (A2–43), (A2–47), and (A2–48), because

$$\bar{u}v = \bar{x}\Lambda_{-}\Lambda_{+}x = 0, \quad \bar{v}u = \bar{x}\Lambda_{+}\Lambda_{-}x = 0.$$

The projection operators $\Lambda_{\pm}$ separate the four-dimensional spinor space into two-dimensional subspaces. Each of these subspaces can be separated further by a second pair of projection operators, $\Sigma_{\pm}$. We choose these conveniently as follows:

$$\Sigma_{\pm}(\mathbf{p}) = \frac{\pm \boldsymbol{\sigma} \cdot \mathbf{p} + |\mathbf{p}|}{2|\mathbf{p}|}. \qquad (A2-50)$$

where $\boldsymbol{\sigma} = (\sigma_1, \sigma_2, \sigma_3)$ is a set of three matrices with the components

$$\sigma_k = -i\gamma_l\gamma_m, \quad (k, l, m = \text{cyclic permutation of } 1, 2, 3). \quad \text{(A2–51)}$$

The 4×4 matrices σ_k satisfy the relations

$$\sigma_k^2 = 1, \quad (k = 1, 2, 3),$$

$$\sigma_k\sigma_l = -\sigma_l\sigma_k = i\sigma_m, \quad (k, l, m = \text{cyclic permutation of } 1, 2, 3),$$

which are formally indentical with the relations (A2–13). Indeed, if we choose the special representation (A2–14) for the γ_μ, the matrices (A2–51) are of the form

$$\begin{pmatrix} \sigma_k & 0 \\ 0 & \sigma_k \end{pmatrix},$$

where the σ_k are the 2×2 matrices defined by (A2–12).

The matrices σ_k occur often in connection with expressions of the type

$$\boldsymbol{\gamma} \cdot \mathbf{P}\, \boldsymbol{\gamma} \cdot \mathbf{Q},$$

where P and Q are any two three-vector operators not containing γ-matrices and $\boldsymbol{\gamma}$ is the set of three matrices

$$\boldsymbol{\gamma} = (\gamma^1, \gamma^2, \gamma^3).$$

We have

$$\boldsymbol{\gamma} \cdot \mathbf{P}\, \boldsymbol{\gamma} \cdot \mathbf{Q} = \sum_{k,l=1}^{3} \gamma_k \gamma_l P^k Q^l$$

$$= \sum_{k=1}^{3} \gamma_k^2 P^k Q^k + \sum_{k \neq l=1}^{3} \gamma_k \gamma_l P^k Q^l$$

$$= \mathbf{P} \cdot \mathbf{Q} + \sum_{k>l} \gamma_l \gamma_{l'} (P^k Q^l - P^l Q^k)$$

$$= \mathbf{P} \cdot \mathbf{Q} + i\boldsymbol{\sigma} \cdot \mathbf{P} \times \mathbf{Q}, \quad \text{(A2–52)}$$

which is a very useful identity.

The projection operators $\Sigma_\pm(\mathbf{p})$, Eq. (A2–50), have properties very similar to those of the $\Lambda_\pm(p)$, namely,

$$\Sigma_\pm(\mathbf{p})\Sigma_\mp(\mathbf{p}) = 0, \quad \Sigma_+(\mathbf{p}) + \Sigma_-(\mathbf{p}) = 1, \quad \Sigma_\pm^2(\mathbf{p}) = \Sigma_\pm(\mathbf{p}). \quad \text{(A2–53)}$$

All $\Sigma_\pm$ commute with all $\Lambda_\pm$,

$$[\Sigma_\pm, \Lambda_\pm] = 0, \quad \text{(A2–54)}$$

for all combinations of signs. We can therefore choose our spinors u, v to be simultaneous eigenspinors of $\Lambda_\pm$ and $\Sigma_\pm$. Thus, we define the two pairs of spinors* $u_\pm, v_\pm$ (generally denoted by u_r, v_r with $r = \pm$),

$$\Sigma_r(\mathbf{p})u_r(\mathbf{p}) = 0, \quad \Sigma_{-r}u_r = u_r, \tag{A2-55}$$

$$\Sigma_{-r}(\mathbf{p})v_r(\mathbf{p}) = 0, \quad \Sigma_r v_r = v_r. \tag{A2-56}$$

The Σ_r also satisfy a relation analogous to Eq. (A2-48),

$$\Sigma_\pm{}^+ A = A\Sigma_\pm, \tag{A2-57}$$

from which follow the orthogonality relations

$$\bar{u}_r u_{-r} = \bar{v}_s v_{-s} = 0. \tag{A2-58}$$

The spinors u_r, v_s are the solutions of homogeneous defining equations and are therefore not entirely determined. For instance, we can impose a normalization condition by the requirement that $\bar{u}_r u_r$, $\bar{v}_s v_s$ shall have certain values. Whether these quantities are positive or negative depends on the choice of the arbitrary sign in the matrix A. However, once such a sign is adopted, the absolute sign of these four expressions is determined. We shall arbitrarily fix the sign of A such that

$$\bar{u}_+ u_+ > 0.$$

To determine the sign of $\bar{u}_- u_-$, we construct a matrix F which transforms u_+ into u_-, v_+ into v_- and vice versa:

$$Fu_+ = e^{i\alpha}u_-, \quad Fu_- = e^{-i\alpha}u_+,$$
$$Fv_+ = e^{i\beta}v_-, \quad Fv_- = e^{-i\beta}v_+. \tag{A2-59}$$

These relations impose on F the following conditions:

$$F^2 = 1, \quad [F, p] = 0, \quad \{F, \boldsymbol{\sigma} \cdot \mathbf{p}\} = 0. \tag{A2-60}$$

A matrix satisfying these conditions is

$$F = \mathbf{e} \cdot \boldsymbol{\gamma} \gamma_5, \tag{A2-61}$$

where $\mathbf{e}$ is a unit vector perpendicular to $\mathbf{p}$, $\mathbf{e}^2 = 1$, $\mathbf{e} \cdot \mathbf{p} = 0$, and $\boldsymbol{\gamma}$ has the three components γ_1, γ_2, and γ_3. The operator F in Eq. (A2-61) also satisfies

$$F^+ A = AF. \tag{A2-62}$$

The phases α and β in Eq. (A2-59) are arbitrary and depend on the choice of the phase factors in u and v, and the choice of $\mathbf{e}$. Equation (A2-59) allows us to express $\bar{u}_- u_-$ in terms of $\bar{u}_+ u_+$:

$$\bar{u}_- u_- = \bar{u}_+ F^2 u_+ = \bar{u}_+ u_+, \quad \bar{v}_- v_- = \bar{v}_+ F^2 v_+ = \bar{v}_+ v_+. \tag{A2-63}$$

---

* The unsymmetrical definition of the sign indices of u_r and v_r given in Eq. (A2-56) is justified by the physical interpretation. [*Cf.* the discussion following Eq. (3-115).]

Furthermore, we can construct a matrix E which transforms u_r into v_{-r}:

$$Eu_r = re^{i\delta}v_{-r}, \quad Ev_r = re^{-i\delta}u_{-r}. \tag{A2-64}$$

This matrix must satisfy the conditions

$$E^2 = -1, \quad \{E,p\} = 0, \quad [E,\boldsymbol{\sigma}\cdot\mathbf{p}] = 0. \tag{A2-65}$$

A matrix satisfying these conditions is

$$E = \gamma_5 = \gamma_0\gamma_1\gamma_2\gamma_3. \tag{A2-66}$$

This matrix also satisfies

$$E^+A = AE. \tag{A2-67}$$

The relative phases of u and v can always be chosen such that $\delta = 0$ in Eq. (A2-64):

$$Eu_+ = v_-, \quad Ev_+ = u_-. \tag{A2-68}$$

These equations allow us to express $\bar{v}_r v_r$ in terms of $\bar{u}_r u_r$:

$$\bar{v}_r v_r = \bar{u}_{-r} E^2 u_{-r} = -\bar{u}_{-r} u_{-r}. \tag{A2-69}$$

The spinors u_r and v_r are now uniquely determined except for the normalization. We choose this normalization as follows:

$$\bar{u}_+ u_+ = 1. \tag{A2-70}$$

The remaining normalizations follow from this choice and Eqs. (A2-63) and (A2-69). Combined with (A2-58), we can write these equations in the form

$$\bar{u}_r u_s = \delta_{rs}, \quad \bar{v}_r v_s = -\delta_{rs}. \tag{A2-71}$$

The choice (A2-70) also determines the sign of the matrix A and normalizes the amplitudes.

The linear independence of the four spinors follows also from Eq. (A2-71). It implies that*

$$\sum_r (u_r \bar{u}_r - v_r \bar{v}_r) = 1. \tag{A2-72}$$

Several useful relations follow from this equation and Eqs. (A2-55) and (A2-56); we multiply (A2-72) first by $\Lambda_-\Sigma_{-r}$ and then by $\Lambda_+\Sigma_r$ and obtain

$$u_r \bar{u}_r = \Lambda_- \Sigma_{-r}, \quad v_r \bar{v}_r = -\Lambda_+ \Sigma_r. \tag{A2-73}$$

Summing over r, we find, with (A2-53),

$$\sum_r u_r \bar{u}_r = \Lambda_-, \quad \sum_r v_r \bar{v}_r = -\Lambda_+. \tag{A2-74}$$

---

* Note that the expression $u\bar{u}$ is a 4×4 matrix, $(u\bar{u})_{\rho\sigma} = u_\rho \bar{u}_\sigma$. 1 is therefore the 4×4 unit matrix. On the other hand, $\bar{u}u$ is a number and not a matrix.

This result and Eq. (A2–44) lead back to the completeness relation (A2–71).

Equations (A2–73) and (A2–74) contain essentially all the mathematical properties of u and v which are of physical importance. They are clearly independent of the representation of the γ_μ's.

We finally note the effect of charge conjugation on u and v. The charge conjugate spinor is defined by

$$\psi^c = C^*\psi^*, \qquad (A2\text{–}75)$$

where C is given by (A2–27).

When we apply the charge conjugation to the spinor u_r, we obtain the spinor $u_r{}^c = C^*u_r{}^*$. One verifies easily that $u_r{}^c$ satisfies the defining equations of the spinor v_r. Therefore, choosing the phases of u_r and v_r appropriately, we can write

$$u_r{}^c(\mathbf{p}) = v_r(\mathbf{p}), \quad v_r{}^c(\mathbf{p}) = u_r(\mathbf{p}). \qquad (A2\text{–}76)$$

A2–4 A theorem on the traces of γ-matrices.† The actual evaluation of the transition probabilities leads to the problem of evaluating the trace of products of γ-matrices.

The general trace calculation can always be reduced to the evaluation of the expression

$$(a_1 a_2 \cdots a_n) \equiv \tfrac{1}{4} \operatorname{Tr}(\mathbf{a}_1 \mathbf{a}_2 \cdots \mathbf{a}_n), \qquad (A2\text{–}77)$$

where

$$\mathbf{a} \equiv a \cdot \gamma = a_\mu \gamma^\mu, \qquad (A2\text{–}78)$$

and $a_1, a_2, \cdots, a_n$ is an arbitrary set of four-vectors. By suitable specialization of these vectors, we obtain the trace of any product of γ-matrices from the expression (A2–77). This expression may be regarded as a generalization of the ordinary scalar product of two vectors, to which it reduces for $n = 2$, because

$$(a_1 a_2) \equiv \tfrac{1}{4} \operatorname{Tr}(\mathbf{a}_1 \mathbf{a}_2) = a_1 \cdot a_2, \qquad (A2\text{–}79)$$

as we have anticipated with the notation.

In order to evaluate (A2–77), we derive first the following recursion relation:

$$(a_1 a_2 \cdots a_n) = (a_1 a_2)(a_3 \cdots a_n) - (a_1 a_3)(a_2 \cdots a_n) + \cdots$$
$$+ (a_1 a_n)(a_2 \cdots a_{n-1}). \qquad (A2\text{–}80)$$

The plus sign of the last term implies that n is even. This is no restriction because

$$(a_1 a_2 \cdots a_n) = 0, \quad \text{for } n \text{ odd}. \qquad (A2\text{–}81)$$

---

† L. M. Yang, *Phil. Mag.* **42**, 1333 (1951).

The proof of the recursion relation (A2–80) is obtained by observing first that

$$(a_1 a_2 \cdots a_n) + (a_2 a_1 \cdots a_n) = 2(a_1 a_2)(a_3 \cdots a_n).$$

This follows from the anticommutator relation (A2–1). Similarly, we have for the interchange of any two neighboring factors,

$$(a_1 a_2 \cdots a_i a_{i+1} \cdots a_n) + (a_1 a_2 \cdots a_{i+1} a_i \cdots a_n)$$
$$= 2(a_i a_{i+1})(a_1 a_2 \cdots a_{i-1} a_{i+2} \cdots a_n).$$

Successive application of this last relation leads to

$$(a_1 a_2 \cdots a_n) + (a_2 \cdots a_n a_1) = 2(a_1 a_2)(a_3 \cdots a_n)$$
$$- 2(a_1 a_3)(a_2 \cdots a_n) + \cdots + 2(a_1 a_n)(a_2 \cdots a_{n-1}).$$

Now we make use of the invariance of the trace under cyclic interchange of the operators:

$$(a_1 a_2 \cdots a_n) + (a_2 \cdots a_n a_1) = 2(a_1 a_2 \cdots a_n).$$

The last two equations together are equivalent to (A2–80). This proves the recursion relation.

This relation allows us to reduce the evaluation of a trace of n factors to the evaluation of a trace of $(n-2)$ factors. Repeated use of (A2–80) leads to

$$(a_1 a_2 \cdots a_n) = \sum_{\text{all pairs}} \delta_P (a_{i_1} a_{i_2})(a_{i_3} a_{i_4}) \cdots (a_{i_{n-1}} a_{i_n}). \qquad (A2\text{--}82)$$

In this formula the summation is extended over all different selections of pairs $(i_\nu i_{\nu+1})$ with $i_\nu < i_{\nu+1}$. The number of terms in the sum is $(n-1)!! = (n-1)(n-3)\cdots$, and δ_P is the signature of the permutation

$$P = \begin{pmatrix} 1 & 2 & \cdots & n \\ i_1 & i_2 & \cdots & i_n \end{pmatrix}.$$

The number of terms in (A2–82) increases rapidly with n. For instance, for $n = 8$, a case which occurs quite frequently in the lowest-order perturbation calculation, this number is 105. It is therefore often more useful to work with the recursion relation (A2–80) instead of the general expression (A2–82).

By suitable specialization of the vectors a_i, we can calculate the trace of any product of γ-matrices. The following general relation can also be read from the formula (A2–82):

$$(a_1 a_2 \cdots a_n) = (a_n \cdots a_2 a_1). \qquad (A2\text{--}83)$$

The trace remains unchanged if the order of the factors is reversed.

A2-5 Spin sums.

In many applications of the theory it is necessary to sum over the spin states of the initial and final electrons. These sums are always to be carried out on the transition probability, which is (apart from factors) the square of the associated S-matrix element.

Consider, for example, a process involving only one negaton in the initial and final states. The matrix element will be of the form

$$\bar{u}(\mathbf{p}')M(p',p)u(\mathbf{p}).$$

The square of its absolute value is

$$|\bar{u}(\mathbf{p}')Mu(\mathbf{p})|^2 = (\bar{u}(\mathbf{p}')Mu(\mathbf{p}))^*(\bar{u}(\mathbf{p}')Mu(\mathbf{p}))$$

$$= u^+(\mathbf{p})M^+Au(\mathbf{p}')\bar{u}(\mathbf{p}')Mu(\mathbf{p}),$$

since

$$\bar{u} = u^+A$$

and A is Hermitian. The 4×4 matrix M consists of a product of γ-matrices or of a sum of such products. Let us first assume that M is a single product. The structure of the matrix elements in momentum space is such that almost all γ-matrices are multiplied by momentum four-vectors and factors of i. Only the γ's which arise from the corners to which external fields or external photon lines are attached, are not associated with a factor i.

Taking these facts into account and using Eq. (A2-27),

$$\gamma_\mu^+ = -A\gamma_\mu A^{-1},$$

we find, when M contains n factors,

$$A^{-1}M^+A = (-1)^{P_e}\bar{M}. \qquad (\text{A2-84})$$

In this relation $\bar{M}$ means M written in reverse order, i.e., from right to left. For example, when

$$M = ipeike', \quad \bar{M} = e'ikeip.$$

The square of the absolute value of $\bar{u}(\mathbf{p}')Mu(\mathbf{p})$ becomes

$$|\bar{u}(\mathbf{p}')Mu(\mathbf{p})|^2 = (-1)^{P_e}\bar{u}(\mathbf{p})\bar{M}u(\mathbf{p}')\bar{u}(\mathbf{p}')Mu(\mathbf{p}), \qquad (\text{A2-85})$$

where $u(\mathbf{p})$ and $u(\mathbf{p}')$ refer to definite and, in general, different spin states. The summation of the final spin states follows from Eq. (A2-74):

$$\sum_{\text{final spins}} |\bar{u}(\mathbf{p}')Mu(\mathbf{p})|^2 = (-1)^{P_e}\bar{u}(\mathbf{p})\bar{M}\Lambda_-(p')Mu(\mathbf{p}). \qquad (\text{A2-86})$$

If we want to sum over both the initial and the final spin states,

$$\sum_{\text{spins}} |\bar{u}(\mathbf{p}')Mu(\mathbf{p})|^2 = (-1)^{P_e} \text{Tr}\,[\bar{M}\Lambda_-(p')M\Lambda_-(p)]. \qquad (\text{A2-87})$$

When M contains several terms, each of which is a product of a different number of γ's, this result remains unchanged, because the number of external photon lines will always be the same in each term. This follows from the fact that the terms in M which arise from different diagrams or higher radiative corrections always refer to the same physical process, i.e., the same number of external lines.

When one of the u in Eq. (A2–87) is replaced by a v corresponding to pair creation or pair annihilation, the corresponding projection operator Λ_- is replaced by $-\Lambda_+$, as follows from Eq. (A2–74). For positon scattering processes, both Λ_- are replaced by $-\Lambda_+$.

The extension of these considerations to processes involving more than one open electron path is very straightforward. As an example, we consider part of the matrix element for pair production in photon-negaton collisions:

$$\sum_{\text{spins}} |\bar{u}(\mathbf{p}')Mu(\mathbf{p})\bar{u}(\mathbf{p}'')Nv(\mathbf{p}_+)|^2$$
$$= + \operatorname{Tr}[M\Lambda_-(p)\bar{M}\Lambda_-(p')] \operatorname{Tr}[N\Lambda_+(p_+)\bar{N}\Lambda_-(p'')]. \quad \text{(A2–88)}$$

The sign in front is the product of one factor -1 from $-\Lambda_+(p_+)$ and another factor -1 from $(-1)^{P_e}$ with $P_e = 1$. The complete matrix element for this process is discussed in Section 11–4.

A2–6 Polarization sums. Another important summation which occurs frequently in the evaluation of cross sections is the summation over the polarization directions of an incident or outgoing photon.

This task can be accomplished most conveniently by summing covariantly over all four polarization vectors $e_\mu^{(r)}$, $(r = 0, 1, 2, 3)$. The Lorentz condition will always assure that the time-like vector and the space-like vector in the propagation direction cancel exactly, so that effectively this covariant summation is equivalent to a summation over the two transverse polarization directions of the free photon.

The formal proof of the correctness of this procedure follows from Eq. (6–17), which can be written in momentum space,

$$e_\mu^\perp = e_\mu - k_\mu \frac{n \cdot e}{n \cdot k} - \left(n_\mu + \frac{k_\mu}{n \cdot k}\right) \frac{k \cdot e}{n \cdot k}, \quad (n^2 = -1). \quad \text{(A2–89)}$$

The matrix elements always contain $e_\mu^\perp$ multiplied by γ^μ, so that terms in (A2–89) which are proportional to k will never give contributions in a gauge-invariant expression.* When, furthermore, the Lorentz condition is satisfied, the remaining terms will vanish. In this case

$$e_\mu^\perp = e_\mu$$

---

* This statement is proved in Section 9–5.

effectively and, since

$$\sum_r g_{rr} e_\mu^{(r)} e_\nu^{(r)} = g_{\mu\nu},$$

$$\sum_r e_\mu^{\perp(r)} e_\nu^{\perp(r)} = g_{\mu\nu}. \qquad (A2\text{-}90)$$

In a transition probability, this means

$$\sum_{\text{pol}} (\cdots e \cdots e \cdots) = (\cdots \gamma_\mu \cdots \gamma_\nu \cdots) g^{\mu\nu}$$
$$= (\cdots \gamma_\mu \cdots \gamma^\mu \cdots). \qquad (A2\text{-}91)$$

In this form, the polarization sum is extremely simple.

Although this method is sufficient, since all observable effects derive from gauge-invariant expressions, it is of some interest to investigate the polarization sum of a non-gauge-invariant term. The Lorentz condition $k \cdot e = 0$ and $k^2 = 0$ are still assumed to be fulfilled. In that case, only the first two terms of (A2–89) contribute and we find, with (A2–90),

$$\sum_r e_\mu^{\perp(r)} e_\nu^{\perp(r)} = g_{\mu\nu} - \frac{k_\mu k_\nu}{(n \cdot k)^2} - \frac{k_\mu n_\nu - k_\nu n_\mu}{n \cdot k}. \qquad (A2\text{-}92)$$

When the polarization average is required, these sums must be divided by 2, corresponding to the two transverse polarization directions.

APPENDIX A3

A THEOREM ON THE REPRESENTATION OF THE EXTENDED LORENTZ GROUP BY IRREDUCIBLE TENSORS

The fundamental equations of quantum electrodynamics have the invariance properties resulting from the proper Lorentz transformations. In addition to these there exist further invariance properties with respect to two special transformations, namely, the space inversion σ and the time inversion τ. The explicit definition of these transformations is given in Chapter 1 [Eqs. (1-6), (1-8), and (1-9)] and their effect on the field equations is discussed in Sections 5-3 and 5-4.

The proper Lorentz transformations form a group Λ. Any tensor quantity $T_{\mu\nu}\ldots$ which transforms according to the law

$$T'_{\mu\nu}\ldots = a_\mu{}^\rho a_\nu{}^\sigma \cdots T_{\rho\sigma}\ldots \tag{A3-1}$$

furnishes a representation of this group. These tensor representations, in general, are not irreducible. In the theory of group representations we show, however, that any tensor representation is completely reducible and decomposes into irreducible parts. A tensor which transforms according to an irreducible representation of the group Λ is called an irreducible tensor. Examples of irreducible tensors are the vectors A_μ; the skew-symmetrical tensors of rank two, $F_{\mu\nu} = -F_{\nu\mu}$; and the symmetrical tensors $T_{\mu\nu}$ of rank two with vanishing trace $T_\mu{}^\mu = 0$. The irreducible representations of the group Λ are all known and can be constructed explicitly.*

We are interested in the transformation properties of tensors under the extended group Γ which is obtained by adjoining to Λ the two special transformations σ and τ. This means that the group Γ is obtained by combining these two special transformations with all the transformations in Λ. We define a regular tensor as one which, under the extended group, also transforms according to Eq. (A3-1). Thus, for instance, the coordinate vector is by definition a regular tensor. The representation of the extended group by regular tensors is, however, not the only possibility of tensor representations. In order to survey all the possible representations, we need the following properties of the two groups Λ and Γ:

The group Λ is an invariant subgroup of Γ.† It follows that any element

---

* See, for instance, B. L. van der Waerden, *Gruppentheoretische Methoden in der Quantenmechanik*, Springer, Berlin, 1932, §20, pp. 78 ff.

† See, e.g., F. D. Murnaghan, *The Theory of Group Representations*, Johns Hopkins Press, 1938, Chapter XII.

$K \in \Gamma$ may be written in the form

$$K = \varphi L,$$

where $L \in \Lambda$ and φ is one of the four elements $\varphi = 1, \sigma, \tau, \rho = \sigma\tau$. Since the subgroup is invariant, the factor group has the composition law of these four elements. It is thus the so-called "four-group" V characterized by the relations $\sigma^2 = \tau^2 = 1$, $\sigma\tau = \tau\sigma$. This group is Abelian but not cyclic. We write

$$\Gamma/\Lambda = V.$$

Since V is Abelian, its irreducible representations are one-dimensional and are equal to the characters $\chi(\varphi)$ of the group. The four characters $\chi_r(\varphi)$, $(r = 0, \cdots, 3)$ are given in Table A3-1.

TABLE A3-1

Character table of the group V

χ \ φ	1	σ	τ	ρ
χ_0	1	1	1	1
χ_1	1	1	−1	−1
χ_2	1	−1	1	−1
χ_3	1	−1	−1	1

Let us now consider a regular irreducible tensor T transforming under Lorentz transformations K according to an irreducible tensor representation. Let $D(K)$ be the matrix corresponding to K in this representation. Further, let $\Delta(K)$ be the matrix of another irreducible tensor representation of Γ which coincides for the proper transformations Λ with D. Thus

$$D(L) = \Delta(L) \quad \text{for } L \in \Lambda. \tag{A3-2}$$

We can then show that

$$\Delta(K) = \chi_r(\varphi)D(K), \tag{A3-3}$$

where φ stands for one of the elements σ, τ, ρ and is determined by

$$K = \varphi L, \quad (L \in \Lambda).$$

The relation (A3-3) states that there exist, for each irreducible tensor representation of Λ, four irreducible representations of Γ associated with the four characters χ_r, $(r = 0, \cdots, 3)$. We have thus four types of tensors to consider, the regular tensors $(r = 0)$ and three types of *pseudotensors* $(r = 1, 2, 3)$. We distinguish the latter by calling them *pseudotensors of kind r*.*

To prove (A3–3), we observe first that the matrix $E(K)$ defined by

$$E(K) = \Delta(K)D^{-1}(K) \qquad \text{(A3–4)}$$

is a function of the co-sets only. Thus if

$$K = \varphi L \quad \text{with } L \in \Lambda,$$

then

$$E(\varphi L) = \Delta(\varphi L)D^{-1}(\varphi L) = \Delta(\varphi)\Delta(L)D^{-1}(L)D^{-1}(\varphi)$$
$$= \Delta(\varphi)D^{-1}(\varphi) = E(\varphi). \qquad \text{(A3–5)}$$

Next we show that for all K, $E(K) = E(\varphi)$ is a multiple of the unit matrix, because $E(K)$ commutes with all matrices of $D(L)$, $(L \in \Lambda)$. This is clear first for $\varphi = \rho$, since ρ commutes with all K. For $\varphi = \sigma, \tau$, we have

$$L\varphi = \varphi L', \quad L, L' \in \Lambda,$$

since Λ is an invariant subgroup. Thus

$$D(L)E(\varphi) = D(L)\Delta(\varphi)D(\varphi) = D(\varphi)\Delta(L')D(\varphi)$$
$$= D(\varphi)\Delta(\varphi)D(L) = E(\varphi)D(L).$$

From Schur's lemma then follows that

$$E(\varphi) = \lambda(\varphi)1. \qquad \text{(A3–6)}$$

Finally, we show that $\lambda(\varphi)$ is a representation of the factor group V. We have seen already that it is a function of the co-sets only. Now let K_1, K_2 be any two transformations in Γ. Then

$$\Delta(K_1)\Delta(K_2) = \Delta(K_1K_2) = \lambda(\varphi_1\varphi_2)D(K_1K_2)$$
$$= \lambda(\varphi_1)\lambda(\varphi_2)D(K_1)D(K_2).$$

Thus

$$\lambda(\varphi_1)\lambda(\varphi_2) = \lambda(\varphi_1\varphi_2). \qquad \text{(A3–7)}$$

That is, $\lambda(\varphi)$ is a representation of V. It follows that $\lambda(\varphi)$ must be one of the characters

$$\lambda(\varphi) = \chi_r(\varphi)$$

and the theorem (A3–3) is proved.

---

*This classification was given by S. Watanabe, *Phys. Rev.* **84**, 1008 (1951), Appendix.

We summarize the result for convenience. There exist four types of irreducible tensors with respect to the extended Lorentz group Γ. Each irreducible tensor is obtained by multiplying the "regular" tensor defined by (A3-1) by one of the characters $\chi_r(\varphi)$ of the factor group $V = \Gamma/\Lambda$.

For tensors which are reducible there exist many more representations of the extended group, since every irreducible part can belong independently to any one of the four classes. For instance, a symmetrical tensor of second rank with nonvanishing trace can represent sixteen pseudotensors under the extended group Γ.

APPENDIX A4

THE ORDERING THEOREM

The iteration solution of the S-matrix (*cf.* Section 8–1) leads to expressions which contain the chronologically ordered products of field operators. To find matrix elements connecting states of given numbers of free particles, it is necessary to write these expressions as sums of products which are ordered with respect to the emission and absorption operators.

The transition from one type of ordering to another which is involved here can be accomplished with the general ordering theorem.* We distinguish two forms of the theorem corresponding to the two cases of commuting and anticommuting field variables [that is, cases (a) and (b) discussed in Section 1–12].

A4–1 The ordering theorem for commuting fields. We consider a set of operators $\varphi(\alpha,\beta)$ depending on two real parameters α and β. The parameters may vary over a discrete or continuous range of values. We introduce the abbreviations

$$\varphi(\alpha_1,\beta_1) \equiv \varphi_1, \cdots \quad (A4\text{–}1)$$

and assume that the operators satisfy a set of commutation rules

$$[\varphi_2,\varphi_1] = \Sigma_{21}. \quad (A4\text{–}2)$$

Here the Σ_{21} are c-numbers (that is, ordinary numbers multiplying the unit operator) depending on the four-parameter values $\alpha_2\beta_2$, $\alpha_1\beta_1$.

We precede the ordering theroem by a number of definitions and some elementary consequences therefrom.

The A-product of n operators $\varphi_n, \cdots, \varphi_2, \varphi_1$ is defined by

$$A(\varphi_n \cdots \varphi_2\varphi_1) \equiv \varphi_{i_n} \cdots \varphi_{i_2}\varphi_{i_1} \quad (A4\text{–}3)$$

such that

$$\alpha_{i_n} \geq \cdots \geq \alpha_{i_2} \geq \alpha_{i_1}. \quad (A4\text{–}4)$$

The B-product of n operators $\varphi_n, \cdots, \varphi_2, \varphi_1$ is defined by

$$B(\varphi_n \cdots \varphi_2\varphi_1) \equiv \varphi_{j_n} \cdots \varphi_{j_2}\varphi_{j_1} \quad (A4\text{–}5)$$

such that

$$\beta_{j_n} \geq \cdots \geq \beta_{j_2} \geq \beta_{j_1} \quad (A4\text{–}6)$$

---

* G. C. Wick, *Phys. Rev.* **80**, 268 (1950). The special case of commuting fields was previously discussed by A. Houriet and A. Kind, *Helv. Phys. Acta* **22**, 319 (1949).

The above definitions of the A- and B-products are unique and define symmetrical functions of the operators $\varphi_n \cdots \varphi_2\varphi_1$ so long as no two values of the parameter α or β are the same. Ambiguities may arise if two or more parameter values are equal and the commutators do not vanish for these values of the parameters. It is easy to dispose of this difficulty by defining the A- and B-products as symmetrical functions of the operators.

The fact that the A- and B-products are symmetrical functions of their arguments allows a simplification of the notation. We shall sometimes write

$$A^n \equiv A(n, \cdots, 2, 1) \equiv A(\varphi_n \cdots \varphi_2\varphi_1) \tag{A4-7}$$

and, similarly,

$$B^n \equiv B(n, \cdots, 2, 1) \equiv B(\varphi_n \cdots \varphi_2\varphi_1). \tag{A4-8}$$

The derivative $A_{r_p \cdots r_1}{}^n$ of order $p \leq n$ is obtained from an A-product A^n by omitting the p factors $\varphi_{r_p} \cdots \varphi_{r_1}$ in A^n:

$$A_{r_p \cdots r_1}{}^n = A(n, \cdots r_p+1, r_p-1, \cdots r_1+1, r_1-1, \cdots 2, 1). \tag{A4-9}$$

For $p = n$, we define the derivative as 1. For $p = n-1$ it is φ_r, $(r = 1, \cdots, n)$. The number of derivatives of order p is $\binom{n}{p}$. It is clear that (A4-9) is a symmetrical function of the indices $r_p \cdots r_1$. The derivatives of B-products are defined analogously.

We define the *contraction symbol* C_{21} by

$$C_{21} \equiv A(2,1) - B(2,1). \tag{A4-10}$$

It can be expressed in terms of the commutator Σ_{21} and the function $\theta(x)$ introduced in Section A1-3 of this Appendix:

$$C_{21} = [\theta(\alpha_2 - \alpha_1) - \theta(\beta_2 - \beta_1)]\Sigma_{21}. \tag{A4-11}$$

It is apparent from this that C_{21} is a c-number, since Σ_{21} is one. Equation (A4-11) can be verified by taking the difference of the two equations:

$$A(2,1) = \theta(\alpha_2 - \alpha_1)\varphi_2\varphi_1 + \theta(\alpha_1 - \alpha_2)\varphi_1\varphi_2 = \theta(\alpha_2 - \alpha_1)[\varphi_2,\varphi_1] + \varphi_1\varphi_2,$$

$$B(2,1) = \theta(\beta_2 - \beta_1)\varphi_2\varphi_1 + \theta(\beta_1 - \beta_2)\varphi_1\varphi_2 = \theta(\beta_2 - \beta_1)[\varphi_2,\varphi_1] + \varphi_1\varphi_2.$$

We finally introduce the symbol

$$\Pi_{B_{2p}}{}^n = \sum_{\text{pairs}} C_{i_p j_p} \cdots C_{i_1 j_1} B_{i_p j_p \cdots i_1 j_1}{}^n. \tag{A4-12}$$

The summation in this expression is extended over all different selections of p pairs $(i_p j_p) \cdots (i_1 j_1)$. Two selections of pairs are considered equal if they differ only by a permutation of the pairs or an interchange of indices within a pair. The number of different selections of p pairs from n indices is $n!/[(n-2p)!p!2^p]$. This is the number of terms in (A4–12).

The ordering theorem may now be stated in the form of the equation

$$A^n = \sum_{p=0}^{N} \Pi_{B_{2p}}{}^n. \tag{A4–13}$$

The maximum value N of the summation index p in this equation is

$$N = \begin{cases} \dfrac{n}{2}, & \text{for } n \text{ even,} \\ \dfrac{n-1}{2}, & \text{for } n \text{ odd.} \end{cases} \tag{A4–14}$$

A dual form of this theorem may be obtained by interchanging the roles of α and β:

$$B^n = \sum_{r=0}^{N} \Pi_{A_{2p}}{}^n \tag{A4–13}'$$

It is sufficient to prove only one of the two relations, for instance (A4–13). The proof of the ordering theorem is based on the following lemma.

LEMMA. *Let*

$$\alpha_{n+1} \geq \alpha_i, \quad (i = 1, \cdots, n). \tag{A4–15}$$

Then

$$\varphi_{n+1} B^n = B^{n+1} + \sum_{r=1}^{n} C_{n+1,r} B_r{}^n. \tag{A4–16}$$

Proof of lemma. Let the index s be determined by the conditions

$$\beta_{j_r} > \beta_{n+1}, \quad \text{for all } r \geq s, \tag{A4–17}$$

$$\beta_{j_r} \leq \beta_{n+1}, \quad \text{for all } r < s. \tag{A4–18}$$

That is, in the descending series

$$\beta_{j_n} \geq \cdots \geq \beta_{j_2} \geq \beta_{j_1},$$

β_{j_s} is the last of the β's which is larger than β_{n+1}. For any $r \geq s$, we then have

$$\varphi_{n+1} \varphi_{j_r} = A(n+1, j_r) \tag{A4–19}$$

because of (A4–15), and also

$$\varphi_{j_r} \varphi_{n+1} = B(n+1, j_r) \tag{A4–20}$$

because of (A4–17) and $r \geq s$. Thus

$$C_{n+1,j_r} = [\varphi_{n+1}, \varphi_{j_r}], \quad \text{for } r \geq s. \tag{A4–21}$$

On the other hand, for $r < s$, because of (A4–15) and (A4–18), we have

$$A(n+1, j_r) = B(n+1, j_r), \tag{A4–22}$$

and consequently

$$C_{n+1,j_r} = 0, \quad \text{for } r < s. \tag{A4–23}$$

The expression $\varphi_{n+1} B^n$ is transformed into B^{n+1} by interchanging φ_{n+1} with all the φ_{j_r}, $(r \geq s)$. We obtain thus the identity

$$\varphi_{n+1} B^n = B^{n+1} + \sum_{r=n}^{s} [\varphi_{n+1}, \varphi_{j_r}] B_{j_r}{}^n. \tag{A4–24}$$

If we substitute in this (A4–21) and (A4–23), we obtain (A4–16):

$$\varphi_{n+1} B^n = B^{n+1} + \sum_{r=1}^{n} C_{n+1,r} B_r{}^n.$$

This proves the lemma.

The ordering theorem is now proved by induction with respect to n. For $n = 1$ it is trivial and for $n = 2$ it is the definition of the contraction symbol. We assume its validity for n and prove that it follows for $n + 1$.

Now A^{n+1} has on its left the factor with the largest of the $(n + 1)$ α-values. Let this value of α be denoted by α_{n+1}. (It may be necessary to change a previously chosen labeling of the parameters.) We then obtain

$$A^{n+1} = \varphi_{n+1} A^n, \quad \alpha_{n+1} \geq \alpha_i \quad (i = 1, \cdots, n). \tag{A4–25}$$

By assumption, (A4–13) holds for A^n, and therefore

$$A^{n+1} = \sum_{p=0}^{N} \varphi_{n+1} \Pi_{B_{2p}}{}^n. \tag{A4–26}$$

Now each $\Pi_{B_{2p}}{}^n$ is a sum of B-products, and for each of them the condition (A4–15) of the lemma is satisfied. Thus

$$\varphi_{n+1} \Pi_{B_{2p}}{}^n$$
$$= \sum_{\text{pairs}} C_{i_p j_p} \cdots C_{i_1 j_1} (B_{i_p j_p} \cdots {}_{i_1 j_1}{}^{n+1} + \sum_{r=1}^{n} C_{n+1,r} B_{i_p j_p} \cdots {}_{i_1 j_1 n+1 r}{}^{n+1}). \tag{4A–27}$$

By substituting this into (A4–26), we find

$$A^{n+1} = \sum_{p=0}^{N'} \sum_{\text{pairs}} C_{i_p j_p} \cdots C_{i_1 j_1} B_{i_p j_p} \cdots {}_{i_1 j_1}{}^{n+1} \tag{A4–28}$$

or

$$A^{n+1} = \sum_{p=0}^{N} \Pi_{B_{2p}}{}^{n+1}. \tag{A4–29}$$

with

$$N' = \begin{cases} \dfrac{n}{2}, & \text{for } n \text{ even,} \\ \dfrac{n+1}{2}, & \text{for } n \text{ odd.} \end{cases} \qquad (A4\text{--}30)$$

This proves the ordering theorem for commuting fields.

A4–2 The ordering theorem for anticommuting fields. Let us assume the anticommutation rules

$$\{\varphi_2,\varphi_1\} = \Sigma_{21}, \qquad (A4\text{--}31)$$

where Σ_{21} is a c-number.

The ordering theorem can be formulated in close analogy to the case of commuting fields provided only that we adapt the definitions to the sign difference between (A4–31) and (A4–2).

The A-product of n operators $\varphi_n, \cdots, \varphi_2, \varphi_1$ is defined by

$$A(\varphi_n \cdots \varphi_2 \varphi_1) = \delta_P \varphi_{i_n} \cdots \varphi_{i_2} \varphi_{i_1} \qquad (A4\text{--}32)$$

such that

$$\alpha_{i_n} \geq \cdots \geq \alpha_{i_2} \geq \alpha_{i_1}, \qquad (A4\text{--}33)$$

and where δ_P is the signature of the permutation

$$P = \begin{pmatrix} 1, & 2, & \cdots, & n \\ i_1, & i_2, & \cdots, & i_n \end{pmatrix}. \qquad (A4\text{--}34)$$

The B-product is defined by

$$B(\varphi_n \cdots \varphi_2 \varphi_1) = \delta_Q \varphi_{j_n} \cdots \varphi_{j_2} \varphi_{j_1} \qquad (A4\text{--}35)$$

such that

$$\beta_{j_n} \geq \cdots \geq \beta_{j_2} \geq \beta_{j_1} \qquad (A4\text{--}36)$$

and where δ_Q is the signature of the permutation

$$Q = \begin{pmatrix} 1, & 2, & \cdots, & n \\ j_1, & j_2, & \cdots, & j_n \end{pmatrix}. \qquad (A4\text{--}37)$$

If two or more parameter values α or β coincide, the definition is made precise by the requirement that the A- and B-products are antisymmetrical functions of their arguments for all values of the parameters

$$A(\varphi_{r_n} \cdots \varphi_{r_1}) = \delta_R A(\varphi_n \cdots \varphi_1), \qquad (A4\text{--}38)$$

where δ_R is the signature of the permutation

$$R = \begin{pmatrix} 1, & 2, & \cdots, & n \\ r_1, & r_2, & \cdots, & r_n \end{pmatrix}. \qquad (A4\text{--}39)$$

The derivatives of the A- and B-products are defined by

$$A_{r_p \cdots r_1}{}^n = \delta_P A(n, \cdots, r_p + 1, r_p - 1, \cdots r_1 + 1, r_1 - 1, \cdots, 2, 1), \tag{A4-40}$$

where P is the permutation

$$P = \begin{pmatrix} n, & \cdots, & n - p + 1, & n - p, & \cdots, & 1 \\ r_p, & \cdots, & r_1, & n, & \cdots, & 1 \end{pmatrix}. \tag{A4-41}$$

It follows that the derivative is an antisymmetrical function of the indices $r_p \cdots r_1$.

The contraction symbol is defined by (A4–10) or (A4–11) and the symbol $\Pi_{B_{2p}}{}^n$ by (A4–12). These definitions are identical for the two cases.

The ordering theorem for anticommuting fields has the same form as (A4–13). The sign differences which stem from the difference of the commutation rules in the two cases are all absorbed into the definition of the ordered products.

To prove the theorem for the anticommuting case, it is sufficient to prove the lemma (A4–16) for this case.

The proof of the lemma follows. We define the integers again by the properties (A4–17), (A4–18). For any $r \geq s$, we have then

$$\varphi_{n+1}\varphi_{j_r} = A(n + 1, j_r) \tag{A4-42}$$

because of (A4–15), and

$$\varphi_{j_r}\varphi_{n+1} = -B(n + 1, j_r) \tag{A4-43}$$

because of (A4–17) and (A4–35) and $r \geq s$. Consequently,

$$C_{n+1, j_r} = \{\varphi_{n+1}, \varphi_{j_r}\}, \tag{A4-44}$$

which replaces (A4–21) in the present case. On the other hand, for $r < s$, because of (A4–15) and (A4–18), we have

$$A(n + 1, j_r) = B(n + 1, j_r), \quad \text{for } r < s, \tag{A4-45}$$

and consequently

$$C_{n+1, j_r} = 0, \quad \text{for } r < s. \tag{A4-46}$$

Instead of (A4–24), we obtain in the present case

$$\varphi_{n+1}B^n = B^{n+1} + \sum_{r=n}^{s} \{\varphi_{n+1}, \varphi_{j_r}\} B_{j_r}{}^n \tag{A4-47}$$

The alternating signs which we obtain for the summation terms on the right are absorbed in the definition of $B_{j_r}{}^n$. Substitution of (A4–46) and (A4–44) into (A4–47) establishes the lemma and consequently the ordering theorem for the anticommuting case.

A4–3 A generalization of the ordering theorem.

Let us assume that each of the factors in an A-product is itself a B-product of a number of operators, each of which belongs to the same α-value. The ordering theorem is then still true in the form (A4–13) provided only that we define the expression $\Pi_{B_{2p}}{}^n$ by

$$\Pi_{B_{2p}}{}^n = {\sum_{\text{pairs}}}' C_{i_p j_p} \cdots C_{i_1 j_1} B_{i_p j_p \cdots i_1 j_1}{}^n \qquad \text{(A4–12)}'$$

The difference from (A4–12) is in the prime on the summation sign. The summation is thereby restricted to selection of pairs i, j, taken only from different factors in A. The proof involves only minor modifications of the two preceding proofs and is omitted here.

A4–4 The ordering of chronological products.

(a) *The radiation field.* The nth approximation of the S-matrix leads to the chronological product of the free-radiation field operators $a_{\mu_r}(x_r) \equiv a(r)$:

$$P[a(n) \cdots a(2)a(1)] \equiv a(i_n) \cdots a(i_2)a(i_1), \qquad \text{(A4–48)}$$

such that

$$\tau_{i_n} \geq \cdots \geq \tau_{i_2} \geq \tau_{i_1}, \qquad \text{(A4–49)}$$

where

$$\tau_r = -(n \cdot x_r). \qquad \text{(A4–50)}$$

The problem is to write this chronological product as an ordered product with respect to emission and absorption operators, that is, one in which all the absorption operators stand on the right and all the emission operators on the left.

We solve this problem with the general ordering theorem by defining the two parameters α and β for this case in the following manner

$$\alpha = \tau, \qquad \text{(A4–51)}$$

$$\beta = \begin{cases} +1, & \text{for } a^{(+)}, \\ -1, & \text{for } a^{(-)}, \end{cases} \qquad \text{(A4–52)}$$

where the $a^{(\pm)}$ are the emission $(+)$ and absorption $(-)$ parts of the field operators a [*cf.* Eqs. (2–63) and (2–64)]. Since $a^{(\pm)}$ occurs in (A4–48) only in the form $a^{(+)} + a^{(-)}$, we must carry out a summation over all the values of $\beta_n \cdots \beta_1$ in the general ordering theorem. The theorem then still holds in the form given by (A4–13) but with the new definition of the contraction symbol,

$$C(2,1) \equiv \sum_{\beta_2 \beta_1} C_{21}, \qquad \text{(A4–53)}$$

and the B-products,

$$[a(r) \cdots a(2)a(1)] \equiv \sum_{\beta_r \cdots \beta_1} B[a(r) \cdots a(1)]. \qquad \text{(A4–54)}$$

The main problem then is to calculate the contraction symbol $C(2,1)$. To this end, we use the commutation rules (2–65), (2–66):

$$[a^{-}(2), a^{(+)}(1)] = -iD_{+}(2,1), \tag{A4–55}$$

where

$$D_{+}(2,1) = g_{\mu_2\mu_1} D_{+}(x_2 - x_1) \tag{A4–56}$$

and, similarly,

$$[a^{(+)}(2), a^{(-)}(1)] = -iD_{-}(2,1), \tag{A4–57}$$

while

$$[a^{(+)}(2), a^{(+)}(1)] = [a^{(-)}(2), a^{(-)}(1)] = 0.$$

Substitution of these expressions into (A4–11) and (A4–53) gives

$$C(2,1) = -i[-\theta(1,2) D_{-}(2,1) + \theta(2,1) D_{+}(2,1)]$$

The expression in brackets on the right is $D_{1R}(2,1) = D_c(2,1)$ [cf. Eq. (A1–4)].* We find thus the important relation

$$C(2,1) = -iD_c(2,1). \tag{A4–58}$$

(b) *The matter field.* The chronological product of the matter field is a product of n factors, each of which is a bilinear expression of ordered ψ-operators. This is an A-product with the definition (A4–51) of the parameter α including the correct sign factor. Such a product therefore satisfies the hypotheses of the generalized ordering theorem (Section A4–3). Thus the only problem left is the calculation of the contraction symbol. There are four such separate symbols, since we have two different kinds of fields, ψ and $\bar{\psi}$. But it is readily seen that the symbols arising from the same kind of fields vanish.

$$C(\psi(2), \psi(1)) = C(\bar{\psi}(2), \bar{\psi}(1)) = 0 \tag{A4–59}$$

because the corresponding commutators vanish. Thus we need to evaluate only $C(\psi(2), \bar{\psi}(1))$ and $C(\bar{\psi}(1), \psi(2))$. The latter is the negative of the former because of the antisymmetry property of the A- and B-products. The remaining symbol is readily evaluated with (A4–11) and (A4–53):

$$C(\psi(2), \bar{\psi}(1)) = iS_c(2,1). \tag{A4–60}$$

The position of the two contracted factors in the original product is actually irrelevant. We can always rearrange the original chronological product in such a way that the two factors of a contracted pair are adjacent to each other, with ψ on the left and $\bar{\psi}$ on the right. Having done this and then replaced the adjacent pairs by (A4–60), we obtain a chain of interlocking contraction symbols. The chain may be either closed or open. In

---

* This conforms with the notation of Stueckelberg, who first connected this function with the *causal* property of the interaction.

the closed chain, the last operator ψ is contracted with the first $\bar{\psi}$ in the chain. The open chain starts with an operator $\bar{\psi}$ and ends with an operator ψ.

We may summarize the result of this section as follows:

(a) The nth chronological product of the radiation field operators is a sum of ordered products, each containing at most n factors.

A particular one of these ordered products of $n - 2m$ factors is obtained by selecting m pairs of operators from the chronological product and replacing each pair by the contraction symbol (A4–58).

The remaining factors are collected into the ordered product.

(b) The nth chronological product of the matter field operators is a sum of ordered products of m factors $\bar{\psi}$ and m factors ψ ($m \leq n$).

A particular one of these ordered products is obtained by selecting m pairs of operators from different factors, each pair consisting of one operator ψ and one operator $\bar{\psi}$, and replacing each pair by the contraction symbol (A4–60).

The remaining factors are collected into the ordered product.

APPENDIX A5

ON THE EVALUATION OF CERTAIN INTEGRALS

In Chapter 9 we encountered integrals over the intermediate states of the system which play an important role both in the discussion of divergences and in the computations of observable phenomena.

The most convenient method of evaluation for these integrals was given by Feynman.* This method involves the introduction of auxiliary scalar variables, so that the original integration over the four-momentum space of the virtual particles reduces to a standard type of integration which can easily be done. One is then left with the integration over the auxiliary variables, which forms the main task in the actual computations.

There are three advantages to this procedure. One is its explicit covariance, the second is its property to reduce the number of poles to two (see below), and the third is its applicability to all integrals which occur in the evaluation of S-matrix elements, including divergent integrals.

A typical integral is of the form

$$I = \int \frac{F(k) d^4 k}{a_1 a_2 \cdots a_n}, \tag{A5-1}$$

where the a_i are polynomials of second degree in the four-vectors k_μ, and $F(k)$ is a polynomial of degree n'.

The introduction of the auxiliary variables, which we shall call $x_1, x_2, \cdots$, is carried out so that the denominator $a_1 \cdots a_n$ becomes the nth power of a second degree polynomial in k_μ:

$$\frac{1}{a_1 a_2 \cdots a_n} = (n-1)! \int_0^1 dx_1 \int_0^{x_1} dx_2 \cdots$$
$$\times \int_0^{x_{n-2}} dx_{n-1} \frac{1}{[a_1 x_{n-1} + a_2(x_{n-2} - x_{n-1}) + \cdots + a_n(1 - x_1)]^n}. \tag{A5-2}$$

A product of n polynomials requires therefore $n - 1$ auxiliary variables. The identity (A5-2) can be proved by induction as follows. For $n = 1$ no auxiliary variable needs to be introduced. For $n = 2$, we have

$$\frac{1}{a_1 a_2} = \int_0^1 \frac{dx}{[a_1 x + a_2(1 - x)]^2}, \tag{A5-3}$$

---

* R. P. Feynman, *Phys. Rev.* **76**, 769 (1949).

which can be verified by elementary integration. We now introduce the variables u_k

$$x_1 = u_1,$$
$$x_2 = u_1 u_2, \qquad (A5\text{-}4)$$
$$\vdots$$
$$x_{n-1} = u_1 u_2 \cdots u_{n-1}.$$

The identity (A5-2) thereby attains another useful form,

$$\frac{1}{a_1 a_2 \cdots a_n} = (n-1)! \int_0^1 u_1^{n-2} du_1 \int_0^1 u_2^{n-3} du_2 \cdots \int_0^1 du_{n-1}$$
$$\times \frac{1}{[a_1 u_1 \cdots u_{n-1} + a_2 u_1 \cdots u_{n-2}(1 - u_{n-1}) + a_n(1 - u_1)]^n}. \qquad (A5\text{-}5)$$

The integrations over the new variables $u_1 \cdots u_{n-1}$ are interchangeable. Assume that the identity (A5-2) and therefore (A5-5) is true for n polynomials. Then we can write for $n+1$ polynomials,

$$\frac{1}{a_1 a_2 \cdots a_n a_{n+1}} = (n-1)! \int_0^1 u_1^{n-2} du_1 \cdots \int_0^1 du_{n-1} \frac{1}{A^n a_{n+1}}, \qquad (A5\text{-}6)$$

where A^n is the denominator in the integral of (A5-5). But this expression can be rewritten by use of the equation

$$\int_0^1 \frac{u^{n-1} du}{[Au + B(1-u)]^{n+1}} = \frac{1}{nA^n B}, \qquad (A5\text{-}7)$$

which can be proved by elementary methods. We find, with $u = u_0$, $B = a_{n+1}$,

$$\frac{1}{a_1 a_2 \cdots a_n a_{n+1}} = n! \int_0^1 u_0^{n-1} du_0 \cdots \int_0^1 du_{n-1} \frac{1}{[Au_0 + a_{n+1}(1 - u_0)]^{n+1}}.$$
$$(A5\text{-}8)$$

Except for the labeling of the auxiliary variables, this equation is identical with Eq. (A5-5) for $n + 1$ polynomials. The identity (A5-2) is thereby established by induction.

A5-1 Convergent integrals. We shall assume in this section that the integrals to be evaluated are absolutely convergent. The following operations can then be carried out without restrictions. The case of divergent integrals is separately discussed in the following section.

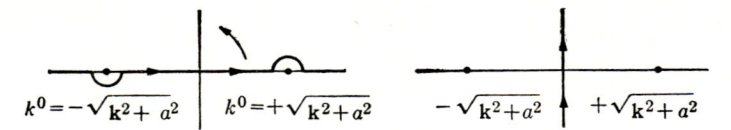

FIG. A5-1. The k^0-integration in the complex k^0-plane.

The denominator in (A5–2) or (A5–5) can always be written in the form

$$A^n = [(k-g)^2 + a^2]^n,$$

where the four-vector g_μ and the scalar a are independent of the integration variable k, but depend on the auxiliary variables x. Since the integral is assumed to be convergent, we can shift the origin of the k-space,

$$k_\mu \to k_\mu + g_\mu,$$

so that A^n becomes

$$A^n = (k^2 + a^2)^n.$$

We next average the integrand over the direction of the four-vector k_μ, which amounts to the substitutions

$$k_\mu k_\nu = \tfrac{1}{4} g_{\mu\nu} k^2,$$

$$k_\mu k_\nu k_\lambda k_\sigma = \tfrac{1}{24}(k^2)^2(g_{\mu\nu}g_{\lambda\sigma} + g_{\mu\lambda}g_{\nu\sigma} + g_{\mu\sigma}g_{\nu\lambda}). \quad (A5-9)$$

With these expressions, the integrals are of the form

$$I_{mn} = \int \frac{(k^2)^{m-2} d^4k}{(k^2 + a^2)^n}. \quad (A5-10)$$

For the evaluation of integrals of this sort, we recall the prescription for handling the poles of the integrand [*cf.* Section 8–3, especially Eqs. (8–18) and (8–19)]. Each of the original denominators a_s must be provided with an additional purely imaginary term $-i\mu_s$. Equation (A5–2) or (A5–5) simply replaces these $-i\mu_s$ by a new term, $-i\mu$, such that the $2n$ simple poles are reduced to 2 poles of nth order which are, after a shift of origin, at the points

$$k^0 = \pm\sqrt{|\mathbf{k}|^2 + a^2}.$$

It is now possible to rotate the path of integration in the complex k^0-plane by $+90°$ without crossing a pole (see Fig. A5–1). The k^0-integration then has the limits $-i\infty$ to $+i\infty$.

We put

$$k^0 = ik^{0\prime}, \quad \mathbf{k} = \mathbf{k}',$$

so that $k^{0\prime}$ is integrated from $-\infty$ to $+\infty$, and we find

$$I_{mn} = i\int \frac{(k'^2)^{m-2}d^4k'}{(k'^2 + a^2)^n}.$$

The integration is now to be carried out over a four-dimensional *spherical* space, $k'^2 = \mathbf{k}'^2 + k_0'^2$. We can introduce four-dimensional polar coordinates κ, φ, θ, χ, such that

$$\int d^4k' = \int \kappa^3 d\kappa \int_0^{2\pi} d\varphi \int_0^{\pi} \sin\theta d\theta \int_0^{\pi} \sin^2\chi d\chi$$

$$= 2\pi^2 \int \kappa^3 d\kappa, \qquad (A5\text{--}11)$$

and we integrate over the angles first;

$$I_{mn} = 2\pi^2 i \int_0^\infty \frac{(\kappa^2)^{m-2}\kappa^3 d\kappa}{(\kappa^2 + a^2)^n}$$

$$= i\pi^2 \int_0^\infty \frac{t^{m-1}dt}{(t + a^2)^n} = \frac{i\pi^2}{(a^2)^{n-m}} \cdot \frac{\Gamma(m)\,\Gamma(n-m)}{\Gamma(n)}.$$

Therefore,

$$\int \frac{(k^2)^{m-2}d^4k}{(k^2 + a^2)^n} = \frac{i\pi^2}{(a^2)^{n-m}} B(m, n-m). \qquad (A5\text{--}12)$$

In the last form we have introduced the Gaussian beta-function $B(m, n-m) = \Gamma(m)\Gamma(n-m)/\Gamma(n)$. The condition for its existence is $n > m > 0$. This is also the condition of convergence for the integrals.

In the following, we call this method of integration in k-space *symmetrical integration*. It consists of an averaging process over all angles [*cf.* Eq. (A5–9)] followed by an integration over k-space in which the angular integration is carried out first.

A5–2 Divergent integrals. The method of handling the convergent integrals serves as a model for a prescription of separating the physically meaningless divergent parts from the finite parts in the divergent integrals. The justification for this particular method of separation is to be sought primarily in the physical interpretation of the result.

In the manipulation of divergent integrals, special attention must be paid to the shift of origin in k-space. The following two examples will illustrate this point.

Consider the integral

$$I_0 = \int \frac{d^4k}{[(k-p)^2 + a^2]^2}.$$

It is logarithmically divergent. We make use of the identity

$$\frac{1}{\alpha^n} - \frac{1}{\beta^n} = -\int_0^1 \frac{n(\alpha - \beta)}{[(\alpha - \beta)z + \beta]^{n+1}} dz \qquad (A5\text{–}13)$$

and find, for $n = 2$,

$$I_0 = \int \frac{d^4k}{(k^2 + a^2)^2} - 2\int d^4k \int_0^1 \frac{(p^2 - 2p \cdot k)dz}{[k^2 + a^2 + (p^2 - 2p \cdot k)z]^3}.$$

In the last integral the k-integration is convergent, so that the integrations may be interchanged, and the method for convergent integrals can be applied. We first shift the origin,

$$k_\mu \to k_\mu + p_\mu z,$$

so that

$$-2\int d^4k \int_0^1 dz \frac{p^2 - 2p \cdot k}{[(k - pz)^2 + a^2 + p^2 z(1 - z)]^3}$$

$$= -2\int_0^1 dz \int \frac{p^2(1 - 2z) - 2p \cdot k}{[k^2 + a^2 + p^2 z(1 - z)]^3} d^4k.$$

The last term in the numerator is odd in k and will vanish upon symmetrical integration. Using (A5–12), we have

$$-2\frac{i\pi^2}{2} \int_0^1 dz \frac{p^2(1 - 2z)}{a^2 + p^2 z(1 - z)} = -i\pi^2 \ln[a^2 + p^2 z(1 - z)]_0^1 = 0.$$

It follows that

$$I_0 = \int \frac{d^4k}{[(k - p)^2 + a^2]^2} = \int \frac{d^4k}{(k^2 + a^2)^2}. \qquad (A5\text{–}14)$$

We can show, in general, that in a logarithmically divergent integral of this type the origin of k can be shifted. This simply follows from the fact that the difference between two logarithmically divergent integrals whose integrands are asymptotically identical is always convergent.

Next we consider a linearly divergent integral,

$$I_1 = \int \frac{k d^4k}{[(k - p)^2 + a^2]^2} = \int \frac{(k + p) d^4k}{(k^2 + a^2)^2} + S_1,$$

where

$$S_1 = -p \int \frac{d^4k}{(k^2 + a^2)^2} + \int k d^4k \left(\frac{1}{[(k - p)^2 + a^2]^2} - \frac{1}{(k^2 + a^2)^2} \right).$$

Equation (A5–13) enables us to write the second integral in S_1 as

$$-2\int \not{k}d^4k \int_0^1 dz \frac{p^2 - 2p\cdot k}{[k^2 + a^2 + (p^2 - 2p\cdot k)z]^3}.$$

This integral is only logarithmically divergent in k-space. Therefore, we can shift the origin of k, $k \to k + pz$:

$$-2\int d^4k \int_0^1 (\not{k} + \not{p}z)\, dz \frac{p^2(1 - 2z) - 2p\cdot k}{[k^2 + a^2 + p^2z(1 - z)]^3}.$$

The denominator is now again reduced to a function of k^2 only, so that we can impose the requirement of symmetrical integration and drop all odd terms in k. This yields, with (A5–9),

$$S_1 = -\not{p}\int \frac{d^4k}{(k^2 + a^2)^2} - 2\not{p}\int d^4k \int_0^1 dz \frac{p^2 z(1-2z)}{[k^2 + a^2 + p^2z(1-z)]^3}$$

$$+ \not{p}\int k^2 d^4k \int_0^1 dz \frac{1}{[k^2 + a^2 + p^2z(1-z)]^3}.$$

The second integral can be integrated by parts over z:

$$-2\not{p}\int_0^1 dz \frac{p^2 z(1-2z)}{[k^2 + a^2 + p^2z(1-z)]^3}$$

$$= \not{p}\,\frac{1}{(k^2 + a^2)^2} - \not{p}\int_0^1 \frac{dz}{[k^2 + a^2 + p^2z(1-z)]^2}.$$

The first term exactly cancels the first term in S_1. We are left with

$$S_1 = \not{p}\int d^4k \int_0^1 dz \left[\frac{k^2}{[k^2 + a^2 + p^2z(1-z)]^3} - \frac{1}{[k^2 + a^2 + p^2z(1-z)]^2} \right]$$

$$= -\not{p}\int d^4k \int_0^1 dz \frac{a^2 + p^2z(1-z)}{[k^2 + a^2 + p^2z(1-z)]^3}.$$

We note that this expression can be put into the form

$$S_1 = -\int d^4k \int dz\, p_\mu \frac{\partial}{\partial k_\mu} \frac{k}{[k^2 + a^2 + p^2z(1-z)]^2}.$$

For this reason, S_1 is sometimes referred to as a *surface term*.

The integration over k is now convergent and gives, according to Eq. (A5–12),

$$S_1 = -\frac{i\pi^2}{2}\not{p}. \tag{A5-15}$$

Therefore, we have the result

$$\int \frac{k d^4 k}{[(k-p)^2+a^2]^2} = \int \frac{(k+p) d^4 k}{(k^2+a^2)^2} - \frac{i\pi^2}{2} p. \qquad (A5\text{--}16)$$

In general, in a linearly divergent integral of this type, a shift of origin, such that the denominator becomes a function of k^2 only, always yields a similar finite additive constant. The argument is completely analogous to the case of logarithmically divergent integrals discussed above.

These examples show how the origin in a divergent integral can be shifted. The resultant integral is now to be treated in a standard way. The sum of terms in the numerator can be separated so that the whole integral separates into a divergent and one or more convergent integrals. The latter are integrated as in A5–1. For the former, symmetrical integration is postulated, such that a divergent integral whose denominator is of the form $(k^2 + a^2)^n$ and whose numerator is odd in k is considered to be zero. The remaining divergent integrals cannot be evaluated further without the introduction of a cutoff function, but there is no need to do so, since the separation of the finite terms has been accomplished.

A5–3 The integral for the photon self-energy part. In Eq. (9–65) the finite part of the photon self-energy was obtained in form of an integral,

$$-k^2 \Pi_f = \frac{2\alpha}{\pi} \int_0^1 x(1-x) \ln\left(1 + \frac{k^2}{m^2} x(1-x)\right) dx. \qquad (A5\text{--}17)$$

To evaluate this integral, we note that the logarithm is to be evaluated in the limit of a vanishing negative imaginary part of k^2. This is equivalent to replacing the first term in the argument of the logarithm by $1 - i\mu$ and going to the limit $\mu \to +0$.

Consider first the function (a real)

$$T(a) = 2 + \lim_{\mu \to +0} \int_0^1 \ln[1 - i\mu + ax(1-x)] dx$$

$$= \frac{1}{2} \lim_{\mu \to +0} \sqrt{1 + \frac{4}{a}(1 - i\mu)} \ln \frac{1 - (1/z_-)}{1 - (1/z_+)}, \qquad (A5\text{--}18)$$

where

$$z_\pm = \frac{1}{2}\left(1 \pm \sqrt{1 + \frac{4}{a}(1 - i\mu)}\right)$$

are the roots of the argument in the logarithm. This function takes on the following forms:

$$T(a) = 2\sqrt{1 + \frac{4}{a}} \sinh^{-1} \frac{\sqrt{a}}{2}, \quad (a \geq 0),$$

$$= 2\sqrt{\frac{4}{|a|} - 1} \sin^{-1} \frac{\sqrt{|a|}}{2}, \quad (-4 \leq a \leq 0),$$

$$= \sqrt{1 - \frac{4}{|a|}} \left(2 \cosh^{-1} \frac{\sqrt{|a|}}{2} - i\pi \right), \quad (a \leq -4). \quad \text{(A5-19)}$$

For small a, $T(a)$ can be expanded:

$$T(a) = 2 + \frac{a}{6} - \frac{a^2}{60} + \frac{a^3}{420} - \cdots. \quad \text{(A5-20)}$$

Consider next the set of functions

$$V_n(a) = \lim_{\mu \to +0} \int_0^1 [x(1-x)]^n \ln[1 - i\mu + ax(1-x)]dx. \quad \text{(A5-21)}$$

We shall be interested especially in V_0 and V_1. V_0 is already known to us:

$$V_0(a) = T(a) - 2. \quad \text{(A5-22)}$$

The function $V_1(a)$ can easily be obtained from $V_0(a)$. In fact, omitting the limiting process for convenience,

$$\int_0^a V_1(b)db = \int_0^1 x(1-x)dx \int_0^a \ln[1 + bx(1-x)]db$$

$$= V_0(a) + aV_1(a) - \frac{a}{6}.$$

This relation can be differentiated and yields

$$V_1(a) = \int_0^a \frac{db}{b} \left(\frac{1}{6} - \frac{dV_0(b)}{db} \right) = \int_0^a \frac{db}{b} \left(\frac{1}{6} - \frac{dT(b)}{db} \right). \quad \text{(A5-23)}$$

We also note the relation

$$\int_0^1 (1 - 2x) \ln[1 + ax(1-x)]dx = 0. \quad \text{(A5-24)}$$

which follows upon substitution of $y = 1 - x$. It implies that V_0 can be expressed in the following two ways:

$$V_0(a) = \int_0^1 \ln[1 + ax(1-x)]dx = 2\int_0^1 x \ln[1 + ax(1-x)]dx. \quad \text{(A5-25)}$$

The evaluation of $V_1(a)$ from Eq. (A5-23) is uniquely determined by $T(a)$ as given in Eqs. (A5-18) and (A5-19). The integration of each case is then elementary.

$V_1(a)$ can also be evaluated directly when a is considered a complex number,

$$\frac{a}{1 - i\mu},$$

and the limit $\mu \to 0$ is taken at the end. Integrating by parts, we find

$$V_1(a) = -\int_0^1 a\left(\frac{x^2}{2} - \frac{x^3}{3}\right)\frac{1 - 2x}{1 + ax(1-x)} dx$$

$$= \frac{1}{\sqrt{1 + (4/a)}} \int_0^1 \left(\frac{1}{2} - \frac{4}{3}x + \frac{2}{3}x^2\right)\left(\frac{1}{x - z_+} - \frac{1}{x - z_-}\right) x^2 dx,$$

where $z_\pm$ are the roots which occurred previously in Eq. (A5-18). The rest is again elementary.

Equation (A5-17) gives

$$-k^2 \Pi_f = \frac{2\alpha}{\pi} V_1\left(\frac{k^2}{m^2}\right) \quad \text{(A5-26)}$$

with

$$V_1(a) = -\frac{1}{6}\left[\frac{5}{3} - \frac{4}{a} - \left(1 - \frac{2}{a}\right)\sqrt{1 + \frac{4}{a}} \ln \frac{\sqrt{1 + (4/a)} + 1}{\sqrt{1 + (4/a)} - 1}\right]. \quad \text{(A5-27)}$$

The functions $T(a)$ and $V_n(a)$ occur frequently in S-matrix elements involving closed loops, and the above remarks illustrate the connection between these functions.

Although the functions $V_n(a)$ are characteristic of the type of integrals one encounters in the evaluation of S-matrix elements in the iteration solution, not all integrals can thus be evaluated in terms of elementary functions. In particular, there occur integrals of the type

$$L(z) = \int_0^z \frac{\ln(1 - \zeta)}{\zeta} d\zeta.$$

These integrals were first studied by Spence* and have recently been tab-

---

* W. Spence, *An Essay on the Theory of the Various Orders of Logarithmic Transcendents*, London and Edinburgh, 1809.

A5–4 The integral for the electron self-energy part. In Section 9–4, Eq. (9–24), we obtained the expression for the finite self-energy part $\Sigma_f(p)$ in the form

$$\Sigma_f(p) = \frac{\alpha}{2\pi} \int_0^1 x(1-x)dx$$

$$\times \int_0^1 dz \, \frac{m(1+x) + (i\not{p}-m)\left[1-x+2z\left(x-\frac{1}{x}\right)\right]}{m^2x^2 + (p^2+m^2)x(1-x)z}.$$

The integration with respect to z can be carried out according to the formulas

$$\int_0^1 \frac{dz}{A+Bz} = \frac{1}{B}\ln\left(1+\frac{B}{A}\right),$$

$$\int_0^1 \frac{z\,dz}{A+Bz} = \frac{1}{B}\left[1 - \frac{A}{B}\ln\left(1+\frac{B}{A}\right)\right].$$

Substituting these expressions into $\Sigma_f(p)$ with $A = m^2x^2$, $B = (p^2+m^2)x(1-x)$, and writing $\rho = (p^2+m^2)/m^2$, we find

$$\Sigma_f(p) = \frac{\alpha}{2\pi m}\left[\frac{I_1}{\rho} + \frac{i\not{p}-m}{m}\frac{1}{\rho}\left(I_2 + 2\frac{I_1}{\rho} + 1 - 2\int_0^1\frac{dx}{x}\right)\right], \quad \text{(A5–28)}$$

where

$$I_1 = \int_0^1 (1+x)\ln\left[1+\rho\left(\frac{1}{x}-1\right)\right] = \frac{\rho}{2(1-\rho)}\left(1 - \frac{2-3\rho}{1-\rho}\ln\rho\right),$$

and

$$I_2 = \int_0^1 (1-x)\ln\left[1+\rho\left(\frac{1}{x}-1\right)\right] = -\frac{\rho}{2(1-\rho)}\left(1 + \frac{2-\rho}{1-\rho}\ln\rho\right).$$

When $\rho < 0$, that is, $p^2 < -m^2$, the phase of the logarithm is obtained by evaluating the integrals with m^2 replaced by $m^2 - i\mu$ ($\mu > 0$). In the limit $\mu \to +0$ the integrals contain imaginary parts independent of μ. For instance,

$$\ln \rho = \ln|\rho| - i\pi, \quad \text{for } \rho < 0.$$

---

* K. Mitchell, *Phil. Mag.* **40**, 351 (1949). Further references can be found in this paper.

APPENDIX A6

A LIMITING RELATION FOR THE δ-FUNCTION

In the general theory of the S-matrix the following limiting relations are used:

$$e^{-i\omega\tau} \lim_{\epsilon \to +0} \frac{1}{\omega + i\epsilon} = \begin{cases} -2\pi i \delta(\omega), & \text{for } \tau \to +\infty, \\ 0, & \text{for } \tau \to -\infty, \end{cases} \quad \text{(A6-1)}$$

and, similarly,

$$e^{-i\omega\tau} \lim_{\epsilon \to +0} \frac{1}{\omega - i\epsilon} = \begin{cases} 0, & \text{for } \tau \to +\infty, \\ 2\pi i \delta(\omega), & \text{for } \tau \to -\infty. \end{cases} \quad \text{(A6-2)}$$

These relations are easily established if we use the integral representation of $1/(\omega + i\epsilon)$:

$$\lim_{\epsilon \to +0} \frac{1}{\omega + i\epsilon} = \lim_{\epsilon \to +0} (-i) \int_0^\infty e^{(i\omega - \epsilon)x} dx = -i \int_0^\infty e^{i\omega x} dx. \quad \text{(A6-3)}$$

Therefore

$$\lim_{\epsilon \to +0} \frac{e^{-i\omega\tau}}{\omega + i\epsilon} = -i \int_0^\infty e^{i\omega(x-\tau)} dx = -i \int_{-\tau}^\infty e^{i\omega y} dy. \quad \text{(A6-4)}$$

The result (A6-1) is now immediately obvious if use is made of the well-known integral representation of the δ-function:

$$2\pi \delta(\omega) = \int_{-\infty}^{+\infty} e^{i\omega\tau} d\tau.$$

In a similar way we can prove (A6-2).

APPENDIX A7

THE METHOD OF ANALYTIC CONTINUATION

This method has proved useful for certain problems, and it offers considerable physical insight even for problems where it is of only academic interest. It has developed essentially on a classical basis and relatively little is known of a strictly quantum field theoretical treatment.

A7–1 The Bohr-Peierls-Placzek relation. Consider a classical monochromatic wave propagating in the x-direction:

$$u_\omega(x,t) = u(\omega)e^{ikx - i\omega t}. \tag{A7-1}$$

For an electromagnetic wave in vacuo,

$$k = k_0 = \frac{1}{\lambda_0} = \omega,$$

whereas in a medium of index of refraction $n(\omega)$,

$$k = k(\omega) = \frac{1}{\lambda} = nk_0 = \omega n(\omega).$$

It is therefore convenient to define the dispersion function

$$f(\omega) = k - k_0 = \omega(n-1) \tag{A7-2}$$

so that $f(\omega) = 0$ in vacuo. The above wave can now be written

$$u_\omega(x,t) = u(\omega)e^{if(\omega)x - i\omega(t-x)}.$$

For an absorptive medium the index of refraction is complex:

$$n = n_1 + in_2, \quad f = f_1 + if_2. \tag{A7-3}$$

The imaginary parts are related to the absorption coefficient

$$\alpha(\omega) = N\sigma(\omega),$$

where N is the number of atoms per unit volume in the medium and $\sigma(\omega)$ is the *total* cross section, i.e., the sum of the scattering and the absorption cross section:

$$\sigma(\omega) = \sigma_s(\omega) + \sigma_a(\omega).$$

The relation between n_2 and $\sigma(\omega)$ is

$$n_2(\omega) = \frac{\alpha(\omega)}{2\omega} = \frac{1}{2\omega} N\sigma(\omega), \tag{A7-4}$$

so that the intensity of the wave decreases by $1/e$ over a distance $x = 1/\alpha$. It follows that

$$f_2(\omega) = \tfrac{1}{2}\alpha(\omega) = \tfrac{1}{2}N\sigma(\omega). \tag{A7-5}$$

The index of refraction $n(\omega)$ is related to the forward scattering amplitude $a(\omega)$ by†

$$n(\omega) = 1 + \frac{2\pi}{\omega^2} N a(\omega). \tag{A7-6}$$

Since this amplitude is necessarily complex,

$$a = a_1 + ia_2,$$

we obtain from Eq. (A7-4) and (A7-6)

$$\sigma(\omega) = \frac{4\pi}{\omega} a_2(\omega). \tag{A7-7}$$

This connection between the total cross section and the imaginary part of the forward scattering amplitude is known as the *Bohr-Peierls-Placzek relation*.‡ It can be shown to be a direct consequence of the law of conservation of probability. In fact, it simply says that whatever is missing in the forward beam is either scattered or absorbed.

It is clear, therefore, that the Bohr-Peierls-Placzek relation is very generally valid. It can be proved on the basis of S-matrix theory as follows.

The S-operator

$$S = 1 + R$$

is unitary, so that

$$SS^* = 1 + R + R^* + RR^* = 1, \quad -RR^* = R + R^*.$$

The corresponding matrix element between two states a and b is

$$\sum (b|R|c)(c|R^*|a) = -2 \operatorname{Re} (b|R|a).$$

On the other hand, we defined the matrix element $(b|M|a)$ in Eq. (14-86) by

$$(b|R|a) = \delta(E_b - E_a)(b|M|a),$$

so that the above relation can be written in the form

$$\sum_c (b|M|c)(c|M^*|a)\delta(E_b - E_a)\delta(E_c - E_a) = -2 \operatorname{Re} \delta(E_b - E_a)(b|M|a);$$

---

† This relation is valid for $|n - 1| \ll 1$ and follows from the interference effect between the incident and the scattered wave. Special cases of it are known from classical electron theory. [Cf. also the fine-print sections following Eq. (A7-26).]

‡ N. Bohr, R. Peierls, and G. Placzek, *Nature* **144**, 200 (1939).

or, after dropping the factor $\delta(E_b - E_a)$ which corresponds to a multiplicative time interval, and putting $b = a = i$, $c = f$,

$$S_f \delta(E_f - E_i)|(f|M|i)|^2 = -2\,\text{Re}\,(i|M|i). \tag{A7-8}$$

This relation can be inserted into the general expression for the total cross section, Eq. (14–91), and we find for the scattering of a photon

$$\sigma = -(2\pi)^2 2\,\text{Re}\,(i|M|i).$$

The initial state can thereby be assumed to be a single state, so that no average over initial substates is necessary. There is no loss of generality involved here.

The forward scattering cross section is

$$\frac{d\sigma(\omega,0)}{d\Omega} = (2\pi)^2 |(i|M|i)|^2 \omega^2,$$

so that

$$|a(\omega)|^2 = |2\pi\omega(i|M|i)|^2.$$

The correct phase factor is $-i$, as will be shown below; therefore,

$$a(\omega) = -2\pi i \omega (i|M|i).$$

The total cross section becomes

$$\sigma(\omega) = -\frac{4\pi}{\omega}\,\text{Re}\,(ia(\omega))$$

$$= \frac{4\pi}{\omega}\,\text{Im}\,a(\omega)$$

$$= \frac{4\pi}{\omega} a_2(\omega).$$

This is exactly the relation (A7-7). We conclude that the unitarity of the S-matrix (which is the only assumption that entered this calculation) corresponds to the conservation of probability.

It remains to prove the relation between the forward scattering amplitude and the matrix element $(i|M|i)$. For this purpose, it is only necessary to take the Fourier transform of the wave matrix, Eq. (7–25), since T_+ describes the scattered wave. In the forward direction and for large r we easily find for photon scattering*

$$\frac{1}{(2\pi)^{3/2}} \int T_+(\mathbf{k}',\mathbf{k}) e^{i\mathbf{k}'\cdot\mathbf{r}} d^3 k'$$

$$= -\frac{i}{\sqrt{2\pi}} \int_0^\infty \omega' d\omega' \left[\frac{e^{i\omega' r}}{r} T_+(\omega',0;\mathbf{k}) - \frac{e^{-i\omega' r}}{r} T_+(\omega',\pi;\mathbf{k})\right].$$

---

* P. A. M. Dirac, *Quantum Mechanics*, Oxford, Clarendon Press, 1947, p. 196.

We can now use Eq. (7–34)$_+$ and find, with the help of Eqs. (A6–1) and (A6–2), that the scattered wave is

$$-\frac{e^{i\omega r}}{r}\sqrt{2\pi}\,\omega G_+(\mathbf{k},\mathbf{k}).$$

On the other hand, the incident wave is

$$\frac{1}{(2\pi)^{3/2}}\int \delta(\mathbf{k}'-\mathbf{k})e^{i\mathbf{k}'\cdot\mathbf{r}}d^3k' = \frac{1}{(2\pi)^{3/2}}e^{i\mathbf{k}\cdot\mathbf{r}}.$$

The scattering amplitude is now defined as the coefficient of $e^{i\omega r}/r$ in the asymptotic form of the scattered wave, when the incident wave has intensity 1. Therefore,

$$a(\omega) = -(2\pi)^2\omega G_+(\mathbf{k},\mathbf{k}).$$

The relation between the S-matrix and G_+ is given in Eq. (7–67)$_+$, so that, in general,

$$(f|M|i) = -2\pi i(f|G_+|i).$$

Combining these two equations, we obtain

$$a(\omega) = -2\pi i\omega(i|M|i),$$

as stated above.

A7–2 The principle of limiting distance. The principle which is closely related to the principle of causality can be stated as follows. Any disturbance $u(x,t)$ which satisfies

$$u(0,t) = 0, \quad (t < 0), \tag{A7–9}$$

must also satisfy

$$u(x,\tau + x) = 0, \quad (\tau < 0, x > 0). \tag{A7–10}$$

This means physically that a disturbance which had not reached $x = 0$ before time t cannot reach the point x at a time prior to $t = x$.

This principle is obviously satisfied in vacuo, where $f(\omega) = 0$. A disturbance

$$u(x,t) = \int_{-\infty}^{\infty} u_\omega(x,t)d\omega = \int_{-\infty}^{\infty} u(\omega)e^{ikx-i\omega t}d\omega \tag{A7–11}$$

yields

$$u(0,t) = \int_{-\infty}^{\infty} u(\omega)e^{-i\omega t}d\omega,$$

and with $k = \omega$,

$$u(x,\tau + x) = \int_{-\infty}^{\infty} u(\omega)e^{-i\omega\tau}d\omega.$$

Therefore, Eq. (A7–10) is indeed valid, provided Eq. (A7–9) is satisfied.

The question now arises under what conditions the principle of limiting distance is valid when $n \neq 1$.

A7-3 The fundamental theorem on analytic continuation.

This theorem has been stated in many different forms. One of the first formulations is due to Kramers and Kronig,[†] who established the relationship between the principle of limiting distance and the dispersion relation. But some of these ideas go much further back, to Cauchy, Sommerfeld, Brillouin, and many others. The most general formulation has been given recently by Toll.[‡] In this form the theorem is a statement about the equivalence of the principle of limiting distance, the dispersion relation, and the analytic character of $f(\omega)$.

Before we state this theorem, we must define $u(\omega)$ and $f(\omega)$ also for $\omega < 0$. To this end, we note first that $u(x,t)$ in Eq. (A7–11) must be real. For $x = 0$, this requires

$$u(-\omega) = u^*(\omega). \qquad (A7\text{–}12)$$

For $x \neq 0$, we find, in combination with Eq. (A7–2), that

$$f(-\omega) = -f^*(\omega). \qquad (A7\text{–}13)$$

We assume now that the absorption coefficient is always positive or zero,

$$f_2(\omega) \geq 0 \quad \text{for } \omega > 0 \text{ and real,} \qquad (A7\text{–}14)$$

which implies that there are no energy sources, but only energy sinks. We want to assume further that the total cross section does not increase with ω stronger than logarithmically or, slightly more generally, we assume that

$$\int_0^\infty \frac{f_2(\omega)\,d\omega}{1+\omega^2} < \infty. \qquad (A7\text{–}15)$$

The last assumption which we want to make, and this assumption is slightly more restrictive than Toll's formulation, is that for complex ω,

$$\omega = \omega_1 + i\omega_2,$$

the function $f(\omega_1 + i\omega_2)$ approaches $f(\omega_1)$ in the limit $\omega_2 \to +0$ uniformly over every finite interval in ω_1:

$$\lim_{\omega_2 \to +0} f((\omega_1 + i\omega_2)) \to f(\omega_1) \text{ uniformly in every finite } \Delta\omega_1. \qquad (A7\text{–}16)$$

[†] H. A. Kramers, *Estratto dagli Atti del Congresso Internazionale de Fisici Como* (publ. by Nicolo Zanichelli, Bologna, 1927); R. de Kronig, *Ned. Tyd. Nat. Kunde* **9**, 402 (1942) and *Physica* **12**, 543 (1946).

[‡] J. S. Toll, *The Dispersion Relation for Light and Its Application to Problems Involving Electron Pairs*, Thesis, Princeton University, 1952. This work also contains a rather complete list of references to earlier work.

Toll's formulation of the fundamental theorem can now be stated as follows.*

Let Eqs. (A7–13) through (A7–15) be valid for $\omega = \omega_1$, and assume Eq. (A7–16) to hold. Then the following statements are equivalent:

(I) The principle of limiting distance is valid.

(II) The function $f(\omega_1)$ is almost everywhere (in the Lebesgue sense) the boundary value of an analytic function of the complex variable ω, which is regular and of nonnegative imaginary part in the plane $\omega_2 > 0$.

(III) The dispersion function $f(\omega_1)$ satisfies the following generalized dispersion relation for real and nonnegative constant A everywhere except possibly on a set of Lebesgue measure zero,

$$f(\omega_1) = A\omega_1 + \frac{\omega_1}{\pi} \lim_{\omega_2 \to +0} \int_0^\infty \frac{2f_2(\omega')d\omega'}{\omega'^2 - \omega^2}. \qquad (A7\text{--}17)$$

We note that this last relation can be written in the form

$$n(\omega_1) = 1 + A + \lim_{\omega_2 \to +0} \frac{1}{\pi} \int_0^\infty \frac{\alpha(\omega')d\omega'}{\omega'^2 - \omega^2}, \qquad (A7\text{--}18)$$

or, using Eq. (A7–6),

$$a(\omega_1) = A\,\omega_1^2 + \lim_{\omega_2 \to +0} \frac{2}{\pi} \omega_1^2 \int_0^\infty \frac{a_2(\omega')d\omega'}{\omega'(\omega'^2 - \omega^2)}. \qquad (A7\text{--}19)$$

Before we proceed to a proof we want to remark on the constant A. It is the contribution to the refractive index of an absorption *line* at infinite frequency. This means that it is *not* the contribution of the absorption as ω_1 approaches infinity; that contribution is included in the integral. It is rather an additional effect, best thought of as an absorption line. However, when $n(\omega_1)$ approaches unity as ω_1 approaches infinity, this pathological line cannot occur. We are thus justified in putting $A = 0$ for all problems of physical interest.

Proof of equivalence of (I) and (II).† We shall first prove that (II) is a consequence of (I) and then show the inverse. For this purpose we make use of a theorem by Titchmarsh.‡

---

* Note that Toll's original formulation does not require assumption (A7–16), but this loss of generality is of little physical importance.

† The general proof of the fundamental theorem (Toll, *loc. cit.*) is simplified in the following by a few minor assumptions.

‡ E. C. Titchmarsh, *Introduction to the Theory of Fourier Integrals*, Oxford, 1937, Theorem #95.

TITCHMARSH THEOREM: Each of the following two conditions is both necessary and sufficient, so that a complex function $F(x)$ of the real argument x, which is Lebesgue square integrable over the domain $-\infty$ to $+\infty$ should be for almost all x the limit as y approaches zero from above of a function $F(x + iy)$ which is regular throughout $y > 0$ and satisfies

$$\int_{-\infty}^{\infty} |F(x+iy)|^2 dx \le K. \qquad (A7\text{-}20)$$

Condition A: If we write $F(x+iy) = F_1(x,y) + iF_2(x,y)$, then F_1 and F_2 are conjugate Hilbert transforms on $y = 0$,

$$F_1(x) = \frac{1}{\pi} P \int_{-\infty}^{\infty} \frac{F_2(x')dx'}{x' - x},$$

$$F_2(x) = -\frac{1}{\pi} P \int_{-\infty}^{\infty} \frac{F_1(x')dx'}{x' - x}. \qquad (A7\text{-}21)$$

Condition B: The Fourier transform $E(t)$ of $F(x)$ is zero for $t < 0$.

If we identify $E(t)$ with $u(x, \tau + x)$ and t with τ, and assume that the principle of limiting distance is valid, Titchmarsh's theorem tells us that $u(\omega)$ can be extended analytically into $\omega_2 > 0$. Since $f_2(\omega_1) \ge 0$ implies

$$|u(\omega_1)e^{if(\omega_1)x}|^2 = |u(\omega_1)|^2 e^{-2f_2(\omega_1)x} \le |u(\omega_1)|^2, \quad (x > 0),$$

the function ue^{ifx} is also square integrable. Assume that we choose $u(\omega)$ to be analytic and positive definite in the upper half plane, then the principle of limiting distance requires that e^{ifx} be extendable into an analytic function in $\omega_2 > 0$ for any $x > 0$. Hence,

$$f(\omega) = \frac{1}{ix} \ln(e^{if(\omega)x})$$

is analytic provided $e^{if(\omega)x} \ne 0$ everywhere in $\omega_2 > 0$. That this latter is indeed fulfilled follows from its expansion in the vicinity of that zero, $\omega = \omega_0$:

$$e^{if(\omega)} = c(\omega - \omega_0)^P + \cdots,$$

where P is a positive integer and c is a constant. Choosing $x = x'/2P$, we see that

$$e^{ifx} = c(\omega - \omega_0)^{x'/2} + \cdots$$

and cannot be analytic for all x. Hence, e^{ifx} has no zeros and $f(\omega)$ is analytic in $\omega_2 > 0$. It remains to prove that the principle of limiting distance also implies $f_2(\omega) \ge 0$ for $\omega_2 > 0$. Consider Eq. (A7-20), which states that

$$\int |u(\omega_1 + i\omega_2) e^{if(\omega_1 + i\omega_2)x}|^2 d\omega_1 \le K. \qquad (A7\text{-}22)$$

Assume that $f_2(\omega) < 0$ for some $\omega = \Omega$ such that $-f_2(\Omega) \geq \epsilon > 0$. Then Eq. (A7–22) says that for a small interval δ about Ω_1 we must have

$$\int_{\Omega_1-(\delta/2)}^{\Omega_1+(\delta/2)} |u(\omega)|^2 e^{2x\epsilon} d\omega_1 \leq K.$$

But this condition is obviously not satisfied for arbitrarily large x. Therefore $f_2(\omega) \geq 0$. This proof is incomplete without verification that K is independent of x. Parseval's relation, combined with the principle of limiting distance, gives

$$\int_{-\infty}^{\infty} |u(\omega) e^{ixf(\omega)}|^2 d\omega_1 = \int_{0}^{\infty} u^2(x,\tau+x) e^{-2\tau\omega_2} d\tau$$

$$\leq \int_{0}^{\infty} u^2(x,\tau+x) d\tau$$

$$= 2\pi \int_{-\infty}^{\infty} |u(\omega_1)|^2 e^{-2f_2(\omega_1)x} d\omega_1$$

$$\leq 2\pi \int_{-\infty}^{\infty} |u(\omega_1)|^2 d\omega_1 = 2\pi K,$$

which proves that K can be chosen independent of x. This concludes the proof that (II) follows from (I).

To show that (I) follows from (II), we assume $u(\omega)$ to be square integrable. Then, if (II) holds, Titchmarsh's theorem tells us that $u(x,t+x) = 0$ for $t < 0$, by condition B; therefore (I) follows from (II). This completes the proof that (I) is equivalent to (II).

Proof of the equivalence of (II) and (III). This proof is rather difficult unless one assumes that $f(\omega)$ is square integrable. In that case, condition A of Titchmarsh's theorem states that

$$f_1(\omega_1) = \frac{1}{\pi} P \int_{-\infty}^{\infty} \frac{f_2(\omega') d\omega'}{\omega' - \omega_1},$$

$$f_2(\omega_1) = -\frac{1}{\pi} P \int_{-\infty}^{\infty} \frac{f_1(\omega') d\omega'}{\omega' - \omega_1}.$$

(A7–23)

The first of these relations can be modified by use of

$$f_2(-\omega_1) = f_2(\omega_1),$$

which follows from Eq. (A7–13); we find

$$f_1(\omega_1) = \frac{2\omega_1}{\pi} P \int_{0}^{\infty} \frac{f_2(\omega') d\omega'}{\omega'^2 - \omega^2}$$

or

$$n_1(\omega_1) = 1 + \frac{1}{\pi} P \int_0^\infty \frac{\alpha(\omega')d\omega'}{\omega'^2 - \omega^2}. \quad (A7\text{--}24)$$

This is exactly the real part of Eq. (A7–18) with $A = 0$. Thus, provided we assume square integrability of $f(\omega_1)$, Titchmarsh's theorem establishes the equivalence of (II) and (III).

For the general proof the reader is referred to Toll's thesis (*loc. cit.*).

A7–4 Applications. The above essentially classical argument can now be reasonably applied to quantum mechanical and field theoretical problems. The principle of limiting distance seems to be a very basic principle, so that there appears to be no reason why it should not be valid in the quantum domain. In fact, in our quantum electrodynamics the ordering of the operators and the resulting form of the propagation function D_c assures the validity of the principle of limiting distance.* The fundamental theorem therefore appears to be also valid in quantum electrodynamics. Apart from this, in many cases the validity of (II) can be seen directly from the explicit expressions for the problem at hand, or they can be inferred from (III), which can be established in some cases. But, in general, it must be kept in mind that the nice theorem above only proves an equivalence, and the validity of at least one of the three points must first be established.

Equation (A7–18) with $A = 0$ can be written

$$n_1(\omega_1) + i n_2(\omega_1) = 1 + \frac{1}{\pi} P \int_0^\infty \frac{\alpha(\omega')d\omega'}{\omega'^2 - \omega^2} + i \frac{\alpha(\omega_1)}{2\omega_1},$$

so that we obtain, with Eq. (A7–4),

$$n_1(\omega_1) - 1 = \frac{1}{\pi} P \int_0^\infty \frac{\alpha(\omega')d\omega'}{\omega'^2 - \omega_1^2}, \quad (A7\text{--}25)$$

or, with Eq. (A7–6),

$$a_1(\omega_1) = \frac{2\omega_1^2}{\pi} P \int_0^\infty \frac{a_2(\omega')d\omega'}{\omega'(\omega'^2 - \omega_1^2)}. \quad (A7\text{--}26)$$

Instead of using Eq. (A7–6), one can also argue as follows. The fundamental theorem is proved for $f(\omega)$ provided $f(\omega)$ satisfies the conditions (A7–13) to (A7–16). Let us consider the function $2\pi N a(\omega)/\omega$. This function, which is originally known only for $\omega = \omega_1 > 0$, can easily be defined for $\omega_1 < 0$, so that it fulfills Eq. (A7–13). Since

$$2\pi N \frac{a_2(\omega_1)}{\omega_1} = f_2(\omega_1) = \frac{1}{2} N\sigma(\omega_1) > 0$$

---

* E. C. G. Stueckelberg, *Helv. Phys. Acta* **19**, 241 (1946); M. Fierz, *Helv. Phys. Acta* **23**, 731 (1950).

by the Bohr-Peierls-Placzek relation, Eq. (A7–14) is fulfilled and Eq. (A7–15) is fulfilled too, provided that the total cross section does not diverge faster than logarithmically. It remains to assume the validity of (A7–16) to conclude that the principle of limiting distance is equivalent to the analyticity of the forward scattering amplitude or, more exactly, of $a(\omega)/\omega$, in the sense of (II) in the fundamental theorem. Equation (A7–6) can then be concluded when one adds that $n_1(\omega_1) \to 1$ for $\omega_1 \to \infty$.

Equation (A7–26) combined with the Bohr-Peierls-Placzek relation, Eq. (A7–7), is a powerful tool for the calculation of forward and small-angle scattering amplitudes.

In quantum electrodynamics, the scattering cross sections are always much smaller than the associated absorption cross sections; compare, for example, the scattering of light by light with the pair production by two-photon collisions, the scattering of light by a Coulomb field with pair production by photons in a Coulomb field, the resonance scattering of light with the photoelectric cross section. From these examples it becomes clear that σ_s is always of twice the order in α as in σ_a, at least in lowest order. Consequently, the total cross section $\sigma(\omega)$ is, to a sufficiently good approximation, equal to σ_a. Furthermore, it follows that σ_a is always much easier to calculate than σ_s.

The above equations give us now a means of obtaining the differential *scattering* cross section in the forward direction,

$$\frac{d\sigma_s(\omega, \theta = 0)}{d\Omega} = |a_1 + ia_2|^2,$$

from the easily calculable total absorption cross section σ_a. Since $\sigma \simeq \sigma_a$, we find, from Eqs. (A7–7) and (A7–26),

$$a_2(\omega) = \frac{\omega}{4\pi} \sigma_a(\omega) \tag{A7–27}$$

and

$$a_1(\omega) = \frac{1}{2\pi^2} \omega^2 P \int_0^\infty \frac{\sigma_a(\omega') d\omega'}{\omega'^2 - \omega^2}, \tag{A7–28}$$

where we omitted the subscript 1 on ω, with the understanding that ω is real.

In this way the forward scattering of light by light (Section 13–2) has been calculated from the Breit-Wheeler formula (Section 13–3) by Toll (*loc. cit.*), and Delbrück scattering (Section 15–8) in the forward direction was obtained from pair production in a Coulomb field (Section 15–7) by Toll (*loc. cit.*) and by Rohrlich and Gluckstern.* The same method can be

---

* F. Rohrlich and R. Gluckstern, *Phys. Rev.* **86**, 1 (1952).

combined with the impact parameter method to obtain a good approximation for the differential scattering cross section at high energies and small angles.* The nuclear recoil distribution in pair production by photons was calculated by use of some of the above relations by Jost, Luttinger, and Slotnick.† For other applications the reader is referred to Toll's thesis.

---

* For Delbrück scattering this was carried through by H. A. Bethe and F. Rohrlich, *Phys. Rev.* **86,** 10 (1952).

† R. Jost, J. M. Luttinger, and M. Slotnick, *Phys. Rev.* **80,** 189 (1950).

APPENDIX A8

NOTATION

Although all symbols and notations are explained when first introduced and, at suitable places, when symbols recur, some general remarks on symbols and notation are given here for the benefit of those who use this book mainly as a reference.

1. The mechanical units are the centimeter, the velocity of light, and $1/2\pi$ times Planck's constant of action, $\hbar$. The electromagnetic quantities are measured in the Heaviside-Lorentz system [cf. Eq. (1–2)].

2. The relativistic metric tensor is real with $g^{ii} = 1$, $g^{00} = -1$, $g^{\mu\nu} = 0$ $(\mu \neq \nu)$. Greek indices run from 0 to 3, Latin indices run from 1 to 3, unless otherwise specified. Scalar products $A_\mu B^\mu \equiv A \cdot B \equiv A_1 B_1 + A_2 B_2 + A_3 B_3 - A_0 B_0$.

3. The star as a superscript, A^*, means "complex conjugate" when A is a c-number; it means Hermitian conjugate when A is an operator in Hilbert space. Only in spinor space is it important to introduce the complex conjugation (γ_μ^*) as well as the adjoint operation (γ_μ^+). The transposition, $\tilde{\gamma}_\mu$, connects these two operations [cf. Eq. (A2–26)]. The Hermitian conjugate of quantities like ψ which are both operators in Hilbert space and vectors in spinor space is denoted by ψ^*.

4. Throughout the theory as well as in the applications all physically meaningful expressions are independent of any particular representation of the gamma-matrices and are treated accordingly.

5. The frequently occurring product $\gamma_\mu A^\mu$, where A^μ is a four-vector, is indicated by use of the italic bold-face $\boldsymbol{A}$. For the differentiation symbol $\partial_\mu \equiv \partial/\partial x^\mu$ we have $\gamma_\mu \partial^\mu \equiv \boldsymbol{\partial}$. It is convenient to have a separate symbol for this product when the differentiation is carried out to the *left*: $\boldsymbol{\mathfrak{d}}$ [cf. Eq. (3–51)]. The symbol ∂ is sometimes used to denote the scalar product $n_\mu \partial^\mu$. Regular bold-face type indicates three-vectors, like **p** and **k**. Regular italic is used for four-vectors and the vector index is sometimes omitted (e.g., one refers to a negaton of momentum p, positon of momentum q, photon of momentum k). The energy of a real electron is denoted by ϵ (ϵ_+ and ϵ_- when electrons of both charge occur, $p^0 = \epsilon_-$, $q^0 = \epsilon_+$); the energy of a real photon is denoted by ω, i.e., $|\mathbf{k}| = \omega$.

6. The continued confusion in much of the current literature between state vectors and operators is avoided by use of the last letter of the Greek alphabet for state vectors. In the Heisenberg picture the operators of the electron field, photon field, and the state vector are denoted by Ψ, A_μ, and Ω; in the free interaction picture they are ψ, a_μ, and ω; in the bound inter-

action picture they are ψ, a_μ, and ω (*cf.* Section 14–2). An unquantized (external) electromagnetic field is denoted by φ_μ.

7. All D-functions, Δ-functions, and S-functions are defined with the same numerical coefficient in their Fourier integral representation (*cf.* Appendix A1). In some cases they therefore differ by trivial factors from those often found in the literature. Since this difference is particularly important for the propagation functions usually denoted by D_F and S_F, we use the symbols D_c and S_c for these functions. "Finite" functions in the sense of renormalization theory have a subscript f, e.g., Σ_f. All renormalized quantities have a subscript zero, e.g., m_0, e_0, D_0', S_0'. In Chapters 9 and 10, which deal exclusively with propagation functions, the subscripts c in D_c and S_c are superfluous and are therefore omitted, since no confusion can arise.

8. Parentheses are used to denote inner products, $(\omega, A\omega)$, and traces, $(\gamma_\mu \gamma_\nu) \equiv \frac{1}{4} \operatorname{Tr} (\gamma_\mu \gamma_\nu)$. Square brackets and curly brackets are used to distinguish commutation and anticommutation relations, respectively.

9. Following the international convention of distinguishing the two types of electrons as negatons and positons, attention is drawn to the fact that wherever the word "electron" is used the statement is meant to be valid for both signs of charge. In Feynman diagrams a distinction is made between electron lines (which go between two corners unless they are external) and electron paths (which refer to the *whole* world line followed along the arrow direction).

10. We refer to "configuration space" meaning either "position" or "momentum" space. "Representation" is used only in the group theoretical sense. We therefore refer to "interaction *pictures*" and "Heisenberg *picture*."

11. As is explained in the introduction, the Hamiltonian formalism is not used and, consistently, the "Hamiltonian" is not introduced. The theory leads naturally to an "interaction operator" which is identical with the "interaction Hamiltonian" in that formalism.

12. Space-like surfaces are in most cases an unnecessary luxury which contribute practically nothing to the generality of the theory. They are therefore replaced by plane surfaces wherever this can be done without loss of covariance.

SUPPLEMENT
FOR THE SECOND EDITION

SUPPLEMENT S1

FORMULATIONS OF QUANTUM ELECTRODYNAMICS

The last twenty years have seen a mushrooming of new and unexpected experimental results in elementary particle physics. This in turn has led to a plethora of formulations of relativistic theories of elementary particle interactions, both strong and weak. Many of these were reformulations of quantum field theories, prepared in the hope that the success of quantum electrodynamics could be duplicated for nonelectromagnetic interactions.

While these goals have not been realized, even remotely, the deeper insight into quantum field theories that has resulted from these efforts has been very valuable indeed; we now have a much better understanding of the mathematical structure of quantum field theory (QFT) and a greater appreciation of the special role played by quantum electrodynamics (QED). This special role is closely connected with the presence in QED of a zero mass field, the electromagnetic field.

Of course, a considerable number of books have been published since our first edition; of those entirely or largely devoted to QED many give excellent expositions including material that was not yet available in 1955. An incomplete list in order of their appearance is given in the Reference section.[1] In addition to these there have appeared many fine books on QFT in which QED is either not treated at all or treated only peripherally. The reason for this is in part due to the mathematical problems one encounters in the seemingly trivial case in which one of the fields happens to have vanishing mass, as indicated above.

Some of the difficulties encountered are of course well known to the students of electrodynamics from the problem of infrared divergences. This problem was treated in Section 16–1 and will be reconsidered from a very different point of view in Supplement S4. In the present section we shall be concerned with various alternative formulations of QFT especially as they affect QED.

In this endeavor we cannot go into detailed expositions, of course, since this is neither feasible here nor necessary. Rather, the characteristic features of the various formulations will be outlined and reference material will be provided in order to enable further and more detailed study.

S1-1 Lagrangian QFT. This is the oldest formulation and developed historically in analogy to the transition from classical to quantum mechanics. The logical sequence of steps are as follows: given a *classical* Lagrangian of a field or a system of interacting fields one begins with an action principle and derives the field equations and conservation laws by means of the calculus of variations (Noether's theorems are here the key). The resulting equations and conservation laws are then regarded as operator equations and form the starting point of a QFT. One must keep in mind that this "derivation" takes place with classical (c-number) fields, since the calculus of variations is not defined for a Lagrangian constructed from operators on a Hilbert space. Consequently, the derived field equations and conservation laws can in general not be regarded as equations for operators without ambiguity of ordering and, usually, other lack of precise mathematical meaning.

Nevertheless, Lagrangian field theory is used extensively as a heuristic tool: it permits intuitive input, it allows an easy way to include invariance properties, and it gives a certain assurance that the theory will have a classical counterpart.

The operator nature of the quantized field is usually characterized by certain commutation relations. These are obtained by passing from the classical Lagrangian to the classical Hamiltonian formalism. The resulting canonical structure is governed by fundamental Poisson brackets of the fields. The transition to a *quantum* field theory is then made as in quantum mechanics by the recipe "(Poisson bracket) $\to -i$ (Commutator)" or in the case of fields with half-odd integer spin "(Poisson bracket) $\to -i$ (Anticommutator)." Obviously, this is not a mathematical derivation. It is a heuristic justification for starting a QFT with certain field equations and commutation relations.[2]

Our Chapters 1 through 3 were at least partly based on Lagrangian field theory.

The zero mass case gives rise to a special difficulty here: the momentum component π^0, canonically conjugate to A^0 is proportional to $\partial_\lambda A^\lambda$ and therefore vanishes in the Lorentz gauge. This leads to the need for imposing the Lorentz condition as a constraint in QED rather than as an operator equation (Gupta method, Section 6-3). Alternative possibilities are of course also available.[3] But quite generally, Lagrangian QFT will be based on a canonical formalism with *constraints*.[4]

Lagrangian QFT and QED are dominated in their outlook by the concept of "quantization." This means that a quantum theory is *not* built from its own assumptions and first principles, but rather from a classical theory which is "quantized" by some prescription. The historical development as an inductive process from classical to quantum physics has here been kept as a permanent crutch. A *derivation* of quantum theory from the less general classical theory is logically impossible. The converse derivation, however, accepted by most physicists, does exist: the classical theory must emerge as a suitable mathematical limit of quantum theory.

S1–2 Axiomatic QFT. The dissatisfaction with the mathematical structure of QED and of QFT in general suggested a careful, rigorous, mathematical study of this subject. For this purpose the first task was to give a mathematically meaningful definition of a quantum field. Under the leadership of Wightman[5] such a study was initiated and soon became known as "axiomatic QFT." The term is obviously a misnomer and has led to misunderstandings: the purpose was *not* to give an axiomatic formulation of the theory (although it may develop into that), but simply to provide definitions which give precise mathematical meaning to basic quantities of the theory. Alternative definitions clearly exist and have in fact been proposed.

Concomitant with these definitions came a judicious use of functional analysis, including the theory of distributions, and a complexification of real Minkowski space which led to the need for the relatively recent theory of several complex variables. These developments started in the early fifties and reached a plateau in the midsixties characterized by two important books on the subject.[6,7]

The first serious attempt to generalize the "axioms" of QFT (sometimes called "Wightman axioms") so as to permit the inclusion of zero mass fields and in particular of the electromagnetic field was made by Wightman and Gårding.[8] The axioms and the necessary generalizations will be listed here but without the full use of the precise *mathematical language* with which the reader may not be familiar.

AXIOM 1 (HILBERT SPACE). *The states are vectors in a separable Hilbert space $\mathcal{H}$ of positive metric (the norm of a vector $\|\Omega\|^2 = (\Omega,\Omega)$ vanishes if and only if $\Omega = 0$).*

AXIOM 2 (OBSERVABLES). *Observables are linear self-adjoint operators on $\mathcal{H}$ and include the ten generators of the Poincaré algebra, P^μ and $M^{\mu\nu}$. There exists a continuous unitary representation of the Poincaré group, $U(L)$, (in the notation of Eq. (1–12)) on $\mathcal{H}$.*

AXIOM 3 (FIELDS). *A field $A(x)$ is an operator valued distribution defined on a suitable test function space $\mathfrak{I}$ such that*

$$A(f) \equiv \int A(x)f(x)d^4x, \quad \text{(formally)} \tag{S1-1}$$

with f in $\mathfrak{I}$, is a linear operator on $\mathcal{H}$.

This axiom is a mathematical statement of the fact that fields have no meaning "at a point" but are experimentally *necessarily* defined as an average over a (small) space-time region. Axiomatic QFT tells us that the idealization to an operator which is a point function is mathematically unsound. A corollary to this fact is the observation that products of any two fields at the same point are mathematically undefined; for example $\bar{\psi}(x)\psi(x)$ does not have a meaning if $\psi(x)$ is a distribution.

AXIOM 4 (POINCARÉ INVARIANCE). *The fields and observables transform covariantly. For example, Eq. (1–13) becomes*

$$U(L)\phi_r(f)U^{-1}(L) = S_r{}^s \phi_s(f(L^{-1}x)), \qquad (S1-2)$$

where $S_r{}^s$ *is a finite dimensional representation of the Lorentz group.*

AXIOM 5 (SPECTRUM). *There exists a unique vacuum state Ω_0 characterized by $U(L)\Omega_0 = \Omega_0$. The spectral decomposition of P^μ has eigenvalues p^μ which all lie inside the future light cone with an isolated minimum mass $m > 0$ above the vacuum (mass gap).*

AXIOM 6 (LOCALITY). *Any two observables whose test functions have spacelike support relative to each other commute. Any two fields with such test functions commute or anticommute.*

This axiom is also known as "microscopic causality." When it is assumed to hold for the vacuum expectation value of a product* it is known as "weak local commutativity."

AXIOM 7 (COMPLETENESS). *Asymptotically, in the distant past and distant future the Fock states span Hilbert spaces $\mathcal{H}_{in}$ and $\mathcal{H}_{out}$ which are identical with the Hilbert space $\mathcal{H}$ of the interacting fields.*

This last axiom was added somewhat later to the theory, but it is an addition mandated by the desire to make contact with QFT as a scattering theory, since that is the way in which QFT was primarily used by most physicists. This requires that the interacting fields must have an asymptotic behavior in the distant past and future such that they become free fields in that limit. They can then be associated with *physical particles* as the quanta of these free fields. No other relation between fields and particles seems available. The S-matrix can then be defined as the mapping of in-states to out-states. This idea was carried through rigorously by Ruelle[9] based on the assumptions suggested earlier by Haag.[10] The same physical goal is achieved in a mathematically less confining way by the LSZ-formulation as will be seen in Section 1–4.

If massless fields are to be included Axioms 1 and 5 are primarily affected. In Axiom 1 the positive definiteness must be given up and an

---

* "Being spacelike" for n points usually means "being on the same spacelike surface." The "weak" in "weak local commutativity" however refers to the weaker condition that the vector $\lambda_1(x_1 - x_2) + \lambda_2(x_2 - x_3) + \cdots + \lambda_{n-1}(x_{n-1} - x_n)$ be spacelike for all $\lambda_i \geq 0$.

indefinite Hilbert space $\mathcal{H}$ must be permitted (See Section 6–3 and the following Section S1–3); the spectrum (Axiom 5) no longer has a mass gap and p^μ can lie *on* the future light cone, $p_\mu p^\mu = 0$. The consequences of this important change, a continuous spectrum from the vacuum on up, will be explored in Supplement S4.

The pragmatic physicist interested primarily in the extraction of predictions for the benefit of the experimenter may be unsympathetic to the complications due to zero mass particles. He will construct QED with a small but nonvanishing photon mass λ and take the limit $\lambda \to 0$ at the very end of his calculations. This is indeed an alternative and was discussed in Section 6–5 and on p. 191. But this approach will not satisfy those who aim for simplicity and beauty in the foundations of physical theory.

The main benefits of axiomatic QFT lie in a deeper mathematical understanding of the divergence difficulties, and in very general proofs of theorems such as those on spin and statistics, the CPT theorem, analyticity properties leading to associated asymptotic behavior of cross sections, etc.[11] The reader is referred to References 5 and 6 where original papers on these theorems are quoted.

Those results of axiomatic QFT which are specifically relevant for QED will be summarized in the following section.

Since the late 1960's most efforts by axiomatic field theorists was devoted to model building: to find a (hopefully physically nontrivial) model field theory which satisfies the Wightman axioms or a similar set. This program started with $1 + 1$ dimensional Minkowski space and over the years built up to $2 + 1$ dimensional models. It became known as *constructive QFT*. The basic approach is to take a (relatively simple) Lagrangian field theory, to put in a sufficient number of cutoffs to ensure mathematical existence of all quantities (thereby losing of course all the invariance properties) and then to make a careful study of the limits as these cutoffs are removed. In this brute force way the program achieved success in the first few years.

Most recently it was possible to clarify the relation of Green functions in $n + 1$ dimensional Minkowski space and Euclidean space of that dimensionality; most work in constructive QFT therefore now starts with a Euclidean QFT. But thus far this program has not reached a stage where it can contribute to QED.

S1–3 Locality, covariance, and indefinite metric. In Chapter 6 alternative methods were presented for the description of the free Maxwell field. It is obvious from Section 6–1 that the Coulomb (or radiation) gauge does not have a local Hamiltonian (6–23): there is a double integral and the nonlocal operator ∂^{-1}. Also, it is only formally covariant, because it depends on the normal n^μ to the spacelike surface. Neither of these features exist in the Gupta method (Section 6–3) but the price paid for covariance is an indefinite metric Hilbert space.

This situation has now been cast into the form of very general and rigorous statements.[12] Let us assume that the Maxwell field $f^{\mu\nu}$ satisfies the free field equations (2–2) and (2–3); that it transforms as an antisymmetric tensor under a unitary representation of the Poincaré group; that the Hilbert space contains an invariant unique vacuum state in the domain of the $f^{\mu\nu}$; that the potentials a^{μ} defined in the usual way by (2–5) are space–time translation covariant, and that they satisfy weak local commutativity, i.e. that

$$(\omega_0, [a^{\mu}(x), a^{\nu}(x')]\omega_0) = 0, \quad (x - x')^2 > 0. \tag{S1-3}$$

Thus we work with Axioms* 2, 3, 4, and 6 but do not assume a positive Hilbert space or a mass gap. Then it follows that $f^{\mu\nu} = 0$, on the physical (positive) subspace of $\mathcal{H}$ i.e., such fields do not exist. Even if one replaces the assumption of weak locality by the apparently innocuous assumption of Lorentz covariance of a^{μ}, the same result, $f^{\mu\nu} = 0$, follows. Obviously, these are strong results.

One now has the following alternatives. Either one keeps covariance and weak locality but one gives up Maxwell's equations as operator equations (Gupta–Bleuler gauge), or one keeps the operator field equations and the usual Hilbert space and instead one gives up weak local commutativity and manifest covariance (Coulomb gauge). In the former case Maxwell's equations will hold only weakly, i.e., as matrix elements of a suitably defined subspace such as in (6–68), but covariance and weak local commutativity are possible. In the latter case Maxwell's equations will hold as operator equations but with potentials that are not covariant and local commutativity will be lost.

It is instructive to see the nonlocality (lack of commutativity for spacelike distances) of a^{μ} in the Coulomb gauge explicitly. The operator equations are

$$a^0 = 0, \quad \nabla \cdot \mathbf{a} = 0, \quad \Box \mathbf{a} = 0, \tag{S1-4}$$

in every Lorentz frame, so that $a^{\mu}(x)$ transforms as

$$a'^{\mu}(x') = \Lambda^{\mu}{}_{\nu} a^{\nu}(x) + \partial^{\mu}\varphi(x,\Lambda) \tag{S1-5}$$

where we used $\Lambda^{\mu}{}_{\nu}$ instead of $a^{\mu}{}_{\nu}$ of (1–6) in order to avoid confusion. The function φ depends on Λ in such a way that the equations (S1-4) are preserved. The commutation relations are

$$[a^k(x), a^l(x')] = -i(\delta^{kl} - \partial^{kl}\nabla^{-2}) D(x - x') \tag{S1-6}$$

---

* Note that Lorentz covariance is assumed of the field strengths, but not of the potentials.

where

$$\nabla^{-2} f(x) \equiv -\frac{1}{4\pi} \int \frac{f(\mathbf{x}')d^3x'}{|\mathbf{x} - \mathbf{x}'|}. \tag{S1-7}$$

This commutation relation is consistent with (S1-4). The second term on the right hand side of (S1-6) is responsible for the nonlocality because it involves

$$-4\pi\nabla^{-2} D(x) = \epsilon(t)\theta(|t| - r) + \frac{t}{r}\theta(r - |t|) \tag{S1-8}$$

and therefore does not vanish outside the light cone ($r > |t|$). It does happen to vanish on the $t = 0$ hyperplane.

The nonlocality of the potentials a^μ in the Coulomb gauge causes no physical problem: the potentials are not observable and the fields $f^{\mu\nu}$ constructed from them *are* local as in the covariant gauge, since they do not depend on the gauge.

The generalization of these strong theorems to an interacting Maxwell field was also carried out and the results are similar.[13] The field equations (4-2) and (4-3) cannot hold as operator equations if the electron field is to be a local field. And if the field equations hold only weakly (as matrix elements on a subspace of $\mathcal{H}$) then $\mathcal{H}$ cannot be positive definite.

The local formulation of quantum electrodynamics thus necessarily involves an indefinite metric for the Hilbert space $\mathcal{H}$. But a local formulation is much better understood mathematically,[8,14] because little is known about nonlocal field theory. Physically, however, this formulation is not attractive; it introduces extra degrees of freedom which only have to be eliminated again and which have no physical meaning; it also requires proofs that the physical subspace has no states of negative norm (sometimes called "ghosts"), that the S-operator is unitary on the physical subspace, etc.

The nonlocal formulation on the other hand seems to present the physical situation directly as it is and is thus preferable. This is quite evident from the Coulomb gauge: the Coulomb field is not quantized in that gauge and this corresponds exactly to the observed situation, *viz.* that the quanta of the electromagnetic field (photons) are restricted to the radiation field. There are no observable Coulomb quanta.

The nonlocal nature of electrodynamics is however much deeper than seems to emerge from the above considerations and it cannot be avoided completely (by use of the Gupta–Bleuler gauge). The following theorem is apparently valid in *any* gauge (although the proof so far is mathematically clean only the Gupta–Bleuler gauge): there are *no local* charged states in quantum electrodynamics.[13,15]

The meaning of this statement is as follows. Given the vacuum state Ω_0 of the electron field, one wants to find a local operator A (presumably a polynomial of the local electron field Ψ) which produces a physical charged state

$$\Omega = A\Omega_0. \tag{S1-9}$$

The charge q is an eigenvalue of the charge operator Q so that $Q\Omega_0 = 0$ and $Q\Omega = q\Omega$. In the Gupta–Bleuler gauge both Ω and Ω_0 must be in the physical subspace of $\mathcal{H}$. The above assertion then states that there is no local operator A with the property (S1-9).

Closely related to this result is the charge superselection rule: every local observable O (i.e. a selfadjoint and gauge invariant local operator) commutes with the charge operator, $[Q,O] = 0$. This "rule" is not an assumption but actually a theorem that is far from trivial to prove.[14] It means that Q belongs to every complete set of observables. It implies that the relative phases of states associated with different total charge are unobservable. Since the above requirements on A imply $[Q,A]\Omega_0 = qA\Omega_0$ it follows that either $q = 0$ or A is not a local observable.

Thus, while the Gupta–Bleuler gauge permits (weakly) local electromagnetic potentials a^μ, it cannot provide local charged states. This nonlocality is intrinsic to the theory. It is therefore only natural that the most physical formulation, the Coulomb gauge, also involves nonlocal potentials. From this point of view the indefinite metric Hilbert space is simply a mathematical device to permit one to work not only with local and covariant field strengths, (which are observables and which enjoy these properties in any gauge), but also with local and covariant potentials (which are not observable).

S1–4 Lehmann–Symanzik–Zimmermann and related formalisms.

There developed soon after axiomatic QFT, and later in parallel with it, a formulation which was much more accessible to most theoreticians and which permitted the incorporation of a great deal more scattering theory at the cost of a less rigid mathematical structure. It became known as the LSZ formulation of QFT after the authors of the fundamental papers on this subject.[16]

Since the axiomatic formulation permits a clean and complete understanding of free fields, the idea is to attempt to expand operators such as Heisenberg fields in terms of normal-ordered (Wick-ordered) products of these; typically,

$$F = \sum_{n=0}^{\infty} \int \cdots \int f(x_1, \cdots, x_n) :a(x_1) \cdots \times a(x_n): \, d^4x_1 \cdots d^4x_n \tag{S1-10}$$

for some free field $a(x)$. The set of coefficient functions $f(x_1, \cdots, x_n)$ for all n will determine F; but F (i.e. its matrix elements) will determine the f's only on the mass shell of the field $a(x)$. The problem is therefore to determine the off-mass shell extension of the f's, because these occur in the matrix elements involving virtual states.

The second idea is the assumed completeness of the "in-fields" and of the "out-fields" as expressed in Axiom 7 of Section S1–2. The S-operator is trivially unitary, since it is just the mapping of $\mathcal{H}_{\text{in}}$ into $\mathcal{H}_{\text{out}} = \mathcal{H}_{\text{in}}$,

$$a_{\text{out}}(x) = S^* a_{\text{in}}(x) S \tag{S1-11}$$

for any complete set of fields $a_{\text{in}}(x)$ (we write $a_{\text{in}}(x)$ for what in general is a set of fields).

These two ideas became linked by the concept of the *interpolating field*: the Heisenberg fields $A(x)$ which satisfy some interacting field equations must approach $a_{\text{in}}(x)$ and $a_{\text{out}}(x)$ asymptotically in the distant past and future. Thus, $A(x)$ interpolates between the distant past and future making mass shells and particle concepts rather vague concepts in the interaction region.

The crucial link is provided by the *asymptotic condition* which links A to a_{in} and a_{out}. If $a_{\text{in}}(x)$ is a neutral free field of mass m, it is defined in terms of $A(x)$ by

$$a_{\text{in}}(x) \equiv \underset{\sigma \to -\infty}{\text{w-lim}} \int \Delta(x - x'; m^2) \overset{\leftrightarrow}{\partial}_\mu{}' A(x') d^3 \sigma^\mu(x'). \tag{S1-12}$$

$\Delta(x, m^2)$ is the invariant Δ-function of Appendix A1, $\overset{\leftrightarrow}{\partial}_\mu \equiv \overset{\rightarrow}{\partial}_\mu - \overset{\leftarrow}{\partial}_\mu$ is the difference between the differentiations to the right and to the left, and "w-lim" indicates the weak limit, i.e., the limit of the matrix elements of the operators in question.

The asymptotic condition permits one to express the S-operator defined by (S1–11) in the form (S1–10): for the above example of a single neutral scalar (necessarily selfinteracting) field,

$$S = 1 + \sum_{n=1}^{\infty} \frac{(-i)^n}{n!} \int \cdots \int K_1 \cdots K_n \varphi(x_1, \cdots, x_n)$$
$$\times :a_{\text{in}}(x_1) \cdots a_{\text{in}}(x_n): d^4 x_1 \cdots d^4 x_n \tag{S1-13}$$

with $K_i = \Box_i - m^2$. The φ-functions can be expressed in terms of the vacuum expectation values of products of the interpolating fields $A(x)$. Define the τ-functions by

$$\tau(x_1, \cdots, x_n) \equiv (\Omega_0, T_+(A(x_1) \cdots A(x_n))\Omega_0) \tag{S1-14}$$

where $T_+ = P$ indicates positive time ordering (increasing from right to left). Then the φ-functions are related to the τ-functions in exactly the same way as the normal-ordered products are related to the time-ordered products.* Thus, the φ-functions and the τ-functions are linear com-

---

* See Appendix A4 and Section S3–3 of Supplement S3 below.

binations of Wightman functions

$$w(x_1, \cdots x_n) \equiv (\Omega_0, A(x_1) \cdots A(x_n)\Omega_0) \qquad \text{(S1-15)}$$

whose coefficients, to be sure, are in general distributions like Δ_c-functions or at least θ-functions. That's where the mathematical questions arise.

The point, of course, is that one works only with coefficient functions (sometimes called "Green functions") instead of the fields themselves. These coefficient functions would have to be determined from equations describing the unitarity of S, its invariance properties, etc.

In their second paper LSZ define another class of functions, the retarded functions or r-functions,

$$r(x;x_1, \cdots, x_n) \equiv (\Omega_0, R(x;x_1, \cdots, x_n)\Omega_0) \qquad \text{(S1-16)}$$

$$R(x;x_1, \cdots, x_n) \equiv (-i)^n \sum_{\text{perm}} \theta(x - x_1)\theta(x_1 - x_2) \cdots$$
$$\theta(x_{n-1} - x_n)[\cdots [A(x), A(x_1)], \cdots, A(x_n)].$$

These functions permit the expansion of the interpolating field in terms of the in-field,

$$A(x) = a_{\text{in}}(x) + \sum_{n=2}^{\infty} \frac{1}{n!} \int \cdots \int K_1 \cdots K_n r(x;x_1, \cdots x_n)$$
$$\times :a_{\text{in}}(x_1) \cdots a_{\text{in}}(x_n): d^4x_1 \cdots d^4x_n \qquad \text{(S1-17)}$$

as was proven by Glaser, Lehmann, and Zimmermann.[16]

The details of this program can be studied from the original literature as well as from various texts.[17] The formulation has the advantage that it can be stated very nicely in a way which is completely independent of perturbation expansions; however, the fundamental equations for the coefficient functions cannot be solved except via this expansion. In that case one recovers the same results as in the usual formulation.

The interesting question is whether one can cast QED into the LSZ-formulation and whether this would be advantageous. When an LSZ-formulation of QED was attempted it was found that the covariant Gupta–Bleuler gauge did not allow the necessary asymptotic conditions to be satisfied and that it therefore led to a formulation which was not manifestly covariant.[18] This difficulty, however, has been removed very recently by Nakanishi.[19] He pointed out that the asymptotic electromagnetic fields need not satisfy the usual commutation relations, (2-28), but more generally,

$$[a_\mu^{\text{in}}(x), a_\nu^{\text{in}}(x')] = -i(g_{\mu\nu} - K\partial_\mu\partial_\nu)D(x - x') \qquad \text{(S1-18)}$$

with K a constant that cannot vanish in general if the gauge is to be covariant. In fact, additional terms arise for gauges different from the

Lorentz–Feynman gauge, such as in the Landau gauge where the photon propagator has a numerator $g_{\mu\nu} - k_\mu k_\nu/k^2$ instead of $g_{\mu\nu}$.

The perturbation expansion of the LSZ-formulation gives of course exactly the same physical predictions as the usually employed theory. But it permits fundamental equations in terms of renormalized quantities which are at least in certain respects on a better mathematical foundation.

Already in the first LSZ paper[16] it was pointed out by Symanzik that the same theory can also be given a functional form by introducing a classical current with respect to which functional differentiation is carried out. This *functional formulation* was first introduced by Feynman and Schwinger[20] and later applied by Hori and by Anderson[22] to QED. One difficulty in this formulation is the appearance of "anticommuting c-numbers." These are the classical currents that must be introduced in connection with Fermi fields. But the functional formulation permits one to cast the equations into a compact form (which however does not make it any easier to solve them) and to establish functional equations for Green functions. Attempts at rigorous justification of this formulation can be found in several references.[22] Texts on the functional formulation of QFT also exist.[23]

Somewhat parallel with the LSZ-formulation there was developed in the Soviet Union a similar formulation by Bogoliubov, Medvedev, and Polivanov.[24] This *BMP-formulation* shares its basic assumptions with the LSZ-formulation, its main new assumption being the *Bogoliubov causality condition*, which for our sample field reads

$$\frac{\delta}{\delta a_{\text{in}}(x)} J(x') = 0 \quad (x - x')^2 > 0 \quad \text{or} \quad x^0 > x^{0'} \quad \text{(S1-19)}$$

$$J(x) \equiv iS^* \frac{\delta S}{\delta a_{\text{in}}(x)}. \quad \text{(S1-20)}$$

with S given by (S1-13). Its causal meaning is obvious. The functional differentiation with respect to free fields can be defined in terms of the algebra of these fields.[25]

Perhaps the most important benefit to elementary particle physics that emerged from these formulations was the development of dispersion relations,[24,26] especially due to the work of Bogoliubov.

Somewhat intermediate between the LSZ and the BMP formulation of QFT is a formulation initiated by Pugh[27] which became known as *asymptotic QFT*. This formalism was applicable only to renormalizable theories and was developed specifically with QED in mind. The hope was to avoid the canonical formalism and to work judiciously with "already renormalized" quantities, so that the perturbation calculation gave directly the renormalized results and renormalization was obviated by the fact that unrenormalized quantities never entered the theory. This idea was further developed in two papers, and the success of the theory demonstrated to second and third order in the ϕ^3 model.[28] But beyond the work by Pugh this theory has so far not been applied to QED.

When LSZ-type theories are used in perturbation expansion they exhibit a tremendous advantage over Lagrangian QFT since they do not require a canonical formulation. Thus, it is possible to express time-ordered products of Heisenberg operators directly in terms of free field operators (at least for renormalizable theories). Without having to go the old road (Heisenberg picture Lagrangian → Hamiltonian → interaction picture) one can prove the following very important relation between Heisenberg fields $A(x)$ and free fields $a(x)$. If $L_1[A]$ is the interaction Lagrangian we define*

$$S = T_+(e^{-i\int L_1[a]d^4x}). \quad (S1\text{-}21)$$

Then

$$T_+(A(x_1) \cdots A(x_n)) = T_+(a(x_1) \cdots a(x_n)S) \quad (S1\text{-}22)$$

This relation was proven by Pugh[27] from asymptotic QFT. Its vacuum expectation value was first given in a different context by Gell-Mann and Low.[29] It will play a fundamental role in the mathematically rigorous schemes of renormalization (see Supplement 2).

When applied to QED the above equation (S1-21) becomes identical with the S-operator (8-8), and (S1-22) becomes

$$T_+(\Psi(x_1) \cdots \Psi(x_n)\bar{\Psi}(y_1) \cdots \bar{\Psi}(y_n)A^{\mu_1}(z_1) \cdots A^{\mu_m}(z_m))$$
$$= T_+(\psi(x_1) \cdots \psi(x_n)\bar{\psi}(y_1) \cdots \bar{\psi}(y_n)a^{\mu_1}(z_1) \cdots a^{\mu_m}(z_m)S). \quad (S1\text{-}23)$$

S1-5 Null plane QED. As first pointed out by Dirac,[30] the usually used pseudoCartesian coordinate system for the description of Minkowski space could in some cases advantageously be replaced by other systems. One of these is the null plane coordinate system in which x^0, x^1, x^2, x^3 are replaced by u, x^1, x^2, v

$$u = \frac{1}{\sqrt{2}}(x^0 - x^3), \quad v = \frac{1}{\sqrt{2}}(x^0 + x^3). \quad (S1\text{-}24)$$

This can be expressed either by a nondiagonal metric tensor $g_{\mu\nu}$ or by two null vectors m^μ and n^μ satisfying

$$m^2 = 0, \quad n^2 = 0, \quad m \cdot n = -1. \quad (S1\text{-}25)$$

One then works with the projections on these vectors. For example, for a fourvector $A^\mu = (A_v, A_1, A_2, A_u)$ where $A_v \equiv -n \cdot A$ and $A_u \equiv -m \cdot A$. In particular, $x_v = u$ and $x_u = v$.

---

* S is assumed to be normalized to that $(0|S|0) = 1$.

This becomes more than a trivial coordinate transformation when the dynamics of the system is described as a development in time from one null plane $u = u_1$ to another one, $u = u_2$, in contradistinction to the usual dynamical development from $t = t_1$ to $t = t_2$. The associated Cauchy problem of the wave equation requires for its uniqueness certain conditions not required in the conventional case.[31] These conditions mean physically that no waves propagate along u = constant planes.

While a Lorentz transformation to a frame that moves with the velocity of light does not exist, the transformation (S1-24) is related to the limit of systems that are boosted to very high momenta. Thus, the null plane coordinate system becomes especially useful for the approximate description of very high energy collisions. In this context the null plane system was found valuable by high energy theorists.[32]

The use of null plane coordinates in QED and the associated reformulation of QED was discovered independently of the above by Neville in the context of the interaction of lasers with electrons.[33] Later, the discovery of scaling in high energy electron scattering led to an independent development of null plane QED (also called "infinite momentum frame QED").[34] One of the characteristic differences of these two formulations is that the former uses four-component electron fields,[31,35] while the latter[34] uses two-component ones, an option made possible by null coordinates.

In null plane QED the commutation relations of the fields must be specified on a null plane because this is the initial plane. It is then natural to work in the *null plane gauge* characterized by $a_v = 0$. The components $a_k(x)$ ($k = 1, 2$) of a free electromagnetic field then satisfy the commutation relations

$$[a_k(x), a_l(x')]_{u=u'} = -\frac{i}{4}\epsilon(v - v')\delta(x_1 - x_1')\delta(x_2 - x_2')\delta_{kl} \quad (k, l = 1, 2).$$

(S1-26)

as follows by restriction of Eq. (2-28) to the null plane $u = u'$. Note that on a spacelike plane this algebra would be Abelian. One sees therefore that various interesting and novel mathematical properties can be expected in this formulation.[36]

The null plane gauge shows considerable similarity to the Coulomb gauge; in particular, it is also nonlocal. A comparison with conventional QED is therefore most convenient when the latter is cast into the Coulomb gauge. Such a comparison is desirable for the following reasons: (1) the formulations are both divergent and yield finite predictions only after renormalization; (2) the associated Cauchy problems are quite different requiring additional conditions in the null plane case; (3) the null plane coordinates are not manifestly covariant, involving a preferred direction; and (4) the gauges are different and are related nontrivially. For these reasons it is not obvious that the two formulations will lead to the same physical predictions to all orders of perturbation expansion.

A study of this comparison requires setting up Feynman rules for diagrams in the null plane formulation. These differ in several respects from the usual diagrams because the simple interaction operator density $-j_\mu a^\mu$ of conventional QED is here replaced by a complicated expression leading to two types of corners in addition to the usual one of Tables 8-1 and 8-2. One of these involves four electron lines and no photon line, the other two electron lines and two photon lines. Finally, the formal expression for the S-operator (8-8) is replaced by one involving integrations over the "null times" u_n rather than over the τ_n.

The result of this study is that the null plane formulation of QED is Poincaré and gauge invariant and gives predictions which are identical to those of the conventional formulation:[37] whereas the unrenormalized S-matrices differ, the renormalized physical S-matrices yield identical scattering amplitudes, life times, and radiative corrections.

References

1. Texts which appeared since the first edition of this book and which are devoted largely or entirely to quantum electrodynamics (this list does not claim to be complete):

    (T) W. Thirring, *Einführung in die Quantenelektrodynamik* [German, Wien: F. Deutike, 1955] translated by J. Bernstein and with additions by the author, Academic Press, New York, 1958.

    (BS) N. N. Bogoliubov and D. V. Shirkov, *Introduction to the Theory of Quantized Fields* [Russian, 1957] translated by G. M. Volkoff, Interscience Publishers, New York, 1959.

    (KA) G. Källen, *Quantenelektrodynamik* [German, 1958] translated by C. K. Iddings and M. Mizushima as *Quantum Electrodynamics*, Springer-Verlag, New York, 1972.

    (AB) A. I. Akhiezer and V. B. Berestetskii, *Quantum Electrodynamics*, Second Edition [Russian, 1959] translated by G. M. Volkoff, Interscience Publishers, New York, 1965.

    (F) R. P. Feynman, *Quantum Electrodynamics*, Benjamin, New York, 1961.

    (KA) D. Kastler, *Introduction a l' électrodynamique quantique* [French], Dunod, Paris, 1961.

    (BD) J. D. Bjorken and S. D. Drell, *Relativistic Quantum Fields*, McGraw Hill, New, York, 1965.

    (G) W. T. Grandy Jr., *Introduction to Electrodynamics and Radiation*, Academic Press, New York, 1970.

    (S) J. Schwinger, *Particles, Sources and Fields*, Addison-Wesley, Reading, MA, 1970. (This book is not quantum electrodynamics in the usual sense, although it deals with the same subject matter. Rather, it is a novel approach which is related to conventional quantum field theory somewhat in the way that the Feynman path integral method is related to field quantization in Hilbert space. The student of this subject should welcome this serious effort by one of the great masters to get away from the traditional approach which—after twenty-five years of effort—has not resolved the basic problem of producing a workable as well as a completely finite theory that yields the desired predictions. Whether this one does succeed is left for Harold* to decide.)

(U) P. Urban, *Topics in Applied Quantumelectrodynamics*, Springer-Verlag, Wien, 1970.

(BB) I. Białynicki-Birula and Z. Białynicka-Birula, *Elektrodynamika kwantowa* [Polish, 1974] translated by Eugene Lapa, Pergamon Press, Oxford, 1975.

2. A classic on this subject is the book by Wentzel (p. 74, footnote), but many recent treatises on QFT make use of it (see, for example Reference 1 above, (BS)).

3. A very general and clear treatment is used by P. A. M. Dirac, *Quantum Mechanics*, 4th Edition, Oxford University Press, 1958.

4. The presence of constraints can be a highly nontrivial matter, as becomes apparent in the problem of quantizing gravitation theory. A careful treatment of the problem of constraints was developed by P. A. M. Dirac, *Canad. J. Math.* **2,** 129 (1950) and *Lectures on Quantum Mechanics*, Belfer Graduate School, Yeshiva University, New York, 1964. See also E. C. G. Sudarshan and N. Mukunda, *Classical Mechanics*, John Wiley and Sons, New York, 1974.

5. One of the cornerstones of the theory was the paper by A. S. Wightman, *Phys. Rev.* **101,** 860 (1956) in which a field was expressed in terms of vacuum expectation values. It should not be forgotten, however, that independently of Wightman and his associates a very similar set of assumptions ("axioms") was proposed by W. Schmidt and K. Baumann, *Nuovo Cim.* **4,** 860 (1956). The latter work is somewhat less general mathematically and in particular does not pay attention to domain questions.

6. R. F. Streater and A. S. Wightman, *PCT, Spin & Statistics, and All That*, Benjamin Inc., New York, 1964. A delightful book which can also be read most profitably by those who are not mathematical sophisticates. It also contains a valuable annotated bibliography.

7. R. Jost, *The General Theory of Quantized Fields*, American Mathematical Society, Providence, R.I., 1965. This condensed and very concise report presupposes a considerable mathematical background since it seems to have been intended more for the mathematician who wants to learn about rigorous QFT than for the physicist who wants to learn how to cast his pragmatic and heuristic QFT into a rigorous language and profit from it.

8. A. S. Wightman and L. Gårding; *Ark. f. Fysik* **28,** 129 (1965).

---

* Harold is the "hypothetical alert reader of limitless dedication" rediscovered by Schwinger. He was first known to Galileo under the name of Sagredo.

9. D. Ruelle, *Helv. Phys. Acta* **35**, 147 (1962).
10. R. Haag, *Phys. Rev.* **112**, 669 (1958) and *Nuovo Cim. Suppl.* **14**, 131 (1959).
11. The exploitation of analyticity properties owes much to A. Martin who summarized this field in *Scattering Theory; Unitarity, Analyticity, and Crossing*, Springer-Verlag, Berlin, 1969.
12. F. Strocchi, *Phys. Rev.* **162**, 1429 (1967) and **D2**, 2334 (1970).
13. B. Ferrari, L. E. Picasso, and F. Strocchi, *Commun. Math. Phys.* **35**, 25 (1974).
14. F. Strocchi and A. S. Wightman, *J. Math. Phys.* **15**, 2198 (1974).
15. R. Haag, *Ann. Physik.* **7**, 29 (1963); A. Swieca, *Cargèse Lectures in Physics*, Vol. 4 (D. Kastler, ed.), Gordon and Breach, New York, 1970; D. Maison and D. Zwanziger, *Nuclear Phys.* **B91**, 425 (1975).
16. H. Lehmann, R. Symanzik, and W. Zimmermann, *Nuovo Cim.* **1**, 205 and **2**, 425 (1955); **6**, 319 (1957). V. Glaser, H. Lehmann, and W. Zimmermann, *Nuovo Cim.* **6**, 1122 (1957).
17. See for example: S. S. Schweber, *An Introduction to Relativistic Quantum Field Theory*, Row, Peterson, and Co., Evanston, Ill., 1961; G. Barton, *Introduction to Advanced Field Theory*, Interscience, New York, 1963. The mathematically minded reader who is interested in the relation between the LSZ-formulation and the Haag–Ruelle theory of References 9 and 10 should consult K. Hepp's lectures in *Axiomatic Field Theory*, (Brandeis Univ. Summer Inst. 1965), Gordon and Breach, New York, 1966.
18. L. E. Evans and T. Fulton, *Nucl. Phys.* **21**, 492 (1960).
19. N. Nakanishi, *Prog. Theor. Phys. (Japan)* **52**, 1929 (1974). Earlier work is quoted therein.
20. R. P. Feynman, *Phys. Rev.* **84**, 268 (1950); J. Schwinger, *Proc. Natl. Acad. Sci. USA.* **37**, 452 (1951). The suggestion goes back to P. A. M. Dirac, *Quantum Mechanics*, 3rd Edition, Oxford Univ. Press, 1947.
21. S. Hori, *Prog. Theor. Phys. (Japan)* **7**, 578 (1952); J. Anderson, *Phys. Rev.* **94**, 703 (1954).
22. J. M. Jauch, *Helv. Phys. Acta* **29**, 287 (1956); F. Rohrlich, paper 1–13 in *Analytic Methods in Mathematical Physics* (R. P. Gilbert and R. G. Newton, eds.), Gordon and Breach, New York, 1970.
23. M. H. Fried, *Functional Methods and Models in Quantum Field Theory*, MIT Press, Cambridge, 1972. Yu. V. Novozhilov and A. V. Tulub, *The Method of Functionals in the Quantum Theory of Fields*, Gordon and Breach, New York, 1961. Functional methods are used extensively in the texts (BS), (S), and (BB) of Reference 1.
24. N. N. Bogoliubov, *Izv. Akad. Nauk SSSR, Ser. Fiz.* **19**, 137 (1955); N. N. Bogoliubov, B. V. Medvedev, and M. K. Polivanov, *Fortsch. d. Physik* **6**, 169 (1958). See also Reference 1, (BS), and B. V. Medvedev, V. P. Pavlov, M. K. Polivanov, and A. D. Sukhanov, *Theor. Math. Phys.* **13**, 939 (1972).
25. F. Rohrlich, *J. Math. Phys.* **5**, 324 (1964) and F. Rohrlich and M. Wilner, *J. Math. Phys.* **7**, 482 (1966).

26. M. L. Goldberger, "Introduction to the theory and application of dispersion relations" in *Dispersion Relations and Elementary Particles* (C. DeWitt and R. Omnes, eds.), John Wiley, New York, 1960. This review includes much of the second paper of Reference 24. An easy book for a first study of this subject matter is R. Hagedorn, *Introduction to Field Theory and Dispersion Relations*, MacMillan, New York, 1964.
27. R. E. Pugh, *Ann. Physics* **23,** 335 (1963), later developed by him in *J. Math. Phys.* **6,** 740 (1965) and **7,** 376 (1966), and by others. See the review article by F. Rohrlich, *Acta Phys. Austriaca Suppl. IV*, 228 (1967).
28. F. Rohrlich, *Phys. Rev.* **183,** 1359 (1969); A. Pagnamenta and F. Rohrlich, *Phys. Rev.* **D 1,** 1640 (1970).
29. M. Gell-Mann and F. Low, *Phys. Rev.* **84,** 350 (1951).
30. P. A. M. Dirac, *Rev. Mod. Phys.* **26,** 392 (1949).
31. R. A. Neville and F. Rohrlich, *Nuovo Cim.* **1A,** 625 (1971).
32. S. Weinberg, *Phys. Rev.* **150,** 1313 (1966); K. Bardacki and G. Segré, *Phys. Rev.* **159,** 1263 (1967); L. Susskind, *Phys. Rev.* **165,** 1535 (1968).
33. R. A. Neville (Ph. D. Thesis, Syracuse University, 1968); R. A. Neville and F. Rohrlich, *Phys. Rev.* **D 3,** 1692 (1971).
34. J. B. Kogut and D. E. Soper, *Phys. Rev.* **D 1,** 2901 (1970); J. D. Bjorken, J. B. Kogut, and D. E. Soper, *Phys. Rev.* **D 4,** 1382 (1971).
35. F. Rohrlich, *Acta Phys. Austriaca* **32,** 87 (1970).
36. H. Leutwyler, J. R. Klauder, and L. Streit, *Nuovo Cim.* **66 A,** 536 (1970), F. Rohrlich and L. Streit, *Nuovo Cim.* **7 B,** 166 (1972). For a review see F. Rohrlich, *Acta Phys. Austriaca Suppl. VIII*, 277 (1971).
37. S. J. Brodsky, R. Roskies, and R. Suaya, *Phys. Rev.* **D 8,** 4574 (1973); J. H. Ten Eyck and F. Rohrlich, *Phys. Rev.* **D 9,** 2237 (1974) and *Phys. Lett.* **46 B,** 102 (1973).

SUPPLEMENT S2

RENORMALIZATION

It is generally stated that the divergences of QED can be successfully removed by renormalization. This is a somewhat misleading statement since it seems to imply that if there were no divergences, renormalization would not be necessary.

However, when physical systems (such as fields) interact one usually obtains expressions (e.g. terms in the matrix elements of the interaction) which are experimentally indistinguishable from expressions describing the free systems (e.g. terms in the free Hamiltonian). This occurs also in theories which are not divergent. But whenever this happens a renormalization procedure is necessary. The fact that this procedure completely removes all divergences (as an accidental by-product, as it were) is the great virtue of "renormalizable" field theories. If the removal of all divergences is not a by-product of the renormalization process, additional terms in the interaction *may* make the theory renormalizable.* However, if an infinite number of such terms would be required the theory is called "nonrenormalizable," since the associated number of phenomenological parameters would be infinite, thus rendering the theory meaningless.

S2–1 Dyson–Salam–Ward renormalization. The original renormalization technique which is known under this name and which was discussed in Chapter 10 has been extended and clarified in several ways. One new development was the *Ward–Takahashi equation*.

Since the Ward identity (10–20), (10–21) has become known, various authors conjectured a generalization.[2] A proof to all orders of the "generalized Ward identity" was given by Takahashi,[3] and it is therefore known under both names.

There are various equivalent forms of this equation. When the Ward identity is written in the form (see page 214)

$$\frac{\partial \Sigma^*(p)}{\partial p^\mu} = i\Lambda_\mu(p,p), \qquad (S2-1)$$

then the Ward–Takahashi equation can be written in a corresponding

---

* An example is QED of spinless charged particles. It is *not* renormalizable. But when an additional interaction of the form $\lambda\phi^4$ is introduced, the theory becomes renormalizable.[1]

way:
$$\Sigma^*(p') - \Sigma^*(p) = i(p_\alpha' - p_\alpha)\Lambda^\alpha(p',p). \tag{S2-2}$$

The Ward identity (2–1) follows by differentiating with respect to p' and then taking the limit $p' \to p$.

An alternative and equivalent form is obtained by use of Eq. (10–10), $\Sigma^* = S^{-1} - S'^{-1}$, yielding

$$-S'^{-1}(p') + S'^{-1}(p) = i(p_\alpha' - p_\alpha)\Gamma^\alpha(p',p). \tag{S2-3}$$

Here we used (10–8), $\gamma_\mu + \Lambda_\mu = \Gamma_\mu$. On multiplication by $S'(p')$ and $S'(p)$ on the left and on the right, respectively, one finds

$$S'(p') - S'(p) = i(p_\alpha' - p_\alpha)S'(p')\Gamma^\alpha(p',p)S'(p). \tag{S2-4}$$

It is important to realize that all these equations hold to all orders in the perturbation expansion. But beyond this, the Ward–Takahashi equation holds also in renormalized form. One only needs to renormalize the unrenormalized equation (S2–4) using (10–48) and (10–49):

$$S_0(p') - S_0(p) = i(p_\alpha' - p_\alpha)S_0(p')\Gamma_0^\alpha(p',p)S_0(p). \tag{S2-5}$$

This result depends crucially on the equality of the renormalization constants for Γ^μ and S'. Thus, the equality $B = L$, Eq. (10–28) was used. But that equality is a consequence of gauge invariance. Therefore, the Ward–Takahashi equation just like the Ward identity is a consequence of gauge invariance, as was conjectured long before the Ward–Takahashi identity was proven.[1] Under certain conditions one can also derive gauge invariance from the Ward–Takahashi equation, *viz.* when one is dealing with a canonical formalism.[4] In that case the Ward–Takahashi equation becomes equivalent to gauge invariance.

Another important addition to the Dyson–Salam–Ward renormalization scheme of the early 1950's was the proof of an important assertion used in this scheme. Although it was possible to make it very plausible[5] no rigorous proof existed until Weinberg's work of 1960.[6]

The separation of divergences by the scheme given in Sections 10–3 and 10–4 is asserted to leave a convergent multiple integral. This separation was based entirely on counting, i.e., on the degree of divergence κ, Eq. (10–3); thus the assertion was that the Dyson–Salam–Ward renormalization scheme gives convergent answers for a given diagram, if—after removal of the divergent parts—its degree of divergence κ is negative and if the degree of divergence of each possible subdiagram is negative. *Weinberg's theorem* proves this assertion and establishes absolute convergence of the renormalized expression.*

---

* In addition, though not relevant in the context of renormalization, this theorem gives the behavior of a renormalized diagram in the high-energy limit of one or more of the external four-momenta.

In Chapter 10 the renormalization scheme involved a Lagrangian in which the bare mass m and bare charge e were used and the "lumping together" of m with δm and of e with δe was carried out in the S-matrix, resulting in a renormalized S-matrix which depends only on the renormalized mass m_0, renormalized charge e_0, and renormalized wave functions.

An obvious suggestion is to *start* the theory with renormalized quantities. This can in fact be done at the cost of having to add "counter terms" (sometimes called "subtraction terms") to the interaction, whose sole purpose is to cancel the terms which arise later on and which require renormalization. These counter terms necessarily contain infinite coefficients.

Thus, one can start with the Lagrangian (4-10)-(4-14) which can be written in the form

$$L = L_{\text{rad}}[A^\mu] + L_{\text{el}}[\Psi,m] + L_1[A^\mu,\Psi,e]. \qquad (S2\text{-}6)$$

Using the renormalization Eqs. (10-40), (10-52), (10-65), and (10-67) it becomes

$$L = ZL_{\text{rad}}[A_0^\mu] + Z'L_{\text{el}}[\Psi_0,m_0] + Z'[A_0^\mu,\Psi_0,e_0] + Z'\delta m\bar\Psi_0\Psi_0. \qquad (S2\text{-}7)$$

Since $Z = 1 - C$ and $Z' = 1 - L$ this can also be written as

$$L = L_{\text{rad}}[A_0^\mu] + L_{\text{el}}[\Psi_0,m_0] + L_1[A_0^\mu,\Psi_0,e_0] + L_c. \qquad (S2\text{-}8)$$

One now has the same functional form as in the original Langrangian but with all arguments replaced by the corresponding renormalized arguments, and with an added piece L_c. This counter term is now seen to consist of a sum of terms each of which contains a factor C, L, or δm which will be found later on to be divergent (as cut-offs are removed). One can apply canonical methods treating $L_1 + L_c$ as the new interaction Lagrangian, the free Lagrangian being identical in form to the one previously used. Of course, the unpleasant terms of L_c result in an interaction Hamiltonian which is equally unpleasant. When one goes to the interaction picture one finds that one must amend the usual Feynman rules since the extra terms give rise to new types of corners. For example, the term $Z'\delta m\bar\Psi_0\Psi_0$ in (S2-7) yields a term $-Z'\delta m\bar\psi_0\psi_0$ in the interaction picture Hamiltonian; this term is described by a corner involving no photon line and two electron lines (one ingoing and one outgoing).

The same result can be achieved much more simply by avoiding the canonical formulation and instead using the relation (S1-23) which requires only the knowledge of $L_1 + L_c$. When the divergences of the S-matrix are then separated (as before) one finds that for each diverging term involving C, L, or δm, there is now a compensating term which arose from the counter term L_c in the Lagrangian. Perfect cancellation results if one identifies the factors C, L, and δm in L_c (which are to begin with unspecified) with the factors of the same name arising in the separation of divergences. The result is a finite renormalized S-matrix which is identical with the one obtained in Chapter 10. This was proven by Matthews and Salam.[7]

S2-2 Bogoliubov-Parasiuk-Hepp-Zimmermann renormalization.

This renormalization scheme, usually referred to as BPHZ is equivalent to the Dyson–Salam–Ward scheme in the sense that it leads to exactly the same renormalized S-matrix. But in its details and especially from the mathematical point of view it is quite different.

The approach by Bogoliubov and Parasiuk[8] is the following. The integrals arising from a particular diagram in perturbation expansion are recognized as (in general) divergent and are made mathematically well defined by some invariant cut-off procedure (e.g. Pauli–Villars regularization, Section 10-9) *before* any integration is carried out. Each propagator (a distribution) is made into an ordinary function. One then carries out the integrations, followed by a separation into terms which would diverge if the cut-off were removed and terms which remain convergent. This separation is of course not unique. But, using the first few terms of a Taylor or MacLaurin expansion in the external momenta as suggested by Dyson* (as e.g. in (9–16)), the separation is fixed by the value of these terms and by the external momenta at which the expansion is carried out, i.e., at which the renormalization constants are defined (e.g. at the mass shell). The result is completely independent of the particular cut-off procedure that is chosen.

Mathematically, it is recognized that the factors in the S-matrix integrals (the electron and photon propagators) are distributions. Products of distributions are in general undefined but can be given an unambiguous meaning by a suitable limiting process. The cut-off plays the role of turning the distributions into ordinary functions whose multiplication is not problematic. The separation of divergences (i.e., of those terms which have no limit as the cut-off is removed) is mathematically simply a regularization procedure which makes out of the product of distributions a well defined distribution (or sum of distributions). Thus, after integration and separation of the "convergent part" one obtains the distribution which is the result of the previously undefined integral of the product.

Bogoliubov and Parasiuk showed how this can be carried through to all orders of the perturbation expansion. Their work was amended and corrected by Hepp.[9] He used the following very simple covariant cut-off for the causal propagators: All causal propagators have the structure

---

* *loc. cit.*, p. 203.

of a polynomial times the distribution

$$\tilde{\Delta}_c(p) = \frac{1}{p^2 + m^2 - i0} = i \int_0^\infty d\alpha e^{-i\alpha(p^2+m^2-i0)} \qquad (S2-9)$$

We wrote $-i0$ instead of $-i\mu$ used in (8–18) and (8–19) in order to indicate that the limit $\mu \to 0$ is meant. The function

$$\tilde{\Delta}_c^{r,\epsilon}(p) = i \int_r^\infty d\alpha e^{-i\alpha(p^2+m^2-i\epsilon)} \qquad (S2-10)$$

with finite r and ϵ becomes the distribution (S2–9) in the limit $r, \epsilon \to 0$. But these limits are taken only after integration and separation of the (in the limit) divergent terms. It is the systematic separation of these terms, consistent to all orders of perturbation expansion which offers the main difficulty in proving the consistency of the BPH renormalization scheme. This scheme culminates in the following fundamental theorem: there exists a regularization scheme so that every truncated* τ-function of Heisenberg fields when expressed by (S1–23) in terms of the τ-function of the free fields, yields after regularization in the limit a Lorentz invariant distribution in p-space which is analytic in the external momenta.

A different mathematical approach to the question of defining the product of distributions was used by Wilson.[10] He expands the product of two local operator valued distributions into a finite sum† of products of c-number distributions and operator valued distributions.

$$A(x)B(y) = \sum_{n=1}^N E_n(x-y)C_n(x) \qquad (S2-11)$$

A, B, and the C_n are operator-valued distributions, the E_n are c-number distributions. This expansion holds for small distances, i.e., when the two points x and y are not too far from one another in Minkowski space. The ultraviolet divergences arise from the limit $x \to y$, i.e., from high momenta in the Fourier transform.

This *operator product expansion* was used by Zimmermann[11] to initiate a similarly motivated regularization process in the BPH renormalization scheme. First he defines *normal operator products* (N-products) by systematically subtracting suitably chosen terms from the ordinary operator product. This corresponds exactly to the regularization of the

---

* In order that a τ-function describe a sum of *connected* Feynman diagrams it must be "truncated," i.e., it must not contain intermediate vacuum states.

† The remaining terms of the sum do not contribute to the ultraviolet divergence problem.

τ-function in the BPH scheme, but it can be done in terms of operator products rather than in terms of vacuum expectation values of operator products. The regularized Feynman integrals are absolutely convergent by Weinberg's theorem.

By means of the N-products one can then achieve what was intended in the Matthews–Salam method, *viz.* to start with a renormalized Lagrangian: one can express the interaction Lagrangian in terms of N-products which are renormalized and have, as it were, all divergences subtracted. A new "effective Lagrangian" thus becomes the starting point, $L = L_0 + L_{eff}$. While one uses renormalized operators in both cases, the difference between this L_{eff} and $L_1 - L_c$ is that L_{eff} contains only *finite* factors while L_c contains the divergent factors L, C, and δm. The transition from the renormalized Lagrangian in the Heisenberg picture to the time-ordered products of Heisenberg operators is again afforded by the relation (S1-23) with L_1 in S replaced by L_{eff}. The divergences are removed in the BPHZ scheme by forming the N-products. Thus, only finite terms are being renormalized.

Finally, one must recall that this renormalization scheme is directed only at ultraviolet divergences. Since QED also contains infrared divergences (see Supplement 4) the theory is here best carried through with a finite photon mass whose limit to zero is taken at the end of the calculation.

S2-3 Analytic renormalization. In 1968 Speer proposed a regularization method of field theoretic propagators which is quite different from the BPHZ techniques.[12] Similar suggestions were made independently by Bollini, Giambiagi, and Gonzalez Dominguez.[13] But the first to suggest analytic regularization was probably Gustafson.[14]

The idea is to replace the typical propagator by an analytic function of a complex variable λ,

$$\tilde{\Delta}_c(p) = \frac{1}{p^2 + m^2 - i0} \to \tilde{\Delta}_c^{\lambda,\epsilon}(p) \equiv \frac{1}{(p^2 + m^2 - i\epsilon)^\lambda}. \quad \text{(S2-12)}$$

One requires at first that Re $\lambda > 2$ for the photon propagator in the Feynman gauge (Re $\lambda > 4$ in other gauges) and Re $\lambda > 3$ for the electron propagator. Assigning a different λ to each propagator one obtains a regularized Feynman integral which depends on $E_i + P_i$ complex variables λ_l ($l = 1, \cdots, E_i + P_i$) with E_i and P_i the number of internal electron and photon lines as in Section 10–1. By means of the representation (S2-9) the multiple integral is carried out and the result is a function analytic in $\mathbb{C}^{E_i + P_i}$ except for singularities at the points $\lambda_l = 1$.

Speer devised a method for removal of these singularities, so that the remainder is absolutely convergent in the limit $\lambda_l \to 1$, $\epsilon \to 0$. For details the reader is referred to his work.[12]

Analytic renormalization can be proven to yield exactly the same renormalized S-matrix as the BPHZ renormalization scheme. It can also be given a counter-term interpretation.

References

1. F. Rohrlich, *Phys. Rev.* **80**, 666 (1950).
2. H. S. Green, *Proc. Phys. Soc.* **66**, 873 (1953); T. D. Lee, *Phys. Rev.* **95**, 1329 (1954); L. D. Landau and I. M. Khalatnikov, *Zh. E.T.F.* **29**, 89 (1955) [*Sov. Phys. JETP* **2**, 69 (1955)]; E. S. Fradkin, *Zh. E.T.P.* **29**, 258 (1955) [*Sov. Phys. JETP* **2**, (1955)].
3. Y. Takahashi, *Nuovo Cim.* **6**, 371 (1957).
4. See e.g., K. Nishijima, *Phys. Rev.* **119**, 485 (1960).
5. A. Salam, *Phys. Rev.* **84**, 426 (1951).
6. S. Weinberg, *Phys. Rev.* **118**, 838 (1960).
7. P. T. Matthews and A. Salam, *Phys. Rev.* **94**, 185 (1954).
8. N. N. Bogoliubov and O. S. Parasiuk, *Acta Math.* **97**, 227 (1957); O. S. Parasiuk, *Ukrainskii Math. J.* **12**, 287 (1960); see also (BS) of Supplement 1, Reference 1.
9. K. Hepp, *Comm. Math. Phys.* **2**, 301 (1966).
10. K. Wilson, *Phys. Rev.* **179**, 1499 (1969).
11. W. Zimmermann's article in *Lectures on Elementary Particles and Quantum Field Theory*, Vol. 1, 1970 Brandeis University Summer Institute, M.I.T. Press, Cambridge, Mass., 1970.
12. E. Speer, *J. Math. Phys.* **9**, 1404 (1968) and *Generalized Feynman Amplitudes*, Princeton University Press, 1969. See also F. Guerra and M. Maranaro, *Nuovo Cim.* **60**, A 759 (1969); F. Guerra, *ibid.* **1**, A, 523 (1971).
13. C. G. Bollini, J. J. Giambiagi, and A. Gonzalez Dominguez, *Nuovo Cim.* **31**, 550 (1964).
14. T. Gustafson, *Ark. Astron. Fys.* **34**, A(2) (1947).

SUPPLEMENT S3

COHERENT STATES

The coherent state representation of Hilbert space is well known in quantum optics where it is used for systems with a finite number of degrees of freedom.[1] The representation must be extended to an infinite number of degrees of freedom for the purpose of field theory. We shall here very briefly review the case of a finite number of degrees of freedom before we proceed to the field theoretic situation.[2]

S3–1 A finite number of degrees of freedom. Given the algebra of K annihilation operators a_k ($k = 1, 2, \cdots K$) and their adjoints, the obvious generalization of Section 2–7,

$$[a_k, a_l] = 0, \quad [a_k, a_l^*] = \delta_{kl}, \tag{S3–1}$$

we seek the eigenvectors of these annihilation operators. Each a_k^* generates from the vacuum, defined by $a_k|0\rangle = 0$ for all k, a set of Fock states $\{|n_k\rangle\}$ defined by

$$|n_k\rangle \equiv \frac{a_k^{*n_k}}{\sqrt{n_k!}} |0\rangle. \tag{S3–2}$$

They provide a complete orthonormal basis of a Fock space, $\mathfrak{F}_1$. The states for different k are orthonormal

$$\langle n_l'|n_k\rangle = \delta_{kl}\delta_{n'n}.$$

The direct products of these states

$$|\{n_k\}\rangle \equiv |n_1\rangle|n_2\rangle \cdots |n_K\rangle \equiv |n_1, n_2, \cdots, n_K\rangle \tag{S3–3}$$

provide a complete orthonormal basis of a Fock space, $\mathfrak{F}_K$.

The eigenvectors of an annihilation operator a_k can be parametrized by a complex number $z_k = x_k + iy_k$ and expressed in terms of the basis of $\mathfrak{F}_1$ as

$$|z_k\rangle = e^{-\frac{1}{2}|z_k|^2} \sum_{n_k=0}^{\infty} \frac{z_k^{n_k}}{\sqrt{n_k!}} |n_k\rangle. \tag{S3–4}$$

One easily verifies their eigenvector property,

$$a_k|z_k\rangle = z_k|z_k\rangle. \tag{S3–5}$$

The Hilbert space spanned by the $\{|z_k\rangle\}$ is called coherent state space, $\mathfrak{C}_1$, and is isomorphic to $\mathfrak{F}_1$. The vectors $|z_k\rangle$ are normalized, but are not mutually orthogonal,

$$\langle z_k'|z_k\rangle = \exp\left[-\tfrac{1}{2}|z_k - z_k'|^2 + i\,\mathrm{Im}\,(z_k'^* z_k)\right]. \tag{S3-6}$$

Their overcompleteness nevertheless permits a resolution of the identity

$$1 = \int |z_k\rangle d\mu_k \langle z_k| \qquad d\mu_k \equiv \frac{dx_k dy_k}{\pi}. \tag{S3-7}$$

Parenthetically it should be remarked that it is sometimes convenient to change this measure to an equivalent one with Gaussian weight. This is done by omitting the exponential factor in the definition (S3-4) and by replacing $d\mu_k$ by $d\mu_k' \equiv \exp(-|z_k|^2) d\mu_k$ in (S3-7). The result is a Hilbert space of entire analytic functions[3] because (omitting the subscript k) any Fock vector $|f\rangle = \sum_n c_n |n\rangle$ yields

$$f(z) \equiv \langle f|z\rangle' = \sum_n c_n z^n / \sqrt{n!}.$$

If z_k ranges over the whole complex plane the set is actually overcomplete as mentioned before. For example, a grid of points in the complex plane suffices if it is not spaced too far apart. One can prove[4] that when one chooses the grid in the form

$$z_k = \gamma(\lambda m + i\, n/\lambda) \qquad m,n \text{ integers}, \lambda > 0, \tag{S3-8}$$

the set is overcomplete for $\gamma < \sqrt{\pi}$, exactly complete for $\gamma = \sqrt{\pi}$, and undercomplete (incomplete) for $\gamma > \sqrt{\pi}$.

The direct product space $\mathfrak{C}_K$ spanned by

$$|\{z_k\}\rangle \equiv |z_1, z_2, \cdots, z_K\rangle \tag{S3-9}$$

can be expressed in terms of the basis vectors of $\mathfrak{F}_K$ (S3-3) in complete analogy to (S3-4),

$$|\{z_k\}\rangle = \sum_{\{n_k\}} \prod_{l=1}^{K} e^{-\frac{1}{2}|z_l|^2} \frac{z_l^{n_l}}{\sqrt{n_l!}} |\{n_k\}\rangle. \tag{S3-10}$$

The corresponding resolution of the identity is

$$\int |\{z_k\}\rangle d\mu(\{z_k\}) \langle\{z_k\}| = 1 \tag{S3-11}$$

with

$$d\mu(\{z_k\}) = \prod_{k=1}^{K} \frac{dx_k dy_k}{\pi}. \tag{S3-12}$$

---

* This overcompleteness is evident from Eq. (S3-8) below.

A very useful expression for $|\{z_k\}\rangle$ is obtained from (S3-10) by means of (S3-2),

$$|\{z_k\}\rangle = e^{-\frac{1}{2}\sum_{k=1}^{K}|z_k|^2} e^{\sum_{l=1}^{K} z_l a_l^*} |0\rangle \qquad \text{(S3-13)}$$

$$= e^{\sum_{1}^{K}(z_l a_l^* - z_l^* a_l)} |0\rangle. \qquad \text{(S3-14)}$$

In the last step the Campbell–Baker–Hausdorff formula was used[1]

$$e^{A+B} = e^A e^B e^{-\frac{1}{2}[A,B]} \qquad \text{(S3-15)}$$

which holds for any two operators A and B whose commutator, $[A,B]$, commutes with both A and B. It is obvious that the operator in (S3-14) is unitary.

Of special interest is of course the action of a creation operator, a_k^*, say, on $\mathcal{C}_K$. The algebra (S3-1) is satisfied if

$$a_k^* |z_1, \cdots, z_k, \cdots, z_K\rangle = \left(\tfrac{1}{2} z_k^* + \frac{\partial}{\partial z_k}\right) |z_1, \cdots, z_K\rangle, \qquad \text{(S3-16)}$$

and this is indeed the only consistent choice for a_k^*.

A most important property of the coherent state representation is that the diagonal elements determine the off-diagonal ones. For if $F(a^*,a)$ is any polynomial in $\{a_k\}$ and $\{a_k^*\}$ the matrix elements

$$\langle z_1', \cdots, z_K' | F(a_1^*, \cdots, a_K^*; a_1, \cdots, a_K) | z_1, \cdots, z_K\rangle$$

differ from a product of entire functions in each of the pairs of complex planes z_k and $z_k'^*$ only by the factor

$$\exp[\tfrac{1}{2} \sum_{1}^{K} (|z_k'|^2 + |z_k|^2)].$$

Since an entire function in z_k and $z_k'^*$ vanishes everywhere if it vanishes on the subdomain $z_k = z_k'$ the above-mentioned property follows.*

But the diagonal elements are given by (using an abbreviated notation)

$$\langle\{z_k\}| F(\{a_k^*\}, \{a_k\}) |\{z_k\}\rangle = F\left(\{z_k^*\}, \left\{z_k + \frac{\partial}{\partial z_k^*}\right\}\right) \qquad \text{(S3-17)}$$

which is one of the most important relations of the coherent state representation. It gives the representation of a function of the $2K$ operators a_k, a_k^* ($k = 1, 2, \cdots K$) in terms of a function of $2K$ complex numbers z_k and z_k^* ($k = 1, 2, \cdots K$). The proof of (S3-17) follows from the

---

* From $f(z'^*, z) = \sum_{m=0}^{\infty} \sum_{n=0}^{\infty} c_{mn} z'^{*m} z^n$ follows that $f(z^*, z) = 0$ implies $c_{mn} = 0 \; \forall_{m,n}$.

observation that for each degree of freedom Eqs. (S3–16) and (S3–6) yield

$$\left(\frac{z_k'}{2} + \frac{\partial}{\partial z_k'^*}\right) (\langle z_k'|z_k\rangle f(z_k')) = \langle z_k'|z_k\rangle \left(z_k + \frac{\partial}{\partial z_k'^*}\right) f(z_k').$$

A special case of (S3–17) is the diagonal matrix element of a normal-ordered (Wick-ordered) function in which all creation operators stand to the left of all annihilation operators, which we shall indicate in the conventional way by

$$F = :F:, \qquad \langle\{z_k\}|F(\{a_k^*\},\{a_k\})|\{z_k\}\rangle = F(\{z_k^*\},\{z_k\}). \qquad \text{(S3–18)}$$

S3–2 Coherent states of the radiation field. We are now ready to consider the limit $K \to \infty$. As is evident from (S3–13) there is no difficulty in that limit as long as

$$\sum_{k=1}^{\infty} |z_k|^2 < \infty, \qquad \text{(S3–19)}$$

since then the expansion of $|z_1, z_2, \cdots, z_K\rangle_{K\to\infty}$ in Fock states exists and will continue to be unitarily equivalent to the corresponding Fock space $\mathfrak{F}_\infty$.

However, if (S3–19) does not hold, so that the normalization factor in (S3–13) vanishes, the coherent states cannot be expressed in terms of Fock states. In this case the scalar product $\langle\{n_k\}|\{z_k\}\rangle$ between any Fock state and any coherent state vanishes in the limit $K \to \infty$; the two bases become orthogonal to one another. There will still exist a state $|0\rangle$ in $\mathfrak{C}_\infty$ but it will not be the vacuum state in the sense that it will not be characterized by $a_k|0\rangle = 0$ for all k. Closely connected with this fact is the absence of a number operator $N = \sum_1^K a_k^* a_k$: if (S3–19) does not hold in the limit then also N will not exist in that limit as an operator on the space $\mathfrak{C}_\infty$, i.e., all expectation values will be $\langle N\rangle = \infty$.

How, then, is $\mathfrak{C}_\infty$ defined, if it cannot be defined in terms of a Fock space? The answer to this question is not a simple one; it was first given by von Neumann.[5] He showed that there exist spaces which are much larger than Fock space and which can be defined by infinite tensor products. The commutation relations (S3–1) of the creation and annihilation operators are still defined on such a space, as well as their enveloping algebra. The expression (S3–14) still has a meaning on this space in the limit $K \to \infty$. In fact,

$$U(z^*,z) \equiv \exp\left[\sum_1^\infty (z_l a_l^* - z_l^* a_l)\right] \qquad \text{(S3–20)}$$

continues to be a unitary operator. It is in terms of just this operator that the coherent states can then be given a mathematical meaning in the infinite degree of freedom case.[6]

In order to exploit the properties of the coherent-state representation which were reviewed in Section S3-1 and which hold at least for those infinite degrees of freedom systems for which (S3-19) is valid, it is convenient to replace the plane wave expansion of the radiation field (Section 2-6) by an expansion in a complete set of wave packets, $\{f_\alpha{}^\mu(x)\}$, so that

$$a^\mu(x) = \sum_\alpha (f_\alpha{}^\mu(x)a_\alpha + f_\alpha{}^{\mu *}(x)a_\alpha{}^*). \qquad \text{(S3-21)}$$

The $f_\alpha{}^\mu(x)$ are positive energy solutions of the wave equation $\Box f_\alpha{}^\mu(x) = 0$. Their completeness is thus given in the covariant form by

$$\sum_\alpha f_\alpha{}^\mu(x) f_\alpha{}^{\nu *}(x') = -ig^{\mu\nu} D_+(x - x') \qquad \text{(S3-22)}$$

with D_+ defined as in Section 2-6. One can recover the plane wave expansion (2-62) from (S3-21) by replacing the index α by a continuum index $\mathbf{k}$ and a polarization index (r) so that the sum over α becomes an integral over d^3k and a sum over polarizations. One must then replace $f_\alpha{}^\mu(x)$ by

$$\frac{e_{(r)}{}^\mu}{(2\pi)^{3/2} \sqrt{2\omega}} e^{ik \cdot x}$$

where $\mathbf{k}$ is on the positive light cone, $|\mathbf{k}| = \omega = k^0$. a_α then becomes $a_{(r)}(\mathbf{k})$. The linear manifold of solutions spanned by the $f_\alpha{}^\mu(x)$ can be given a covariant inner product

$$\langle f_\alpha{}^\lambda, g_{\beta\lambda} \rangle \equiv i \int (f_\alpha{}^{\lambda *}(x) \partial_\mu g_{\beta\lambda}(x) - g_{\beta\lambda}(x) \partial_\mu f_\alpha{}^{\lambda *}(x)) d^3\sigma^\mu(x). \qquad \text{(S3-23)}$$

The integral is over any spacelike plane, since it is independent of the specific choice of such a plane as long as $f_\alpha{}^\mu$ and $g_\beta{}^\nu$ are solutions of the wave equation. The form of (S3-23) is of course closely related to the Cauchy solution (2-16) and was in fact used previously, e.g. in (2-47). This inner product satisfies $\langle f,g \rangle^* = \langle g,f \rangle$.

The norm associated with the product (S3-23) is of course *not* positive definite. For the positive energy solutions we can require $\langle f_\alpha{}^\lambda, f_{\beta\lambda} \rangle = \delta_{\alpha\beta}$ but for the negative energy solutions, $f_\alpha{}^{\mu *}$ one has $\langle f_\alpha{}^{\lambda *}, f_{\beta\lambda}{}^* \rangle = -\delta_{\alpha\beta}$. The two are orthogonal, $\langle f_\alpha{}^\lambda, f_{\beta\lambda}{}^* \rangle = 0$. The operators a_α are now given by

$$a_\alpha = \langle f_\alpha{}^\mu, a_\mu \rangle. \qquad \text{(S3-24)}$$

When the commutation relations

$$[a_\alpha, a_\beta{}^*] = \delta_{\alpha\beta}, \quad [a_\alpha, a_\beta] = 0 \qquad \text{(S3-25)}$$

are imposed, the fields $a^\mu(x)$ will satisfy the expected covariant commutation relations $[a^\mu(x), a^\nu(x')] = -ig^{\mu\nu}D(x - x')$, Eq. (2-28). The operators a_α and $a_\alpha{}^*$ therefore obviously play the same role as the Fourier transforms of $a^{\mu(-)}$ and $a^{\mu(+)}$, (2-63) and (2-64).

The important relation (S3-17) for the representation of an operator can now be generalized to the radiation field:

$$\langle\{z_\alpha\}|F(a_\mu{}^{(+)}(x); a_\nu{}^{(-)}(x'))|\{z_\alpha\}\rangle = F\left(z_\mu{}^{(+)}(x); z_\nu{}^{(-)}(x') + \sum_\alpha f_{\alpha\nu}(x')\frac{\partial}{\partial z_\alpha{}^*}\right) \quad \text{(S3-26)}$$

where $z_\mu(x)$ and z_α are the c-number analogues of $a_\mu(x)$ and a_α following exactly the expansion (S3-21).

$$z^\mu(x) = \sum_\alpha (f_\alpha{}^\mu(x) z_\alpha + f_\alpha{}^{\mu*}(x) z_\alpha{}^*). \quad \text{(S3-27)}$$

If F depends only on $a_\mu(x) = a_\mu{}^{(-)}(x) + a_\mu{}^{(+)}(x)$ this result simplifies to

$$\langle\{z_\alpha\}|F(a^\mu(x))|\{z_\alpha\}\rangle = F\left(z^\mu(x) + \sum_\alpha f_\alpha{}^\mu(x)\frac{\partial}{\partial z_\alpha{}^*}\right).$$

But since the derivative acts only on functions of $z^\mu(x)$ it follows from (S3-27) and (S3-22) that this can be written as

$$\langle\{z_\alpha\}|F(a^\mu(x))|\{z_\alpha\}\rangle = F\left(z^\mu(x) - i\int D_+(x - x')d^4x'\frac{\delta}{\delta z_\mu(x')}\right) \quad \text{(S3-28)}$$

which is the fundamental relation for the coherent state representation of a function of the radiation field.

A further simplification arises when F is normal ordered, because then, according to (S3-18), the integral on the right side is absent and one has the beautiful result

$$\langle\{z_\alpha\}|F(a^\mu(x)|\{z_\alpha\}\rangle = F(z^\mu(x)) \text{ for } F(a^\mu) = {:}F(a^\mu){:} \quad \text{(S3-29)}$$

In all these equations it is understood that $z^\mu(x)$ is defined by (S3-27) in terms of the z_α with respect to which the matrix element is taken.

The unitary operator (S3-20) can also be generalized to apply to a field. Substitution of (S3-24) for a_α and of the analogous equation for z_α yields the simple result

$$U(z^*, z) = e^{\sum_\alpha (z_\alpha a_\alpha{}^* - z_\alpha{}^* a_\alpha)} \quad \text{(S3-30)}$$
$$= e^{\langle z_\mu, a^\mu \rangle}.$$

Finally, it is important to ensure the consistency of this covariant coherent-state representation with the Gupta–Bleuler gauge condition:[7]

this condition, Eq. (6-67), becomes here

$$\partial^\mu a_\mu{}^{(-)}(x)|\{z_\alpha\}\rangle = 0.$$

It will be satisfied provided the $\{f_\alpha{}^\mu(x)\}$ are restricted by $\partial^\mu f_{\mu\alpha}(x) = 0$.

S3-3 Application to ordering theorems. The potential $a^\mu(x)$ occurs in the expression for the S-matrix under the integral expressing the interaction operator H, Eq. (4-43). More generally, we shall consider integrals of the form

$$a(\varphi) \equiv \int a^\mu(x)\varphi_\mu(x)d^4x \qquad (S3\text{-}31)$$

where $\varphi^\mu(x)$ is a set of c-number functions which are sufficiently well behaved to ensure the selfadjointness of the operator $a(\varphi)$. The use of $a(\varphi)$ instead of $a^\mu(x)$ not only corresponds to the actual physical situation but also will avoid possible ambiguities when we consider functions of $a^\mu(x)$, rather than of functionals, $a(\varphi)$. The appearance of δ-functions, D-functions, and other distributions makes it clear that many theorems will not be applicable point-wise, i.e., for each x. They will hold only for functions of functionals $a(\varphi)$.

The most important ordering theorem is the one that relates time-ordered to normal-ordered products. Now the time-ordered product defined in (8-6) has the representation, according to (S3-28)

$$\langle\{z_\alpha\}|P(a^\mu(x)a^\nu(x'))|\{z_\alpha\}\rangle$$

$$= \theta(x - x')\left(z^\mu(x) - i\int D_+(x - x'')d^4x'' \frac{\delta}{\delta z_\mu(x'')}\right)z^\nu(x')$$

$$+ \theta(x' - x)\left(z^\nu(x') - i\int D_+(x' - x'')d^4x'' \frac{\delta}{\delta z_\nu(x'')}\right)z^\mu(x)$$

$$= z^\mu(x)z^\nu(x') - iD_c(x - x')$$

according to (A1-16) with $D_c \equiv D_{1R}$. Proceeding in this way one proves by induction that

$$\langle\{z_\alpha\}|PF(a(\varphi))|\{z_\alpha\}\rangle$$

$$= F\left(z(\varphi) - i\int \varphi^\mu(x)d^4x D_c(x - x')d^4x' \frac{\delta}{\delta z^\mu(x')}\right). \qquad (S3\text{-}32)$$

We can now use the CBH formula (S3-15) to obtain

$$\langle\{z_\alpha\}|Pe^{i\lambda a(\varphi)}|\{z_\alpha\}\rangle = e^{i\lambda z(\varphi) + \frac{1}{2}i\lambda^2 D_c(\varphi,\varphi)}$$

where

$$D_c(\varphi,\varphi) = \int \varphi^\mu(x)d^4x D_c(x - x')d^4x' \varphi_\mu(x').$$

If we employ (S3-29) and observe that a relation that holds for all matrix elements of an operator (the diagonal elements in the coherent state representation are just as good as all elements, as we proved in S3-1), will also hold for the operator itself, we find a closed form of Wick's ordering theorem[8]

$$Pe^{i\lambda a(\varphi)} = :e^{i\lambda a(\varphi)}:e^{\frac{i}{2}\lambda^2 D_c(\varphi,\varphi)} \tag{S3-33}$$

This is the special case of the ordering theorem proven in Section A4-1 which relates time-ordered to normal-ordered products. The products of n factors $a(\varphi)$ are related by comparing the coefficients of λ^n on both sides of (S3-33):

$$Pa^n(\varphi) = \sum_{k=0}^{(\frac{n}{2},\frac{n-1}{2})_I} (2k-1)!! \binom{n}{2k} (-iD_c(\varphi,\varphi))^k :a(\varphi)^{n-2k}: \tag{S3-34}$$

where $(\frac{1}{2}n, \frac{1}{2}n - \frac{1}{2})_I$ is the integer of the two numbers, $\frac{1}{2}n$ and $\frac{1}{2}n - \frac{1}{2}$; $(2k-1)!! \equiv (2k-1)(2k-3)(2k-5)\cdots$, and $(-1)!! \equiv 0$.

Of course the ordering theorem for electron fields cannot be derived in this fashion because the coherent state representation exists nontrivially only for Bose fields.

In a similar way one can derive the ordering theorems

$$e^{i\lambda a(\varphi)} = :e^{i\lambda a(\varphi)}:e^{\frac{i}{2}\lambda^2 D_+(\varphi,\varphi)} \tag{S3-35}$$

and

$$:e^{i\lambda a(\varphi)}::e^{i\mu a(\psi)}: = :e^{i\lambda a(\varphi)+i\mu a(\psi)}:e^{i\lambda\mu D_+(\varphi,\psi)}. \tag{S3-36}$$

The coefficients of $\lambda^m \mu^n$ provide the reordering theorem for the product of two normal-ordered products to be expressed as a sum of normal-ordered products:[2]

$$:a^m(\varphi)::a^n(\psi): = \sum_{k=0}^{(m,n)_<} k![-iD_+(\varphi,\psi)]^k \binom{m}{k}\binom{n}{k} :a^m(\varphi)a^n(\psi): \tag{S3-37}$$

Here $(m,n)_<$ indicates the smaller of the two integers m and n.

The relations between the normal-ordered product and the time-ordered and ordinary products, Eqs. (S3-33) and (S3-35), can be combined by eliminating the normal-ordered product. Since $D_c - D_+ = D_A$ according to (A1-9), we find the relation between a time ordered and an ordinary product,

$$Pe^{i\lambda a(\varphi)} = e^{i\lambda a(\varphi)}e^{\frac{i}{2}\lambda^2 D_A(\varphi,\varphi)} \tag{S3-38}$$

We conclude this discussion with an important application of the closed form of Wick's theorem to the S-operator. The Dyson form of the S-operator, (8–7) and (8–8), can be written

$$S = Pe^{-i\int H(\tau)d\tau} = Pe^{i\int j_\mu(x)a^\mu(x)d^4x} \tag{S3-39}$$

The symbol P indicates here a time ordering of the a^μ and of the $\psi(x)$ and $\bar\psi(x)$ contained in $j_\mu(x)$. Under the P symbol the operators can be manipulated as if they were c-numbers. Thus, with (S3-31) we can write in obvious notation

$$S = P^{(\psi)}(P^{(a)}e^{ia(j)})$$

and we can treat j like an ordinary function under the $P^{(\psi)}$ symbol. With (S3-33) this becomes

$$S = P^{(\psi)}(:e^{-i\int H(\tau)d\tau}:e^{\frac{i}{2}\int j_\mu(x)d^4x D_c(x-x')d^4x' j^\mu(x')}). \tag{S3-40}$$

The normal ordering here refers only to the photon field a^μ.

This expression for the S-operator contains a clean separation into internal and external photon lines. It shows that all internal photon lines appear as propagators $D_c(x-x')$ while all external photon lines occur exactly as in the interaction operator H attached to one $\bar\psi$ and one ψ. Equation (S3-40) gives a closed form for the operator describing all possible electron transitions for a given number of external photon lines.

References

1. R. J. Glauber, *Phys. Rev.* **131**, 2766 (1963). J. R. Klauder and E. C. G. Sudarshan, *Fundamentals of Quantum Optics*, Benjamin, N.Y., 1968. W. H, Louisell, *Quantum Statistical Properties of Radiation*, John Wiley & Sons. New York, 1973.
2. F. Rohrlich, *Coherent State Representation and Quantum Field Theory*, paper 1–13 in *Analytic Methods in Mathematical Physics* (R. P. Gilbert and R. G. Newton, eds.), Gordon and Breach, New York, 1970.
3. V. Bargmann, Comm. *Pure Appl. Math.* **14**, 187 (1961); *Proc. Nat. Acad. Sci. USA* **48**, 199 (1962).
4. V. Bargmann, P. Butera, L. Girardello, and J. R. Klauder, *Rep. Math. Phys.* **2**, 221 (1971).
5. J. von Neumann, *Comp. Math.* **6**, 1 (1938).
6. The existence of representation spaces of the canonical commutation relations that are not unitarily equivalent to Fock space, a situation which occurs only for an infinite number of degrees of freedom, has been mentioned earlier (p. 39, footnote). They were investigated by L. Gårding and A. S. Wightman, *Proc. Nat. Acad. Sci. USA* **40**, 617 and 622 (1954). See also K. O. Friedrichs, *Mathematical Aspects of the Quantum Theory of Fields*, Interscience Publishers, New York, 1953, who called these representations myriotic. The coherent

state representation is a special case of a much larger class of important myriotic representations called continuous representations. These were studied in a series of papers by J. R. Klauder, *J. Math. Phys.* **4,** 1055 and 1058 (1963); **5,** 177 (1964), and by J. R. Klauder and J. McKenna, *ibid.* **5,** 878 (1964) and **6,** 68 (1965).

7. For a more detailed discussion see J. Gomatam, *Phys. Rev.* **D3,** 1292 (1971), R. E. Pugh, *Ann. Phys.* **30,** 422 (1964).

8. See footnote, p. 445.

SUPPLEMENT S4

INFRARED DIVERGENCES

Our treatment of infrared divergences in Section 16–1 concentrated on the three questions raised on p. 390, i.e., it was concerned with the physical understanding of these apparent divergences and with the proof of their exact cancellation to all orders. This proof has since been carried out in different ways paying more attention to the overlapping infrared divergences.[1]

However, this solution to the infrared problem is now superseded by a much deeper understanding of this difficulty. That a full understanding was lacking as late as the early 1960's can also be seen from the inability until that time to compute a transition probability *amplitude* that is infrared-divergence free. The cancellation of infrared divergences could be carried out only in the transition probability or cross section.

One notes first that the infrared problem occurs already on the non-relativistic and classical level: the Rutherford cross section (15–8) diverges in the forward direction and no total cross section exists. Although it has been known for a long time that the electron wave function in a Coulomb field is asymptotically *not* a plane wave and is modified by a logarithmic term as in (15–9), this indication did not lead to the detection of the cause of the mathematical difficulty until 1964.[2]

S4–1 Dollard's discovery. In the perturbation approach to scattering it is assumed that the interaction vanishes asymptotically so that in the limit as $|t| \to \infty$ the particles can be described by the free field equations. This assumption underlies the interaction picture and the associated iteration solution of the S-matrix, Section 8–1. It also underlies the existence of the limits $(7-44)_\pm$ of the wave operators $\Omega_\pm(\tau)$ in the foundations of scattering theory.

Dollard[2] proved on the level of nonrelativistic quantum mechanics that the long range of the Coulomb field *precludes* that assumption because the interaction H does not vanish fast enough. A screened Coulomb field permits it, and so does a finite photon mass. However, scattering theory and a corresponding iteration solution for the unscreened Coulomb field can be "saved" in the following way: one does not separate the total Hamiltonian H_{tot} into the usual $H_0 + H$ of (7–39) but into $H_0' + H'$ such that H' *does* vanish asymptotically so as to permit the existence of a suitably defined limit $\Omega_\pm(\pm \infty)$. Such a new separation is of course not unique. For example, for Coulomb scattering of a charge e by the field (15–3) one can add to H_0 any operator which asymptotically approaches $\alpha Z/(vt)$ where v is the asymptotic velocity of the particle. The same operator must then be subtracted from H to form H'.

When the existence of the asymptotic limit of the wave operators is thus ensured, the resulting iteration solution of the S-matrix obtained from the S-operator (7–70) is no longer infrared divergent. This spurious divergence is thus eliminated from the *amplitudes*. The price one pays is of course obvious: a much more complicated "free particle" solution based on H_0'.

The generalization of these results to relativistic quantum electrodynamics was carried out most successfully by Kulish and Fadde'ev[3] and later by Zwanziger.[4] Their work is in substantial agreement with earlier attempts, especially by Chung[5] and by Kibble.[6] We shall follow here the ideas of Kulish and Fadde'ev but work relative to the interactive picture rather than the Schrödinger picture which they used. The results will then be related to those of Zwanziger who used a very different method.

S4-2 A new picture. We start in the interaction picture as developed in Section 4–3 because it is from this picture that the easily applicable iteration solution of Chapter 8 is derived. In this picture the fields a^μ and ψ are free fields, i.e., they are characterized by a time dependence governed by the free Hamiltonian H_0. The state vectors $\omega(\tau)$ have a time dependence governed by the interaction operator $H(\tau)$, Eq. (4–43). According to the ideas emerging from Dollard's paper discussed above we wish to change the time dependence of both the field operators and the state vectors, i.e., we want to go into a new picture. This picture will be characterized by state vectors $\omega'(\tau)$ with a time dependence

$$i\dot{\omega}'(\tau) = H'(\tau)\omega'(\tau) \tag{S4-1}$$

and the new interaction operator $H'(\tau)$ must be such that it vanishes asymptotically: the usual interaction $H(\tau)$ does *not* vanish asymptotically but approaches some operator $H^{as}(\tau)$. The transformation to the new picture is accomplished in exact analogy to the transformation from the Heisenberg to the interaction picture, Section 4–3. If $\varphi(x)$ is a typical field in the interaction picture we are seeking a transformation $U(\tau)$ such that the field in the new picture $\varphi'(x,\tau)$ is related to $\varphi(x)$ by

$$\varphi'(x,\tau) = U(\tau)\varphi(x)U^{-1}(\tau) \tag{S4-2}$$

and the new state vectors are related by

$$\omega'(\tau) = U(\tau)\omega(\tau). \tag{S4-3}$$

The instant τ_0 at which the two pictures coincide is arbitrary and will be discussed later.

It now follows that $H'(\tau)$ of (S4–1) is related to its counterpart in the interaction picture according to (S4–2). But that counterpart should be $H(\tau) - H^{as}(\tau)$, since this is guaranteed to vanish asymptotically. Thus,

$$H'(\tau) = U(\tau)(H(\tau) - H^{as}(\tau))U^{-1}(\tau)$$

Substitution of this and (S4–3) into (S4–1) leads to an equation for $U(\tau)$

$$i\dot{U}(\tau) = -U(\tau)H^{as}(\tau) \tag{S4–4}$$

which, in view of the initial condition $U(\tau_0) = 1$ has the unique solution

$$U(\tau) = T_-\left(\exp\left[i\int_{\tau_0}^{\tau} H^{as}(\tau')d\tau'\right]\right). \tag{S4–5}$$

Here T_- is a time ordering which orders operators with time increasing from left to right; the usual time ordering, $P \equiv T_+$ orders them with time increasing from right to left, as in (8–6).

We must now find H^{as} before we can evaluate $U(\tau)$. It is obvious from the plane wave expansion (2–62) that $a^\mu(x)$ is not approaching a different form for large t, thus,

$$a_\mu^{as}(x) = a_\mu(x). \tag{S4–6}$$

Not so for the current of the interaction picture which we write in normal ordered form as

$$j_\mu(x) = -ie:\bar{\psi}(x)\gamma_\mu\psi(x):. \tag{S4–7}$$

The decomposition (3–87) of $\psi(x)$ into creation part, $\psi^{(+)}(x)$, and annihilation part, $\psi^{(-)}(x)$, separates $j_\mu(x)$ into four terms, $j_\mu^{(--)}$, $j_\mu^{(++)}$, $j_\mu^{(+-)}$, and $j_\mu^{(-+)}$. It will be shown below that asymptotically only $j^{(--)}$ and $j^{(++)}$ survive; the argument will be as follows: Each of the four terms of j_μ contains the time t in the form $e^{i\alpha t}$. Asymptotically, only those terms will survive for which α can approach zero within the range of integration over momentum space. Thus a term of the form $\bar{u}(\mathbf{p}')\gamma_\mu u(\mathbf{p})$ which occurs in $j^{(--)}$ has a factor $\exp[i(\epsilon' - \epsilon)t]$ and will survive; but $\bar{u}(\mathbf{p}')\gamma_\mu v(\mathbf{p})$ which occurs in $j_\mu^{(-+)}$ has a factor $\exp[i(\epsilon' + \epsilon)t]$ and will not survive.

In order to find j_μ^{as} one must look at H^{as} because one must define j_μ^{as} as that current which produces the correct asymptotic interaction. Consider, therefore,

$$-\int j_\mu^{(--)}(x)a^{\mu(-)}(x)d^3x = \frac{iem}{(2\pi)^{3/2}}\int\frac{d^3p'd^3pd^3k}{\sqrt{\epsilon'\epsilon}\sqrt{2\omega}}\delta(\mathbf{p} - \mathbf{p}' + \mathbf{k}) \tag{S4–8}$$

$$\sum_{r'}\sum_r \bar{u}_{r'}(\mathbf{p}')\gamma\cdot a(\mathbf{k})u_r(\mathbf{p})a_{r'}{}^*(\mathbf{p}')a_r(\mathbf{p})\exp[i(\epsilon' - \epsilon - \omega)t].$$

For large t only very small ω will contribute because only for those can $\epsilon' - \epsilon - \omega$ become small. Because of the δ-function,

$$\epsilon' - \epsilon - \omega = \sqrt{(\mathbf{p}+\mathbf{k})^2 + m^2} - \sqrt{\mathbf{p}^2 + m^2} - \omega$$
$$= \mathbf{k} \cdot \nabla_{\mathbf{p}}\epsilon - \omega + O(\omega^2).$$

At this point the covariant generalization of (3–101)

$$i\bar{u}_r(\mathbf{p})\gamma^\mu u_{r'}(\mathbf{p}) = i\bar{v}_r(\mathbf{p})\gamma^\mu v_{r'}(\mathbf{p}) = \delta_{rr'} p^\mu/m \tag{S4-9}$$

can be used where p^μ is on the positive mass shell. The integral then becomes (with the help of the inversion of (2–63))

$$\frac{e}{(2\pi)^{3/2}} \int \frac{d^3p}{\epsilon} \int \frac{d^3k}{\sqrt{2\omega}} \sum_r a_r^*(\mathbf{p}) a_r(\mathbf{p}) p \cdot a(\mathbf{k}) \exp[i(\mathbf{k}\cdot\mathbf{p}/\epsilon - \omega)t]$$
$$= e \int \frac{d^3p}{\epsilon} \sum_r a_r^*(\mathbf{p}) a_r(\mathbf{p}) p_\mu \int e^{i\mathbf{k}\cdot\mathbf{p}t/\epsilon} \cdot \frac{d^3k}{(2\pi)^3} \cdot \int a^{\mu(-)}(x) e^{-i\mathbf{k}\cdot\mathbf{x}} d^3x.$$

Thus one finds the asymptotic form

$$-\int j_\mu^{(--)}(x) a^{\mu(-)}(x) d^3x \to -\int j_\mu^{\mathrm{as}(-)}(x) a^{\mu(-)}(x) d^3x$$

with

$$j_\mu^{\mathrm{as}(-)}(x) = \int d^3p\, \rho^{(-)}(\mathbf{p}) \frac{p_\mu}{\epsilon} \delta_3(\mathbf{x} - \mathbf{v}t)$$
$$\mathbf{v} \equiv \mathbf{p}/\epsilon, \quad \rho^{(-)}(\mathbf{p}) = -e \sum_r a_r^*(\mathbf{p}) a_r(\mathbf{p}).$$

Similarly, the integrals over $j_\mu^{(--)} a^{\mu(+)}$ and $j_\mu^{(++)} a^{\mu(\pm)}$ are obtained in the small ω limit. The final result is

$$H^{\mathrm{as}}(\tau) = -\int j_\mu^{\mathrm{as}}(x) a^\mu(x) d\sigma(\tau) \tag{S4-10}$$

with

$$j_\mu^{\mathrm{as}}(x) = \int d^3p\, \rho(\mathbf{p}) \frac{p_\mu}{\epsilon} \delta_3(\mathbf{x} - \mathbf{v}t)$$
$$= \int d^3p\, \rho(\mathbf{p}) v_\mu \delta_{\sigma(v)}(x - v\tau_v). \tag{S4-11}$$

In this expression ρ is the charge density operator in momentum space

$$\rho(\mathbf{p}) = e \sum_r (b_r^*(\mathbf{p}) b_r(\mathbf{p}) - a_r^*(\mathbf{p}) a_r(\mathbf{p})) \tag{S4-12}$$

and the invariant three-dimensional hypersurface delta function is defined in terms of the surface delta function δ^μ of (1–108) by

$$\delta_{\sigma(v)}(x) = v_\mu \delta^\mu(x), \tag{S4-13}$$

the surface normal arises here naturally as the four-velocity direction, $v^\mu = p^\mu/m = (\gamma, \gamma\mathbf{v})$. We also note for future reference that with $\tau_v \equiv -v \cdot x$,

$$\delta_{\sigma(v)}(x - v\tau_v) = \int_{-\infty}^{\infty} \delta_4(x - v\tau')d\tau' = \frac{1}{\gamma} \delta_3(\mathbf{x} - \mathbf{v}t) \cdot \frac{d^3x}{d\sigma(v)}$$

The manifest covariance of $j_\mu{}^{\text{as}}$ is now evident since $Q = \int d^3p \rho(\mathbf{p})$ is the invariant charge operator already encountered in (3–106). The structure of $j_\mu{}^{\text{as}}$ is that of a classical charge density moving uniformly: a free particle of velocity v^μ with proper time $\bar\tau$ has the world line $x^\mu = x_0{}^\mu + v^\mu\bar\tau$, but for large $\bar\tau$ the initial position is irrelevant and $\bar\tau = -v \cdot x = \tau_v$. Thus τ_v has the physical meaning of asymptotic proper time.

The derivation of $j_\mu{}^{\text{as}}$ emphasizes two points. First, it makes apparent the role of the soft photon limit in the asymptotic behavior of the current; secondly, it shows that one cannot take the limit of $j_\mu(x)$ itself, but only of $H(\tau)$; i.e., only the integrated current (integrated over the photon field a^μ) permits this limit.

The asymptotic currents in the distant future and in the distant past can be distinguished as follows. One defines

$$j_\mu{}^{\text{as}}(x;\,{}^{\text{out}}_{\text{in}}) \equiv \int d^3p \rho(\mathbf{p}) v_\mu \int_{-\infty}^{\infty} \theta(\pm\tau')\delta_4(x - v\tau')d\tau' \\ = \int d^3p \rho(\mathbf{p}) v_\mu \theta(\pm\tau_v)\delta_\sigma(x - v\tau_v) \quad \text{(S4–14)}$$

so that

$$j_\mu{}^{\text{as}}(x) = j_\mu{}^{\text{as}}(x;\text{in}) + j_\mu{}^{\text{as}}(x;\text{out}) \quad \text{(S4–15)}$$

The Fourier transform of (S4–14), defined as in (16–15), is obtained now by means of the identities preceding (16–20),

$$j_\mu{}^{\text{as}}(x;\,{}^{\text{out}}_{\text{in}}) = \frac{1}{(2\pi)^{3/2}} \int j_\mu{}^{\text{as}}(k;\,{}^{\text{out}}_{\text{in}}) e^{ik\cdot x} d^4k \\ j_\mu{}^{\text{as}}(k;\,{}^{\text{out}}_{\text{in}}) = \frac{\mp i}{(2\pi)^{3/2}} \int \frac{p_\mu}{p \cdot k} \rho(\mathbf{p}) d^3p. \quad \text{(S4–16)}$$

The classical limit, $s_\mu(k)$, (16–16), results for the specific case of an electron of momentum p^μ for $t < 0$ changing to momentum p'^μ for $t > 0$,

$$s_\mu(k) = \int [(p'|j_\mu{}^{\text{as}}(k,\text{out})|p'') + (p''|j_\mu{}^{\text{as}}(k,\text{in})|p)] d^3p'' \\ = \frac{ie}{(2\pi)^{3/2}} \left(\frac{p_\mu'}{p' \cdot k} - \frac{p_\mu}{p \cdot k}\right).$$

The Fourier transform of the whole operator $j_\mu{}^{\mathrm{as}}$ follows easily from (S4–11)

$$\begin{aligned}j_\mu{}^{\mathrm{as}}(k) &= \frac{1}{(2\pi)^{3/2}} \int e^{-ik\cdot x} d^4x \int d^3p \rho(\mathbf{p}) v_\mu \int \delta(x - v\tau') d\tau' \\ &= \frac{1}{(2\pi)^{3/2}} \int d^3p \rho(\mathbf{p}) v_\mu \delta(k\cdot v)\end{aligned} \qquad (\text{S4–17})$$

This expression shows that the asymptotic current is conserved, $k^\mu j_\mu{}^{\mathrm{as}}(k) = 0$, i.e., also $\partial^\mu j_\mu{}^{\mathrm{as}}(x) = 0$, and that $j_\mu{}^{\mathrm{as}}(k)$ is limited to space-like momenta k^μ, since v^μ is timelike.

The $j_\mu{}^{\mathrm{as}}(x)$ are operators because ρ is an operator. Nevertheless, they commute with one another at *all* space–time points,

$$[j_\mu{}^{\mathrm{as}}(x), j_\nu{}^{\mathrm{as}}(x')] = 0 \quad \text{for all } x, x'. \qquad (\text{S4–18})$$

Consequently, the commutator of $H^{\mathrm{as}}(\tau)$ with itself at two different times can be evaluated from the commutator of the $a_\mu(x)$, (2–28),

$$[H^{\mathrm{as}}(\tau), H^{\mathrm{as}}(\tau')] = -i \int j_\mu{}^{\mathrm{as}}(x) d\sigma(\tau) D(x - x') d\sigma(\tau') j_{\mathrm{as}}{}^\mu(x').$$

This commutator commutes with $H^{\mathrm{as}}(\tau'')$ for all values of τ', τ'', and τ because of (S4–18). It is clear from this that $j_\mu{}^{\mathrm{as}}$ can be treated like a c-number in algebraic operations.

The transformation $U(\tau)$, (S4–5), which leads from the interaction picture to the new picture according to (S4–3) is now known explicitly,

$$U(\tau) = T_-(e^{-i\int_{\tau_0}^{\tau} j_\mu{}^{\mathrm{as}}(x') a^\mu(x') d^4x'});$$

the integration extends over a four-dimensional volume bounded by the two hyperplanes τ and τ_0. Its adjoint is

$$U^*(\tau) = P(e^{i\int_{\tau_0}^{\tau} j_\mu{}^{\mathrm{as}}(x') a^\mu(x') d^4x'}) = U^{-1}(\tau). \qquad (\text{S4–19})$$

The transformation is therefore formally unitary. Further evaluation is facilitated by the ordering formula (S3–38) which relates a time-ordered to an ordinary product. The operator $a(j^{\mathrm{as}})$ is here defined as in (S3–31) but with finite time limits. Thus,

$$U(\tau)^* = e^{\frac{i}{2} D_A(j^{\mathrm{as}}, j^{\mathrm{as}})_\tau} e^{ia(j^{\mathrm{as}})_\tau} \equiv e^{i\Phi(\tau)} e^{R(\tau)}, \qquad (\text{S4–20})$$

and since the two exponentials commute and R is formally antihermitian,

$$U(\tau) = e^{-i\Phi(\tau)} e^{-R(\tau)}. \qquad (\text{S4–20})'$$

The transformation thus consists of two commuting factors. The first one does not involve real photons and will later be related to the Coulomb field effect. It gives a phase factor. The second factor does involve real photons (i.e., the field operators) and will be the center of interest in the later discussion.

We now come to the important question of the choice of τ_0. It is one of the key points in the argument. While $|\tau|$ is obviously large since we want to describe asymptotic behavior, this is not so for τ_0. In fact, for symmetry reasons between asymptotic in and out states, one might even be tempted to choose $\tau_0 = 0$.

Such a choice however would ignore that $H^{as}(\tau)$ has physical meaning only asymptotically, for large $|\tau|$. Thus, one is interested in modifying the separation of H_{tot} into $H_0' + H'$ *only for large* $|\tau|$, i.e., for distant interactions. If $U(\tau)$ were to contain contributions from smaller interaction regions, this fact would appear in $\omega'(\tau)$ of (S4–3) *even for large τ*. In that case, the $|\tau| \to \infty$ limit would differ from the classical limit. For example,[3] the linearity of $R(\tau)$ in the photon creation and annihilation operators results in a lack of commutativity of $R(\tau)$ with the total momentum operator whose radiation field part is (2–38). This noncommutativity disappears in the limit $|\tau| \to \infty$ only when $U(\tau)$ contains no contributions from τ_0. $U(\tau)$ cannot depend on τ_0.*

If this argument is accepted the transformation (S4–2), (S4–3) can have only heuristic value: there is no time at which $\omega'(\tau)$ and $\omega(\tau)$ coincide. In fact, this is exactly the foreboding of the result to be derived below: that the space of the $\omega'(\tau)$ is a coherent state space and has very little to do with the Fock space spanned by the $\omega(\tau)$; $U(\tau)$ is *not* a unitary operator in Fock space. Our arguments are therefore necessarily only of a formal nature.

S4–3 The asymptotically modified fields.

The new fields a_μ' and ψ' can now be obtained explicitly by means of the transformation (S4–2). To this end we cast $\Phi(\tau)$ and $R(\tau)$ into a more convenient form. The Coulomb phase $\Phi(\tau)$ becomes,† after a lengthy and not very illuminating calculation,[3]

$$\Phi(\tau) \equiv \frac{1}{2} D_A(j^{as}, j^{as})_\tau = -\frac{1}{2} \int^\tau d\tau'' \int^{\tau''} d\tau' j_\mu^{as}(x') d\sigma' D(x' - x'') d\sigma'' j^\mu_{as}(x'')$$

$$= \frac{1}{8\pi} \int d^3p \int d^3p' \frac{:\rho(\mathbf{p})\rho(\mathbf{p'}):}{v(p,p')} \int^\tau \frac{d\tau'}{|\tau'|}$$

$$= \frac{1}{2} \int d^3p :\rho(\mathbf{p})\varphi(p,\tau):$$

(S4–21)

---

* The present discussion suggests that the contribution to the integral (S4–5) from the lower limit, τ_0, must vanish.

† In this computation divergent selfenergy terms have been ignored. They would be removed by a mass renormalization.

In this expression $v(p,p')$ is the Lorentz invariant relative velocity[7] of two particles of equal mass m and momenta p and p',

$$v(p,p') = \sqrt{1 - \left(\frac{m^2}{p \cdot p'}\right)^2} \; ; \qquad (S4\text{--}22)$$

it is related to the flux F of (8–50) by $F = p \cdot p' v(p,p')$. The quantity $\varphi(p,\tau)$ is a transform of the Coulomb phase $\Phi(\tau)$ and can be called a "Coulomb phase density"; it is given by

$$\varphi(p,\tau) = Q(p)\epsilon(\tau) \ln |\tau/h| \qquad (S4\text{--}23)$$

with h a constant and

$$Q(p) = \frac{1}{4\pi} \int \frac{\rho(\mathbf{p}')d^3p'}{v(p,p')}. \qquad (S4\text{--}24)$$

The momenta are always on the positive mass shell. Finally, the lower limit of the τ' integration in (S4–21) is omitted in accordance with the discussion at the end of the preceding Section.

The close relationship between the operator Φ of (S4–21) and the classical Coulomb phase is apparent. For a specific nonrelativistic system the latter is given by (15–9) and (15–10). In fact, the above derivation confirms the conjecture by Dalitz (see p. 330). When matrix elements are taken one sees that $\Phi(\tau)$ depends on *all* the charges of the physical system. Correspondingly, the states $\omega'(\tau)$ will describe each electron as *dependent on all other electrons* in the system: the infinite range of the Coulomb potential has completely destroyed the asymptotic free-particle behavior of the interaction picture.

The transform used to define $\varphi(p,\tau)$ in (S4–21) can also be used for $R(\tau)$,

$$R(\tau) = \int d^3p \rho(\mathbf{p}) r(p,\tau). \qquad (S4\text{--}25)$$

From the definition (S4–20) follows

$$R(\tau) = i \int_0^\tau d\tau' \int j_\mu^{as}(x') a^\mu(x') d\sigma'$$
$$= \int d^3p \rho(\mathbf{p}) v_\mu i \int_0^\tau d\tau' d\sigma' \delta_{\sigma(v)}(x' - v\tau_v') a^\mu(x').$$

The surface δ-function here refers to a hypersurface with normal v^μ while the integration is over σ' with arbitrary normal n^μ. Since

$$\delta_{\sigma(v)}(x' - v\tau_v') = \frac{1}{|n \cdot v|} \delta_{\sigma(n)}(x' - n\tau')$$

where $\tau_v' = -v \cdot x'$ and $\tau' = -n \cdot x'$ we find for $r(p,\tau)$

$$r(p,\tau) = i \int^\tau d\tau' \frac{v \cdot a(n\tau')}{|v \cdot n|}$$
$$= \frac{1}{(2\pi)^{3/2}} \int \frac{d^3k}{\sqrt{2\omega}} \frac{1}{|v \cdot n|} \left(\frac{v \cdot a(\mathbf{k})}{v \cdot k} e^{ik \cdot n\tau} - \frac{v \cdot a^*(\mathbf{k})}{v \cdot k} e^{-ik \cdot n\tau} \right). \quad \text{(S4-26)}$$

Substitution into (S4-25) shows that $R(\tau)$ can be written in terms of $j_\mu^{as}(k; \text{in})$ and $j_\mu^{as}(k; \text{out}) = j_\mu^{as}(k, \text{in})^*$ in the suggestive form

$$R(\tau) = -i \int \frac{d^3k}{\sqrt{2\omega}} \frac{1}{|n \cdot v|} (j_\mu^{as}(k, \text{in})a^\mu(\mathbf{k})e^{ik \cdot n\tau} + j_\mu^{as}(k, \text{out})a^{\mu *}(\mathbf{k})e^{-ik \cdot n\tau}). \quad \text{(S4-27)}$$

The fields in the new picture now follow from (S4-2). Since the asymptotic currents commute, (S4-18), the Coulomb phase factor drops out and one has

$$a_\mu'(x,\tau) = e^{-R(\tau)} a_\mu(x) e^{R(\tau)} = a_\mu(x) - [R(\tau), a_\mu(x)].$$

There are no repeated commutators in this expansion because $R(\tau)$ commutes with

$$[R(\tau), a_\mu(x)] = i \int^\tau d^4x' j_\nu^{as}(x') [a^\nu(x'), a_\mu(x)] = \int^\tau d^4x' j_\mu^{as}(x') D(x' - x).$$

Since x is on the hypersurface $\sigma(\tau)$ only $D_A(x' - x) = D_R(x - x')$ contributes. Thus,

$$a_\mu'(x,\tau) = a_\mu(x) + \int^\tau D_R(x - x') j_\mu^{as}(x') d^4x', \quad \text{(S4-28)}$$

with the obvious physical meaning that the field in the new picture is the retarded field due to j_μ^{as} which reduces in the distant past to the free field $a_\mu(x)$. The integration can be carried out explicitly with the result[8]

$$a_\mu'(x,\tau) = a_\mu(x) + \frac{1}{(2\pi)^3} \int \frac{d^3k}{\omega} d^3p \rho(\mathbf{p}) \frac{v_\mu}{v_n \cdot k |v \cdot n|} \cos[k \cdot (x + v_n\tau)]$$
$$v_n^\mu \equiv v^\mu v \cdot n + n^\mu [(v \cdot n)^2 - 1]. \quad \text{(S4-29)}$$

The derivation of this equation requires a separation of x' and v into components parallel and perpendicular to the surface normal n.

The integral in (S4-29) is convergent at the lower limit (infrared limit); but it seems to diverge for large frequencies.* However, one

---

* Actually, it does not diverge when treated as an operator-valued distribution and when integrated over suitable test functions before the high-frequency limit is taken.

must keep in mind that the asymptotic operator j_μ^{as} has been derived under the assumption of small ω (see for example the discussion following (S4–8)). The null vectors k can thus be integrated in a consistent way only over a small sphere of the three dimensional space of $\mathbf{k}$ (with measure d^3k/ω) which surrounds the origin. The upper bound of $\omega = |\mathbf{k}|$ will be related to the experimental error in energy $\Delta\epsilon$ of the specific system observed, as was already indicated at the end of Section 16–1. Thus, the integral is convergent at *both* ends.*

The transformation of the electron field is slightly more complicated. Both factors in $U(\tau)$, (S4–20), contribute. Since they commute, each one can be treated separately. First, one notes that

$$\left[\rho(\mathbf{p}'), \begin{pmatrix} a_r(\mathbf{p}) \\ b_r^*(\mathbf{p}) \end{pmatrix}\right] = e\delta_3(\mathbf{p}' - \mathbf{p}) \begin{pmatrix} a_r(\mathbf{p}) \\ b_r^*(\mathbf{p}) \end{pmatrix}. \tag{S4–30}$$

Then, with $r(p,\tau)$ defined in (S4–26)

$$[R(\tau),\psi(x)] = \frac{e}{(2\pi)^{3/2}} \sum_r \int \sqrt{\frac{m}{\epsilon}}\, d^3p\, r(p,\tau)[a_r(\mathbf{p})u_r(\mathbf{p})e^{ip\cdot x}$$
$$+ b_r^*(\mathbf{p})v_r(\mathbf{p})e^{-ip\cdot x}]. \tag{S4–31}$$

Since this expression is again linear in a_r and b_r^* the n times repeated commutator

$$[R(\tau),\psi(x)]_n = [R(\tau),[R(\tau),\psi(x)]_{n-1}]$$

becomes simply the same expression as $\psi(x)$ but with a factor $[er(p,\tau)]^n$ in the integrand. Thus,[8]

$$e^{-R(\tau)}\psi(x)e^{R(\tau)} = \psi(x) + \sum_{n=1}^{\infty} \frac{(-1)^n}{n!} [R(\tau),\psi(x)]_n$$
$$= \frac{1}{(2\pi)^{3/2}} \sum_r \int \sqrt{\frac{m}{\epsilon}}\, d^3p\, e^{-er(p,\tau)}[a_r u_r e^{ip\cdot x} + b_r^* v_r e^{-ip\cdot x}]. \tag{S4–32}$$

The Coulomb factor in $U(\tau)$ yields a similar transformation. One computes first from (S4–21) to (S4–24)

$$\left[\Phi(\tau), \begin{pmatrix} a_r(\mathbf{p}) \\ b_r^*(\mathbf{p}) \end{pmatrix}\right] = e{:}\varphi(p,\tau) \begin{pmatrix} a_r(\mathbf{p}) \\ b_r^*(\mathbf{p}) \end{pmatrix}{:}.$$

Since $\Phi(\tau)$ commutes with $\varphi(p,\tau)$ this expression is effectively again linear in a_r and b_r^* and repeated commutators just produce additional factors of

---

* An invariant cut-off must be a function of $v \cdot k$.

$e\varphi(p,\tau)$. Thus, similar to (S4–32),

$$e^{-i\Phi(\tau)}\psi(x)e^{i\Phi(\tau)} = \frac{1}{(2\pi)^{3/2}}\sum_r \int \sqrt{\frac{m}{\epsilon}}\, d^3p : e^{-ie\varphi(p,\tau)}[a_r u_r e^{ip\cdot x} + b_r^* v_r e^{-ip\cdot x}] : .$$
(S4–33)

The electron field in the new picture, i.e., the free electron field modified for the asymptotic behavior characterized by H^{as} now becomes

$$\psi'(x,\tau) = \frac{1}{(2\pi)^{3/2}}\sum_r \int \sqrt{\frac{m}{\epsilon}}\, d^3p [e^{-er(p,\tau)-ie\varphi(p,\tau)} a_r u_r e^{ip\cdot x} + b_r^* v_r e^{-ip\cdot x} e^{-er-ie\varphi}].$$
(S4–34)

These fields, a_μ' and ψ', (S4–29) and (S4–34), depend on the point x as well as on the normal n to the spacelike hypersurface on which x lies: $\tau = -n\cdot x$. Since we are interested only in the correct asymptotic behavior (large $|\tau|$) it is easy to *eliminate* the dependence on the surface normal and make the fields depend on the point x only. As we have seen in the expression for the asymptotic current, (S4–11), which *is* a point function, there occurs the proper time τ_v of the asymptotically quasiclassical free particle which moves along $x = v\tau_v$. It is therefore natural to choose the hypersurface so that the normal at x is just the tangent to the classical particle that pierces the plane at that point.* Thus, we choose $n = v$ so that $\tau = -n\cdot x = -v\cdot x$. This surface normal is integrated and appears explicitly only in expectation values. We can now write instead of the normal dependent field $a_\mu'(x,\tau)$ the point function field

$$a_\mu'(x) = a_\mu(x) - \frac{1}{(2\pi)^3}\int \frac{d^3k}{\omega}\, d^3p\, \rho(\mathbf{p})\frac{v_\mu}{v\cdot k}\cos[k\cdot(x+vv\cdot x)], \quad (S4\text{–}35)$$

since for $v = n$ one has $v_n^\mu = -v^\mu$. Similarly, for the normal dependent field $\psi'(x,\tau)$ we can write the point function field

$$\psi'(x) = \frac{1}{(2\pi)^{3/2}}\sum_r \int \sqrt{\frac{m}{\epsilon}}\, d^3p [D(p,x)a_r u_r e^{ip\cdot x} + b_r^* v_r e^{-ip\cdot x} D(p,x)] \quad (S4\text{–}36)$$

where $p = mv$,
$$D(p,x) = e^{-er(p,x)-ie\varphi(p,x)} \quad (S4\text{–}37)$$

$$r(p,x) = \frac{1}{(2\pi)^{3/2}}\int \frac{d^3k}{\sqrt{2\omega}}\left(\frac{v\cdot a(\mathbf{k})}{v\cdot k} e^{-ik\cdot vv\cdot x} - \frac{v\cdot a^*(\mathbf{k})}{v\cdot k} e^{+ik\cdot vv\cdot x}\right)$$

$$\varphi(p,x) = Q(p)\epsilon(-p\cdot x)\ln\left|\frac{v\cdot x}{h}\right|.$$
(S4–38)

---

* The shape of the hypersurface of the Hamiltonian time development is thus fixed asymptotically by the state of the physical system.

Asymptotically, the fields (S4–35) and (S4–36) will give the same results as (S4–29) and (S4–31), respectively. The result (S4–36) was obtained by Zwanziger[4] from very different considerations. The integrals over **k** space are again convergent at *both* ends.

The "distortion operator" $D(p,x)$ which represent the "asymptotic distortion" of the plane wave contains a great deal of information. The operator φ gives the Coulombic interaction with all the other charges in the system; $\exp[ip \cdot x - ie\varphi(p,x)]$ is the generalization of (15–9). The operator r describes the soft photon cloud surrounding each physical charge even when it is "free." The physical electron now emerges as a very complicated structure, even asymptotically, because it is never "free" in the usual, more naive sense of the term.

One can attempt to derive the analogue of the photon and electron propagators of the usual theory, as it obtains in the new picture. For the photon this is trivial because the second term in (S4–35) does not contribute,

$$-ig_{\mu\nu}D_c'(x-x') = (0|T_+(a_\mu'(x)a_\nu'(x')|0) = (0|T_+(a_\mu(x)a_\nu(x')|0) \\ = -ig_{\mu\nu}D_c(x-x'). \quad \text{(S4–39)}$$

The propagator is the same as in the interaction picture.

For the electron the situation is more complicated. The modification of the usual propagator, $iS_c(x-x')$ is obtained from

$$iS_c'(x-x') = (0|T_+(\psi'(x)\bar\psi'(x'))|0). \quad \text{(S4–40)}$$

One finds,* apart from a multiplicative renormalization constant[4,6]

$$S_c'(x) = \frac{1}{(2\pi)^4}\int d^4p e^{ip\cdot x}\frac{i\gamma\cdot p - m}{p^2 + m^2 - i0}\left(\frac{2m^2}{p^2 + m^2 - i0}\right)^{\alpha/\pi}\Gamma\left(1+\frac{\alpha}{\pi}\right)$$

$$= \frac{i}{(2\pi)^3}\int \frac{d^3p}{2\epsilon} e^{ip\cdot x}(-ip\cdot x)^{\alpha/\pi}(i\gamma\cdot p - m) \quad \text{(S4–41)}$$

valid for t very large and in momentum space near the mass shell $p^2 + m^2 = 0$. In the second expression p is on the positive mass shell; Γ is the Gaussian gamma function and $-i0$ in the denominator is a symbolic way to indicate the limit from the negative imaginary part of m^2 of the causal function replacing the possibly confusing notation $-i\mu$ of (8–19).

The derivation of (S4–41) involves a divergent renormalization because the distortion operator leads to an ultraviolet divergent integral in the propagator which appears as a multiplicative factor and must be

---

* This result neglects the Coulomb phase. Its inclusion changes the exponent α/π to $\alpha/\pi + i\gamma$ where γ depends on the invariant relative velocities between the charged particles in the system. Equation (S4–40) then describes the propagation of an electron in the presence of all the other charges at great distance.

removed by wave function renormalization. A clean derivation within the general framework of renormalization theory has so far not been accomplished.

The result (S4–41) shows that a free charged particle is *not* correctly characterized as a pole in the associated propagator. The inevitable photon cloud surrounding such a particle is responsible for modifying this pole into a branch point. The consequence is that a physical charged particle has a "smeared out" mass shell, or equivalently an energy-momentum relation which is not sharp. We shall see later that the associated one-particle states are not exact momentum eigenstates, nor exact eigenstates of $P_\mu P^\mu$ the mass operator. Of course, this situation is characteristic of all particles in interaction with a zero-mass field such as the electromagnetic field. The term "infraparticles" has been used for such particles.[9]

Physically, the inexactness of the mass of the physical electron is a direct consequence of its long-range interaction which does not vanish fast enough asymptotically. The distant charge is therefore still slightly in interaction. This has two consequences: it is always capable of emitting or absorbing (sufficiently soft) photons, and it is still in Coulombic interaction with all the other charged particles.

S4–4 The new S-matrix. As was seen in Chapters 7 and 8 the conventional S matrix in the interaction picture is (see e.g. (7–14) and (8–8))

$$S_{fi} = \lim_{\tau_2 \to +\infty} \lim_{\tau_1 \to -\infty} (\omega(\tau_2), V(\tau_2,\tau_1)\omega(\tau_1)), \qquad (S4–42)$$

the S-operator being

$$S = V(\infty,-\infty) = \lim_{\tau_2 \to +\infty} \lim_{\tau_1 \to -\infty} T_+(e^{-i\int_{\tau_1}^{\tau_2} H(\tau)d\tau}). \qquad (S4–43)$$

In the picture characterized by (S4–2) and (S4–3) this becomes

$$S_{fi} = \lim_{\tau_2 \to \infty} \lim_{\tau_1 \to -\infty} (\omega'(\tau_2), V'(\tau_2,\tau_1)\omega'(\tau_1)) \qquad (S4–44)$$

where the S operator is given by

$$S' = \lim_{\tau_2 \to \infty} \lim_{\tau_1 \to -\infty} V'(\tau_2,\tau_1)$$

$$V'(\tau_2,\tau_1) = U(\tau_2)V(\tau_2,\tau_1)U^{-1}(\tau_1). \qquad (S4–45)$$

The two ways of computing the S-matrix, (S4–42) and (S4–44), give identical physical predictions; but while the former yields infrared *divergent* amplitudes, the latter is convergent. The expression for the operator simplifies because the radiation operator $R(\tau)$ does not contribute; from (S4–25) and (S4–26) we see that in the limit

$$R(\pm \infty) = \int d^3p \rho(\mathbf{p}) r(p, \pm \infty)$$

$$r(p, \pm \infty) = \pm \frac{1}{(2\pi)^{3/2}} \int \frac{d^3k}{\sqrt{2\omega}} v \cdot a(\mathbf{k}) 2\pi i \delta(k \cdot v) - \text{h.c.} = 0,$$

as follows from Appendix A6, Eqs. (A6–1) and (A6–2), and from the fact that $k \cdot v$ cannot vanish.* Thus, with (S4–21) and (S4–23),

$$S' = \lim_{\tau_2 \to \infty} \lim_{\tau_1 \to -\infty} e^{-\frac{i}{2}\int d^3p : \rho(\mathbf{p}) Q(p) : \ln|\tau_2/h|} V(\tau_2, \tau_1) e^{+\frac{i}{2}\int d^3p' : \rho(\mathbf{p}') Q(p') : \ln|\tau_1/h|} \quad (S4–46)$$

and only the Coulomb phases modify the conventional S-operator.

Let us now turn our attention to the Hilbert space of the state vectors $\omega'(\tau)$. Here the Coulomb phase factors are of little interest while the radiation operators produce a qualitative change of Fock space in the sense that $\exp[-R(\tau)]$ transforms a Fock vector into a coherent state vector. In addition, as we shall see, the resultant coherent state space $\mathcal{C}_\infty$ is *not* unitarily equivalent to Fock space; it corresponds to the case discussed in the previous Section, S3, where condition (S3–19) is not satisfied.

In order to prove these assertions one expresses $R(\tau)$ as follows:

$$R(\tau) = (\zeta(\tau), a) - (\zeta(\tau), a)^* \quad (S4–47)$$

$$(\zeta(\tau), a) \equiv \frac{1}{(2\pi)^{3/2}} \int \zeta_\mu^*(k, \tau) a^\mu(\mathbf{k}) \frac{d^3k}{\sqrt{2\omega}} \quad (S4–48)$$

and $\zeta_\mu(k, \tau)$ is defined on the positive light cone by

$$\zeta_\mu(k, \tau) = \frac{1}{(2\pi)^{3/2}} \int d^3p \rho(\mathbf{p}) \frac{v_\mu}{v \cdot k |v \cdot n|} e^{-iv \cdot k\tau}. \quad (S4–49)$$

This operator is of course closely related to the asymptotic current. In fact

$$\lim_{\tau \to \pm \infty} \zeta_\mu(k, \tau) = \frac{1}{(2\pi)^{3/2}} \int d^3p \rho(\mathbf{p}) v_\mu \delta(v \cdot k) = j_\mu^{\text{as}}(k)$$

as in the above argument leading to $R(\pm \infty) = 0$; here $v \cdot k$ is given a small imaginary part of the correct sign to ensure convergence for $\tau \to \pm \infty$. But it has nonvanishing value only for spacelike k while (S4–48) forces k on the null cone, so that $(\zeta(\pm \infty), a) = 0$.

---

* In $\exp(ik \cdot v\tau)$ for $\tau \to +\infty$ $(-\infty)$ one must assume a small positive (negative) imaginary part of $k \cdot v$.

$\zeta_\mu(k,\tau)$ is diagonal when operating on states of negatons and positons of given momenta and spins. Let such a state be given by $v_j{}^\mu$, r_j ($j = 1, \cdots, m$) for negatons and by $u_j{}^\mu$, s_j ($j = 1, \cdots, n$) for positons. If this state is abbreviated by $|mn)$ then the expectation value is

$$(mn|\zeta^\mu(k,\tau)|mn) \equiv z_{mn}{}^\mu(k,\tau)$$
$$= \frac{-e}{(2\pi)^{3/2}} \left(\sum_{j=1}^{m} \frac{v_j{}^\mu}{v_j \cdot k |v_j \cdot n|} e^{-iv_j \cdot k\tau} - \sum_{j=1}^{n} \frac{u_j{}^\mu}{u_j \cdot k |u_j \cdot n|} e^{-iu_j \cdot k\tau} \right) \tag{S4-50}$$

It is a sum of classical currents of uniformly moving charges. The corresponding radiation operator is

$$R_{mn}(\tau) = (z_{mn}, a) - (z_{m,n}, a)^*,$$

the $z_{mn}{}^\mu(k,\tau)$ being c-numbers.

One should not forget that a^μ must satisfy a subsidiary condition. In the covariant formulation this condition is given by the Gupta relation $\partial \cdot a^{(-)}\omega = 0$, (6-67). It ensures that at least in matrix elements only the transverse components contribute. Effectively, then, $a^\mu(\mathbf{k}) = \sum_r e_{(r)}{}^\mu a_{(r)}(\mathbf{k})$ where $r = 1, 2$ are the two transverse polarization directions as in Section A2-6. Therefore, effectively, $\zeta_\mu{}^* a^\mu = \sum_r \zeta_{(r)}{}^* a_{(r)}$ with $\zeta_{(r)}{}^* \equiv e_{(r)}{}^\mu \zeta_\mu{}^*$, and only the transverse part, $\zeta_\mu{}^\perp$ of the ζ_μ contributes,

$$(mn|(\zeta(\tau),a)|mn) = (mn|(\zeta^\perp(\tau),a)|mn).$$

In the following we shall anticipate this fact and replace ζ by $\zeta^\perp$ which is a spacelike fourvector.

The transverse operator $\zeta_\mu{}^\perp$ can also be constructed without reference to the polarization directions. The null vector κ^μ is defined relative to a given null vector $k^\mu = (\omega, \mathbf{k})$ by $\kappa^\mu \equiv (\omega, -\mathbf{k})$. Let $v_\perp{}^\mu \equiv v^\mu + \frac{1}{2}\kappa^\mu v \cdot k/\omega^2$, then $k \cdot v_\perp = 0$ and $v_\perp{}^2 > 0$, so that $v_\perp{}^\mu$ is spacelike. Then $\zeta_\mu{}^\perp$ is given by*

$$\zeta_\mu{}^\perp(k,\tau) = \frac{1}{(2\pi)^{3/2}} \int d^3p\, \rho(\mathbf{p}) \frac{v_\mu{}^\perp}{v \cdot k |v \cdot n|} e^{-iv \cdot k\tau}$$

The radiation operator R now gives a transformation e^{-R} of Fock space which has exactly the structure of (S3-30). When one normal orders it as in (S3-13) one finds

$$e^{-(\zeta^\perp,a)+(\zeta^\perp,a)^*} = e^{\frac{1}{2}[(\zeta^\perp,a)^*,(\zeta^\perp,a)]} e^{(\zeta^\perp,a)^*} e^{-(\zeta^\perp,a)}.$$

---

* This construction of $\zeta_\mu{}^\perp$ can be carried out in each Lorentz frame but is not manifestly covariant.

But the exponent of the normalization factor is

$$[(\zeta^\perp,a)^*,(\zeta^\perp,a)] = -\frac{1}{(2\pi)^3}\int \frac{d^3k}{2\omega}\,\zeta_\mu^{\perp*}(k,\tau)\zeta_\perp^\mu(k,\tau)$$

$$= -\frac{1}{(2\pi)^3}\int d^3p\rho(\mathbf{p})\int d^3p'\rho(\mathbf{p}')\int \frac{d^3k}{2\omega}$$

$$\frac{v_\perp \cdot v_\perp{}'}{v\cdot k v'\cdot k |v\cdot n||v'\cdot n|}\cos[(v-v')\cdot k\tau].$$

This integral is *divergent* (at the $\omega \to 0$ limit*) and positive definite since $v_\perp \cdot v_\perp{}' > 0$. Consequently, the normalization factor vanishes and we have exactly the case of a coherent state space $\mathfrak{C}_\infty$ in which the condition (S3–19) is not satisfied, as asserted above.

Nevertheless, the operator $\exp(-R(\tau))$, with $R(\tau)$ given by (S4–47), is a unitary operator in a suitably defined space. This space can be proven to be a separable Hilbert space[10] for any given electron field matrix element, $R_{mn}(\tau)$.

As special examples of states of physical electrons (including their photon cloud) we have the one-negaton state

$$\omega'(\tau;\mathbf{p},r) = e^{-R(\tau)}a_r^*(\mathbf{p})\omega_0$$

and the negaton–positon state

$$\omega'(\tau;\mathbf{p},r;\mathbf{q},s) = e^{-R(\tau)-i\Phi(\tau)}a_r^*(\mathbf{p})b_s^*(\mathbf{q})\omega_0.$$

The latter contains a contribution from the Coulomb interaction.

The states defined in this way and the S-operators (S4–46) permit the definition and explicit computation of an S-matrix. Perturbation calculations have verified that this S-matrix is indeed free of infrared divergences. In fact, a set of Feynman rules has been developed taking account of the modified mass-shell behavior (S4–41). A carefully prescribed order of limits is here necessary.[4] However, for *practical* calculations this method does not at the present time seem to offer advantages over the older perturbation techniques, especially as streamlined recently.[11]

What has not been achieved so far is a rigorous derivation of these results which would include a consistent treatment of the renormalization terms, a general proof of the unitarity of the new S-matrix and a nonperturbative proof of the absence of infrared divergences. It is obvious that the present treatment of the infrared problem comes as an afterthought. The coherent state space is not introduced into the theory from the beginning. Nor is the necessary infrared renormali-

---

* The upper limit is convergent for the same reason as in (S4–29).

zation carried through and shown to be consistent with ultraviolet renormalization.

References

1. D. R. Yennie, S. C. Frautschi, and H. Suura, *Ann. Physics* **13**, 379 (1961), and other references given there.
2. J. D. Dollard, *J. Math. Phys.* **5**, 729 (1964).
3. P. P. Kulish and L. D. Fadde'ev, *Theor. Math. Phys.* **4**, 745 (1971). [This is an English translation of *Teor. Matem. Fiz.* **4**, 153 (1970) which is in Russian.]
4. D. Zwanziger, *Phys. Rev.* **D 7**, 1082 (1973); *Phys. Rev. Lett.* **30**, 934 (1973); *Phys. Rev.* **D 11**, 3481 and 3504 (1975).
5. V. Chung, *Phys. Rev.* **140 B**, 1110 (1965).
6. T. W. B. Kibble, *Phys. Rev.* **173**, 1527; **174**, 1882; **175**, 1624 (1968); *J. Math. Phys.* **9**, 315 (1968).
7. A. Aurilia and F. Rohrlich, *Am. J. Phys.* **43**, 261 (1975).
8. Similar expressions were recently also given by N. Papanicolaou, *Ann. Physics* **89**, 423 (1975), but without the normal dependence.
9. The spectral properties of infraparticles was investigated by B. Schroer, *Fortschr. d. Phys.* **11**, 1 (1963).
10. G. Reents, *J. Math. Phys.* **15**, 31 (1974).
11. G. Grammer and D. R. Yennie, *Phys. Rev.* **D 8**, 4332 (1973).

SUPPLEMENT S5

PREDICTIONS AND PRECISION EXPERIMENTS

It has become commonplace to emphasize the excellent agreement of quantum electrodynamics with precision experiments and in the same breath to complain about its mathematical inadequacies. But just how good is this agreement when viewed through our rapidly developing technology and the concomitant rapid improvement of our experimental accuracy? Can we ascertain validity limits of this theory on the basis of sufficiently precise observations?

This question has been raised many times but to date no validity limits have been found. On the contrary, we have come close to the limits of our ability to find such validity limits in the near future and may have to wait until we shall have solved the problem of the nature of strong and weak interactions: quantum electrodynamic predictions have been confirmed to an accuracy at which the interplay of hadronic effects with purely electromagnetic ones can no longer be neglected. We have thus reached the limit of unambiguous predictions because our knowledge of hadronic effects is only of a phenomenological nature: we have so far still no fundamental theory of strong interactions. Just to give one example: we cannot predict from first principles the charge or magnetic moment distribution of a proton.

There is another, less serious limitation on our present ability to ascertain the validity limits of the theory: the fine structure constant α is not known to sufficiently high accuracy from experiments which are independent of quantum electrodynamics. The best value of α is obtained from a weighted least square analysis of *all* available data.[1] This value is not much more accurate than the value α_E that would be obtained from a comparison of any one precision experiment with its quantum electrodynamic prediction. It is therefore more reasonable in such cases to judge the agreement with experiment by making a comparison of the values of α and α_E.

Accurate predictions require very complex higher-order calculations which can be handled most successfully by means of computer programs. A whole field of specialization is growing up around this technique.[2] The results quoted below are heavily dependent on this new field.

S5-1 The anomalous magnetic moment. As of this writing the best experimental value[3] for the magnetic moment of the electron is*

$$(\mu|\mu_0)_{\exp} = 1.001\ 159\ 656\ 7\ (35). \qquad (S5\text{--}1)$$

---

* The number in parenthesis indicates the error in the last two figures.

It was therefore necessary to compute terms to order $(\alpha/\pi)^3$. This is a difficult task and was a joint effort of several groups. There are a total of 72 diagrams of that order: 50 diagrams involving no closed-loop subdiagrams, 12 involving second-order and 4 involving fourth-order vacuum polarization subdiagrams, and 6 involving photon–photon scattering subdiagrams. All were evaluated numerically (many by more than one method) and some also analytically. The final theoretical result is[4,5]

$$(\mu/\mu_0)_{th} = 1 + \frac{1}{2}\left(\frac{\alpha}{\pi}\right) - 0.328\ 478\left(\frac{\alpha}{\pi}\right)^2 + 1.195\ (26)\left(\frac{\alpha}{\pi}\right)^3. \quad (S5\text{-}2)$$

The error here is due to error limits in numerical integration.

One obtains from (S5-1) and (S5-2)

$$1/\alpha_e = 137.035\ 49\ (42)$$

while the presently accepted best value for this constant is[1]

$$1/\alpha = 137.036\ 04\ (11) \quad (S5\text{-}3)$$

a considerable improvement over (15-75). The agreement is therefore very good indeed.

The above calculations can easily be adapted to the computation of the anomalous magnetic moment of the muon. This particle is treated as a heavy electron but permitting closed loop subdiagrams of electron lines as well as muon lines. However, the $(\alpha/\pi)^3$ corrections cannot be determined reliably by comparison with experiment because the hadronic corrections have been estimated to be of nearly the same order of magnitude (about 25% of it) and are obviously less reliably known. Nevertheless, it is remarkable that to the present high accuracy the muon is correctly described as a "heavy electron."

S5-2 The hyperfine structure of the hydrogen ground state.

The most accurate physical measurement known lies in this field of energy level differences: the hyperfine splitting of the ground state of hydrogen. The most recent value is[6]

$$\Delta\nu_H = 1420.405\ 751\ 766\ 7\ (10)\ \text{MHz} \quad (S5\text{-}4)$$

an accuracy of better than one part in 10^{12}. Unfortunately, the theoretical predictions are accurate only to a few parts per million (ppm) because of the difficulty of computing higher-order corrections. Thus, this experiment is 10^6 times more precise than the predictions!

One can obtain a rough idea of the present state of the theory of the hydrogen hyperfine structure from the general expression*

$$\Delta \nu_H = \frac{16}{3} \alpha^2 Ry \left(\frac{\mu_e}{\mu_0}\right)\left(\frac{\mu_P}{\mu_0}\right)\left(1 + \frac{3}{2}\alpha^2\right)\frac{1}{\left(1 + \frac{m_e}{m_P}\right)^3}\left(1 + \alpha^2 Q + \frac{m_e}{m_P} R\right).$$

(S5–5)

μ_0 is the Bohr magneton of the electron, μ_e and m_e the magnetic moment and mass of the electron, μ_P and m_P those of the proton. The leading term is the well known Fermi formula.† This nonrelativistic formula must be corrected by using Dirac instead of Schroedinger wave functions (Breit correction $1 + 3\alpha^2/2$). Also the finite mass of the proton must be taken into account.

But beyond that there are three types of corrections (of order α and higher): the anomalous magnetic moment of the electron (which is contained in μ_e), nuclear recoil corrections (indicated by R and proportional to m_e/m_P) and other radiative corrections (indicated by Q). The latter include the correction previously given (last term of Eq. (15–74)) as well as binding-energy effects and other radiative effects which are smaller by another factor α/π but contain $\ln^2 1/\alpha$ and $\ln 1/\alpha$ terms. These were computed by Brodsky and Erickson.[7] The recoil terms can be written

$$R = \frac{\alpha}{\pi}\left(R_1 + \frac{(g_P - 2)^2}{2g_P} R_2\right) + \frac{1}{\left(1 + \frac{m_e}{m_P}\right)^2}\left(\alpha^2 \ln \frac{1}{\alpha} R_3 + \alpha^2 R_4\right).$$

The R_i have been computed‡ for $i = 1$ to 3. R_4 is completely unknown and is thus an indicator of the present accuracy of the prediction.

The presently best value for α, (S5–3), yields for the last bracket in (S5–5) together with the Breit correction,

$$1 - 0.000\ 056\ 6\ (32),$$

i.e., about 3 ppm accuracy. The large error is due to the uncertainties in the estimates of the proton structure (its polarizability). The accuracy of the remaining input data, μ_P/μ_0 and m_e/m_P is not any better. Obviously one uses this very accurate experiment not as a check of the theory but as a means to determine these constants.

Finally one observes that twenty years of very complex calculations to find the value of the last bracket in (S5–5) barely gives us the first two nontrivial significant figures of the corrections, mainly because of unknown hadronic effects.

---

* Hadronic corrections are estimated to be negligible at present accuracy, but their uncertainty is included in the error estimates.
† For a derivation see for example Bethe and Salpeter (p. 360, last footnote).
‡ See references 1 and 5 for details.

S5-3 The Lamb–Retherford shift in hydrogen.

The most famous test of quantum electrodynamics is of course the Lamb–Retherford shift in hydrogen which we computed to lowest order in Section 15–3. While it is today by far not the most accurate test of the theory, its present state will be discussed briefly, partly for its historical importance, partly to show how much better this effect is known twenty years later, and partly as an example of an often encountered state of temporary uncertainty in the comparison between theory and experiment. The reader should realize that this comparison is often not clear cut either because different measurements disagree with one another or because different theorists in their independent computations obtain different results for these very complex calculations. Sometimes it takes years until such discrepancies are resolved.

We shall focus on $\Delta E_H = 2^2S_{1/2} - 2^2P_{1/2}$ in hydrogen but we shall retain the factor Z corresponding to hydrogenlike atoms in order to keep an easy distinction between binding effects (αZ) and radiative corrections (α). We recall that we are computing terms in a double expansion, as discussed on p. 320, expanding in increasing number of virtual photon line insertions (α) as well as in powers of the external field (αZ).

The result (15–66) with $n = 2$ and $c_{1\frac{1}{2}} = -1$ yields the lowest order relativistic shift

$$\Delta E_H^{(0)} = \frac{\alpha m}{6\pi}(\alpha Z)^4 \left[\ln\frac{1}{(\alpha Z)^2} + \ln\frac{k_0(2,1)}{k_0(2,0)} + \frac{19}{30} + \frac{1}{8}\right] \quad \text{(S5-6a)}$$

The term 19/30 includes additive terms 1/2 and −1/5 which are the contributions of the second-order anomalous electron magnetic moment (contributing +68 MHz) and of second-order vacuum polarization (−27 MHz), as discussed on p. 358. To this should be added the binding energy correction of 7.14 MHz given in (15–69) (of order αZ relative to $\Delta E_H^{(0)}$) and mass corrections (of order m_e/m_P) to yield the shift correct to order $\alpha(\alpha Z)^5 m_e$ and $\alpha(\alpha Z)^4(m_e/m_P)m_e$

$$\Delta E_H^{(1)} = 1057.696\ (6)\ \text{MHz}. \quad \text{(S5-6b)}$$

The improved values of $\ln[k_0(n,l)/Ry]$ were used here which are more accurate than (15–67),[8] as well as the presently best value of α, (S5–3).

A great deal of effort has been devoted to additional small corrections, improving over the ones listed in Table 15–1, p. 359, so that the theoretical error is now down to about 10 kHz. These corrections are as given in Table S5–1.

The best theoretical value (including an error estimate on not yet calculated terms) is thus[5]

$$\Delta E_H^{(2)} = 1057.991\ (12). \quad \text{(S5-6c)}$$

TABLE S5–1 (Adapted from Reference 5)

Corrections in MHz to $\Delta E_H^{(1)}$ for the hydrogen $2^2S_{1/2} - 2^2P_{1/2}$ separation (r_0 and R_P are the classical electron radius, e^2/m, and the root-mean-square proton charge radius)

Order relative to $\Delta E_H^{(0)}$	Effect	Value
α	Radiative corrections incl. magnetic moment and vacuum polarization	0.103
$(\alpha Z)^2$ and higher	Binding energy	-0.372 (5)
Zm_e/m_P	Proton recoil	0.359
$(R_P/r_0)^2$	Proton charge distribution	0.125 (6)
	Total	0.215 (8)

However, a different method of including the binding energy effects (αZ-type terms) was recently used by Mohr[9] who finds

$$\Delta E_H^{(M)} = 1057.864 \ (14) \tag{S5-7}$$

in disagreement with the above value. We shall now see that the present experimental situation is also unclear.

The old direct measurements of the $2^2S_{1/2} - 2^2P_{1/2}$ separation ΔE_H (see footnote, p. 345) more recently gave way to three measurements of the $2^2P_{3/2} - 2^2S_{1/2}$ transition. Since the theoretical value of the fine structure splitting $\Delta E_{FS} = 2^2P_{3/2} - 2^2P_{1/2}$ is well known, these measurements can be regarded as the difference $\Delta E_{FS} - \Delta E_H$ so that ΔE_H can be determined from them. With the theoretical value for ΔE_{FS} assumed to be exact[1] (its error is less than ten times smaller than the experimental error),

$$\Delta E_{FS} = 10969.035 \text{ MHz} \tag{S5-8}$$

one finds from the three measurements the values

$$\Delta E_H^{(a)} = 1057.658 \ (26) \text{ MHz}^{10} \tag{S5-9a}$$

$$\Delta E_H^{(b)} = 1057.785 \ (63) \text{ MHz}^{11} \tag{S5-9b}$$

$$\Delta E_H^{(c)} = 1057.862 \ (42) \text{ MHz}^{12} \tag{S5-9c}$$

A very recent direct measurement using a method of line narrowing[13] yields the preliminary result

$$\Delta E_H^{(d)} = 1057.893 \ (20) \text{ MHz.}^{14} \tag{S5-9d}$$

There is obviously a considerable discrepancy between the first and the remaining ones of these measurements.

When we now compare these observations with the predictions $\Delta E_H^{(2)}$ and $\Delta E_H^{(M)}$ we see that the latter prediction agrees very well with all but the lowest observation, $\Delta E_H^{(a)}$. The prediction $\Delta E_H^{(2)}$ agrees very well with $\Delta E_H^{(c)}$ and $\Delta E_H^{(d)}$ and is within twice the error of $\Delta E_H^{(b)}$. Thus, one concludes that, while it does not seem likely that there is a discrepancy, the present experimental and theoretical results do not permit the conclusion that there is complete agreement.

S5-4 Energy levels in positronium.

The positon–negaton bound state was discussed in Section 12-5. But a relativistic calculation of its energy levels to the high accuracy of present experiments ($\lesssim 10$ ppm) requires a systematic approximation procedure for the quantum dynamics of relativistic two-body systems. Two general approaches of the quantum relativistic two-body problem exist, the Bethe–Salpeter equation[15] and the quasipotential method.[16] Of these the former is older and has been pushed to much higher precision.

A treatment of the Bethe–Salpeter equation for positronium which permits systematic approximations to higher order has only recently been developed. Feldman, Fulton, and Townsend devised a simplified equation[17] which permits such a treatment. Older perturbation calculations were relatively *ad hoc*, although they were very successful in their goal. Only a systematic procedure promises to permit the theory to keep up with the steady improvement in the accuracy of observed positronium energy levels. This is important because positronium is a *pure* electrodynamic system: unlike hydrogen, the hadronic effects of the proton size and of the charge and magnetic moment distributions are absent.

The $n = 1$ levels show a separation $^3S_1 - \,^1S_0$ which corresponds to the hyperfine structure of the ground state of hydrogen. Its calculation began in 1952 when Karplus and Klein[18] calculated the first two terms of the presently available theoretical value

$$\Delta \nu_{\text{Pos}} = \frac{1}{2} \alpha^4 m \left[\frac{7}{6} - \left(\frac{16}{9} + \ln 2 \right) \frac{\alpha}{\pi} + \frac{1}{2} \alpha^2 \ln \frac{1}{\alpha} - 1.937 \alpha^2 \right]. \quad \text{(S5-10)}$$

The next term is a sum of several contributions which were obtained in the last five years.[19] They can be separated into two groups, those coming from recoil and retardation[19a] ($\frac{3}{4}\alpha^2 \ln 1/\alpha$) and those from virtual annihilation[19b] ($-\frac{1}{4}\alpha^2 \ln 1/\alpha$). The last term was computed by Fulton.[20]

The above result is of course related to the hyperfine structure result for hydrogen, $\Delta \nu_H$, (S5-5). In fact, if one replaces m_P and μ_P by m_e and μ_e, and if one adds to this the contribution due to virtual pair annihilation (the proton–negaton system cannot annihilate) one does indeed obtain (S5–10).* However, some of the recoil corrections of positronium were known before the corresponding ones in hydrogen. This is primarily because corrections proportional to the anomalous part of the magnetic moment of the positive particle are small for positronium (positon) but large for hydrogen (proton).†

With our present best value of α, (S5–3), (S5–10) yields

$$\Delta \nu_{\text{Pos}}{}^{\text{th}} = 203.385\ 9\ \text{GHz}.$$

Two independent recent experiments, more accurate than any of the previous ones yield

$$\Delta \nu_{\text{Pos}}{}^{\text{exp}} = 203.386\ (3)\ \text{GHz}^{21} \qquad \text{(S5–11a)}$$

and

$$\Delta \nu_{\text{Pos}}{}^{\text{exp}} = 203.386\ (1)\ \text{GHz.}^{22} \qquad \text{(S5–11b)}$$

in beautiful agreement with the theory. In fact, the agreement is almost too good: some terms of order α^2 in the square bracket of (S5–10) have not yet been computed. We thus expect these to be small.

Beyond the $n = 1$ levels there have been observed most recently also the $n = 2$ levels of positronium, a considerable experimental achievement.[23] One found not only the Ly α line 2×1215 Å, but also its structure and in particular the 2^3S_1–2^3P_2 level separation (the analogue of the Lamb shift in hydrogen), viz.[24]

$$\Delta E_{\text{Pos}}{}^{\text{exp}} = 8628.4 \pm 2.8\ \text{MHz}. \qquad \text{(S5–12)}$$

The theoretical value[25] consists of a lowest order separation of $7413 - (-981) = 8394$ MHz and a correction to that of order α, $232 - 1 = 231$ MHz, where we have indicated the shifts of each of the two levels. This gives a prediction of

$$\Delta E_{\text{Pos}}{}^{\text{th}} = 8625\ \text{MHz} \qquad \text{(S5–13)}$$

thus confirming the correction term to 1.5 ± 1.2 percent.

In conclusion it should be remarked that the life times of parapositronium ($^1S_0$) and orthopositronium ($^3S_1$) as computed on p. 286 have since been computed to higher accuracy[5] and agree with the corresponding more precise measurements.

S5–5 Muonium hyperfine structure. Muonium is a hydrogenlike structure consisting of a positive muon and a negaton. Since the muon

---

* But (S5–5) cannot be expected to hold in higher order. Not all corrections are proportional to the magnetic moments which themselves contain radiative corrections. This holds in positronium from $\alpha^6 m$ on.

† The Bethe–Salpeter equation has of course also been used for hydrogen, but it plays a much more important role in positronium.

is known from its magnetic moment to be correctly described as a "heavy electron" this structure is also "purely electrodynamic," just as positronium.

The hyperfine structure of muonium can be obtained from the corresponding result for hydrogen, (S5–5), by the replacement of μ_P and m_P by μ_μ and m_μ. But the recoil terms must be known to higher accuracy because m_e/m_μ is larger than m_e/m_P. The fact is, therefore, that $\Delta \nu_{\mu e}$ is not known theoretically to as high an accuracy as $\Delta \nu_H$ (if we ignore the hadronic uncertainty in $\Delta \nu_H$).

The best value of $\Delta \nu_{\mu e}$ obtained from a series of measurements of the Chicago group and the Yale group[1] is

$$\Delta \nu_{\mu e}^{\text{exp}} = 4463.303\ 8\ (14)\ \text{MHz} \qquad (S5\text{–}14a)$$

and a very recent measurement gives[26]

$$\Delta \nu_{\mu e}^{\text{exp}} = 4463.301\ 1\ (16)\ \text{MHz} \qquad (S5\text{–}14b)$$

Comparing these results with theory yields a value for α (the magnetic moments are taken from direct measurements)

$$1/\alpha_{\mu e} = 137.036\ 34\ (21) \qquad (S5\text{–}15a)$$

and

$$1/\alpha_{\mu e} = 137.036\ 20\ (21) \qquad (S5\text{–}15b)$$

respectively. These are in good agreement with (S5–3).

References

1. E. R. Cohen and B. N. Taylor, *J. Phys. Chem. Ref. Data* **2**, 663 (1973); B. N. Taylor, W. H. Parker, and D. N. Langenberg, *Rev. Mod. Phys.* **41**, 375 (1969). These papers are the most important references to the steadily changing results of the field of determination of fundamental constants.
2. J. A. Campbell, *Act. Phys. Austriaca* **XIII**, 595 (1974); *Progress in Elementary Particle Physics* (P. Urban, ed.), Springer-Verlag, 1974. Another recent review with many pertinent references was given by A. Visconti in *Renormalization and Invariance in Quantum Field Theory*, (E. R. Caianiello, ed.), Plenum Press, New York 1974, p. 329.
3. J. C. Wesley and A. Rich, *Phys. Rev.* **A 4**, 1341 (1971) and *Rev. Mod. Phys.* **44**, 250 (1972), corrected by S. Granger and G. W. Ford, *Phys. Rev. Lett.* **28**, 1479 (1972).
4. T. Kinoshita and P. Cvitanovic, *Phys. Rev. Lett.* **29**, 1534 (1972) and P. Cvitanovic and T. Kinoshita, *Phys. Rev.* **D 10**, 4007 (1974). References to the work of others can be found in these summary papers.

5. B. E. Lautrup, A. Petermann, and E. de Rafael, *Physics Rep.* **3** C(4) (1972). This review contains a wealth of detailed computational results for the comparison of quantum electrodynamics with experiments and also a very extensive bibliography.

6. L. Essen, R. W. Donaldson, M. J. Bangham, and E. G. Hope, *Nature* **229**, 110 (1971). Their result is in excellent agreement with that of H. Hellwig, R. F. C. Vessot, M. W. Levine, P. W. Zitzewitz, D. W. Allan, and D. J. Glaze, *IEEE Trans. Instr. Meas.* **IM19**, 200 (1970).

7. S. J. Brodsky and G. W. Erickson, *Phys. Rev.* **148**, 26 (1966).

8. Chas. Schwartz and J. J. Tiemann, *Ann. Phys.* **2**, 178 (1959).

9. P. J. Mohr, *Phys. Rev. Lett.* **34**, 1050 (1975). This calculation should be compared with that by G. W. Erickson, *Phys. Rev. Lett.* **27**, 780 (1971) on whose work (S5–6) is based.

10. S. L. Kaufman, W. E. Lamb, Jr., K. R. Lea, and M. Leventhal, *Phys. Rev.* **A 4**, 2128 (1971).

11. T. W. Shyn, T. Rebane, R. T. Robiscoe, and W. L. Williams, *Phys. Rev.* **A 3**, 116 (1971).

12. B. L. Cosens and T. V. Vorburger, *Phys. Rev.* **A 2**, 16 (1970).

13. C. W. Fabian and F. M. Pipkin, *Phys. Rev.* **A 6**, 556 (1972).

14. S. R. Lundeen and F. M. Pipkin, *Phys. Rev. Lett.* **34**, 1368 (1975).

15. First suggested by E. E. Salpeter and H. A. Bethe, *Phys. Rev.* **84**, 1232 (1951), it has a very large literature devoted to it (see N. Nakanishi, *Prog. Theor. Phys. (Kyoto) Suppl.* **43**, 1 (1969)). It is discussed in D. Lurié, *Particles and Fields*, Interscience, 1968, in S. S. Schweber, Supplement S1, Reference 17, and in other texts.

16. A. A. Logunov and A. N. Tavkhelidze, *Nuovo Cim.* **29**, 380 (1963); A. A. Logunov, A. N. Tavkhelidze, I. T. Todorov, and O. A. Khrustalev, *ibid.* **30**, 134 (1963); a local formulation was given by I. T. Todorov in *Properties of Fundamental Interactions*, Vol. 9, Part C (A. Zichichi, ed.), Compositori Bologna, 1973.

17. G. Feldman, T. Fulton, and J. Townsend, *Ann Physics* **82**, 501 (1974). This work is based on the paper by W. Kummer, *Nuovo Cim.* **31**, 219 (1964) and **34**, 1840 (1964).

18. R. Karplus and A. Klein, *Phys. Rev.* **87**, 848 (1952).

19. (a) T. Fulton, D. Owen, and W. W. Repko, *Phys. Rev.* **A 4**, 1802 (1971) and *Phys. Rev. Lett.* **26**, 61 (1971); (b) D. Owen, *Phys. Rev. Lett.* **30**, 887 (1973); R. Barbieri, P. Christillin, and E. Remiddi, *Phys. Rev.* **A 8**, 2266 (1973) and *Phys. Lett.* **43B**, 411 (1973).

20. T. Fulton, *Phys. Rev.* **A 7**, 377 (1973). Annihilation and recoil corrections of this same order have so far not been computed.

21. This is a weighted average of two measurements by A. P. Mills and G. H. Bearman, *Phys. Rev. Lett.* **34**, 246 (1975).

22. P. O. Egan, E. R. Carlson, V. W. Hughes, and M. Yam, *Bull. Am. Phys. Soc.* **20**, 703 (1975), corrected value reported at the meeting.

23. K. F. Canter, A. P. Mills, and S. Berko, *Phys. Rev. Lett.* **34,** 177 (1975).
24. A. P. Mills, S. Berko, and K. F. Canter, *Phys. Rev. Lett.* **34,** 1541 (1975).
25. T. Fulton and P. Martin, *Phys. Rev.* **95,** 811 (1954).
26. D. E. Casperson, T. W. Crane, V. W. Hughes, P. A. Souder, R. D. Stambaugh, P. A. Thompson, H. F. Kaspar, H.-W. Reist, H. Orth, G. zu Pulitz, and A. B. Denison, *Bull. Am. Phys. Soc.* **20,** 702 (1975).

AUTHOR INDEX

Akhieser, A., 296, 384, 492
Abraham, M., 34
Adler, S., 389
Albert, A. A., 67
Allan, D. W., 538
Anderson, J. L., 489, 494
Arago, D. F. J., 44
Arnous, E., 409
Ashkin, A., 261
Ashkin, J., 330, 371
Aurilia, A., 529

Bangham, M. J., 538
Baranger, M., 360
Barbieri, R., 538
Bardacki, K., 495
Barton, G., 494
Baumann, K., 163, 396, 493
Beaman, G. H., 538
Belinfante, F. J., 49, 69, 110, 114, 134
Berestetskii, V. B., 492
Bergmann, P. G., 19
Bersohn, R., 360
Berko, S., 539
Beth, R. A., 41
Bethe, H. A., 163, 269, 330, 349, 358, 360, 368, 370, 371, 372, 375, 386, 396, 408, 475, 538
Bhabba, H. J., 257
Bialynicka-Birula, Z., 389, 493
Bialynicki-Birula, I., 389–493
Bjorken, J. D., 492
Bleuler, K., 103, 108, 110
Bloch, C., 8
Bloch, F., 397
Boccaletti, D., 389
Boekelheide, I. F., 241
Bogoliubov, N. N., 489, 492, 494, 499, 502
Bohr, N., 80, 377, 466
Bollin, C. G., 501, 502
Bolsterli, M., 389
Bopp, F., 71, 224

Borowitz, S., 414
Borsellino, A., 251
Braunbeck, W., 397
Breit, G., 300, 368
Breuner, S., 388
Brodsky, S. J., 495, 532, 538
Brown, L. M., 245, 358, 396
Bukhvostov, A. P., 389
Burton, W. K., 24
Busch, H., 34
Butera, V., 511

Caianiello, E. R., 55
Carlson, B. C., 331
Casperson, D. E., 539
Cavanaugh, P. E., 241
Chandrasekhar, S., 42
Chang, T. S., 9
Cheng, H., 388, 387
Christillin, P., 538
Chung, V., 514, 529
Clark Jones, R., 42
Clifford, W. K., 66, 67
Coester, F., 49, 97, 114, 116, 134, 142, 190
Cohen, E. R., 539
Compton, A. H., 229
Conway, A. W., 34
Corinaldesi, E., 81, 396
Corson, D. R., 332
Cosens, B. L., 538
Costantini, V., 388, 389
Cowan, C. E., 269
Crane, T. W., 539
Critchfield, C. L., 55
Cvitanovic, P., 537

Dalitz, R. H., 330, 378, 520
Davies, H., 372
Dayhoff, E. S., 345
De Benedetti, S., 269
deBroglie, L., 34, 113
Delbrück, M., 379

Demeur, M., 309
Denison, A. B., 539
de Rafael, E., 538
Dirac, P. A. M., 5, 7, 8, 9, 15, 28, 38, 50, 71, 97, 103, 125, 269, 286, 467, 490, 493, 494, 495
Dodo, T., 102
Dollard, J. D., 513, 529
Donaldson, R. W., 538
Drell, S. D., 492
Dyson, F. J., 134, 146, 175, 176, 187, 203, 213, 218, 290, 325, 360, 420, 499

Egan, P. O., 538
Ehlotzky, F., 388
Einstein, A., 3
Eliezer, C. J., 71
Elton, L. R. B., 339
Elwert, G., 371
Epstein, P. S., 34
Epstein, S. T., 413
Erickson, G. W., 532, 538
Essen, L., 538
Euler, H., 294, 295
Evans, L. E., 494

Fabian, C. W., 538
Fadde'ev, L. D., 514, 529
Falkoff, D. L., 42
Fano, U., 42
Feldman, D., 225
Feldman, G., 535, 538
Fermi, E., 28, 532
Ferrari, B., 494
Ferretti, B., 134
Feshbach, H., 330, 331
Feynman, R. P., 146, 196, 224, 245, 314, 337, 358, 360, 396, 454, 489, 492, 494
Fierz, M., 8, 41, 397, 473
Fock, V. A., 28, 56
Foldy, L. L., 69
Ford, G. W., 537
Fradkin, E. S., 502
Franz, W., 388
Frautschi, S. C., 529
French, J. B., 358
Fresnel, A. J., 44

Fried, M. H., 494
Friedrichs, K. O., 511
Fubini, A., 134
Fukuda, H., 358
Fukuda, N., 407
Fulton, T., 494, 535, 538, 539
Furry, W. H., 58, 159, 306, 372

Gårding, L., 481, 493, 511
Géhéniau, G., 309
Gell-Mann, M., 134, 490, 495
Giambiage, J. J., 501, 502
Girardello, L., 511
Glaser, V., 488, 494
Glauber, R. J., 397, 511
Glaze, D. J., 538
Gluckstern, R., 368, 379, 384, 474
Goldberger, M. L., 134, 494
Gomatam, J., 512
Gonzalez Dominguez, A., 501, 502
Gora, E., 408
Gordon, W., 343
Goto, K., 142
Grammar, G., 529
Grandy, W. T., 492
Granger, S., 537
Green, H. S., 502
Gupta, S. N., 103, 224, 407, 414
Gustafson, T., 501, 502

Haag, R., 482, 493, 494
Hagedorn, R., 594
Hamermesh, M., 134
Hamilton, J., 142
Hanson, A. O., 332, 340
Haug, R., 251, 262
Hayakawa, S., 97
Heisenberg, W., 28, 33, 34, 84, 120, 125
Heitler, W., 103, 135, 163, 368, 370, 371, 375, 406, 408, 409
Hellwig, H., 538
Henriot, E., 34
Hepp, K., 494, 499, 502
Hill, E. L., 45
Hodes, I., 262
Hönl, H., 71
Hope, E. G., 538
Hori, S., 489, 494

Hough, D. V. C., 371
Houriet, A., 443
Hu, N., 97
Hughes, V. W., 538, 539
Hull, M. H., 368
Hulme, H. R., 378
Humblet, M. J., 34
Hurst, C. A., 218

Imamura, T., 102
Ivanenko, D., 273

Jaeger, J. C., 378
Jarlskog, G., 389
Jauch, J. M., 24, 45, 49, 57, 97, 164, 390, 494
Jeffreys, B. S., 27
Jeffreys, H., 27
Jordan, P., 28, 41, 56, 67, 419
Joseph, J., 250
Jost, R., 320, 377, 396, 475, 493

Kahane, S., 389
Källén, G., 167, 169, 170, 492
Kanei, E., 71
Karplus, R., 183, 201, 290, 345, 360, 361, 535, 538
Kaspar, H. F., 539
Kaufmann, S. L., 538
Kessler, P., 388
Khalatnikov, I. M., 502
Khrustalev, O. A., 538
Kibble, T. W. B., 514, 529
Kind, A., 443
Kinoshita, T., 396, 537
Kirkpatrick, P., 371
Klauder, J. R., 495, 511, 512
Klein, A., 360, 361, 535, 538
Klein, O., 56, 229, 312
Koba, Z., 84, 97
Kockel, B., 163, 298
Koenig, S., 345
Kogut, J. B., 495
Kohn, W., 320, 414
Konnecker, W. R., 269
Kramers, H. A., 469
Kraus, J., 388
Kristensen, P., 169

Kroll, N., 183, 201, 345, 358, 360, 361
Kronig, R. de, 469
Kulish, P. P., 514, 529
Kusch, P., 345

Lamb, W. E., 251, 263, 345, 358, 372, 538
Landau, L., 8, 280, 286, 502
Landé, A., 224
Langenberg, D. N., 537
Lautrup, B. E., 538
Lea, K. R., 538
Lee, H. C., 67
Lee, T. D., 502
Lehmann, H., 488, 494
Leutwyler, H., 495
Leventhal, M., 538
Levine, M. W., 538
Lifshitz, E. M., 273
Lippman, B. A., 134
Logunov, A. A., 538
Lomont, J. S., 110
Lorentz, H. A., 71, 410
Louisell, W. H., 511
Low, F., 409, 490, 495
Lundeen, S. R., 538
Lurie, D., 538
Luttinger, J. M., 377, 475
Lyman, E. M., 340

Ma, S. T., 134, 409
Macdonald, J. E., 42
Maison, D., 494
Majorana, E., 58
Mandl, F., 237, 239
Martin, A., 494
Martin, P., 539
Massey, H. W. S., 123, 319, 330, 331
Matthews, P. T., 114, 190, 320, 499, 502
Maximon, L. C., 372
May, M. M., 368, 375
McKenna, J., 512
McKinley, W. A., 330, 331
McManus, H., 169, 224
Medvedev, B. V., 489, 494
Michel, L., 280
Mills, A. P., 538, 539

Mitchell, K., 463
Mitter, H., 330
Miyamoto, Y., 97, 358
Miyazima, T., 407
Mohr, P. J., 538
Møller, C., 120, 131, 134, 166, 169, 252
Morch, R., 389
Mork, K. J., 251
Morpurgo, G., 397
Moses, H. E., 122
Mott, N. F., 123, 257, 319, 330, 331
Mueller, H., 42
Mukunda, N., 493
Murnaghan, F. D., 38, 441

Nakanishi, N., 488, 494, 538
Neuman, M., 290, 320
Neumann, J. v., 6, 31, 39, 121, 506, 511
Neville, R. A., 491, 495
Nishijima, K., 502
Nishina, Y., 229, 378
Nordsieck, A., 397
Novobatsky, K. F., 110
Novoshilov, Yu, V., 494

Oisi, Y., 84, 97
Oppenheimer, J. R., 172, 396, 408
Ore, A., 273, 286
Orth, H., 539
Owen, D., 538

Page, L. A., 261
Pagnamenta, A., 495
Pais, A., 17, 71, 225, 320, 410, 413
Papanicolaou, N., 529
Papapetrou, A., 71
Parasiuk, O. S., 499, 502
Parker, W. H., 537
Pauli, W., 3, 8, 23, 28, 33, 58, 84, 89, 103, 135, 224, 316, 397, 419, 425, 430
Pavlov, V. P., 494
Peierls, R., 8, 377, 466
Peng, H. W., 142, 408
Perrin, F., 42
Petermann, A., 218, 538

Picasso, L. E., 494
Pipkin, F. M., 538
Pirenne, J., 122, 280, 286, 407
Placzek, G., 377, 466
Planck, M., 171
Podolsky, B., 28, 224
Poincaré, H., 42
Polivanov, M. K., 489, 494
Pollock, F., 361
Pomeranchuk, I., 384
Pontrjagin, L., 13
Powell, J. L., 273, 286
Poynting, J. M., 34
Pradell, A. G., 345
Primakoff, H., 269
Pryce, M. H. L., 71
Pugh, R. E., 489, 490, 494, 512

Racah, G., 58, 89, 370, 377
Ravenhall, D. G., 280
Rayski, J., 169, 224
Rebane, T., 538
Reents, G., 529
Reist, H. W., 539
Remiddi, E., 538
Repko, W. W., 538
Retherford, R. C., 345
Rich, A., 537
Rivier, D., 224, 420
Roberts, K. V., 9
Robertson, H. H., 339
Robiscoe, R. T., 538
Rohrlich, F., 205, 224, 274, 379, 384, 386, 390, 414, 474, 475, 494, 495, 511, 529
Roskies, R., 495
Rosenfeld, L., 34, 80
Ruark, A. E., 274
Ruelle, D., 482, 493

Sakata, S., 224, 378
Salam, A., 203, 320, 499, 502
Salpeter, E. E., 360, 538
Sasaki, M., 84, 97
Sauter, F., 370, 371
Schafroth, M. R., 245, 396
Schiff, L. I., 346
Schiller, R., 19

Schmidt, W., 493
Schönberg, M., 71
Schroer, B., 529
Schur, I., 427
Schweber, S. S., 39, 494, 538
Schwartz, C., 538
Schwinger, J., 9, 16, 19, 24, 36, 97, 134, 190, 304, 309, 325, 339, 342, 345, 358, 360, 392, 420, 423, 489, 492, 494
Scott, M. B., 340
Segrè, E., 373
Segre, G., 495
Serber, R., 358
Serpe, J., 34
Sheppey, G. C., 388
Shirkov, D. V., 492
Shyn, T. W., 538
Skyrme, T. H. R., 237, 239
Slotnick, M., 377, 475
Sokolov, A., 273
Soleillet, P., 42
Sommerfeld, A., 34, 371
Soper, D. E., 495
Souder, P. A., 539
Speer, E., 501, 502
Spence, W., 462
Stambaugh, R. D., 539
Stehn, J. R., 358
Stokes, G. G., 41
Streater, R. F., 493
Streit, L., 495
Strocchi, F., 494
Stueckelberg, E. C. G., 114, 119, 134, 224, 420, 473
Suaya, R., 495
Sudarshan, E. C. G., 493, 511
Suh, K. S., 251
Sukhanov, A. D., 494
Sunakawa, S., 102
Susskind, L., 495
Suura, H., 529
Swieca, A., 494
Symanzik, K., 494

Takagi, S., 71
Takahashi, Y., 57, 496, 502
Tamm, I., 229

Tanaka, K., 134, 231
Tanikawa, Y., 15
Tati, T., 84, 97
Tavkhelidze, A. N., 538
Taylor, B. N., 537
Ten Eyck, J. H., 495
Thirring, W., 218, 246, 397, 492
Thomas, L. H., 224
Thompson, P. A., 539
Thomson, J. J., 235
Tiemann, J. J., 538
Tiomno, J., 55, 87
Titchmarsh, E. C., 470
Todorov, I. T., 538
Toll, J. S., 384, 469–473
Tolman, R. C., 46
Tomonaga, S., 9, 84, 97, 358, 378
Touschek, B. F., 24
Townsend, J., 535, 538
Triebwasser, S., 345
Tulub, A. V., 494

Uehling, A. E., 195, 358
Uhlenbeck, G. E., 17, 225
Urban, P., 330, 493
Utiyama, R., 15, 71, 102

Valatin, J. G., 56, 97, 110
Vessot, R. F. C., 538
Villars, F., 224, 309,, 415
Volterra, V., 9
Vorburger, T. V., 538
Votruba, V., 249

Waerden, B. L. van der, 45, 50, 441
Waller, I., 172, 408
Ward, J. C., 203, 213
Wataghin, G., 169
Watanabe, S., 15, 91, 95, 96, 443
Weinberg, S., 495, 497, 502
Weinmann, E., 397
Weiss, P., 9
Weisskopf, V., 172, 298, 358, 408
Weizsäcker, C. F. V., 287
Weneser, J., 360
Wentzel, G., 71, 74, 190, 493
Wesley, J. C., 537
Wessel, W., 71

Weyl, H., 425
Wheeler, J. A., 120, 251, 263, 280, 286, 300, 372
Wick, G. C., 87, 93, 445
Wiedemann, L., 371
Wightman, A. S., 39, 67, 87, 93, 481, 493, 494, 511
Wigner, E. P., 55, 56, 67, 87, 89, 91, 93, 408
Williams, E. J., 287
Williams, W. L., 538
Wilner, M., 494
Wilson, K., 500, 502
Wilson, R. R., 379, 408
Wolfenstein, L., 280
Woodward, W. M., 261
Wu, T. T., 388, 387

Yam, M., 538
Yang, C. N., 55, 74, 87, 167, 280, 282
Yang, L. M., 436
Yennie, D. R., 8, 529
Yukawa, H., 8

Zienau, S., 409
Zimmermann, W., 488, 494, 500
Zitzewitz, P. W., 538
zu Pulitz, G., 539
Zwanziger, D., 494, 514, 524, 529

SUBJECT INDEX

All listings refer to page numbers. In case of several listings, those involving the basic discussion of the item are given first. References to footnotes are indicated by the letter n.

Action integral, 16
Action principle, 14–18
Adjoint spinor, 53
Analytic continuation method, 465–475
Angular momentum paradox, $34n$
Annihilation, electron pair (*see* Pair annihilation)
  operator (*see* Operator)
  positronium (*see* Positronium)
Anticommutation rules, 56–58
Asymptotic condition, 487
Auxiliary condition (*see* Subsidiary condition)
Auxiliary fields (*see* Regulators)
Average excitation energy, 355

Baker–Campbell–Hausdorff formula, 505
Bethe–Heitler formula, for bremsstrahlung, 369–370
  for pair production, 375
Bethe–Salpeter equation, 535
Bhabha scattering, 257–261, 252
Bogoliubov causality, 489
Bohr–Peierls–Placzek relation, 466–467, 377
Born approximation, 319–321
Bose–Einstein statistics, 22–23
Bound interaction picture, 306–308
  diagrams, 318
Breit correction, 532
Bremsstrahlung, 364–373
  in electron-electron collisions, 261–263
  extreme relativistic limit, 371
  nonrelativistic limit, 370

Canonical momentum tensor (*see* Momentum tensor)

Canonical operator (*see* Operator)
Canonical transformation, 10–11
Canonical variation, 15
Causality (*see* Locality)
Charge conjugation (*see also* Invariance), 94–96, 54, 142–143, 228
  for coupled fields, 73
  symmetry under, 64
Charge-current density (*see* Current density)
Charge parity, 143
Charge renormalization (*see also* Renormalization), 187, 197, 175
  spurious, 202
Chronological product, 144, 145
  ordering of, 451–453, 509–511
Classical limit of quantum electrodynamics, 302
Classical theory of point charge, 71
Closed loop, 159–161
  diagrams, 147
  divergences, 174
  theorem on nonrelativistic limit, 160
Coherent states, 503–512
Commutation relations (*see* Commutation rules)
Commutation rules, 22–24
  bound interaction picture, 308–312
  current operator, 58
  interacting fields, 73–76
  invariance under charge conjugation, 58
  radiation field, 28–30
  subsidiary condition, 74–75
Commutators (*see* Commutation rules)
Compton scattering, 229–235, 150, 157–158
  double, 235–241, 229
  radiative corrections, 241–247
Conservation laws, 20–22

Conservation laws, *cont.*
  charge, 73
  parity, 139
Contraction symbol, 446
Convergence of iteration solution, 144n
Convergence of renormalized S-matrix, 218
Corners of diagrams (*see* Diagrams)
Coulomb gauge, 484, 491
Coulomb interaction, 97–100
Coulomb phase, 519 520
Coulomb scattering, 327–332
  radiative corrections, 332–342
  radiative corrections irrespective of radiation loss, 342
Covariants, bilinear, 53–55
Creation operator (*see* Operator)
Cross section, 163–167
  differential, 167
Current, orbital, 343
  polarization, 343
Current density, 55
  measurement, 81
  operator, 72

Deceleration radiation (*see* Bremsstrahlung)
Decomposition into plane waves (*see* Plane waves)
Delbrück scattering, 379–388, 134, 298
Delta functions, Dirac, 464
  explicit expressions, 424
  homogeneous, 420
  inhomogeneous, 420
  integral representations, 421
  relations between, 421
  surface, 23, 27, 421
Density of final states, 165
Density matrix, 46
Destruction operator (*see* Operator)
Detailed balance (*see* Principle)
Diagrams, 146–150
  bound interaction picture, 318
  closed loop (*see* Closed loop)
  corners, 146
  cut-off dependent, 209n
  equivalent, 148
  improper, 207

  irreducible, 206–209
  momentum space, 151–159
  proper, 207
  reducible, 206, 207
  skeleton, 207
  vacuum, 176
Differential cross section, 167
Dirac equation, 50–51
  boundary value problem, 51–52
  relativistic invariance, 52–53
Divergences, 170–202
  classification, 173–175
  closed loop, 174
  degree of, 205
  infrared (*see* Infrared divergences)
  overlapping, 212
  primitive, 203–206
  separation in irreducible parts, 210–211
  separation in reducible parts, 211–218
  skeleton, 212
Double Compton scattering (*see* Compton scattering)

Einstein–Bose (*see* Bose–Einstein)
Electron (*see also* Negaton *and* Positon)
  classical point, 171
  classical radius, 233
  magnetic moment, 342–345
Electron line, 146–147
Electron mass, 184–185
Electron path, 147
Electron resonance scattering (*see* Rayleigh scattering)
Electron self-energy, 178–188
  integral, 463
Electron self-stress, 410–415
Energy levels, in external fields, 322–326, 533
  in hydrogen-like atoms, 345–361
Energy-momentum tensor (*see* Momentum tensor)
Equations of motion, 18
External field, 303
  cross sections, 323–324
  energy levels, 325–326

External field approximation, 303–306
  S-matrix, 318–321

Fermi–Dirac statistics, 22–23
Feynman diagrams (*see* Diagrams)
Feynman–Dyson diagrams (*see* Diagrams)
f-field, 224
Field, localizable, 8
  interpolating 487
  longitudinal, 97
  transverse, 48, 99
Field equations, 18
  coupled fields, 69–73
  electron field, 50–51
  radiation field, 25–26
Field operators, representation of, 65–67
Field variables (*see* Variables)
Fine-structure constant, 180
  value, 361, 531
Fluctuations, charge, $81n$
  current, $81n$
  effect on energy levels, 346
Flux density, 166
Fock space, 503
Furry's theorem, 159
  generalized, 160

Gamma matrices, 425–429
  standard representation, 427
Gauge-independent interaction, 110–113
Gauge integral, 84
Gauge invariance (*see also* Invariance)
  photon self-energy, 189–190
Gauge-invariant operators, 85 (*see also* Operators)
Gauge transformations, 83–86, 26, 138–139
  first kind, 138–139
  Gupta method, 108, 484–486, 508
  infinitesimal, 84, 33
  Landau, 489
  null plane, 491
  restricted, 26
Generating operator, gauge transformations, 84

infinitesimal canonical transformations, 11
infinitesimal displacement for radiation field, 33
infinitesimal Lorentz transformations, 20
Graphs (*see* Diagrams)
Green's function, 167

Hamiltonian (*see* Interaction operator)
Hamiltonian mechanics, 7
Heisenberg fields, 74, 117
Heisenberg picture, 76, 10, 14
  evaluation of S-matrix, 167–169
Heitler's integral equation, 135, 406
Heitler's subtraction procedure, $408n$
Hermitian conjugate, 6 (*see also* Notation)
Hermitian operator (*see* Operator)
Hyperfine structure, radiative corrections, 361, 531
Hypermaximal operator (*see* Operator)

Incident wave, 123
Indefinite metric, 103, 483
Infinitesimal transformations (*see the respective transformations*)
Infraparticle, 525
Infrared divergence, 390–405, 174, 513–529
  Compton effect, 239
  electron-electron collisions, 256
  two- and three-photon annihilation, 264, 274
Instant parameter, 9
Integrals, evaluation of, 454–463
  convergent, 455–457
  divergent, 457–460
Integration, symmetrical, 457
Interaction operator, 117
Interaction picture, 76–78
Invariance properties of coupled fields, 82–96
  charge conjugation, 94–95
  gauge transformations, 83–86
  Lorentz transformations, 82–83
  scale transformation, 96
  space inversion, 86–88

Invariance properties of coupled fields, cont.
  time inversion, 88–93
Invariance properties of S-matrix, 136–143
  charge conjugation, 142–143
  gauge transformation, 138–139
  Lorentz transformations, 138
  phase transformations, 139
  space inversion, 139
  time inversion, 140–142
Invariant functions, 419
Invariant relativistic velocity, 520
Iteration approximation (see Iteration solution)
Iteration solution, 144–145, 117, 170
  as asymptotic expansion, 176

Klein-Nishina formula, 234
Klein's paradox, 312

Lagrangian, 16
  coupled fields, 72
  electron field, 56
  radiation field, 28
Lamb shift (see Energy levels)
Lie group, 13
Lifetime, 165
  positronium, 285–286, 536
Light-by-light scattering (see Photon-photon scattering)
Lines, in diagram, 146–147
Line shape, 408
Line width, natural, 408–410, 354
Locality, 482–485
Localizability, 7
Localizable field, 8
Lorentz condition (see Subsidiary condition)
Lorentz group, 441–444
Lorentz transformation, 3, 138 (see also Invariance properties)
  infinitesimal, 12

Magnetic moment of electron, 342–345, 530
Mass, electromagnetic, 184–185
  mechanical, 185
  observable, 184, 219
Mass renormalization (see Renormalization)
Maxwell's equations, 25–26
Measurability of fields, 78–81
Measurement, of charge-current density, 81
  of first kind, $8n$
Metric tensor, 3
  indefinite, 103
Møller scattering, 252–257, 149, 157
Momentum operators, 13–14, 19–20
  (see also Operators)
  for radiation field, 31–32
  for spinor field, 59
Momentum tensor, canonical, 20, 83
  symmetrical, 21
Muonium, 536

Negatons, 65
Negaton-negaton scattering (see Møller scattering)
Negaton-positon scattering (see Bhabha scattering)
Neutral vector meson, 113–116
  longitudinal, 116
Nonlinear interaction, 298
Nonlocal field theories, 169
Nonrelativistic limit, 316–318
Normal operator product, 500
Notation, 476–477
Number operator (see Operator)
Numerical factor of S-matrix elements, 155 (see also S-matrix element)

Observable, 8, 85
  first and second kind, 85
Occupation number, 40, 47
One-particle processes, 147
One-quantum annihilation (see Pair annihilation)
Operator (see also Generating operator)
  annihilation, 37
  canonical, 105
  creation, 37
  current density, 72

distortion, 524
electron number, 64
gauge-invariant, 85
Hermitian, 6
hypermaximal, 121n
linear, 5
photon number, 40
product expansion, 500
self-adjoint in Hilbert space with indefinite metric, 105
spin, 65
unitary, 6
Ordered product, 47, 506
for anticommuting fields, 63
Ordering theorem, 445–453, 148, 509–511
anticommuting fields, 449–450
commuting fields, 445–449
generalized, 451

Pair annihilation (*see also* Positronium)
one-quantum, 379
three-quantum, 270–274
two-quantum, 263–270
Pair production, 373–379
in photon-electron collisions, 247–251
in photon-photon collisions, 299–301
Parity operator, 88, 139
Phase transformations, 139
Photoelectric process, 372
Photon-electron scattering (*see* Compton scattering)
Photon number operator (*see* Operator)
Photon-photon scattering, 287–298
cross section, 294–298
Photon rest mass, 113
Photon scattering by Coulomb field (*see* Delbrück scattering)
Photon self-energy, 188–197
diagram, 188
part, 188, 206
integral, 460–463
Plane wave, decomposition of electron field, 60–65
decomposition of radiation field, 35–37

normalization for electron field, 63, 434–435
normalization for radiation field, 35–37
solutions of Dirac equations, 431–436
Poincaré parameters, 43
Polarization, 44–46
degree of, 44, 46
opposite, 44
partial, 44, 46
Polarization effect, 346
Polarization sums, 439–440
Positons, 65
Positon-positon scattering (*see* Møller scattering)
Positronium, 274–286, 252
annihilation, 282–286
charge-parity, 274
energy levels, 535
lifetime, 285
ortho-, 536
para-, 536
selection rules, 274–282
space-parity, 274
spin-exchange operator, 275
three-photon annihilation, 286
two-photon annihilation, 285–286
Potentials, retarded, 305
Potential scattering (*see* Delbrück scattering)
Principle, of detailed balance, 142n
of limiting distance, 468
of reciprocity, 141–142
Propagation functions,
electron in external field, 312–318
finite, 218, 524
nonrelativistic limit, 317–318
one-particle theory, 316
true, 208

Quantization, second, 51, 480
Quantum Field Theory,
asymptotic, 489
axiomatic, 481–483
constructive, 483
functional formulation, 489
Lagrangian, 480
LSZ formulation, 486

Quantum Field Theory, *cont.*
  null plane, 490–492
  renormalizable, 496
Quasipotential method, 535

Racah operator, 95
Radiation damping, 405–408
Radiation field operator, 29
  longitudinal component, 101
  representations, 37–39
Radiative corrections (*see specific process*)
Radiative transitions, 361–364
Ray representation, 12n
Rayleigh scattering, 387–388, 379
Reciprocity, principle of, 141–142
Regulators, 223–227
Renormalization, 173, 203, 496–502
  analytic, 501
  charge, 219–221, 187, 197, 202, 175
  Dyson–Salam–Ward, 496–499, 501
  external field, 321–322
  mass, 219, 184–185, 175
  wave function, 221–222, 186–187
Rest mass of photon, 113
Retarded functions, 488
Rutherford scattering, 329, 257, 513

Scalar field, auxiliary, 111
Scalar photon, 115
Scale transformations, 96
Scattered wave, 123
Scattering (*see specific process*)
Scattering matrix (*see S*-matrix)
Scattering states, 124
Schroedinger equation, 77, 118–119
Schroedinger picture, 120
Schroedinger wave function, 316
Schur's lemma, 427
Screening, in bremsstrahlung, 372
  in pair production, 378
Selection rules, in positronium, 274–282
Self-adjoint (*see* Operator)
Self-energy, 172
Self-energy of electron, 178–188
  part, 179

Self-energy of photon, 188–197
  part, 188
Self-stress of electron, 410–415, 172, 223
Self-stress operator, 412
SE-parts, 206
S-function, 424
Shape factor, 224
Sign factor in S-matrix element, 155
Skeleton, diagram, 207
  divergence, 212
  integral, 212
S-matrix, 117–143, 119
  definition, 131–136
  evaluation, 144–169
  evaluation in Heisenberg picture, 167–169
  external field approximation, 318–321
  invariance, 136–143 (*see also* Invariance properties)
  physical meaning, 132
  preliminary definition, 117–120
S-matrix element, sign and numerical factors, 155
S-operator, 131, 119, 487
  unitary, 132
Space inversion, 86, 139, 3
  invariance under, 86–88 (*see also* Invariance properties)
Space-like surface, 9
Space parity, 139
Spence functions, 462
Spin operator, 65
  photon, 40–46
Spin sums, 438–439
Spinor wave functions, large and small components, 316
State vector, construction of, 101–103
Stokes operators, 45
Stokes parameters, 42
Subsidiary condition, 26, 29–30, 71, 77, 100
  commutation rules for, 74–75
  conservation of, 138
  Gupta method, 107
  interacting fields, 110
  momentum representation, 100

Substitution law, 161–163, 257, 261, 264, 299, 379
Superselection rule, 93, 142
Surface term, 459

Tau-functions, 487, 500
Tensors, regular, 443
Thomson scattering, 235, 246–247
Time inversion, 88–93, 4, 140–142
Time-ordered products (see Chronological product)
Time-reversal operator, 91–93, 89n
Time-reversed state vector, 91, 141
Traces of gamma-matrices, 436–437
Transformation matrix, 15, 119
Transition probability, 163, 119
 amplitude, 163

Uehling term, 195, 358
Uncertainty relation, 79
Unitary deficiency, 128
Unitary operator, 6 (see also Operator)
Units, 1–2

Vacuum, definition of,
 for electron field, 67
 for external field, 312
 for photon field, 46–49
Vacuum diagrams, 176
Vacuum expectation values, 48, 488
Vacuum fluctuation, 176–178

Vacuum polarization, 217, 195, 358
 fourth rank tensor, 288
Variables, dynamical, 7
 field, 8
Vector meson, neutral, 113–116
Vector potential, 25
Vector space, linear, 5
Vertex (see Corner)
Vertex part, 197–202
 finite, 218
 three-photon, 215
 true, 208
V-parts, 206–207 (see also Vertex part)

Ward's identity, 213, 496–497
Ward–Takahashi equation, 496–497
Wave function, 316
Wave function renormalization, 186–187, 221–222 (see also Renormalization)
Wave matrix, 120–126
Wave operator, 126–128
 integral representation, 129–131
Weak local commutativity, 482, 484
Weinberg theorem, 497, 501
Wentzel potentials, 74n
Width of spectral lines (see Line width)
Wightman functions, 488

Zero point energy, 172

Texts and Monographs in Physics

Editors: W. Beiglböck,
J. L. Birman,
E. H. Lieb, T. Regge,
W. Thirring

W. Greiner, B. Müller, J. Rafelski

Quantum Electrodynamics of Strong Fields

1985. 260 figures. Approx. 600 pages. ISBN 3-540-13404-2

G. Ludwig

Foundations of Quantum Mechanics II

Translated from the German by C. Hein
1985. Approx. 54 figures. Approx. 300 pages.
ISBN 3-540-13009-8

R. M. Santilli

Foundations of Theoretical Mechanics I

The Inverse Problem in Newtonian Mechanics
Corrected 2nd printing. 1984. XIX, 266 pages.
ISBN 3-540-08874-1

N. Straumann

General Relativity and Relativistic Astrophysics

1984. 81 figures. XIII, 459 pages. ISBN 3-540-13010-1

G. Ludwig

Foundations of Quantum Mechanics I

Translated from the German by C. A. Hein
1983. XII, 426 pages. ISBN 3-540-11683-4

R. M. Santilli

Foundations of Theoretical Mechanics II

Birkhoffian Generalization of Hamiltonian Mechanics
1983. XIX, 370 pages. ISBN 3-540-09482-2

F. J. Ynduráin

Quantum Chromodynamics

An Introduction to the Theory of Quarks and Gluons
1983. XI, 227 pages. ISBN 3-540-11752-0

Springer-Verlag
Berlin
Heidelberg
New York
Tokyo

Texts and Monographs in Physics

Editors: W. Beiglböck,
J. L. Birman,
E. H. Lieb, T. Regge,
W. Thirring

R. G. Newton
Scattering Theory of Waves and Particles
2nd edition. 1982. 35 figures. XX, 743 pages.
ISBN 3-540-10950-1

R. D. Richtmyer
Principles of Advanced Mathematical Physics II
1981. 60 figures. XI, 322 pages. ISBN 3-540-10772-X

P. Ring, P. Schuck
The Nuclear Many-Body Problem
1980. 171 figures. XVII, 716 pages. ISBN 3-540-09820-8

M. D. Scadron
Advanced Quantum Theory
and Its Applications Through Feynman Diagrams
Corrected 2nd printing. 1981. 78 figures. XIV, 386 pages
ISBN 3-540-10970-6

W. Rindler
Essential Relativity
Special, General, and Cosmological
Revised printing of the 2nd edition. 1980. 44 figures.
XV, 284 pages. ISBN 3-540-10090-3

H. M. Pilkuhn
Relativistic Particle Physics
1979. 85 figures, 39 tables. XII, 427 pages. ISBN 3-540-09348-6

R. D. Richtmyer
Principles of Advanced Mathematical Physics I
1978. 45 figures. XV, 422 pages. ISBN 3-540-08873-3

C. Truesdell, S. Bharatha
The Concepts and Logic of Classical Thermodynamics as a Theory of Heat Engines
Rigorously Constructed upon the Foundation Laid by S. Carnot and F. Reech
1977. 15 figures. XXII, 154 pages. ISBN 3-540-07971-8

Springer-Verlag
Berlin
Heidelberg
New York
Tokyo